PHENOLIC COMPOUNDS - NATURAL SOURCES, IMPORTANCE AND APPLICATIONS

Edited by **Marcos Soto-Hernández, Mariana Palma-Tenango** and **María del Rosario García-Mateos**

Phenolic Compounds - Natural Sources, Importance and Applications
http://dx.doi.org/10.5772/67213
Edited by Marcos Soto-Hernández, Mariana Palma-Tenango and María del Rosario García-Mateos

Contributors

Alfredo Aires, Yana Brandão, Mohand Benachour, Shagufta Perveen, Areej Al-Taweel, Magdalena Sobiesiak, Mohamed Roby, Federico Casassa, Alexey Gogotov, Carlos Velasco-Santos, Ana Laura Martínez-Hernandez, Veronica Saucedo-Rivalcoba, Eduardo E. Perez-Ramirez, Miguel De La Luz-Asunción, Thuan Minh Huynh, Udo Armbruster, Andreas Martin, Yolanda Salinas-Moreno, Carolina García-Salinas, José Luis Ramírez-Díaz, Ivone Alemán De La Torre, Norma Francenia Santos-Sánchez, Raúl Salas-Coronado, Beatriz Hernández-Carlos, Joseoziel LLaguno-Guilberto, Julia Martín, Mª José Navas, Ana María Jiménez, Agustín G. Asuero, Fernanda Cosme, Berta Gonçalves, Eunice Bacelar, António Inês, António M. Jordão Alice Vilela, Li Fan, Claudine Dubé, Shahrokh Khanizadeh, Najat Kamal, Rosa Galvez-Cloutier, Gerardo Buelna, Rino Dubé, Pankaj Chowdhury, Sharmistha Nag, Ajay Ray, Messai A. Mamo, William Anku, Penny Govender

Published by InTech
Janeza Trdine 9, 51000 Rijeka, Croatia

Publishing Process Manager Maja Bozicevic
Technical Editor SPi Global
Cover InTech Design team

First published March, 2017
Printed in Croatia
Legal deposit, Croatia: National and University Library in Zagreb

Additional hard copies can be obtained from orders@intechopen.com

Phenolic Compounds - Natural Sources, Importance and Applications, Edited by Marcos Soto-Hernández, Mariana Palma-Tenango and María del Rosario García-Mateos
p. cm.
Print ISBN 978-953-51-2957-8
Online ISBN 978-953-51-2958-5

For EU product safety concerns:
IN TECH d.o.o., Prolaz Marije Krucifikse Kozulić 3, 51000 Rijeka, Croatia,
info@intechopen.com or visit our website at intechopen.com.

Meet the editors

Dr. Marcos Soto-Hernández is a pharmacist from the National University of México and completed his PhD degree at the University of Wales, Cardiff, UK, and now is a full professor at the Colegio de Postgraduados where he conducts research in phytochemistry and bioactivity natural products. He has established collaboration with research groups in the UK, the Netherlands, and Spain and other groups in México; he has received several awards locally and abroad. At present, his main line of research is related with bioguided isolation of secondary metabolites with its importance in medicine and agriculture and the potential of the local aromatic plants is part of his recent research.

Dr. Mariana Palma-Tenango is an engineer agronomist from the Universidad Autónoma Chapingo; holds a PhD degree in Plant Physiology from the Colegio de Postgraduados, México; and has teaching duties in the National Autonomous University of Mexico. She is an assistant professor of Phytochemistry at the Colegio de Postgraduados, and she is a reviewer of several journals and an editor. She has participated in the organization of meeting and symposiums in México and is a supervisor of master and PhD degree students. Her research line is phytochemistry, medicinal, and aromatic plants.

Dr. María del Rosario García-Mateos is a chemist of the University of Puebla, México, and holds a PhD degree from the Colegio de Postgraduados. Her research line is related with the nutraceutical properties and biological activity of horticultural crops. She has supervised master and PhD thesis, has published more than 71 papers, and has been a reviewer of several scientific journals. She has established collaboration with many groups national and international.

Contents

Preface

Phenolic compounds as a large class of metabolites found in plants have attracted attention since long time ago due to their properties and the hope that they will show beneficial health effects when taken as dietary supplements. In this broad group of compounds, we find flavonoids, phenolic acids, and stilbenoids, being flavonoids, as the most isolated compounds. They can be found in fruits, roots, leaves, seeds and flowers, and their concentration varies depending on many biotic or abiotic factors or development stage, agricultural practices, or edafo-climatic or geographic conditions. This book presents the state of the art of some of the natural sources of phenolic compounds, for example, medicinal plants, grapes, or blue maize, as well as the modern methods of extraction, quantification, and identification, and there is a special section discussing the treatment, removal, and degradation of phenols, an important issue in those phenolics derived from the pharmaceutical or petrochemical industries.

Dr. Marcos Soto-Hernández
Department of Botany
Postgraduate College
Texcoco, México

Dr. Mariana Palma-Tenango
Science Faculty
National Autonomous University of Mexico
Mexico City, México

Dr. María del Rosario García-Mateos
Department of Plant Science
Institute of Horticulture
Chapingo Autonomous University
Texcoco, México

Section 1

Source and Extraction

Chapter 1

Chemical Structure of Phenols and Its Consequence for Sorption Processes

Magdalena Sobiesiak

Additional information is available at the end of the chapter

http://dx.doi.org/10.5772/66537

Abstract

Sorption of phenolic compounds is a very complex process and many factors influence it. At the beginning, detailed chemical structure of phenols is presented with its consequence for physical properties, for example, values of melting and boiling points, solubility in water, pKa and Log P. Also influence of activating and deactivating substituents on the properties is explained. On this basis, interaction with the most frequently used sorbents, for example, chemically modified silicas, polymers and porous carbons, is described. Both sorbents characteristics including physical (porosity) and chemical properties (functional groups) and experimental conditions such as concentration of solutes, contact time, temperature, solvent effects and presence or absence of oxygen are taken into account. The explanations of irreversible adsorption and oxidative coupling phenomena are included. The mechanisms of phenolic compounds sorption are described.

Keywords: phenol's properties, phenol's sorption, phenol's interactions, structure, adsorbents

1. Introduction

There is a huge variety of phenolic compounds in our natural environment. Some like eugenol, thymol, pyrogallol, guaiacol or pyrocatechol are formed in natural way, but a vast majority of them are introduced as a consequence of industrial, agricultural and communal activities of humans. While the former are neutralized as the result of natural processes, the latter pose a serious risk to the environment.

Quantitative isolation of phenol and its derivatives from environmental matrices is usually difficult because of considerably different properties and low concentration. It was a stimulus to development of research on process of sorption and search for efficient sorptive materials.

This chapter is devoted to review of current state of knowledge on sorption process of phenolic compounds. Many different types of sorbents are used for phenols in chromatographic columns and solid phase extraction devices. Their efficiency is diverse and depends on many factors. As the most important chemical structure of adsorbate, a type of sorbent and its porosity as well as properties of solvent (or eluent) should be mentioned. Also other properties such as pH, temperature and presence of oxygen influence the process. For this reason, sorption of phenols is a very complex phenomenon. Although many researchers try to explain the mechanism of interaction of phenols with adsorbents, it is still an open problem.

Many groups of researchers tried to solve it using different scientific methods, for example, chromatography [1, 2], spectroscopy (UV, mass spectrometry [MS], Fourier transform infrared spectroscopy [FTIR]) [3–5], thermal analysis [6], and computer simulations [7]. Based on the obtained results, some authors proposed explanations of phenomena and mechanisms accompanying the sorption of phenol. Understanding the mechanism is important from scientific point of view. Elucidation of this process is essential for reasons of utility and finding an answer to the questions how to improve efficiency of sorption phenolic compounds in the processes of aqueous environment remediation. Effective removal of these species from industrial and urban waste water helps to protect aquatic ecosystem from toxic impact of phenols on the living organisms, which is an important aim all over the world.

2. Chemical structure and properties of phenols

Phenols are compounds possessing one or more hydroxyl groups (–OH) directly connected to the aromatic system (e.g. phenyl, naphthyl) [8, 9]. **Figure 1** presents some examples of phenols with their systematic and common names.

All carbon atoms forming aromatic ring are sp^2 hybridized. Therefore, phenyl has hexagonal planar structure with all bond angles 120° and delocalized π-electrons distributed over the ring. The C-O bond is formed from Csp^2-Osp^3 and the O-H bond is formed from Osp^3-H1s. Two other orbitals of oxygen atom are occupied by two nonbonded electron pairs. For this reason, hydroxyl functional group C-O-H has a bent shape with almost the tetrahedral bond angle of 109.5° as it is represented in **Figure 2**. Oxygen is more electronegative than carbon and hydrogen that makes both the C-O and O-H bonds polar [10].

Moreover, electron pairs of oxygen atom are conjugated with aromatic system that causes partial transfer of negative charge from oxygen to the ring and delocalization of the charge. This effect additionally strengthens polarization of O-H bond. In consequence, phenol gains acidic character and ability to form phenoxide (phenolate) ion [9, 10]. Both phenol and its conjugate base are resonance stabilized. Dispersion of the negative charge over the molecule can be illustrated with the resonance structures or as a resonance hybrid as in **Figure 3**.

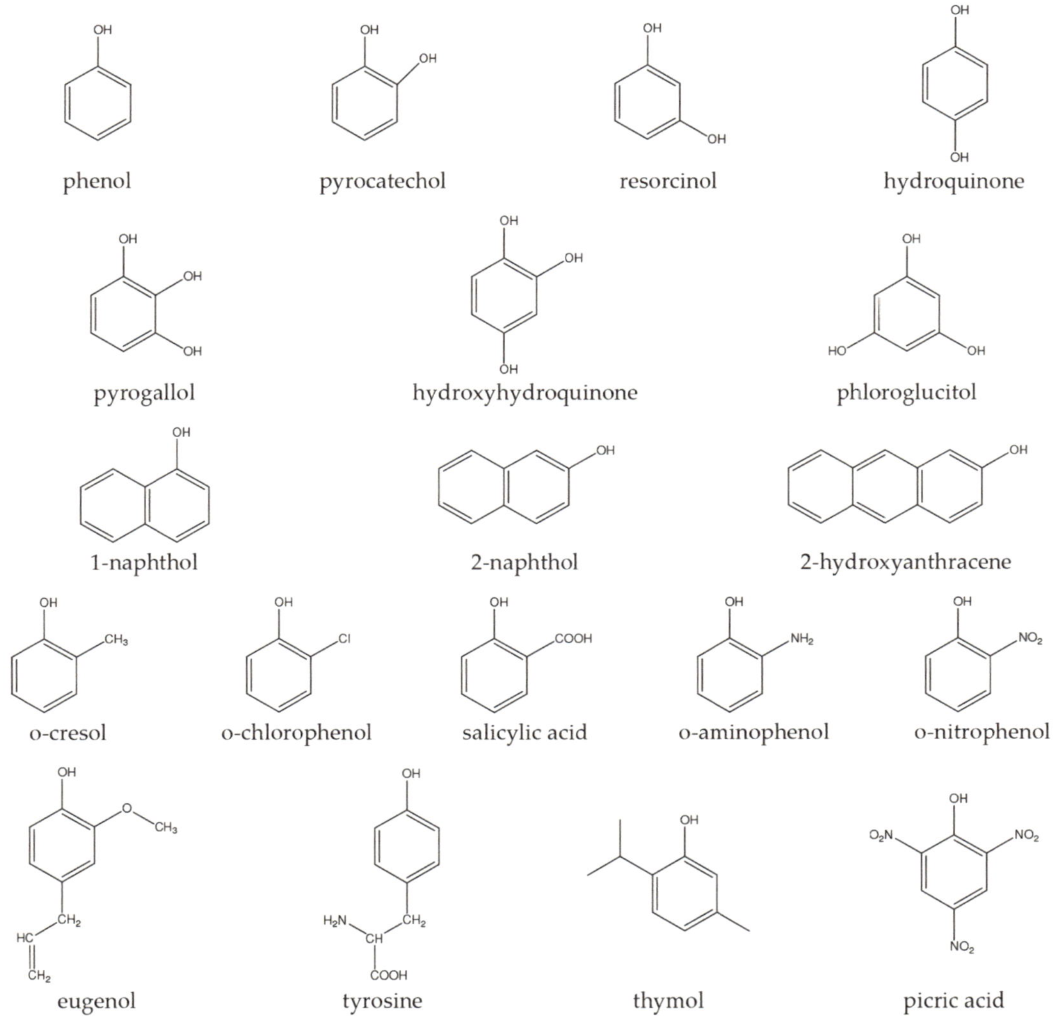

Figure 1. Examples of phenolic compounds.

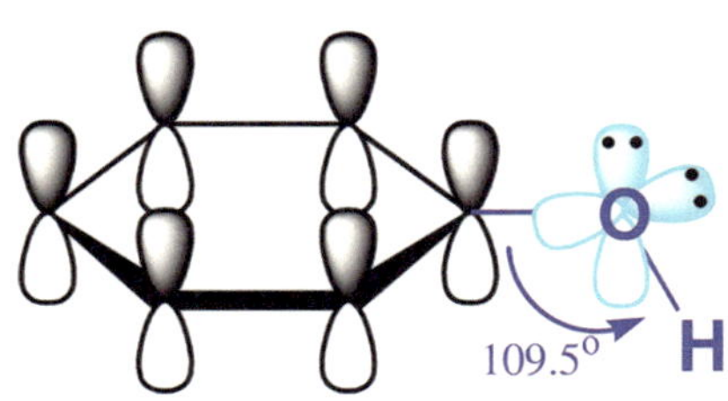

Figure 2. The structure of phenol.

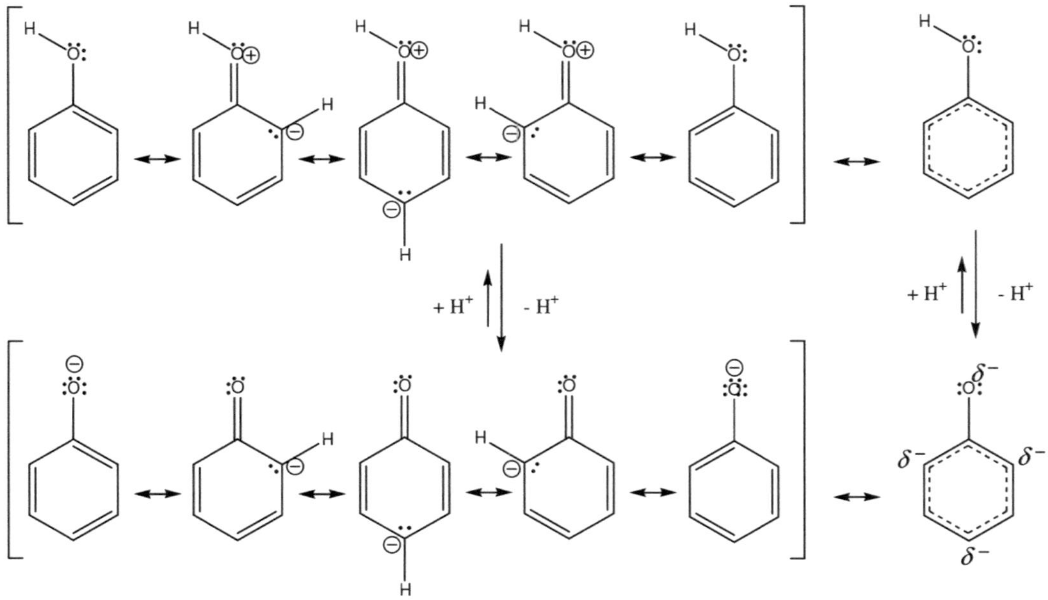

Figure 3. Resonance stabilized structures of phenol and phenoxide representing dispersion of negative charge and its resonance hybrid. Based on Refs. [9, 10].

All the above-mentioned structural features give the phenol specific physical and chemical properties. Phenol is a peculiar smelling, water-soluble crystalline solid, with low melting and high boiling point values. These attributes can be reinforced or weakened, if a substituent is connected to the aromatic ring. A chemical character (electron withdrawing or donating), the way of substitution—position and a number of functional groups attached to the ring—can alter properties of the phenol significantly. Some properties such as melting (MP) and boiling points (BP), density (D), water solubility (w.s.), acidity in water (pKa) and octanol-water partition coefficient (Log P) of phenol and its derivatives are presented in **Tables 1** and **2** [11, 12]. Comparison of these data allows to evaluate the impact of substituents on change of the properties.

Many monohydroxyl derivatives (alkylphenols, alkoxyphenols, halogenophenols, 2- and 3-nitrophenols) similar to phenol are crystalline solids with melting points below 100°C and boiling points ca. 200°C. Other nitrophenols, aminophenols, pentachlorophenol, hydroxybenzoic acids and species possessing two or more hydroxyl groups in a molecule have very high values both melting and boiling points. While some phenols are sublime, others decompose before reaching the boiling point. So, high temperatures are the consequence of intermolecular hydrogen bonds, which are formed between the molecules, thereby justifying water solubility of these compounds. The examples of intermolecular hydrogen bonds that phenols can form with their own molecules and with molecules of water are shown in the **Figure 4**.

Compound Chemical name IUPAC name	Case no.	M [g mol^{-1}]	MP [°C]	BP [°C]	d [g cm^{-3}]	p.s.	w.s. [g L^{-1}]	pKa	LogP	μ [D]
Benzenol Phenol	108-95-2	94.11	40.91	181.7	1.071	c.s.	83	9.99	1.46	1.61
Pyrocatechol Benzene-1,2-diol	120-80-9	110.11	105	245.5	1.344	c.s.	461	9.45 12.8	0.88	nda
Resorcinol Benzene-1,3-diol	108-46-3	110.11	109.8	280	1.278	c.s.	1400	9.30 11.06	0.80	2.07
Hydroquinone Benzene-1,4-diol	123-31-9	110.11	171	287	1.330	c.s.	72	10.85	0.59	1.4
Pyrogallol Benzene-1,2,3-triol	87-66-1	126.11	133	309	1.453	c.s.	507	9.01	0.97	nda
Hydroxyhydroquinone Benzene-1,2,4-triol	533-73-3	126.11	141	nda	nda	c.s.	nda	nda	nda	nda
Phloroglucinol Benzene-1,3,5-triol	108-73-6	126.11	217 –219	S. D.	nda	c.s.	10	8.45	nda	nda
1-Naphthol Naphthelene-1-ol	90-15-3	144.17	96	288	1.095	c.s.	0.86	9.34	2.85	1.40
2-Naphthol Naphthelene-2-ol	135-19-3	144.17	121.6	285	1.28	c.s.	0.75	9.51	2.70	1.53
o-Cresol 2-Methylphenol	95-48-7	108.14	31.0	191.0	1.047	c.s.	25.9	10.287	1.95	1.60
m-Cresol 3-Methylphenol	108-39-4	108.14	12.2	202.2	1.034	l	22.2	10.09	1.96	1.44
p-Cresol 4-Methylphenol	106-44-5	108.14	34.77	201.9	1.0185	c.s.	21.5	10.26	1.94	1.64
m-Xylenol 2,4-Dimethylphenol	105-67-9	122.16	25.4 –26	211.5	0.965	c.s	7.87	10.60	2.30	nda

Compound Chemical name IUPAC name	Case no.	M [g mol⁻¹]	MP [°C]	BP [°C]	d [g cm⁻³]	p.s.	w.s. [g L⁻¹]	pKa	LogP	μ [D]
Mesitol 2,4,6-trimethylphenol	527-60-6	136.19	73	220	nda	s	1.01	10.88	2.73	nda
2-Ethylphenol	90-00-6	122.16	18	204.5	1.015	l	5.34	10.20	2.47	nda
3-Ethylphenol	620-17-7	122.16	−4	218.4	1.028	l	nda	9.9	2.40	nda
4-Ethylphenol	123-07-9	122.16	46	217.9	1.011	c.s.	4.90	10.0	2.58	nda
Guaiacol 2-Methoxyphenol	8000-58-9	124.14	28-32	205	1.129	l/ c.s.	18.7	9.98	1.32	nda
3-Methoxyphenol	150-19-6	124.13	−18 −16	244	1.131	l	nda	nda	nda	nda
4-Methoxyphenol	150-76-5	124.14	54-57	243	1.55	c.s.	40.0	10.05	1.41/ 1.34	nda
o-Aminophenol 2-Aminophenol	95-55-6	109.13	170 −174	153 11 mmHg	1.328	c.s.	20	4.72 9.71	0.62	nda
m-Aminophenol 3-Aminophenol	591-27-5	109.13	123	164 at 11mmHg	1.195	c.s.	26.3	4.37 9.815	0.21	1.83
p-Aminophenol 4-Aminophenol	123-30-8	109.13	187.5	284 D.	1.29	c.s.	1.6	5.48 10.46	0.04	nda
2,4-Diaminophenol	95-86-3	124.14	205 D.	nda	nda	c.s.	275	nda	nda	nda

M, molar mass; d, density; w.s., water solubility; MP, melting point; p.s., physical state; c.s., crystalline solid; l,– liquid; pKa acidity in water; BP, boiling point; D., decomposes; E., explodes; S, sublimes; Log P, partition coefficient; μ, dipole moment in benzene as solvent; nda, no data available.

Table 1. Comparison of physical and chemical properties of phenol and its derivatives with electron-donating substituents.

http://dx.doi.org/10.5772/66537

Compound **Chemical name** **IUPAC name**	Case no.	M [g mol^{-1}]	MP [°C]	BP [°C]	d [g cm^{-3}]	p.s.	w.s. [g L^{-1}]	pKa	Log P	μ [D]
o-Nitrophenol 2-Nitrophenol	88-75-5	139.11	44–45	216	1.49	c.s.	2.5	7.230	1.79	3.10
m-Nitrophenol 3-Nitrophenol	554-84-7	139.11	96.8	194 70 mmHg	1.485	c.s.	13.55	8.360	2.00	3.90
p-Nitrophenol 4-Nitrophenol	100-02-7	139.11	113 −114	279 D.	1.479	c.s.	15.6	7.15	1.91	5.05
2,4-Dinitrophenol	51-28-5	184.11	114.8	S.	1.6683	c.s.	2.790	4.09	1.67	nda
Picric acid 2,4,6-Trinitrophenol	88-89-1	229.10	122 −123	300 E.	1.763	c.s.	12.7	0.42	1.44	1.75
Salicylic acid 2-Hydroxybenoic acid	69-72-7	138.12	159	211 20 mmHg	1.443	c.s.	2.240	2.98 13.6	2.26	2.63[d]
4-Hydroxybenoic acid	99-96-7	138.12	214.5 D.	nda	1.46	c.s.	5.000	4.54	1.58	2.73[d]
o-Fluorophenol 2-Fluorophenol	367-12-4	112.10	14–16	151-152	1.246	l	37.7	8.82	nda	1.84[d]
m-Fluorophenol 3-Fluorophenol	372-20-3	112.10	13.5	178	1.213	l	nda	9.28	nda	nda
p-Fluorophenol 4-Fluorophenol	371-41-5	112.10	48	185	1.22	c.s.	80	9.96	nda	2.08
o-Chlorophenol 2-Chlorophenol	95-57-8	128.56	9.8	174.9	1.264	l	11.3	8.52	2.15	1.39 2.11[d]
m-Chlorophenol 3-Chlorophenol	108-43-0	128.56	33.5	214	1.245	c.s.	26	9.12	2.5	2.15
p-Chlorophenol 4-Chlorophenol	106-48-9	128.56	42.8	220	1.2238	c.s.	24	9.41	2.39	2.15
2,4-Dichlorophenol	120-83-2	163.00	45	210	1.4	c.s.	4.5	7.89	3.06	1.59
2,4,6-Trichlorophenol	88-06-2	197.45	69.5	249	1.4901	c.s.	0.500	6.23	3.69	1.62

Compound Chemical name IUPAC name	Case no.	M [g mol^{-1}]	MP [°C]	BP [°C]	d [g cm^{-3}]	p.s.	w.s. [g L^{-1}]	pKa	Log P	μ [D]
2,3,4,5,6-Pentachlorophenol	87-86-5	266.34	174	309–310 D.	1.978	c.s.	0.014	4.70	5.12	nda
o-Bromophenol 2-Bromophenol	95-56-7	173.00	5.6	194.5	1.492	l	2.23	8.45	2.35	1.36 2.36^{d}
m-Bromophenol 3-Bromophenol	591-20-8	173.00	33	236.5	nda	c.s.	23	9.03	2.63	nda
p-Bromophenol 4-Bromophenol	160-41-2	173.00	66.4	238	1.840	c.s.	14	9.17	2.59	2.12
o-Iodophenol 2-Iodophenol	533-58-4	220.01	39–43	186 100 mmHg	nda	c.s.	nda	8.47	nda	nda
m-Iodophenol 3-Iodophenol	626-02-8	220.01	40	D.	nda	c.s.	nda	8.88	nda	nda
p-Iodophenol 4-Iodophenol	540-38-5	220.01	94	D.	nda	c.s.	nda	9.2	nda	nda

M, molar mass; d, density; w.s., water solubility; MP, melting point; p.s., physical state; c.s., crystalline solid; l, liquid; pKa, acidity in water; BP, boiling point; D., decomposes; E., explodes; S, sublimes; Log P, partition coefficient; μ, dipole moment in benzene as solvent; d, dioxane; nda, no data available.

Table 2. Comparison of physical and chemical properties of phenolic compounds with electron withdrawing substituents.

http://dx.doi.org/10.5772/66537

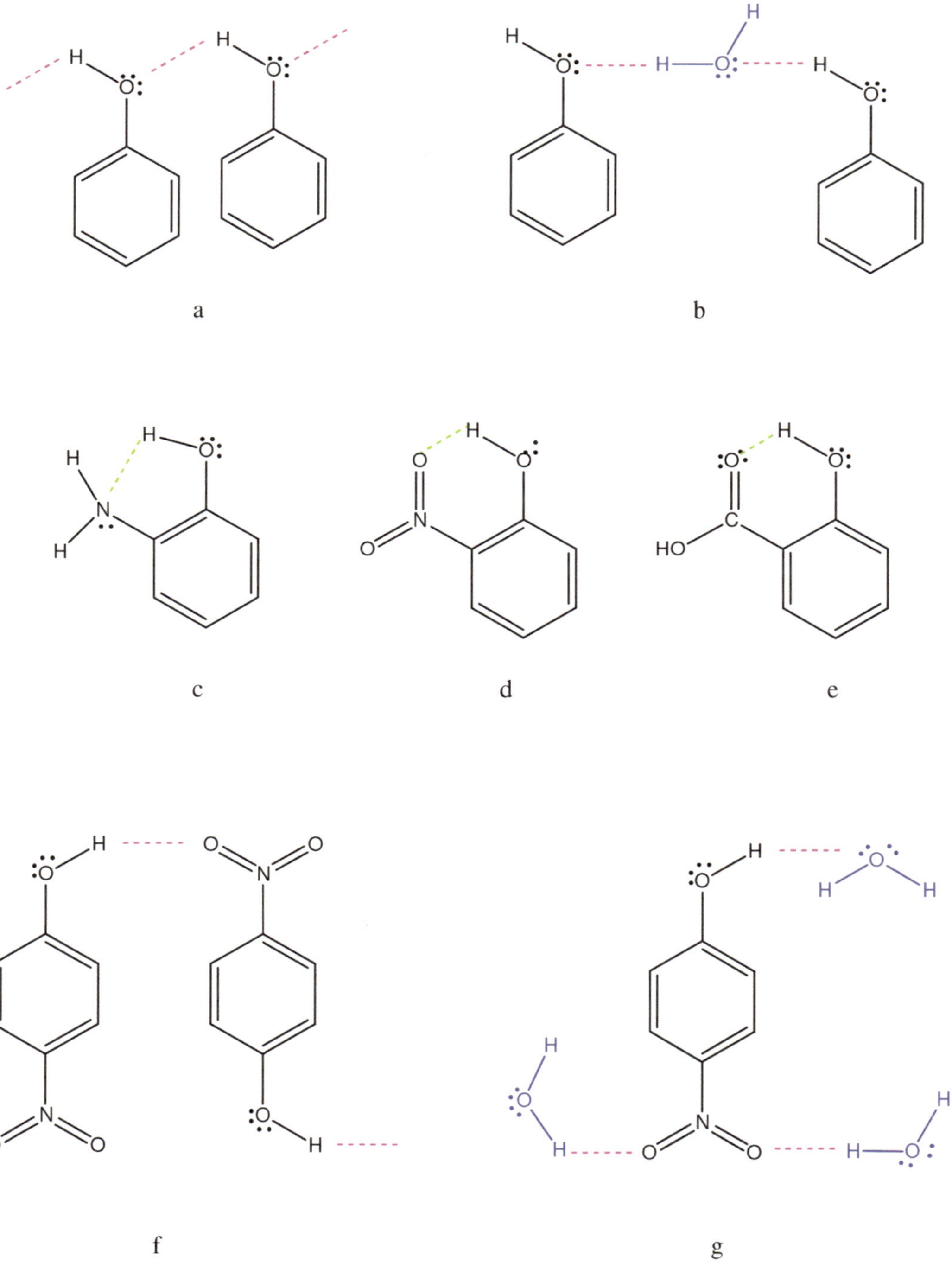

Figure 4. The intermolecular(purple dotted lines) and intramolecular hydrogen bonds (green dotted lines) formed by phenol and its derivatives. On the basis of Ref. [9].

The possibility to form hydrogen bonds suggests that phenols should have good water solubility, but this is not the rule. The unsubstituted phenol is relatively well soluble in water (83 g L^{-1}), while its substituted derivatives are not. For most of them, solubility does not exceed 30 g L^{-1}. Alkyl and halogen groups enhance hydrophobic character of aromatic ring resulting in decrease of water solubility. For phenols possessing functional groups with strong polar character, differences in ability to dissolve are more pronounced. Even for isomers of the same compound, they can be significant. The close proximity of the –OH group to substituents such as $-NO_2$, $-NH_2$ and -COOH lead to formation of intramolecular hydrogen bonds. In this way coordinated the hydroxyl group becomes less active in the solvation process. Due to this fact, solubility of ortho-isomer is lower than meta or para ones (e.g. nitrophenols, hydroxylbenzoic acids).

Very interesting behavior demonstrates benzenediols and benzenetriols. Increasing number of hydroxyl groups implies their solubility should be better in comparison to phenol. It is indeed, but hydroquinone and phloroglucinol are exceptions. Solubility of these two compounds is lower and amount 72 and 10 g L^{-1}, respectively. This apparently abnormal behavior is the result of the presence of hydrogen bonds whose strength are additionally enhanced by symmetry of the molecules. The adjacent molecules form a kind of network in which each of them is strongly bonded with others by infinite chains of hydrogen bonds (**Figure 5**) [13–15]. In this way, a compact structure is created that prevents the penetration of solvent/water molecules into the interior. For this reason, dissolution process is significantly hindered. In

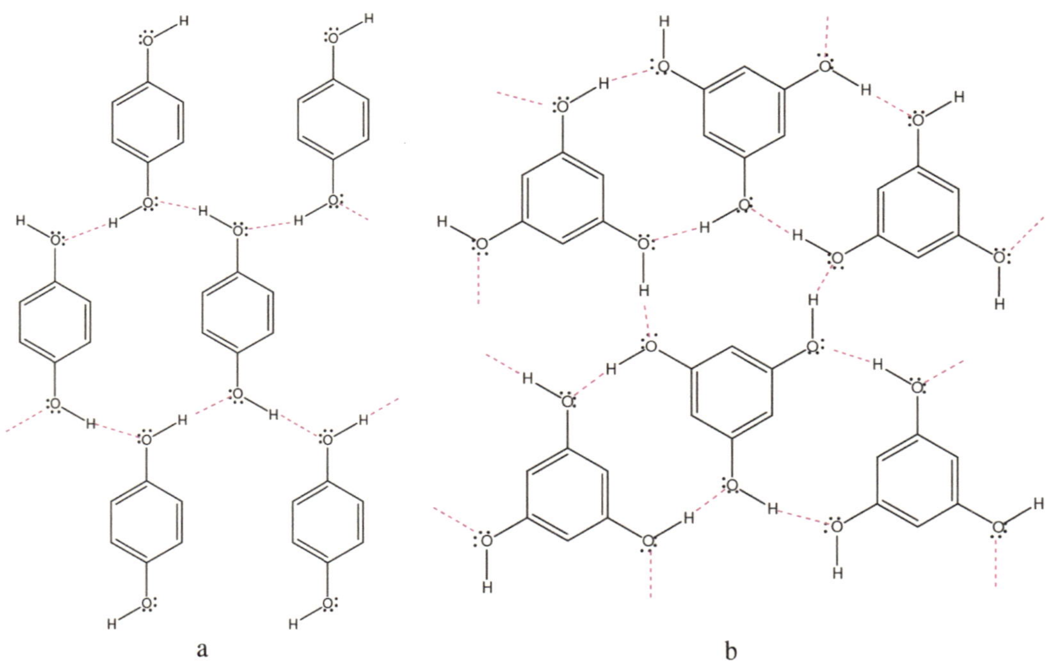

Figure 5. Hydrogen bonds enhanced by symmetrical structure of the molecules in (a) hydroquinone and (b) phloroglucine. Based on Refs. [13–15].

the similar way, slight solubility of other phenols (4-aminophenol/para-isomers, multi-substituted phenols) can be explained.

Acidity values (pKa) of most phenols are in the range of 8–10, which means they are acids stronger than water (pKa of water 15.7) but weaker than carbonic acid (pKa of 6.4). As it was mentioned above, acidity of phenol is the result of delocalizing of negative charge over the aromatic ring and resonance effect. Dipole moment is sensitive to this kind of changes. Entering *an electron donating* (-CH_3, -C_2H_5, -OCH_3, -NH_2, -OH) *or electronwithdrawing* (-COOH, -NO_2, -F, -Cl, -Br, -I) *group* to the ring causes changes in its electron density. As a consequence, altering of dipole moment of the molecule and its acidity is observed. For this reason, alkylphenols are less acidic and nitrophenols are more acidic than phenol. This effect is particularly well visible in case of picric acid. Three symmetrically situated nitro groups exert so strong electron-withdrawing effect that the pKa value for this phenol decreases up to of 0.42 and is comparable with acidity of mineral acids, for example, $HBrO_3$ (pKa = −0.69), H_3PO_4 (pKa = 2.12) or strong organic acids CF_3COOH (pKa = 0.2) and CCl_3COOH (pKa = 0.6). Similar effect is also observed for chlorophenols. The higher is a number of chlorine substituents the more acidic is character of phenol (decrease of pKa value) particularly in case of tri- and pentachlorophenols.

3. The way of interaction with sorbents

Great varieties of physicochemical properties make phenols very difficult compounds to adsorb or extract, especially they occur at extremely low level of concentration. Although so many different sorbents are commercially available, none of them is universal. For this reason, researchers are constantly working to develop new, better and more efficient materials for phenols sorption processes. All accessible sorbents due to their origin and chemical structure can be divided as the following classes:

1. Natural and mineral based (clay, silicas, zeolites) [16–23]
2. Polymer based [2, 17, 18, 20, 21, 24–28]
3. Carbons and carbonaceous materials (porous, activated, graphitized) [1, 17–20, 29–35]
4. Bioadsorbents: chitin, chitosan, peat, biomass [16]
5. Waste materials or by-products from agriculture or industry [16, 36–39]
6. Nanocomposites [40] and hybrid materials [41]

Independently of the above division, in each class, the materials can have hydrophobic or hydrophilic character due to the lack or presence of polar functional groups.

Retention of phenols from aqueous solution by nonpolar reversed phase adsorbent, for example, silica sorbents (RP-C_8, RP-C_{18}, RP-cyclohexyl, RP-phenyl), is the result of apolar Van der Waals forces that play the key role in reversed phase mechanisms. Differences in the polarity and solubility of the phenols between the aqueous and the solid apolar phase cause the mass transfer leading to partition process. For this reason, the efficiency of sorbent is related to the octanol-water partition coefficient, Log P (**Tables 1** and **2**), which is a measure of hydrophobicity (or lipophilicity) of the compound. The smaller is the value of Log P, the more limited is sorbent efficacy towards the compound. Conversely, increase in Log P provides its better affinity to the sorbent.

In sorption of phenols using hydrophobic polymers (e.g. PS-DVB (copolymer styrene-divinylbenzene) or hypercrosslinked resins) and graphitized carbons, the possible interactions are also hydrophobic ones. The partition process occurs due to combination of apolar Van der Waals and π–π electrons interactions of the aromatic rings of the adsorbate and adsorbent. Therefore, the retention is the result of the reversed phase mechanism, too [18].

Unfortunately, for very polar phenols, the described interactions in many cases are insufficient to carry out the process of sorption efficiently. In order to facilitate the retention, two methods can be applied:

- Introduction of polar groups (acetyl, hydroxymethyl, benzoil, o-carboxybenzoil, carboxylic, sulfonic, amino [24]) to the hydrophobic skeleton of already existing sorbent [27, 42]. This method is applicable to preparation of chemically modified polymers, activated or grafted carbons [42].
- Synthesis of a new sorbent using compounds with polar moieties. In this way, silica-based materials: RP-cyano, RP-hydroxyl, RP-amine [43], polar polymers [2, 21, 25–28] and chemically activated or porous carbons [21, 27–29, 39] can be produced.

Resulting materials still retain the high capacity towards less hydrophilic compounds and gain possibility to specifically interact with polar molecules due to stronger interactions including: dipole-dipole, hydrogen bonding or even ionic if sulfonic or carboxylic or amine groups are present. Additionally, polar surface is more wettable and consequently support mass transfer of the more polar species from aqueous solution to the sorbent [18], thus influencing the selectivity of sorption process and the capacity of the sorbent [26].

In a situation when a solution consists of phenols with significantly different properties, one sorbent may be insufficient to trap all the compounds quantitatively. In such cases, the possibility to achieve high retention and selectivity for all analytes is the preparation of device containing mixed or layered sorbents with different chemical characters [18, 20, 21].

4. Influence of different factors on process of phenols sorption

Process of sorption of phenolic compounds is very complex and many factors exert influence on its efficiency. Some of them were already described above. Others are presented below. The factors are arranged into three groups: the first, connected with adsorbents; the second, associated with adsorbate properties; and the last, related to environmental conditions. Nevertheless, one should remember that none of the factors is working as separate. All of them make a system in which each element is closely linked with others, and changing one of them affects the rest.

4.1. Adsorbent properties

4.1.1. Physical properties of an adsorbent

Many studies were carried out to find the correlation between specific surface area of sorbent and uptake of phenolic compounds. It is clear that as higher is the value of specific surface area, the higher are microporosity and the larger number of active sites available to interact with adsorbate. Consequently, extraction efficiency increases [23]. That is why, for hyper-crosslinked polymers, higher sorption capacity is observed in comparison with traditional polymeric sorbents [18, 21].

On the other hand, there are many works indicating lack of correlation between the value of specific surface area and ability to adsorb phenols [16, 24, 25, 27, 33, 41]. However, specific surface area is still put into characteristic in order to define the morphology of the adsorbent and for comparative purposes.

In general, an adsorption process consists of three following steps:

- transport of the adsorbates from the bulk solution to the adsorbent exterior surface,
- diffusion of molecules to the pore of adsorbent, although a small amount of adsorbate is retained on the external surface and
- proper adsorption of the solutes on the interior surface of the micropores and in capillary spaces of the adsorbent. This final step is assumed to be rapid and leads to state of equilibrium [16, 33]. A driving force of that process is an adsorption potential near the walls of micropores [23].

Consequently, adsorption of phenols is controlled by diffusion processes which are the slowest stage of the overall process. Therefore, characteristic features of porous structure like type of pores, their shape, size, a way of their connection and a size distribution are the most important factors determining the adsorption process and deciding on proper application [31, 44].

In case of sorption processes realized in a dynamic mode (e.g. chromatography, solid phase extraction), size, shape and uniformity of sorbent particles also play significant role. These parameters have an impact on a length of diffusional path, availability of pores for adsorbed molecules, mass transfer resistance and contact time. Not meaningless is the geometry of a device (e.g. column, cartridge, sampler) as its length and diameter decide on the rate and efficiency of the process.

4.1.2. Chemical properties of an adsorbent

Chemical composition of the adsorbent surface determines the mechanism of phenols adsorption. As it is described above, the retention of phenols on the solid surface is the result of specific and nonspecific interactions. The main role is played by strong dipole-dipole interactions and hydrogen bonds between phenolic compounds and surface functional groups. Equally important are π electrons interactions of the aromatic systems of the adsorbate and adsorbent (if contains) [26, 32].

Thus, retention of the polar phenols is enhanced by the presence of polar moieties on the sorbent surface, while sorption of the hydrophobic molecules and those forming strong intramolecular hydrogen bonds is not influenced by the polarity of an adsorbent surface [25].

The studies show that functional groups being active in retention of phenolic species are mainly located in larger micropores, and the increase in their concentration on the surface leads to the favoring of adsorption in larger pores. Consequently, the competition takes place between filling of the smallest micropores and the adsorption on active sites located in larger micropores [33].

The moderately and slightly acidic oxygen groups are as considered as the most important ones influencing the mechanism of phenols adsorption. The vital surface groups for this process are bases and carbonyls [18, 33, 45]. According to Su et al., the increase in concentration of carbonyl groups on the surface provides more sites for the donor-acceptor interactions resulting in the improvement in phenol adsorption [23].

In addition, other polar moieties, for example nitrile [25], amine [24] or bicarbonate [27] on the sorbent surface, can positively influence the uptake of phenol and its derivatives by the formation of hydrogen bonds.

Many researchers observe decrease in adsorption of phenol with the increase in surface acidity and with surface oxidation [33, 46, 47]. This phenomenon can be explained by

- strong interactions of phenols with the surface groups that lead to the creation of polymorphic forms of phenol [33],
- hydration process of polar groups (e.g. carboxyl) leading to creation of water clusters, which occupy active sites. The formed water complexes block the pore entrances and thus reduce surface available for sorption resulting in decrease of phenol adsorbability [23, 48].

In case of adsorption from solutions, the role of surface functionalities increases more and maybe even dominant [49]. The surface of adsorbent can be treated as a huge molecule covered with uniform (e.g. polymers) or various (e.g. active carbons) in chemical character functional groups. If they are able to dissociate, ions are released to the solution. Their presence affects the acid-base equilibrium of the solvent. In order to characterize this process, point of zero charge (PZC) [44, 50], contents of acidic and basic groups [47] and their pKa [45, 50], or pH of sorbent slurry [48], are determined.

The ions on surface, those in solution and adsorbate molecules interact with one another by electrostatic attraction or repulsion. Confirmation of this phenomenon was presented in Tamon and Ozaki's studies, who found that sorption characteristic of phenols in aqueous solution depends on electronic states of adsorbent surface and adsorbate [51]. The same phenomenon controls adsorption of acids on surface with basic groups and bases on the acidic ones of ion exchangers or porous carbons [32].

4.2. Adsorbate properties

Adsorption of small organic molecules, especially containing functional groups, is affected by surface chemistry. The bigger is the molecule, the weaker is the effect due to the steric hindrance. As phenols possess in their molecules aromatic ring and at least one polar group (-OH), both these parts are active in competition for high-energy adsorption sites. The strength of retention of the molecule depends on size of molecule (the number of carbon atoms) and specific interaction caused by polar moiety [32]. The presence of substituents in the ring changes physical, chemical and energetic characteristics of the molecule, resulting in additional strengthening or weakening impact on the sorption process. It was also found that solutes with electron-donating groups show a tendency to irreversible adsorption, while those with electron-withdrawing groups do not [51].

Thus, for example, chloro or nitro groups which reduce the electron density of the aromatic ring cause growth of molecule hydrophobicity and decrease in water solubility. In consequence, interactions between the adsorbed molecules and hydrophobic surface of adsorbent become stronger and an increase in adsorption capacity is observed [17, 46, 49, 52, 53]. For polar adsorbents, the effect will be opposite [25, 35].

4.2.1. Size of molecule

The molecular size of the adsorbate has a crucial meaning for adsorption process. The rate of mass transfer is controlled by diffusion of adsorbates in porous structure of the adsorbent. However, availability of the pores strictly depends on size of the molecule. The smallest molecules penetrate almost all pores, while for the larger ones, not all of them are accessible. This elimination of large molecules from entering the pores describes size exclusion mechanism. As a result, porous materials act as a molecular sieve. An increase in adsorption resulting from the rise of adsorption energy is supposed if the pore size is about twice the size of kinetic diameter of adsorbate molecule [46].

The species with branched or bulky substituents, which are too large to reach interior of micropores, are retained in mesopores [31]. Desorption of such solutes can be hindered due to being trapped inside pores.

4.2.2. Multicomponent effect

For multicomponent solutions additional effects become apparent. If the mixture consists of solutes having molecules of similar size, a competition between them for the active sites takes place. Compounds possessing higher affinity to adsorbent interact stronger with its surface and displace molecules interacting weaker [22, 27]. Hence, it follows a selectivity of adsorbent towards the specific compounds.

On the other hand, if molecules that are significantly different in size are present in the mixture, the larger of them can block entrances of some pores effectively preventing adsorption of the smaller species [31].

Both the effects adversely affect the sorption and reduce the overall efficiency of the process

4.2.3. Initial concentration and contact time

Increase in the phenols concentration in solution leads to an increase in the amount of adsorbed compounds. This can be attributed to a rise of the driving force of the concentration gradient with the rise in the initial phenols concentration and different arrangements of solute molecules on the surface of adsorbent [30]. At low concentrations, phenol molecules are lying flat in parallel with the surface. As a result, the surface occupied per molecule corresponds to the largest of its dimensions [54, 55]. At higher concentrations, accessibility of the surface for the individual molecules is considerably less. Consequently, the molecules have to change their position for upright to allow closer packing in parallel to each other. Then the area occupied in conversion to a single molecule is smaller that is observed as rise in adsorption density [55].

Different studies show that the contact time required for phenols solution to reach the equilibrium is in the range of 1–24 h [30, 56, 57]. However, the uptake of phenols is the most rapid during the initial 30 min [56, 57]. It was also observed that the higher initial concentration, the longer equilibrium times were needed [30, 56]. Shaarani and Hameed explained the run of the process as follows: "initially the adsorbate molecules have to first encounter the boundary layer effect and then diffuse from the boundary layer film onto adsorbent surface and then finally, they have to diffuse into the porous structure of the adsorbent. This phenomenon takes relatively long contact time" [30].

Moreover, it was noted that at relatively large phenol adsorption values, when interactions between adsorbate molecules predominate, phenol uptake is higher for the more oxidized surfaces, whereas, in case of diluted phenol solutions, the competition between phenol molecules and water leads to substantial reduction in adsorption of phenol on oxidized surface. "Competition between water adsorbing in pores and phenol is mainly responsible for the changes in adsorption" [33].

4.3. Environmental conditions

4.3.1. Water and other solvents

Water molecules interact very weakly with the hydrophobic surface unless their concentration in adsorbed mixture is low (e.g. sorption from gas phase). If the contents of water rise like in case of sorption from aqueous solution or in head space techniques, the influence of water interaction with the surface is no longer negligible. Water begins to interfere with adsorption of organic compounds. Due to the better affinity to the surface, organic solutes are able to replace preadsorbed water. That impacts the kinetics and efficacy of the process. The effect is even stronger, if the surface is hydrophilic. The polar functional groups are predominantly adsorption centers capable of forming the hydrogen bonding. Interactions between water molecules lead to formation of clusters, which block entrances of pore as well as "the condensation of water in micropores at much lower humidity than that at which it happens on a fully hydrophobic surface" can take place [32, 33]. In consequence, adsorption of phenols diminishes.

From practical point of view both adsorption and desorption processes are equally important. Adsorption of phenols is usually carried out from aqueous solution. However, to desorb retained compounds, organic solvent is necessary. It should be chosen so that its properties (solvating power, good wettability of adsorbent, miscibility, chromophoric nature and purity) provided quantitative, quick and simple process of recovery.

For elution of adsorbed phenolic compounds, the most frequently used solvents are methanol [58], acetonitrile [59], acetone [18], ethyl acetate [18], tetramethylammonium hydroxide [17] and others [43]. If a single eluent is ineffective, a mixture of solvents can be considered. The volume of applied eluent depends on its eluting strength, polarity of the phenols, amount and surface chemistry of adsorbent, and type of device [18].

Bruzzoniti et al. in their paper presented efficiency of different eluents in desorption of phenolic compounds from graphitized carbon black. They noted the effectiveness of eluent was affected by the value of pKa phenolic solute. Phenols with pKa about 7 were effectively eluted with tetramethylammonium hydroxide, while for phenols with higher pKa, methanol was better [17].

For newly synthesized sorptive materials, studies on eluent selection and optimization of desorption process should be conducted [18, 59].

4.3.2. Electrolytes: ionic strength

The presence of electrolytes, for example salts in the solution, affects the strength of electrostatic interaction between phenolic solutes and the surface of the adsorbent. If these interactions are attractive and surface concentration of adsorbates is low, then the rise in ionic strength of solution causes decrease in uptake. But if they are repulsive or concentration of adsorbates on the surface is high, increase in ionic strength of the solution enhances the adsorption [60].

The presence of salts can influence the adsorption of phenols also by the effect of salting out. The electrolytes dissolved in the solution reduce or even destroy the hydration layer of organic compound causing decrease in its solubility, which affects favorably on adsorption.

Kuśmiderek and Świątkowski have compared the influence of an electrolyte on the adsorption of 4-chlorophenol on active carbon and multiwalled carbon nanotubes (MWCNTs). They used three different salts: Na_2SO_4, NaCl, $NaNO_3$ at various concentrations. The presented results proved higher uptake of 4-chlorophenol on the active carbon in the presence of salts in all cases, whereas no significant difference was observed for adsorption on MWCNTs. It was also indicated that the desorption of the 4-chlorophenol increased with the increase in the ionic strength of the solution [60].

Dissimilar results were obtained by Mukherjee and De for sorption of catechol on mixed matrix membrane in the presence of NaCl. They justified their outcome by weakening forces between sorbent and phenols resulting in decrease of the uptake. They also defined this phenomenon as the shielding effect [22].

4.3.3. pH

The change in pH affects chemistry of both the adsorbate and the adsorbent by shifting the equilibrium dissociation process of solutes and surface functional groups of the sorbent towards ionized or unionized form. Thus, adsorption of phenols is limited by the acid-base characteristic of the adsorbent and its microporosity, which subsequently influences kinetics and effectiveness of the overall process. Higher values of uptake are observed for the compounds in undissociated form [30, 61]. Therefore, most of phenols such as weak acids are better adsorbed from neutral or acidic solutions. At lower pH values, oxonium ions are present in solution and they prevent dissociation of surface acidic groups. In these conditions, formation of hydrogen bonds and π-π interactions between phenolic adsorbates and sorbent surface is privileged. Accordingly, adsorption capacity is the highest [19]. The same applies to phenols, their transition point between acidic and basic form is associated with their pKa value. Below this value dominates acidic form and above it the conjugated base-phenoxide ion takes advantage.

The uptake of phenols decreases at higher pH, when the molecules turn into phenolate ions and the surface of the adsorbent is negatively charged. The presence of electrostatic repulsive forces impedes adsorption [16, 57]. Thus, the more acidic the surface of sorbent, the lower the adsorbability of phenols.

However, some researchers report the in certain pH range, adsorption for most phenols initially increases with the rise in pH to reach a certain value and then decreases with further increase in pH [16, 56]. This phenomenon can be explained by the fact that, in the presence of oxonium ions the surface functional groups of adsorbent gain a positive charge. As a result, adsorption of water increases and cluster formation takes place. Consequently, some of more active sites are not available. Second, already adsorbed molecules of phenols can block the entrances to fine pores and cause decrease in uptake. This effect is stronger at the lowest pH values [33].

4.3.4. Temperature

Due to the fact that process of physical adsorption is exothermic in nature, increase of temperature causes decrease in phenol's uptake [56]. Moreover, the rise in temperature increases the rate of the sorption process and leads to the disappearance of the effects associated with surface chemical composition [33].

In contrast, chemisorption is an endothermic process. Therefore, Shaarani and Hameed observed the increase of temperature slightly increases the uptake of 2,4-dichlorophenol at a higher concentration. They also found that the change of temperature alters the equilibrium capacity of the adsorbent for particular adsorbates [30].

4.3.5. Oxic and anoxic conditions

Influence of oxygen for process of phenols sorption was at first perceived for carbon adsorbents; hence, the facts presented here have been developed only based on the results obtained for these materials.

When phenols are adsorbed under oxic and anoxic conditions, differences in the uptake can be observed. The sorption capacities of some sorbents are higher under oxic conditions that imply promoting action of molecular oxygen in this process [3, 49]. The increase in adsorption is attributed to the oxidative coupling reaction leading to formation of timers or more substituted derivatives of phenols (multimers) that are attached to the surface of adsorbent [33, 55].

http://dx.doi.org/10.5772/66537

The course of the process can be explained as follows: the molecules of phenols and those of oxygen diffuse to the surface of carbon, where they are preadsorbed. The adsorbed oxygen molecules form superoxo ions that yield rearrangements and further reactions with surface functional groups and adsorbed phenols leading to creation of various surface compounds. Thus, described reaction occurs only on the surface of carbon and is catalyzed by it [33].

Studies on this phenomenon revealed that

- the small pore size precludes oligomerization, thereby the presence of molecular oxygen has no significant effect on adsorption [3];

- narrow pore size distribution (PSD) reduces the oligomerization of phenols to some extend while the wider PSD favors it [1];

- 2-methylphenol shows oligomerization under oxic condition. The oligomerization of 2-chlorophenol depends on the overall competitive effect of 2-nitrophenol and 2-methylphenol [1];

- surface concentration of carboxylic groups diminishes the oxidative coupling in the whole range of coverages [62];

- the lower ability of carbon to adsorb ions on the surface, the more significant differences in adsorption determined at oxic and anoxic conditions [33];

- oxidative polymerization (spontaneous and/or electrochemical) of chlorophenol is responsible for a lack of anodic peak during voltammetric cyclization [5].

4.3.6. Irreversible sorption of phenols

Irreversible adsorption of phenols is closely related to process of chemical adsorption involving chemical reaction between adsorbates and surface functional groups of adsorbent [55]. As it was described above, the hydrogen bonding arising in the system adsorbate-active centers of adsorbent strengthens interactions and sorptive capacity, but also can cause difficulties related to irreversible sorption [18]. It can be also considered that phenol molecules or phenoxide radicals can react with active sites on a carbon surface, which leads to creation of covalent bond between them [33].

Tessmer et al. investigated this problem and found that "grater structural ordering and delocalized electrons on the carbon surface may increase the basicity of the carbon but do not enhance its ability to promote irreversible adsorption. The presence of oxygen containing basic groups (chromene-type or pyrone-type) is likely the key factor promoting irreversible adsorption" [48].

Other studies demonstrate that lactone surface groups are responsible for both irreversibility of phenol adsorption due to chemical reaction and decrease in the difference between adsorption in oxic and anoxic conditions [33].

And another one proves that irreversible adsorption to carbon surfaces does not depend on oxidative coupling reaction, but oxidative coupling may enhance irreversible adsorption when experimental conditions are conductive (pH of solution high enough, microporosity of adsorbent) [63].

The discrepancies in this matter indicate that it is a topic still open for discussion and further research.

5. Conclusions and future prospects

A lot of scientific research was performed in order to study and understand the process of sorption of phenolic compounds. This chapter presents the complexity of this process and shows how many different factors have influence on it. The starting point is chemical structure and overview of physical and chemical properties of phenol and its ring-substituted derivatives. This characteristic is the basis to explain the behavior of the molecules in solution and during sorption. Next, the way phenols interact with the most popular sorbents are shortly described. Finally, factors influencing the sorption process were characterized.

Factors promoting the sorption:

- matching of the size of adsorbate molecules to the porous structure of adsorbent;

- the retention process of phenolics is mainly due to reversed phase mechanism, but in case of polar sorbents, it is also supported by specific attractive interaction (dipole-dipole, donor-acceptor and ionic);

- elongation of contact time of a phenols solution with an adsorbent;

- salting effect;

- pH. Sorption characteristic depends on electronic states of both an adsorbent surface and an adsorbate. Phenols uptake is higher for compounds in unionized form as formation of hydrogen bonds and π-π interactions are favored;

- increase in temperature in case of chemisorption process.

Factors hindering the sorption:

- polar functional groups of the sorbent acts as adsorption centers. Their presence in larger pores can limit access to micropores (pore blocking effect) and lead to competition between the process of micropore filling and adsorption on active sites;

- steric effects and size exclusion mechanism;

- pore blocking caused by adsorption of large molecules or water clusters;

- multicomponent effect and displacement phenomena;

- shielding effect;

- increase in temperature in case of physisorption process.

Irreversibility of phenols adsorption can be the result of

- a chemical reaction between adsorbate and surface functional groups;
- an oxidative coupling process in oxic conditions;
- trapping molecules (especially large or branched ones) in pores of adsorbent;
- the presence of electron donating substituents in phenol molecule enhances the effect of irreversible adsorption.

Although so many works were devoted to studies on phenols sorption, the general mechanism of the process is still not fully understood and explained. There are some issues and unsolved problems requiring further investigations:

- There is still no consensus in the matter, what kind of the adsorbent functional groups has favorable and which undesirable influence on adsorption of phenols.
- Process of irreversible adsorption due to oxidative coupling requires additional studies confirming the proposed in literature mechanisms.
- Most of the studies concern the process of sorption of single compounds. Only a few research studies relate to multicomponent system, in which competitive adsorption take place. This aspect is particularly important from environment protection point of view. Interferents, such as ions from salts or organic contaminants of natural or anthropogenic origin, impede quantitative analysis of phenols in waste water.
- Relatively little studied are processes of cyclic adsorption—desorption, reproducibility of the adsorption properties and regeneration of adsorbent, which are extremely important for utility reasons.

Author details

Magdalena Sobiesiak

Address all correspondence to: magdalena.sobiesiak@umcs.pl

Department of Polymer Chemistry, Faculty of Chemistry, Maria Curie-Skłodowska University, Lublin, Poland

References

[1] Lu Q, Sorial GA: A comparative study of multicomponent adsorption of phenolic compounds on GAC and ACFs. J Hazard Mater 2009; **167**(1–3):89–96. DOI: 10.1016/j.jhazmat.2008.12.086

[2] Sobiesiak M: Bead-shaped porous polymers containing bismaleimide – their physicochemical characteristics and sorption properties towards chlorophenols. Polish J Appl Chem. 2011;**LV**(1):25–32.

[3] Lu Q, Sorial GA: Adsorption of phenolics on activated carbon--impact of pore size and molecular oxygen. Chemosphere 2004;**55**(5):671–679. DOI: 10.1016/j.chemosphere.2003.11.044

[4] Rodil R, Quintana JB, Concha-Graña E, López-Mahía P, Muniategui-Lorenzo S, Prada-Rodríguez D. Emerging pollutants in sewage, surface and drinking water in Galicia (NW Spain). Chemosphere 2012;**86**(10):1040–1049. DOI: 10.1016/j.chemosphere.2011.11.053\

[5] Biniak S, świątkowski A, Pakuła M, Sankowska M, Kuśmiderek K, Trykowski G: Cyclic voltammetric and FTIR studies of powdered carbon electrodes in the electrosorption of 4-chlorophenols from aqueous electrolytes. Carbon 2013;**51**:301–312. DOI: 10.1016/j.carbon.2012.08.057

[6] Velasco LF, Ania CO: Understanding phenol adsorption mechanisms on activated carbons. Adsorption 2011;**17**(1):247–254. DOI: 10.1007/s10450-011-9322-x

[7] Terzyk AP, Gauden PA., Zieliński W, Furmaniak S, Wesołowski RP, Klimek KK: First molecular dynamics simulation insight into the mechanism of organics adsorption from aqueous solutions on microporous carbons. Chem Phys Lett 2011;**515**(1–3):102–108. DOI: 10.1016/j.cplett.2011.08.093

[8] Gold Book of IUPAC [Internet]. Available from: http://goldbook.iupac.org/P04437.html

[9] Morrison RT, Boyd R N. Chemia Organiczna. 4th ed. Warszawa: Państwowe Wydawnictwo Naukowe; 1985. 966p.

[10] Smith JG. Organic chemistry. 3rd ed. New York: McGraw-Hill; 2011. 1285p.

[11] PubChem database. Available from: https://www.ncbi.nlm.nih.gov/pccompound.

[12] Poradnik fizykochemiczny, WNT, Warszawa, 1974 – values of dipole moments.

[13] Maartmann-Moe. The crystal and molecular structure of phloroglucinol. Acta Crystallogr. 1965;**19**;155–157. DOI:10.1107/S0365110X65002992

[14] Gorbitz CH, Kaboli M, Lovell M, Vestli K: Benzene 1,3,5-triol at 105 K. Acta Crystallogr Sect E. 2008;**64**:48–52. DOI: 10.1107/S1600536808030638

[15] Oswald IDH, Motherwell WDS: Formation of quinol co-crystals with hydrogen-bond acceptors. Acta Crystallogr Sect B. 2005;**61**:46–57. DOI: 10.1107/S0108768104028605

[16] Ahmaruzzaman M: Adsorption of phenolic compounds on low-cost adsorbents: a review. Adv Colloid Interface Sci. 2008;**143**(1–2):48–67. DOI: 10.1016/j.cis.2008.07.002

[17] Bruzzoniti MC, Sarzanini C, Mentasti E: Preconcentration of contaminants in water analysis. J Chromatogr A. 2000;**902**:289–309.

[18] Rodríguez I, Llompart MP, Cela R: Solid-phase extraction of phenols. J Chromatogr A. 2000;**885**(1–2):291–304. Available from: http://www.ncbi.nlm.nih.gov/pubmed/10941678

[19] An F, Du R, Wang X, Wan M, Dai X, Gao J: Adsorption of phenolic compounds from aqueous solution using salicylic acid type adsorbent. J Hazard Mater. 2012;**201–202**:74–81. DOI: 10.1016/j.jhazmat.2011.11.037

[20] Hennion MC: Solid-phase extraction: method development, sorbents, and coupling with liquid chromatography. J Chromatogr A. 1999;**856**(1–2):3–54. Available from: http://www.ncbi.nlm.nih.gov/pubmed/10526783

[21] Fontanals N, Marcé RM, Borrull F: New materials in sorptive extraction techniques for polar compounds. J Chromatogr A. 2007;**1152**(1–2):14–31. DOI: 10.1016/j.chroma.2006.11.077

[22] Mukherjee R, De S: Adsorptive removal of phenolic compounds using cellulose acetate phthalate-alumina nanoparticle mixed matrix membrane. J Hazard Mater. 2014;**265**:8–19. DOI: 10.1016/j.jhazmat.2013.11.012

[23] Su F, Lv L, Hui TM, Zhao XS: Phenol adsorption on zeolite-templated carbons with different structural and surface properties. Carbon 2005;**43**:1156–1164. DOI: 10.1016/j.carbon.2004.12.0341.

[24] Pan BC, Zhang X, Zhang WM, Zheng JZ, Pan BJ, Chen JL, Zhang QX: Adsorption of phenolic compounds from aqueous solution onto a macroporous polymer and its animated derivative: isotherm analysis. J Hazard Mater. 2005;**121**(1–3):233–241. DOI: 10.1016/j.jhazmat.2004.10.027

[25] Trochimczuk AW, Streat M, Kolarz N: Highly polar polymeric sorbents Characterization and sorptive properties towards phenol and its derivatives. React Funct Polym. 2001;**46**:259–271

[26] Drechny D, Trochimczuk AW: Synthesis and some sorptive properties of highly crosslinked cyanomethyl styrene/divinylbenzene copolymers. React Funct Polym. 2006;**66**:323–33. DOI: 10.1016/j.reactfunctpolym.2005.10.024

[27] Sobiesiak M, Podkościelna B, Podkościelny P: New functionalised polymeric microspheres for multicomponent solid phase extraction of phenolic compounds. Adsorption. 2016;**22**(4–6):653–662. DOI: 10.1007/s10450-015-9749-6

[28] Sobiesiak M, Podkościelna B: SPE of phenols on new amine – based copolymers. Key Eng Mater. 2014;**605**:173–176. DOI: 10.4028/www.scientific.net/KEM.605.173

[29] Jaroniec M, Gorka., Choma J, Zawiślak A: Synthesis and properties of mesoporous carbons with high loadings of inorganic species. Carbon 2009;**47**; 3034–3040. DOI: 10.1016/j.carbon.2009.06.059

[30] Shaarani FW, Hameed BH: Batch adsorption of 2,4-dichlorophenol onto activated carbon derived from agricultural waste. Desalination 2010;**255**(1–3):159–164. DOI:10.1016/j.desal.2009.12.029

[31] Nakagawa K, Namba A, Mukai SR, Tamon H, Ariyadejwanich P, Tanthapanichakoon W: Adsorption of phenol and reactive dye from aqueous solution on activated carbons derived from solid wastes. Water Res. 2004;**38**(7):1791–1798. DOI: 10.1016/j.watres.2004.01.002

[32] Bandosz T, Ania CO: Surface chemistry of activated carbons and its characterization. In: Activated Carbon Surfaces in Environmental Remediation. 1st ed. New York: Elsevier Ltd; 2006. p. 159–229.

[33] Terzyk AP: Further insights into the role of carbon surface functionalities in the mechanism of phenol adsorption. J Colloid Interface Sci. 2003;**268**(2):301–329. DOI: 10.1016/S0021-9797(03)00690-8

[34] Sobiesiak M: Nanoporous carbons obtained by carbonization of copolymers impregnated by salts. Adsorption 2012;**19**(2–4):349–356. DOI: 10.1007/s10450-012-9457-4

[35] Gawdzik B, Sobiesiak M, Puziy AM, Poddubnaya OI: carbon sorbents derived from porous polymers for off-line preconcentration of chlorophenols from water. J Liq Chromatogr Relat Technol 2004;**27**(6):1027–1041. DOI: 10.1081/JLC-120030176

[36] Srihari V, Das A: The kinetic and thermodynamic studies of phenol-sorption onto three agro-based carbons. Desalination 2008;**225**(1–3):220–234. DOI: 10.1016/j.desal.2007.07.008

[37] Bratek W, Swiątkowski A, Pakuła M, Biniak S, Bystrzejewski M, Szmigielski R: Characteristics of activated carbon prepared from waste PET by carbon dioxide activation. J Anal Appl Pyrolysis. 2013;**100**:192–198. DOI: 10.1016/j.jaap.2012.12.021

[38] Sobiesiak M: Preparation of porous carbons from polymeric precursors modified with acrylated kraft lignin. IOP Conf Ser Mater Sci Eng. 2016;**123**:012052–012057. DOI: 10.1088/1757-899X/123/1/012052

[39] Choma J, Marszewski M, Osuchowski L, Jagiello J, Dziura A, Jaroniec M: Adsorption properties of activated carbons prepared from waste CDs and DVDs. ACS Sustainable Chemistry & Engineering 2015;**3**;733–742 DOI: 10.1021/acssuschemeng.5b00036

[40] Jancar J, Douglas JF, Starr FW, Kumar SK, Cassagnau P, Lesser AJ, Sternstein SS, Buehler MJ: Current issues in research on structure e property relationships in polymer nanocomposites. Polymer 2010;**51**(15):3321–3343. DOI: 10.1016/j.polymer.2010.04.074

[41] Podkościelna B, Sobiesiak M: Synthesis and characterization of organic–inorganic hybrid microspheres. Adsorption 2016;**22**(4–6):631–638. DOI: 10.1007/s10450-015-9741-1

[42] Stein A, Wang Z, Fierke MA: Functionalization of porous carbon materials with designed pore architecture. Adv. Mater. 2009,**21**,265–293. DOI: 10.1002/adma.200801492

[43] Simpson, NJK: Solid-Phase Extraction – Principles. Techniques and Applications. 1st ed. New York: Marcel Dekker. Inc., 2000. 514p.

[44] Yan L, Fitzgerald M, Khov C, Schafermeyer A, Kupferle MJ, Sorial G A: Elucidating the role of phenolic compounds in the effectiveness of DOM adsorption on novel tailored activated carbon. J Hazard Mater 2013;**262**:100–105. DOI: 10.1016/j.jhazmat.2013.08.022

[45] Montes-Morán MA, Suárez D, Menéndez JA, Fuente E: On the nature of basic sites on carbon surfaces: an overview. Carbon 2004;**42**(7):1219–1225. DOI: 10.1016/j.carbon.2004.01.023

[46] Derylo-Marczewska A, Buczek B, Świątkowski A. Applied Surface Science Effect of oxygen surface groups on adsorption of benzene derivatives from aqueous solutions onto active carbon samples. Appl Surf Sci 2011;**257**(22):9466–9472. DOI: 10.1016/j.apsusc.2011.06.036

[47] Salame II, Bandosz TJ: Role of surface chemistry in adsorption of phenol on activated carbons. J Colloid Interface Sci. 2003;**264**:307–312, DOI: 10.1016/S0021-9797(03)00420-X

[48] Tessmer C: Impact of oxygen-containing surface functional groups on activated carbon adsorption of phenols. Environ Sci. 1997;**31**(7):1872–1878. DOI: 10.1021/es960474r

[49] Dąbrowski A, Podkościelny P, Hubicki Z, Barczak M: Adsorption of phenolic compounds by activated carbon — a critical review. Chemosphere 2005;**58**:1049–1070. DOI: 10.1016/j.chemosphere.2004.09.067

[50] Puziy AM, Poddubnaya OI, Sobiesiak M, Gawdzik B. Structural and surface heterogeneity of phosphorus-containing polyimide-derived carbons: effect of heat treatment temperature. Adsorption 2013;**19**(2–4):717–722. DOI: 10.1007/s10450-013-9497-4

[51] Tamon H, Okazaki M: Desorption characteristics of aromatic compounds in aqueous solution on solid adsorbents. J Colloid Interface Sci 1996;**187**(179):181–187. Available from: http://www.sciencedirect.com/science/article/pii/S0021979796902003

[52] Radovic L, Silva I, Ume J, Menendez J: An experimental and theoretical study of the adsorption of aromatics possessing electron-withdrawing and electron-donating functional groups by chemically modified activated carbons. Carbon 1997;**35**(9):1339–1348. Available from: http://www.sciencedirect.com/science/article/pii/S0008622397000729

[53] Terzyk AP: Molecular properties and intermolecular forces--factors balancing the effect of carbon surface chemistry in adsorption of organics from dilute aqueous solutions. J Colloid Interface Sci 2004;**275**(1):9–29. DOI: 10.1016/j.jcis.2004.02.011

[54] Furuya E, Chang H, Miura Y, Noll K: A fundamental analysis of the isotherm for the adsorption of phenolic compounds on activated carbon. Sep Purif. 1997;**1**(96):69–78. Available from: http://www.sciencedirect.com/science/article/pii/S1383586696010015,

[55] Grant TM, King CJ: Mechanism of Irreversible Adsorption of Phenolic Compounds by Activated Carbons. Ind Eng Chem Res. 1990;**29**:264–271. DOI: 10.1021/ie00098a017

[56] Khare P, Kumar A: Removal of phenol from aqueous solution using carbonized Terminalia chebula-activated carbon: process parametric optimization using conventional method and Taguchi's experimental design, adsorption kinetic, equilibrium and thermodynamic study. Appl Water Sci 2012;**2**(4):317–326. DOI: 10.1007/s13201-012-0047-0

[57] Srivastava VC, Swamy MM, Mall ID, Prasad B, Mishra IM: Adsorptive removal of phenol by bagasse fly ash and activated carbon: equilibrium, kinetics and thermodynamics. Colloids Surfaces A Physicochem Eng Asp. 2006;**272**(1–2):89–104. DOI: 10.1016/j.colsurfa.2005.07.016

[58] Sobiesiak M, Gawdzik B, Puziy AM, Poddubnaya OI: polymer-based carbon adsorbents obtained from copolymer of 4,4′-Bis(maleimidodiphenyl)methane and divinylbenzene for use in SPE. Chromatographia 2006;**64**(3–4):175–181. DOI: 10.1365/s10337-006-0798-6

[59] Feng QZ, Zhao LX, Yan W, Lin JM, Zheng ZX: Molecularly imprinted solid-phase extraction combined with high performance liquid chromatography for analysis of phenolic

compounds from environmental water samples. J Hazard Mater. 2009;**167**(1–3):282–288. DOI: 10.1016/j.jhazmat.2008.12.115

[60] Kuśmierek K, Świątkowski A: The influence of an electrolyte on the adsorption of 4-chlorophenol onto activated carbon and multi-walled carbon nanotubes. Desalin Water Treat. 2015;**56**:2801–2816. DOI: 10.1080/19443994.2014.965221

[61] Terzyk AP, Ćwiertnia MS, Wiśniewski M, Gauden PA., Rychlicki G, Szymański GS: Impact of the interaction with the positive charge in adsorption of benzene and other organic compounds from aqueous solutions on carbons. Appl Surf Sci. 2007;**253**(8):4006–4009. DOI: 10.1016/j.apsusc.2006.08.042

[62] Terzyk AP: The impact of carbon surface chemical composition on the adsorption of phenol determined at the real oxic and anoxic conditions. Appl Surf Sci. 2007;**253**(13):5752–5755. DOI: 10.1016/j.apsusc.2006.12.039

[63] Cesar A, Pimenta DO, Kilduff JE: Oxidative coupling and the irreversible adsorption of phenol by graphite. J Colloid Interface Sci. 2006;**293**:278–289. DOI: 10.1016/j.jcis.2005.06.075

Chapter 2

Phenolic Compounds from the Natural Sources and Their Cytotoxicity

Shagufta Perveen and Areej Mohammad Al-Taweel

Additional information is available at the end of the chapter

http://dx.doi.org/10.5772/66898

Abstract

Natural phenolic compounds are considered as one of the important secondary metabolites for their chemopreventive and chemotherapeutic effects in cancer. These compounds show potent activities for cancer prevention and its treatment. There are many phenolic compounds present in medicinal and edible plants such as, flavonoids, bioflavonoids, stilbenes, chalconoids, chromones, phenylpropanoids, curcuminoids, coumarins, tannins, lignans, neolignans, anthraquinones, quinones, xanthones, phenolic acids and their glycosides and many more. The antioxidant potential of phenolic compounds is almost bolded in the treatment and prevention of cancer. Mono phenolic, polyphenolic and phenolic acids compounds from a large variety of plants, foods, spices, insects, fungus, beverages, lichens, algae and mammals have been shown to inhibit or attenuate the initiation, progression and spread of cancers in cells in vitro and in animals in vivo. In this chapter, we try to cover general view and the recent literature to summarize structural information and cytotoxic effects of phenolic compounds on different cancer cell lines from medicinal herbs and plants.

Keywords: natural phenolic constituents, cytotoxic activities, cancer cell lines

1. Introduction

Natural products offer opportunities for innovation in drug discovery and play a major role for cancer cure. A considerable number of antitumor agents currently used in the clinic are of natural origin. For instance, over half of all anticancer prescription drugs approved internationally between the 1940s and 2006 were natural products or their derivatives. Among them, plants have been the chief source of natural compounds used for medicine [1]. Many traditional and folk medicinal plants have been use for cancer therapy throughout the world and they work very well for the prevention of cancers. Phenolic compounds play very important role for the treatment and prevention on different types of cancers which are a major health problem around the world. These secondary metabolites associated with the health benefits of human derived from consuming high levels of fresh and dried fruits and vegetables. These compounds with hydroxyl bearing aromatic ring skeleton exhibit wide range of different biological activities including: anti-inflammatory, antioxidant, cytotoxicity, antimicrobial, anti-allergic. According to WHO, 8.2 million people die each year from cancer and it is estimated 13% of all deaths worldwide [2]. In this chapter, we discuss the most recent literature on phenolic constituents from natural sources and their anticancer activities on different cancer cell lines (**Table 1**).

Source	Compound name	Class of compounds	Cytotoxic activity	Refernces
Artocarpus heterophyllus	Artocarpin (**1**), cudraflavone C (**2**), 6-prenylapigenin (**3**), kuwanon C (**4**), norartocarpin (**5**), albanin A (**6**), cudraflavone B (**7**)	Isoprenoid-substituted flavonoids	B16 melanoma cells	[3]
Apple, grapes, tomato green tea, pine and many other	Kaempferol (**8**)	Flavonoid	p53 or PLK-1, in MCF-7 breast cancer and HeLa cervical cancer cells, U-2 OS human osteosarcoma cell, A549 lung cancer cell, Miapaca-2, Panc-1 and SNU-213 human pancreatic cancer cells	[4]
Dorstenia mannii	Dorsmanin F (**9**)	Flavanone	LeukaemiaCCRF-CEM, MDA-MB-231-BCRP, CEM/ADR5000 cells	[5]
Morus mesozygia	Artocarpesin (**10**), cycloartocarpesin (**11**)	Flavonoid	Induced apoptosis in CCRF-CEM leukemia cells	[6]
Oriental tobacco, *Nicotiana tabacum*	6,7-dimethoxy-4'hydroxy-8-formylflavon (**12**), 4',7-dihydroxy-8-formyl-6-methoxyflavon (**13**)	Flavonoid	Human tumor (NB4, A549, SHSY5Y, PC3 and MCF7) cell lines	[7]
Sophora flavescens	Isoxanthohumol (**14**)	Flavonoid	Breast cancer (MCF-7), ovarian cancer (A-2780), prostate cancer (DU145 and PC-3) and colon cancer (HT-29 and SW620) cells, human cytochrome P450 (CYP1A2).	[8]
Kushen *Sophora flavescens*	Kushecarpin D (**15**)	Novel flavonoid	Human umbilical vein endothelial cell line (ECV304), antiangiogenic activity, together with its antiproliferative effect on endothelial cells without causing apoptosis	[9]
Scutellaria barbata	Luteolin (**16**)	Flavonoid	HepG2 and Bel7402 cells, human hepatocellular carcinoma cells, prostate cancer (PCa), assessing the PC3 and LNCaP cells, MCF-7 human breast cancer cells, PC12 cell line	[10–12]

Source	Compound name	Class of compounds	Cytotoxic activity	Refernces
Propolis and in honey	Chrysin (**17**)	Flavonoid	Oropharyngeal KB, mammary LM3, anaplastic thyroid KAT18, anaplastic thyroid HTh7, pancreatic PANC-1, liver H22, gastric SGC-7901, colon HT-29, cervical Hela, melanoma A375, oesophageal OE33, lung A549, colon DLD-1, rectal SW837, breast MDA-MB-231, glioma U87-MG, esophageal squamous KYSE-510, leukemia U937, prostate PC-3, hepatocellular HepG2, acute T-lymphoblastic leukemia CEM, neuroblastoma SH-SY5Y, squamous cell carcinoma FaDu, breast MCF-7, oral SCC-9, prostate DU-145, leukemia K562, cell line PC12	[13]
Macrothelypteris torresiana	DICO (**18**)	Novel nonaromatic B-ring flavonoid	Human hepatoma HepG2 cells	[14]
Mulberry tree *Morus* species	Morusin (**19**)	Prenylated flavonoid	Human hepatoma SK-Hep1 cells	[15]
Artocarpus communis	Artocarpin (**20**)	Prenylated flavonoid	HepG2 and PLC/PRF/5 hepatoma cells, human cutaneous SCC cell line HSC-1, human T47D breast cancer cells	[16, 17]
Daphne genkwa Sieb	Genkwanin (**21**)	Flavonoid	HT-29 and SW-480 human colorectal cancer cells	[18]
Strawberries, apples, persimmons, grapes, onions and cucumbers	Fisetin (**22**)	Flavonoid	Human malignant melanoma cells A375 (CLR- 1619) and RPMI-7951 (HTB-66), SK-MEL-28 cells, A375 and SK-MEL-28 cells, RPMI-7951 cells	[19]
Clerodendrum inerme and various medicinal herbs	Hispidulin (**23**)	Flavonoid	Human HepG2 hepatocarcinoma cell line and the mouse L929 fibroblast cell line, human HEK293 fibroblast cell lines, human renal cancer cell (HRCC) lines 786-0 and Caki-1, PANC-1 cancer cells from human pancreas	[20, 21]
Fruits, vegetables, leaves and grains	Quercetin (**24**)	Flavonoid	HepG2 cells, Human GL-15 glioblastoma cells, breast cancer cells MCF-7 and MCF-7/dox	[22, 23]
All citrus fruits and *Cordia obliqua*	Hesperetin (**25**)	Flavonoid	Human esophageal squamous cell carcinoma Eca109 cells, human breast cancer MCF-7 cells, prostate cancer PC-3 cells, HT-29 human colon adenocarcinoma cell, human cervical cancer cell line HeLa cells	[24]
Hypericum perforatum	Apigenin (**26**)	Flavonoid	MCF-7 breast carcinoma cells, breast tumor cell line MDA-MB231, human cervical carcinoma HeLa cells, human colon carcinoma cell, human leukemia cells, A549 lung cancer cell, human ovarian carcinoma HO-8910PM cells, human prostate cancer PC-3 cell, thyroid carcinoma cell lines, UCLA NPA-87-1 (NPA), cell line PC12	[25, 26]

Source	Compound name	Class of compounds	Cytotoxic activity	Refernces
Azadirachta indica	Quercetin-3-*O*-β-D-glucopyranoside (**27**)	Flavonoid	Protect cells from H_2O_2-induced cytotoxicity	[27]
Polygonum amplexicaule	Amplexicaule A (**28**)	Flavonoid glycoside	Human breast cancer cell lines MCF-7 and MDA-MB-435	[28]
Wheat bran	Triticuside A (**29**)	Flavonoid glycoside	Human breast cancer cells (MCF-7 and MDA-MB-231)	[29]
Citrus fruits	Naringin (**30**)	Flavonoid diglycoside	Breast cancer (TNBC), MDA-MB-231, MDA-MB-468 and BT-549 cells, K562 (human leukemia cell line), Raji (human Burkitt's lymphoma cell line) and NK-92MI	[30, 31]
Filamentous bacterium streptomyces	Quercetin-3-*O*-α-L-rhamnopyranosyl-(1→6)-β-D-glucopyranoside (**31**)	Flavonoid diglycoside	Human lung cancer A549 cells through p53 and cytochrome c	[32]
Cirsium setidens, Aster scaber	Pectolinarin (**32**) and astragalin (**33**)	Flavonoid glycoside	Human brain neuroblastoma SK-N-SH cells	[33]
Sophora japonica	Troxerutin (**34**)	Flavonoid diglycoside	Human prostate cancer radioresistant (DU145) and radiosensitive (PC3) cells	[34]
Bupleurum flavum, Artemisia capillaris	Vicenin (**35**)	Flavonoid glycoside	Hepatocellular carcinoma HEP-G2 cells, human prostate carcinoma LNCaP, PC- 3 & DU-145	[35, 36]
Soybean	Genistein (**36**)	Isoflavone	CIP2A in MCF-7-C3 and T47D breast cancer cells, human prostate cancer cell PC3-M, PC3 and DU-145 human PCa cell lines, sarcomatoid mammary carcinoma cell line F3II, B16F0 melanoma cell line	[37, 38]
Soybean	Daidzein (**37**)	Isoflavone	LnCaP, DU145 and PC3 human prostate cancer cell lines, MCF-7 breast cancer cell	[39]
Ateleia glazioviana	Biochanin A (**38**)	Isoflavone	Pancreatic cancer cells (Panc1 and AsPC-1)	[40]
Erythrina stricta, E. variegata	Alpinum isoflavone (**39**)	Isoflavon	Leukemia CEM/ADR5000 cells, drug-resistant breast adenocarcinoma MDA-MB-231-*BCRP* cells, CEM/ADR5000 cells, colon carcinoma HCT116	[41]
Pueraria lobata	Puerarin (**40**)	Isoflavone-C-glucoside	Myeloid leukemia cell lines, U937, Kasumi-1, HL-60 and NB4 cells, breast cancer MCF-7/adriamycin (MCF-7/adr) cells, colon cancer HT-29 cell	[42]
Toxicodendron vernicifluum	Butein (**41**)	Chalcone	Human colon adenocarcinoma cell line 220.1, human leukemia cells HL-60, $HER2^+$ HCC-1419, HCC-2218 and SKBR-3 breast cancer cells, human PCa (LNCaP, CWR22Rv1 and PC-3), human uveal melanoma cell lines (M17, SP6.5 and C918), HeLa human cervical cancer cell line	[43–47]

Source	Compound name	Class of compounds	Cytotoxic activity	Refernces
Spatholobus suberectus	Isoliquiritigenin (**42**)	Chalcone	Human leukemia cells HL-60 cell, human U373 glioblastoma cells, human hepatoma HepG2) cells, DU145 human prostate cancer cells	[48]
Eugenia aquea	2′,4′-dihydroxy-6-methoxy-3,5-dimethylchalcone (**43**)	Chalcone	Human cell lines liver cancer SMMC-7721 cells, pancreas cancer 8898 cells, tumor of cervix uteri HeLa cells, lung cancer SPC-A-1 cells, high metastatic lung carcinoma 95-D cells and gall bladder carcinoma GBC-SD cells	[49, 50]
Polygonum limbatum	4′-hydroxy-2′,6′-dimethoxychalcone (**44**)	Chalcone	MDA-MB-231-pcDNA3 breast cancer, HCT116 (p53+/+) colon cancer cells/−), the U87MG glioblastoma cells	[51]
Humulus lupus	Xanthohumol (**45**)	Prenylated Chalone	Human breast cancer (MCF-7), colon cancer (HT-29) and ovarian cancer (A-2780) cells	[52, 53]
Piper methysticum	Flavokawain B (**46**)	Chalcone	Breast cancer cell lines, MCF-7 and MDA-MB231, HCT116 human colon cancer	[54]
Artocarpus communis	Isolespeol (**47**)	Chalcone	Liver cell line Hep3B, PLC5, Huh7, human colon cancer HT-29, COLO205 cell, SW 872 human liposarcoma cell	[55]
Ashitaba (*Angelica keiskei*)	Xanthoangelol (**48**), 4-hydroxyderricin (**49**)	Chalcone	Human neuroblastoma (IMR-32) and leukemia (Jurkat) cells, SW 872, HT-29, COLO205, Hep3B, PLC5, Huh7, HepG2, KATO III, human tumor cell lines, HL60 (leukemia), CRL1579 (melanoma), A549 (lung) and AZ521 (stomach)	[56, 57]
Dorstenia barteri, Psoralea corylifolia	Isobavachalcone (**50**)	Chalcone	CCRF-CEM leukemia cells, human neuroblastoma cell lines (IMR-32 and NB-39), Induced apoptosis in CCRF-CEM leukemia cells	[58]
Dorstenia poinsettifolia	Poinsettifolin B (**51**)	Chalcone	Leukaemia CCRF-CEM, MDA-MB-231-BCRP, M/ADR5000 cells	[59]
Citrus kinokuni	2′-hydroxy-3,4,4′,5,6′-pentamethoxychalcone (**52**)	Chalcone	Human breast adenocarcinoma (MCF-7), non-small cell lung cancer (NCI-H460) and melanoma (A375-C5)	[60]
Rhuspyroides	Rhuschalcones II–VI (**53, 54**) (**53–57**)	Bichalcone	HT29 and HCT-116 colon tumor cell lines	[61]
Hibiscus syriacus	Syriacusin A (**58**)	Naphthalene	Human skin fibrobrast cells (CRL-2076), human tumor cell lines UACC62, ACHN, SW620 and SF539	[62]
Rumex nepalensis	Rumexneposide A (**59**)	Naphthalene acylglucoside	A549 (human non-small cell lung cancer), SKBR3, MCF-10A, MCF-7 and H522 cells (nonsmall-cell lung carcinoma)	[63]
Pentas parvifolia and Pentas bussei	Parvinaphthol B (**60**), parvinaphthols C (**61**)	Naphthalene	MDA-MB-231 human triple-negative breast cancer cell line	[64]

Source	Compound name	Class of compounds	Cytotoxic activity	Refernces
Rumex dentatus	Chrysophanol (**62**)	Anthraquinone	MCF-7 breast cancer cell line, gastric cancer 7901 cells, Melanoma A375 cells and oophoroma SKOV-3 cells	[65]
Streptomyces sp. ERINLG-26	9,10-anthraquinone (**63**), 2-hydroxy-9,10-anthraquinone (**64**), 2,3-dihydroxy-9,10-anthraquinone (**65**)	Anthraquinone	A549 lung adenocarcinoma and COLO320 cancer cell line	[66–68]
Rheum palmatum	Chrysophanol 8-*O*-beta-(6′-acetyl) glucopyranoside (**66**)	Anthraquinone glucoside	Human oral squamous cell carcinoma (HSC-2) and salivary gland tumor (HSG) cell lines than against normal human gingival fibroblasts (HGF)	[69]
Cratoxylum maingayi and *C. cochinchinense*	Formoxanthone C (**67**)	Xanthone	Human lung cancer (NCI-H187), MCF-7 (breast adenocarcinoma), KB (human oral cancer), HeLa (human cervical cancer) and HT-29 (colon cancer)	[70, 71]
Mangrove fungus *Phomopsis longicolla*, Fungus *Phomopsis longicolla*, *Rhizhopora mucronata*	Phomoxanthone A (**68**)	Xanthone	Tumour cell lines or of blood cancer cell lines	[72]
Garcinia mangostana	Garcinone C (**69**)	Xanthone	MCF-7, A549, Hep-G2 and CNE cell lines	[73]
Garcinia nobilis	Morusignin I (**70**), 8-hydroxycudraxanthone G (**71**), cudraxanthone I (**72**)	Xanthone	Breast cancer cells transduced with control vector (MDA-MB-231-pcDNA3), Human wild-type HCT116 (p53+/+) colon cancer cells, Human glioblastoma multiforme U87MG cells, Human HepG2 hepatocellular carcinoma cells and normal AML12	[74]
Securidaca longepedunculata	1,6,8-trihydroxy-2,3,4,5-tetramethoxyxanthone (**73**), 1,6-dihydroxy-2,3,4,5,8-pentamethoxyxanthone (**74**)	Xanthone	Human pancreatic cancer cell line, PANC-1	[75]
Garcinia hunburyi	Desoxymorellin (**75**)	Xanthone	HEL (human embryonic lung fibroblasts) and HeLa (Henrietta Lacks cervical cancer)	[76]
Garcinia cantleyana	Cantleyanone (**76**)	Xanthone	Breast cancer (MDA-MB-231 and MCF-7), ovarian cancer (CaOV-3) and HeLa cells	[77]
Garcinia lateriflora	Lateriflorone (**77**)	Xanthone	P388 cancer cell line	[78]
Garcinia morella	Gambogic acid (**78**)	Xanthone	T47D and DLD-1 breast cancer cells	[79]
Garcinia gaudichaudii	Gaudichaudione (**79**)	Xanthone	Parental murine leukemia P388 and P388/doxorubicin-resistant cell lines	[80]
Garcinia cowa	Cowaxanthones G (**80**)	Xanthone	Human cancer cell lines (HeLa, PANC-1 and A549)	[81]

Source	Compound name	Class of compounds	Cytotoxic activity	Refernces
Oncidium baueri	Batatasin III (**81**)	Stilbenoid	U251 (glioma, CNS), MCF-7 (breast), NCI-ADR/RES (ovarian expressing phenotype multiple drugs resistance), 786-0 (renal), NCI-H460 (lung, non-small cell), HT29 (colon), HaCat (human keratinocytes, immortalized non-tumoral cell)	[82]
Scirpus yagara	Sciryagarol I (**82**) and II (**83**)	*Cis*-stilbenoid	HeLa Cell Line human epitheloid cervix carcinoma	[83]
Gnetum macrostachyum	Macrostachyols D (**84**)	Oligostilbenoid	Human cervical carcinoma (HeLa) and human mouth epidermal carcinoma (KB) cell lines	
Cajanus cajan	Cajanotone (**85**), cajaninstilbene acid (**86**), pinosylvin monomethyl ether (**87**), longistylin A (**88**), longistylin C(**89**)	Stilbenoid	Human hepatoma cell line HepG2, human breast adenocarcinoma MCF-7 and human lung cancer cell line A549	[84]
Peanuts	Arachidin-1 (**90**)	Peanut Stilbenoid	Human leukemia HL-60 cells	[85]
Combretum caffrum	Combretastatin A-4 (**91**)	Stilbenoid	Human hepatocellular carcinoma HepG2, SMMC-7721, gastric carcinoma BGC-803 cells, breast carcinoma MDA-MB-435	[86]
Macaranga siamensis	Macasiamenenes A (**92**), macasiamenene K (**93**), Macasiamenene L (**94**), acasiamenenes M (**95**)	Prenylated stilbenoid	MOLT-3 (acute lymphoblastic leukemia)	[87]
Monomeria barbata	1,4,7-trihydroxy-2-methoxy-9,10-dihydrophenanthrene (**96**), 1,3,8-tri (phydroxybenzyl)-4-methoxy-phenanthrene-2,7-diol (**97**)	Phenanthrene	HepG-2, promyelocytic leukaemia HL-60, ovarian carcinoma Skov-3	[88]
Bulbophyllum odoratissimum	Bulbophythrins A (**98**) and B (**99**)	Dimeric phenanthrene	Human leukemia cell lines K562 and HL-60, human lung adenocarcinoma A549, human hepatoma BEL-7402 and human stomach cancer SGC-7901	[89]
Helicteres hirsuta	Pinoresinol (**100**), boehmenan (**101**) and boehmenan H (**102**)	Lignan	LNCaP, Lu1 (human lung cancer), MCF-7 (human breast cancer) and HUVEC (human umbilical vein endothelial) cell lines,	[90]
Sambucus williamsii	Sambucasinol A (**103**), B (**104**), C (**105**)	Lignan	Human cell lines A549, SK-OV-3, SK-MEL-2 and XF498	[91]
Pycnanthus angolensis	Pycnanthulignene A (**106**)	Lignan	CCRF-CEM leukemia cell line, CEM/ADR50 0 0 cells	[41]
Phyllanthus glaucus	Phyllanthusmin C (**107**)	Lignan glycoside	HL-60, MCF-7 and SW480 cells	[92]

Table 1. Cytotoxic phenolics from different natural source.

2. Natural phenolics with anticancer properties

2.1. Flavonoids and their cytotoxic activities

Flavonoids and their glycosides are the main phenolic constituents isolate form many natural sources in a reasonable amount. These phenolic compounds are very well-known for their cytotoxic activity and gave very promising results against different types of tumors. The basic structures of flavonoids are based on a C_6-C_3-C_6 skeleton derived from 1,3-diphenylpropane. Flavonoids are further divided into three subclasses on the basis of their little structural changes, substitutions and degree of hydroxylation and polymerization, such as, flavonoids (2-phenylchromen-4-one), isoflavonoids (3-phenylchromen-4-one) and neoflavonoids (4-phenylchromen with no hydroxyl group substitution at position 2). More than ten thousand flavonoids have been identified from natural sources and it exhibit a number of human health benefits due to their interactions with different cellular targets including antioxidant, anti-inflammatory, antiviral and anticancer properties. Isoprenylated flavonoids isolated from Moraceae family has many cytotoxic activity in different cancer cell lines including MCF-7 human breast cancer, TK-10 human renal cancer and UACC-62 human melanoma cells, A549 human lung cancer, Hep3B human hepatocellular cancer, HT-29 human colorectal cancer cells.

The cell viability of B16 melanoma cells after 3 days culture with compounds Artocarpin, cudraflavone C, prenylapigenin, kuwanon C, norartocarpin, albanin A and cudraflavone B at different concentrations was determined using 3-(4,5-dimethylthiazole-2-yl)-2,5-diphenyltetrazolium bromide MTT colorimetric assays. These compounds 1–7 exhibit potent cytotoxicity in a concentration-dependent manner. The IC_{50} values are 10.3, 9.2, 32.5, 14.2, 7.8, 84.7 and 12.5 μM [3].

Kaempferol is a dietary flavonoid and it has many protective effects for human health, was treated with human breast cancer cell line, MCF-7 and it regulated down the expression of polo-like kinase PLK-1, which has reported to regulate mitotic progression and to be upregulated in several human tumors. This polyphenolic compound significantly reduces cell viabilities of U-2 OS, HOB and 143B cells, but exerts low cytotoxicity on human fetal osteoblast progenitor hFOB cells. It has anticancer effects on Miapaca-2, Panc-1 and SNU-213 human pancreatic cancer cells. In a dose-dependent manner, its decreased viability of pancreatic cancer cells by increasing apoptosis. The anticancer effect of kaempferol mediated by inhibition of EGFR-related Src, ERK1/2 and AKT pathways and it could act as potent pancreatic cancer cells inhibitor [4]. The IC_{50} value of Dorsmanin F is a prenylated flavonoid and it ranged from 5.34 to 1.94 μM towards leukemia CCRF-CEM cells to 33.30–28.92 μM towards MDA-MB-231-BCRP cells, respectively and from 0.20 to 195.12 μM against CCRF-CEM. This compound induced apoptosis in CCRF-CEM leukemia cells, mediated by MMP disruption and increased reactive oxygen spp (ROS) production [5]. Artocarpesin, cycloartocarpesin (unique flavonoid structures) displayed cytotoxic effect on four cell lines with IC_{50} values, respectively, below 106, 50 and 25 μM. The IC_{50} values of these compounds ranged from 23.95 μM for hepatocarcinoma HepG2 cell, 105 μM towards colon carcinoma HCT116 (p53−/−) cells for artocarpesin, from 15.51 μM for leukemia CCRF-CEM cells and 49.83 μM for glioblastoma U87MG.EGFR cells for cycloartocarpesin [6]. 6,7-dimethoxy-4′-hydroxy-8-formylflavon and 4′,7-dihydroxy-

8-formyl-6-methoxyflavon are formylated flavonoid and showed significant cytotoxicity against PC3 and A549 cell lines with IC_{50} values of 2.6 and 1.6 μM, respectively [7]. Isoxanthohumol is known as hops and wines flavonoid and it exhibits antiproliferative activity against human breast cancer MCF-7 cell lines, A-2780 ovarian cancer cell lines, prostate cancer PC-3 and DU145 and colon cancer SW620 and HT-29 cells lines. It inhibits the activation of carcinogens: 2-amino-3-methylimidazol-[4, 5-f]quinoline and aflatoxin B1, AFB1 via human cytochrome P450 CYP1A2 [8]. Kushecarpin D is a Chinese traditional herbal medicine and exhibits antiangiogenic activity demonstrated by its effects on migration, adhesion, tube formation of an endothelial cell line. The antiangiogenic activity, together with its antiproliferative effects on endothelial cells, indicates that kushecarpin D is an excellent candidate for development as chemo preventive molecule to be used in combating tumor development [9]. Luteolin is one of the very important flavonoid found in many foods with lot of health benefits and it prevent cancers strongly, it induce cell cycle arrest and apoptosis in various human cancer cells and it synergize antitumor effects of 5-FU on Bel7402 and HepG2 cells, which can related with apoptosis and regulation of 5-FU metabolism. The IC_{50} value of luteolin on PC_3 and LNCaP cells was 31.44 and 32.05 μM, respectively. It resulted in a marked reduction in cell proliferation in a dose-dependent manner and enhanced the paclitaxel-induced apoptosis in human breast cancer MDA-MB-231 cells by blocking STAT3 [10–12]. Chrysin is found in many flowers and honey and it has preventive effect on cancer induced chemically on xenograft tumor by inducing activity of antioxidant and detoxification enzyme, reducing the activities of cytochrome P450 (CytP450)-dependent monooxygenases, inhibiting cellular proliferation and inducing apoptosis. Chrysin has induced breast cancer resistance protein (BCRP) in Caco-2 cells. Chrysin and many other flavonoids such as fisetin, kaempferol, galangin, myricetin and apigenin reported as potent inhibitors of P-form phenol sulfo transferase mediated sulfation induced carcinogenesis in human hepatoma cell line HepG2. Chrysin kill several histotype cancer cells, including hematological, cervical, liver, colon, lung, breast, nasopharyngeal, glioblastoma, prostate, thyroid and pancreatic cancer [13]. DICO is a nonaromatic B ring flavonoid having potent antitumor activity and inhibited growth of HepG2 cells in different dose/time dependent manners. It induced G2/M cell cycle arrest and apoptosis via a ROS-mediated mitochondrial pathway. It has significant antitumor effect through G2/M cell cycle arrest and apoptosis induction, which suggested DICO has therapeutic potential against tumors [14]. Morusin belongs to the prenylated class of flavonoids and it suppressed signal transducer and activator of transcription 3 STAT3 and nuclear factor-kB/NFkB signaling pathway, which modulate protein expression involved in invasion process. Morusin decreased lung colonization of SK-Hep1 cells in mice. It indicates that morusin possesses antitumor progression potential by suppressing STAT3 and NFkB [15]. Artocarpin showed significant anticancer activities on breast cancer cells (T47D cells) with IC_{50} value on T47D cells was 12.6 μM on concentration-dependent manner. Anticancer effect of artocarpin in HepG2 and PLC/PRF/5 hepatoma cells is mediated through autophagic cell death mechanism and it showed dose-dependent reduction on cancer cell viability after 24 h of treatment and IC_{50} value was calculated to be approximately 15 μM in both cell lines [16, 17]. Genkwanin a methoxy flavonoid, significantly inhibited HT-29 and SW-480 human colorectal cancer cells proliferation and inflammatory cytokine IL-8 secretion. It has a better antitumor activity via enhancing host immunity and decreasing inflammatory cytokine levels [18]. Fisetin is a tetrahydroxy flavonoid having many health benefits and it effects on short- and long-term growth of BRAF-mutated A375, SK-MEL-28 and RPMI-7951 melanoma cancer cells. Results of MTT assay demonstrated that fisetin (10–60 μM) treatment significantly decreased the growth of A375 8.64–61.75%, SK-MEL-28, 6.94–59.79% and RPMI-7951, 11.60–64.11% cells in a concentration-dependent manner [19]. Hispidulin is found in many herbs and significantly inhibited HepG2 cell growth in a time and dose-dependent manner. Hispidulin (at 200 μM) inhibited the growth of HepG2 cells by nearly 50, 70 and 90% after 24, 48 and 72 h of treatment, respectively, which suggest that hispidulin promotes HepG2 cell death through apoptosis. It inhibited cell growth in a dose-dependent manner in (HRCC) lines 786-0 and Caki-1 cell lines, whereas Caki-1 cells were found to be more resistant to hispidulin treatment [20, 21]. Quercetin is a dietary flavonoid (berries flavonoid) showed strong cytotoxicity as 1.49-fold in MCF-7 cells and 1.98-fold in MCF-7/dox cells. It suppressed proliferation and survival of HepG2 cancer cells and induced apoptosis by enhancing the expression of p53 and BAX through downregulation of ROS, PKC, PI3K and COX-2 [22, 23]. Hesperetin a citrus flavonoid exhibited inhibition of cell growth in a concentration and time-dependent manner with IC_{50} at 72 h 200 μM. It could significantly promote apoptosis of Eca109 cells in a dose-/time-dependent manner [24]. Apigenin a trihydroxy flavonoid with potential health benefits exhibited strong growth inhibitory activity in HER2/neu breast cancer cells but was much less effective in inhibiting growth of cells expressing basal levels of HER2/neu. It induces apoptosis in MDA-MB-453 breast cancer cells with involving intrinsic and extrinsic apoptotic pathways. It has shown to downregulate levels of cyclin D1, D3 and cdk4 and increases p27 protein levels in breast cancer cells. Apigenin inhibited human cervical carcinoma HeLa cells growth through an apoptotic pathway. In various human colon carcinoma cells, it resulted cell growth inhibition and G2/M cell cycle arrest. The effects of apigenin on lung cancer cells were evaluated and it inhibited A549 lung cancer cell. It has capability to significantly reduce cell number and induce apoptosis in PWR-1E, LNCaP, PC-3 and DU145 cells [25, 27]. Quercetin-3-*O*-β-D-glucopyranoside also known as isoquercetin and found in mangoes in high amount and it is an interesting dietary compound worth further investigation as a cytoprotective agent. Pretreatment of PC12 cells with nontoxic concentrations of this compound protect cells from H_2O_2 induced cytotoxicity with decrease in generation of reactive oxygen species (ROS). These observations qualify it as cytoprotective dietary compound [27]. Amplexicaule A is found in many herbs and it increased levels of cleaved caspase-3,-8,-9 and PARP, resulted from suppression of MCL-1 and BCL-2 expression in cells. This compound inactivated the Akt/mTOR pathway of breast cancer cells. It influenced strongly on breast cancer cells, most likely by induction of apoptosis [28]. Triticuside A a dietary flavonoid found in wheat and induced apoptosis accompanied by a significant decrease in Mcl-1 and Bcl-2 proteins and by increase in cleavage of caspases-3, -7, -9 and PARP. It suppressed the level of phospho-Akt and its downstream targets, mTOR and P70 S6 kinase. LY294002, a specific inhibitor of PI3K, significantly enhanced the triticuside A induced apoptosis. It may a potentially useful wheat bran component and can used for treatment of breast cancer [29].

Naringin is a diglycoside flavonoid and found in many citrus fruits and it inhibited cell proliferation and promoted cell apoptosis and G1 cycle arrest, accompanied by increased p21 and decreased surviving. Significant inhibitory effects of naringin on the cell proliferation of TNBC cells were observed. MDA-MB-231 and BT-549 cells treatment with naringin (50, 100 and 200 M) for 48 h, significantly increased apoptosis in breast cancer cells. It exerts significant effects on inhibition of breast cancer cells growth in through mediating-catenin pathway. Treatment of Raji with naringin showed maximum sensitivity towards NK cell lysis and the activity was 2.5-fold with naringin treatment [30–32]. Quercetin-3-*O*-α-ʟ-rhamnopyranosyl-(1→6)-β-ᴅ-glucopyranoside induces apoptosis in A549 cancer cells via caspase activation through cytochrome release from mitochondria. It tested against A549 lung cancer cell line, COLO320DM cancer cell line and Vero cell line and it showed prominent cytotoxic activity against A549 lung cancer cell. This compound showed 87.41% activity at dose of 164 μM with IC_{50} value of 82 μM. COLO320DM cancer cell line was maintained in complete tissue culture medium [32]. Pectolinarin and astragalin isolated from edible plants against H_2O_2-induced cell death of human brain neuroblastoma SK-N-SH cells. These compounds showed protective effects against H_2O_2-induced cell death and inhibited ROS generation by oxidative stress [33]. Troxerutin showed strong binding with calf thymus DNA in vitro and DNA interaction was confirmed by CD spectropolarimetry. The mode of binding of troxerutin to DNA was assessed by competing with EtBr or DAPI, known DNA intercalator and minor groove binder, respectively. It induced cytotoxicity in radioresistant DU145 and sensitive PC3 prostate cancer cells. When troxerutin was pretreated with DU145 and PC3 cells and it were exposed to gamma-radiation, the cytotoxicity induced in PC3 and DU145 prostate cancer cells and it was monitored by MTT assay. The toxicity induced by troxerutin 5 μM was more than 40% over control at 24 h, but more than 50% increase was observed at 48 h. Radiation per se induced around 18% cell death in PC3 cells [34]. Vicenin gave IC_{50} values 141.7 μM, 195.7 μM and 369.3 μg mL^{-1} on treatment with PC-3, DU-145 and LNCaP cells. Curcumin was used as positive control with IC_{50} 17.90 μM. It is an active constituent of the medicinal herbs (Tulsi) and it effectively induced antiproliferative, antiangiogenic and proapoptotic effect in CaP cells (PC-3, DU-145 and LNCaP) [35, 36] (**Figure 1**).

2.2. Isoflavonoids and their cytotoxic activities

Isoflavonoids play very important role in human health-promoting natural chemicals. They belong to plants secondary metabolites, mediate diverse biological functions through numerous pathways. Isoflavonoids are phenolic compounds and possess a 3-phenylchroman skeleton that biogenetically derived from 2-phenyl chroman skeleton of flavonoids. Some studies reported that anticarcinogenic activities of dietary soy isoflavonoids play important role for preventing colorectal cancer. Isoflavonoids have shown to possess many biological properties that can account for cancer prevention. Isoflavones exert their effects through numerous pathways with respect to the cancer prevention; and it use mechanisms of action which appear to be various, complementary and overlapping.

Genistein is a soybeans isoflavones and it decrease the risk of breast cancer and induced downregulation of CIP2A in breast cancer cells MCF-7-C3 and T47D, which correlates with its growth inhibition and apoptotic activities. In a model of human prostate carcinoma, treatment with genistein, decreased the metastatic burden, without changing the size of primary tumor and cell detachment were also decreased. According to studies, patients with PCa effectively tolerate genistein and therapeutic modulation of metastasis, specifically MMP-2 (matrix metallopeptidase 2) is possible [39, 40]. Inhibitory effect of daidzein investigated by

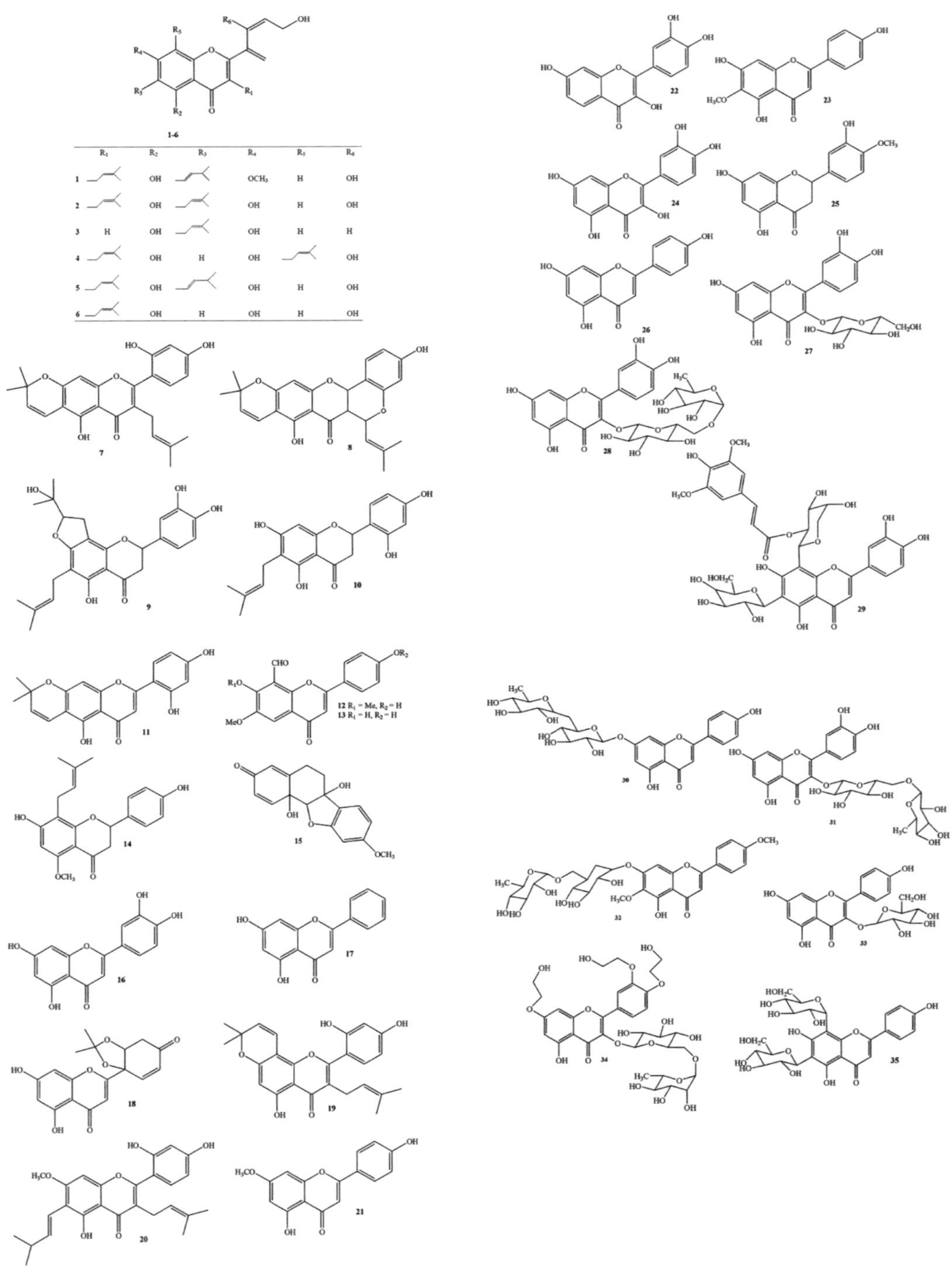

Figure 1. Structures of flavonoids **1–35**.

bacterial flora in intestines, on DU145, LnCaP and PC3 human prostate cancer cell. The results provide a better understanding of the biomolecular mechanisms of this compound as natural anticancer agent and provide base for development of daidzein and its analogs as potent anticancer molecules [39]. Biochanin A is found in nuts and beers and it effects on pancreatic cancer progression and it induced dose-dependent toxicity on pancreatic cancer cells Panc1 and AsPC-1. It reduced colony formation ability of Panc1 cells and induced dose-dependent apoptosis [40]. The IC_{50} value of Alpinum isoflavone ranged from 5.91 μM towards leukemia CEM/ADR5000 cell and to 65.65 μM towards drug-resistant breast adenocarcinoma MDA-MB-231-*BCRP* cells. This sioflavone induced apoptosis of CCRF-CEM cells, mediated by the loss of MMP and increase the ROS production [41]. Puerarin prevents the proliferation of breast cancer cells (MDA-MB-231, HS578T and MCF-7) at 50% of cell growth inhibition with concentration of 46, 71 and 69 μM, respectively. Puerarin was further preventing three different types of breast cancer in the G0/G1 phase of the cell cycle and stimulated apoptosis in these cells [42] (**Figure 2**).

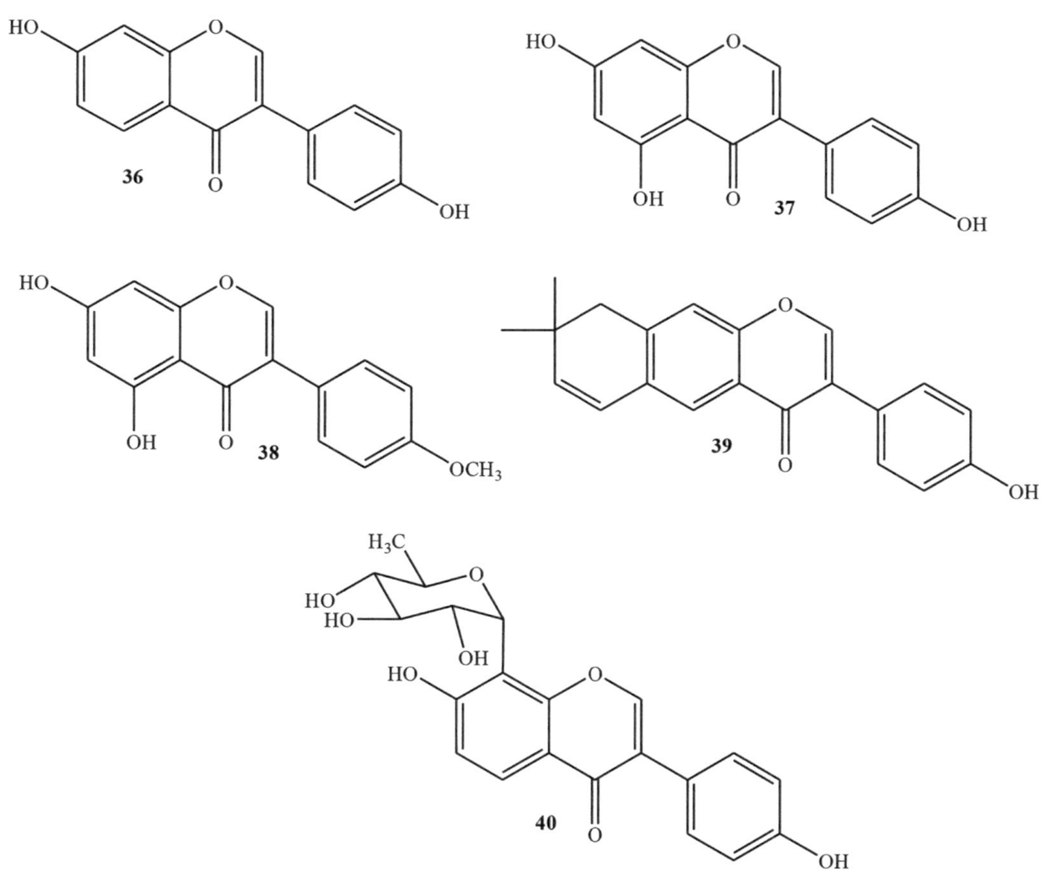

Figure 2. Structure of isoflavonoids **36–40**.

2.3. Chalcone and their cytotoxic activities

Chalcones mainly belongs to flavonoids family and known as open chain flavonoid in which two aromatic rings A and B structurally joined by α,β-unsaturated carbonyl system. Plants containing chalcones have been used traditionally as anti-inflammatory, antioxidant, antimalarial, antimicrobial, antifungal, antitubercular, cytotoxic, antiviral, antitumor and chemopreventive agent. These are very common phenolics specially founds in Leguminosae, Moraceae and Asteraceae families.

Butein is a natural dietary chalcone has many traditional uses and a herbal medicine. It reduced the cell viability of cultured human uveal melanoma cells in a dose-dependent manner with IC_{50} at 13.3 and 15.8 μM in SP6.5 and M17 cell lines, respectively. Similar effects were also found in a highly aggressive and metastatic C918 cell line (IC_{50} 16.7 μM). At 2 μM concentration, it inhibited the incorporation of 14C-labelled thymidine, uridine and leucine into the colon cancer cells whilst 5-fluorouracil (5-FU, a chemotherapeutic drug) at 50 μM concentration. The cytotoxic action of butein was different from 5-FU but may be similar to colchicine, a known HeLa cell inhibitor. Butein inhibit telomerase activity by downregulating hTERT gene expression in human leukemia cells and it causes apoptosis of breast cancer cells, while luminal $HER2^+$ HCC-1419, HCC-2218 and SKBR-3 breast cancer cells. Treatment with butein (10–30 μM) decreased cell viability in LNCaP (29, 42 and 52%), CWR22Rv1 (20, 31 and 42%) and PC-3 (11, 22 and 35%) cells. It inhibited colony formation, cell viability, migration, invasion, induced cell cycle at G2/M stage, cell apoptosis and enhanced caspase-3, -8 and -9 activity in HeLa cells in a dose-dependent manner [43–47]. Isoliquiritigenin is a natural chalcone and used for the treatment of cancer, it is cytostatic and able to overcome the intrinsic resistance of U373 cancer cells to proapoptotic stimuli. After 72 h treatment with 10 μg mL^{-1} of isoliquiritigenin, a typical differentiated morphology observed in HL-60 cancer cells, including the decrease in karyoplasmic ratio and the increase in kidney-shape nuclear cells. This compound is able to induce the monocytic differentiation in leukemia cells. It has potential as a drug for leukemia. HepG2 cells are significantly more resistant to isoliquiritigenin when the activity of p53 was blocked. Isoliquiritigenin inducible p53 plays a key apoptotic role and may do so by regulating the expression of specific target molecules that promotes efficient apoptotic cell death following G2/M-cell cycle arrest. It inhibits the proliferation of prostate cancer cells, via inhibition of ErbB3 signaling and PI3K/Akt pathway [48]. 2′,4′-dihydroxy-6-methoxy-3,5-dimethylchalcone tested on human cell lines including liver cancer SMMC-7721 cells, pancreas cancer 8898 cells, tumor of cervix uteri HeLa cells, lung cancer SPC-A-1 cells, high metastatic lung carcinoma 95-D cells and gall bladder carcinoma GBC-SD cell lines was dose-dependent. It observed that different cells had a different sensitivity for inhibition effect of this chalcone. The IC_{50} values on cytotoxicity were 32.3, 37.2, 37.7, 81.3, 84.6 and 84.8 μM for SMMC-7721, 8898, HeLa, GBC-SD SPC-A-1 and 95-D cells, respectively [49, 50]. 4′-hydroxy-2′,6′-dimethoxychalcone showed IC_{50} values in a range of 2.54 μM against CEM/ADR5000 leukemia cells to 58.63 μM towards hepatocarcinoma HepG2 cells. This compound arrested cell cycle between Go/G1 phase and induced apoptosis via disrupted mitochondrial membrane potential MMP and increased production of reactive oxygen species (ROS) in the studied leukemia cell line [51]. Xanthohumol is a prenylated chalcone found in beers; it caused dose dependent (0.1–100 μM) decrease in growth of MCF-7, HT-29 and A-2780 cancer cell lines. After two-day treatment, the concentrations at which growth of MCF-7 cells was inhibited by 50% (IC_{50}) were 13.3 μM for xanthohumol. After four-day treatment, IC_{50} for xanthohumol was 3.47 μM [52, 53]. Flavokawain B is known as kava chalcone and it induced caspase and mitochondria-dependent apoptosis which characterized by cytochrome c release and Bak translocation to mitochondria. It induces G2/M accumulation, autophagy and apoptosis, leading to HCT116 colon cancer cell growth inhibition. It further induced both MCF-7 and MDA-MB231 and significant G2/M arrest was seen in MDA-MB231 cells [54]. Isolespeol is a geranyl chalcone and it showed inhibitory activity against human liposarcoma cells SW 872 with IC_{50} values of 3.8 μM. Treatment of SW 872 human liposarcoma cells with this compound stimulated increase protein expression of Fas, FasL and p53 [55]. Xanthoangelol and 4-hydroxyderricin inhibit adipocytes differentiation through AMPK and mitogen-activated protein kinase pathways, resulting in the down-expression of adipocyte-specific transcription factors. Xanthoangelol induce apoptotic cell death by activation of caspase-3 in neuroblastoma and leukemia cells through a mechanism that does not involve Bax/Bcl-2 signal transduction. Therefore, this compound may effective drug for the treatment of neuroblastoma and leukemia. 4-Hydroxyderricin showed significant cytotoxicity in four human tumor cell with IC_{50} values 4.8 μM (CRL1579), 5.5 μM (HL60), 4.2 μM (AZ521) and 10.2 μM (A549). 4-Hydroxyderricin induced further the early apoptosis of HL60 cells and it observed as membrane lipid exposure in the flow cytometry [56, 57]. The IC_{50} values for Isobavachalcone is a prenylated chalcone and having wide range of biological activities including anticancer, it ranged from 0.20 μM (towards CCRF-CEMcells) to 195.12 μM (towards leukemia CEM/ADR 5000 cells) for doxorubicin. It induces apoptosis in CCRF-CEM leukemia cells, mediated by caspase activation and the disruption of MMP [58]. Cytotoxicity of Poinsettifolin B assessed against different sensitive and multidrug-resistant cancer cell lines. The IC_{50} values for this compound ranged from 5.34 to 1.94 μM towards CCRF-CEM leukemia cells, to 33.30 and 28.92 μM towards human breast cells MDA-MB-231-BCRP, respectively and from 0.20 μM against human leukemia cell CCRF-CEM, to 195.12 μM against CEM/ADR5000 cells [59]. *2′-hydroxy-3,4,4′,5,6′-pentamethoxychalcone is very potent in inhibiting* breast adenocarcinoma (MCF-7), lung cancer (NCI-H460) and melanoma (A375-C5) cell lines [60]. Rhuschalcones II–VI showed potent cytotoxic activities against HCT-116 and HT29 human colon cancer cells [61] (**Figure 3**).

2.4. Naphthalenes and their cytotoxic activities

Naphthalenes are simplest and most important member of arenas, in which two benzene rings are fused with each other and it based on a C_6-C_4 skeleton. Many naphthalene-based natural products have been isolated from different sources and showed significant biological activities.

Syriacusin A inhibit activity of human neutrophil elastase HNE, a serine protease to degrade extracellular matrix ECM proteins including elastin, with IC_{50} values 8.0, 5.2 and 6.1 μM, respectively [62]. Rumexneposide A showed against broad spectrum activity against human cancer cells lines, including lung and breast cancer cells (A549, H522, MCF-7, MCF-10A, SKBR3) with IC_{50} value 31.0, 15.7, 21.8, 22.8 and 20.7 μM [63]. Parvinaphthol B, parvinaphthols C showed marginal cytotoxicity against the MDA-MB-231 human triple-negative breast cancer cell line with IC_{50} values ranging from 62.3 to 129.6 μM [64].

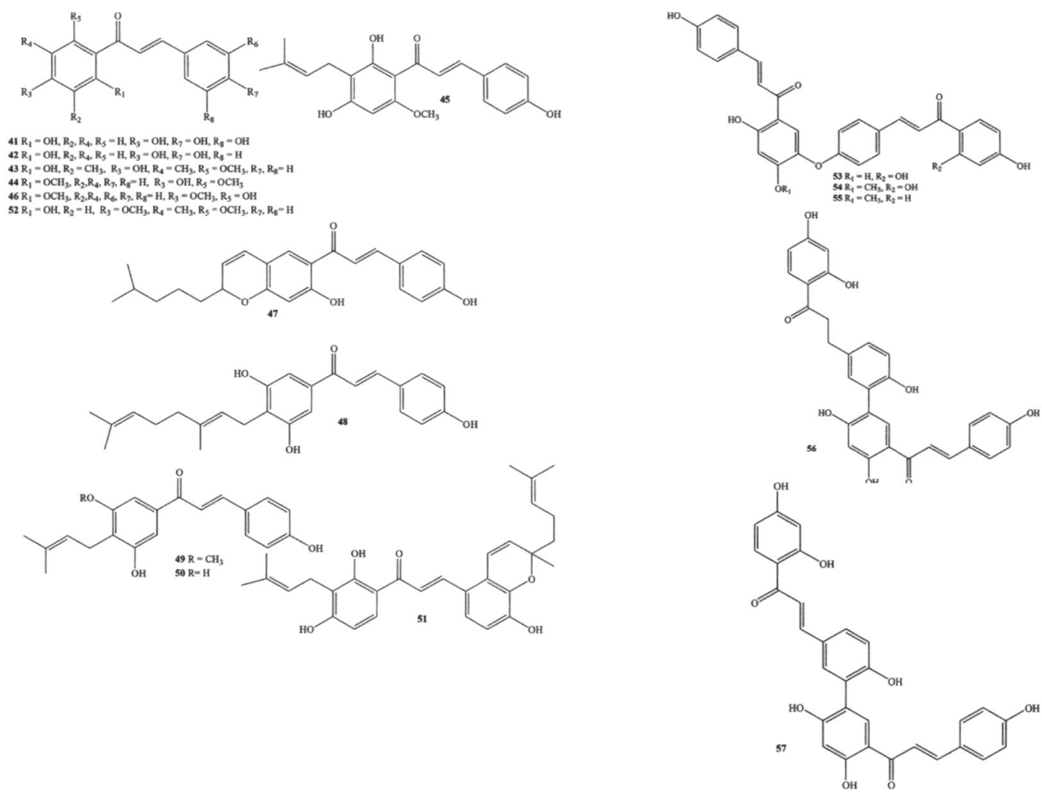

Figure 3. Structure of chalcones **41–57**.

2.5. Anthraquinone and their cytotoxic activities

Anthraquinones are the active constituents of many herbs with a tricyclic C6-C2-C6-based skeleton. They are widely distributed in the *Fabaceae, Liliaceae, Labiate, Polygonaceae* and *Rhamnaceae* families. Anthraquinone found in herbs (*Polgyonum*) in vegetable (cabbage, lettuce and beans).

Chrysophanol was active in the oophoroma SKOV-3 cell line, where the IC_{50} value was 5.62 μM and with IC_{50} value 20.4 μM for MCF-7 breast cancer cell line [65]. 9,10-anthraquinone showed strong cytotoxic activity against COLO320 human colon carcinoma cell line. It gave 79.7% cytotoxicity at 300 μg mL^{-1} concentration of compound with 75 μg mL^{-1} IC_{50} value. The treatment of human colon carcinoma cell line COLO320 with 9,10-anthraquinone significantly decreased the proliferation of cells and enhanced the formation of apoptotic bodies and fragmented DNA. The expressions of p53 and caspase-3 was upregulated from 9,10-anthraquinone in colon adenocarcinoma cells [66]. 2-hydroxy-9,10-anthraquinone showed cytotoxic activity against A549 lung and COLO320 cells lines and it showed 62.7% activity at the dose of 500 μg mL^{-1} with IC_{50} value of 400 μg mL^{-1} against COLO320 cells [67]. 2,3-dihydroxy-9,10-anthraquinone inhibits PI3K/AKT activity after treatment. Also, COX-2 enzyme plays a major role in colorectal cancer and significantly reduced COX-2 enzyme in COLO320 cells.

It involves in apoptotic pathway, mitochondrial function, cell cycle checkpoint and control the over expression gene during the colorectal cancer [66]. Chrysophanol 8-O-beta-(6′-acetyl) glucopyranoside exhibited relatively higher cytotoxic activities against human oral squamous cell carcinoma HSC-2 and salivary gland tumor HSG cell lines than against normal human gingival fibroblasts HGF cells lines [69] (**Figure 4**).

2.6. Xanthone and their cytotoxic activities

Xanthones are three-membered ring compounds with (C6-C1-C6) skeleton and are mainly found in higher plants and microorganisms. Now days, they gained great importance due to their significant pharmacological and biological properties. These types of natural compounds have broad biological profile, such as, antihypertensive, anti-inflammatory, antioxidant antithrombiol, anticancer and antiviral activities. Xanthones are commonly found in Gentianaceae, Moraceae, Guttiferae, Clusiaceae, Polygalaceae in citrus fruits and mangosteen.

Formoxanthone C has α,α,β-trimethylfuran ring and it showed potent cytotoxic activity against NCI-H187 with IC_{50}, 0.22 μg mL^{-1}, which is stronger than elliptecine, a standard drug with IC_{50}, 0.45 μg mL^{-1} [70, 71]. Phomoxanthone A displayed strong anticancer activity, on the treatment against cisplatin resistant (CisR) cancer cell lines or blood cancer cell lines with an IC_{50} values in sub-micromolar concentration and it was up to 100-folds less active against PBMC peripheral blood cells from a healthy donors [72]. Garcinone C was tested for cytotoxicity against MCF-7, A549, Hep-G2 and CNE cell lines by 3-(4,5-dimethylthiazol2-yl)-2,5-diphenyltetrazolium bromide (MTT) colorimetric assay, while doxorubicin used as positive control. Garcinone C showed strong cytotoxicity against all cell lines with IC_{50} values from 4.3 ± 0.1 to 7.1 ± 0.5 μM [73]. Morusignin I 70 and cudraxanthone I inhibited proliferation of all cancer cell lines including sensitive and drug-resistant phenotype. 8-hydroxycudraxanthone G showed activity on many cell lines with IC_{50} values ranged from 16.65 μM against CCRF-CEM leukemia cells to 70.38 μM against HepG2 hepatocarcinoma cells. The IC_{50} values of different cells ranged from 7.15 μM against CCRF-CEM human leukemic cells to 53.85 μM against U87MG.Δ*EGFR* human glioblastoma cells for morusignin I and 2.78 μM against MDA-MB231 BCRP breast cancer cells to 22.49 μM against U87MG human glioblastoma *cell* line for cudraxanthone I [74]. 1,6,8-trihydroxy-2,3,4,5-tetramethoxyxanthone, 1,6-dihydroxy-2,3,4,5,8-pentamethoxyxanthone were tested for preferential cytotoxic activity against human pancreatic cancer cells PANC-1 under nutrient deprived condition. These compound displayed potent cytotoxicity with PC_{50} of 22.8 and 17.4 μM, respectively. They triggered apoptosis-like PANC-1 cell death in NDM with glucose sensitive mode [75]. Desoxymorellin inhibited the growth of HEL human embryonic lung fibroblasts and HeLa Henrietta Lacks cervical cancer cell with a minimum inhibitory concentration MIC of 0.39 mg mL^{-1} [76]. Cantleyanone displayed significant cytotoxicity against breast cancer MDA-MB-231 and MCF-7, ovarian cancer CaOV-3 and HeLa cells with EC_{50} values ranging from 0.22 to 17.17 mg mL^{-1} [77]. Lateriflorone was cytotoxic against P388 cancer cell line with an ED_{50} value 5.4 mg mL^{-1} [78]. Gambogic acid inhibited proliferation of T47D and DLD-1 breast cancer cells with GI_{50} values of 0.04 and 0.03 mm, respectively [79]. Gaudichaudione displayed strong growth inhibitory activity against parental murine leukemia P388 and P388/doxorubicin resistant cell lines at low μM concentrations [80]. Cowaxanthones G showed potent inhibition on cell viability $IC_{50} < 10$ μM, while etoposide was used as positive control [81] (**Figure 5**).

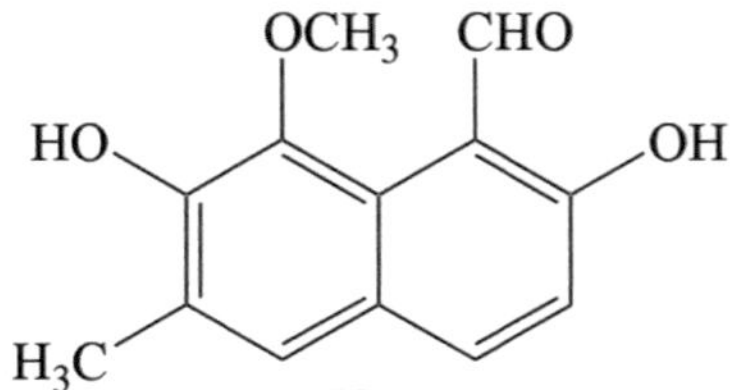

58

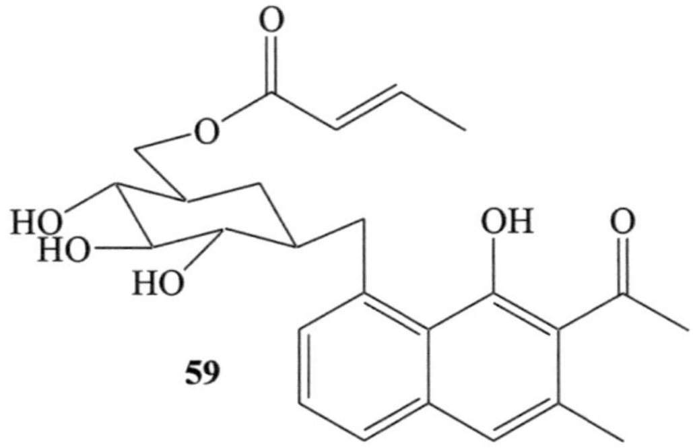

59

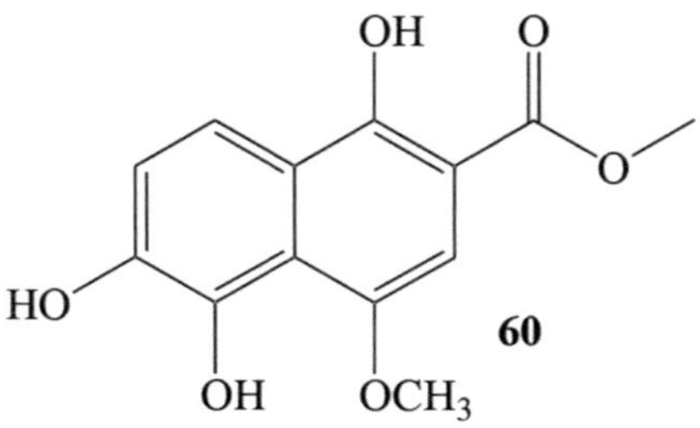

60

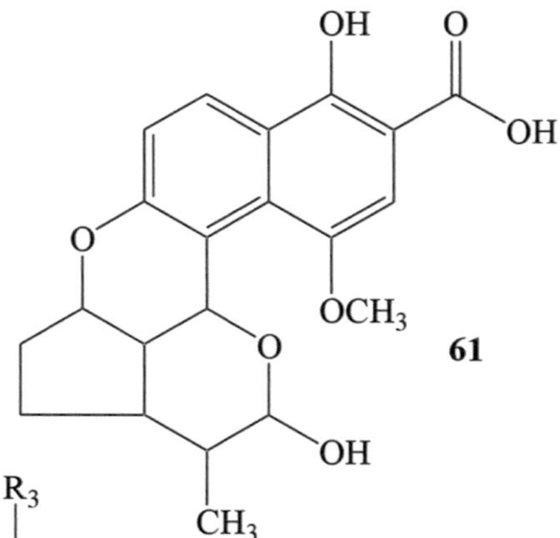

61

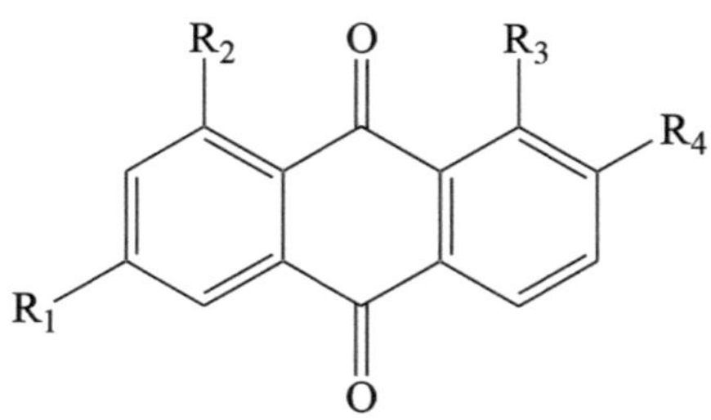

62 : $R_1 = CH_3$, $R_2 = OH$, $R_3 = OH$, $R_4 = H$
63 : $R_1 = H$, $R_2 = H$, $R_3 = H$, $R_4 = H$
64 : $R_1 = H$, $R_2 = H$, $R_3 = H$, $R_4 = OH$
65 : $R_1 = CH_3$, $R_2 = H$, $R_3 = OH$, $R_4 = OH$

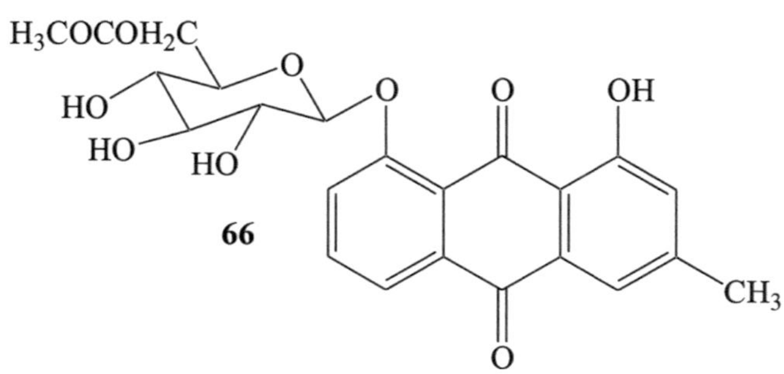

66

Figure 4. Structure of naphthalenes **58–61** and Anthraquinones 62–66.

http://dx.doi.org/10.5772/66898

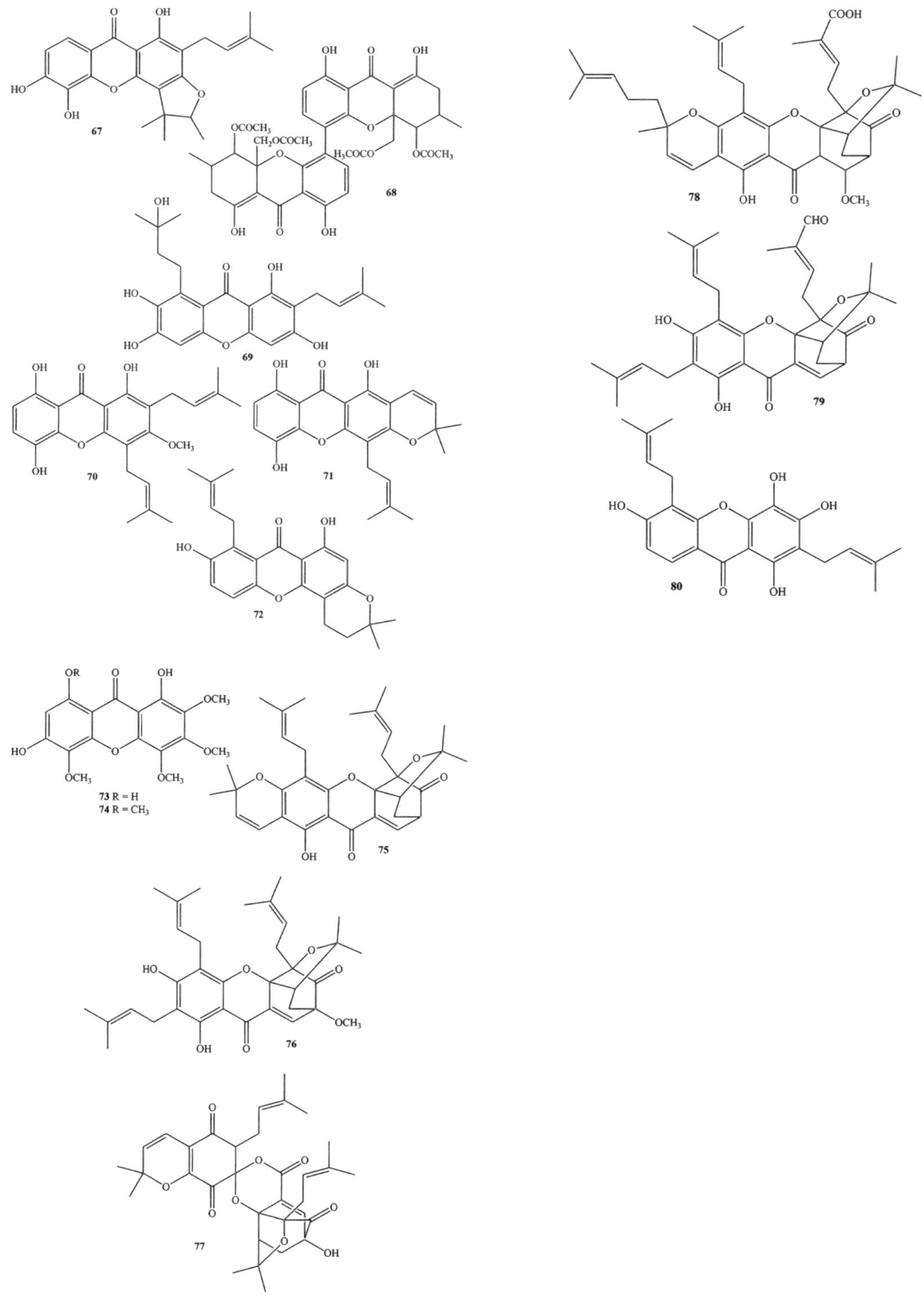

Figure 5. Structure of structure of xanthones **67**–**80**.

2.7. Stilbene and their cytotoxic activities

Stilbenoids are formed by flavonoid biosynthesis pathway and it consists on C6-C2-C6-based skeleton and usually found in peanuts, grapes and wines. Stilbenoid have a great interest on account due to their promising pharmacological activities including anticancer, antimicrobial, antioxidant and anti-inflammatory. Stilbenoid play important role of phytoalexin in plants. They play very important role in the defense of different pathogen.

Batatasin III inhibited growth of all cell lines including U251 glioma, CNS, MCF-7 breast, NCI-ADR/RES (ovarian expressing phenotype multiple drugs resistance), 786-0 renal NCI-H460 lung, non-small cell, HT29 colon, HaCat human keratinocytes, immortalized non-tumoral cell with GI_{50} values close to 30 mg mL^{-1} [82]. The cytotoxicity of Sciryagarol I and II was evaluate by MTT assay against human tumor cell lines MGC803, SMMC7721 and Hela and it showed significant cytotoxicity against Hela cell lines with IC_{50} values 61.21 and 7.21 μM, respectively [83]. Oligostilbene macrostachyol D showed significant cytotoxicity to HeLa cells with IC_{50} value of 4.13 μM. Cajanotone, cajaninstilbene acid, pinosylvin monomethyl ether, longistylin A and longistylin C showed strong cytotoxicities against human hepatoma HepG2, human breast adenocarcinoma MCF-7 and human lung cancer A549 cell lines with IC_{50} values in the range of 3.5–15.5 μM, while doxorubicin used as a positive control [84]. Arachidin-1 induced human leukemia HL-60 cells death with EC_{50} value 4.2 μM and it was more potent than resveratrol and induces cell death in HL-60 cells through only the intrinsic apoptotic pathway [85]. Combretastatin A-4 showed strong cytotoxity against human cancer cell lines: HepG2 with IC_{50} value 9.2, SMMC-7721 with IC_{50} value 12.8, BGC-803 with IC_{50} value 12.2, MDA-MB-231 with IC_{50} value 17.6 μM [86]. Macasiamenenes A2, macasiamenene K, Macasiamenene L, acasiamenenes M exhibited significant cytotoxicity against MOLT-3 cancer cell line with the IC_{50} values in the range of 0.66–9.78 μM, etoposide used as Positive control standard [87] (**Figure 6**).

2.8. Phenanthrene and their cytotoxic activities

Phenanthrenes consist on three fused aromatic rings and found in *Dendrobium* and *Dioscorea* spp.

Cytotoxicities of 1,4,7-trihydroxy-2-methoxy-9,10-dihydrophenanthrene, 1,3,8-tri (phydroxybenzyl)-4-methoxy-phenanthrene-2,7-diol towards liver carcinoma HepG-2, promyelocytic leukaemia HL-60, ovarian carcinoma Skov-3 and epidermoid carcinoma A431 cancer cell lines were determined by MTT method. 1,4,7-trihydroxy-2-methoxy-9,10-dihydrophenanthrene is the strongest one to HepG-2 and Skov-3 with IC_{50} of 15.9 and 124.0 mmol L^{-1}, respectively; 1,3,8-tri (phydroxybenzyl)-4-methoxy-phenanthrene-2,7-diol is the strongest one to HL-60 and A431 with IC_{50} of 34.9 and 46.4 mmol L^{-1}, respectively [88]. Biphenanthrenes bulbophythrins A and B were evaluated in vitro for their inhibitory ability against human leukemia cell lines K562 and HL-60, human lung adenocarcinoma A549, human hepatoma BEL-7402 and human stomach cancer SGC-7901, using cisplatin as a positive control. Bulbophythrins A exhibited some selectivity against HL-60 and BEL-7402 with IC_{50} values of 1.27×10^{-3} and 1.22×10^{-3} μmol mL^{-1}, respectively, whereas bulbophythrins B was most active against A549 with IC_{50} value of 1.18×10^{-3} μmol mL^{-1} [89] (**Figure 7**).

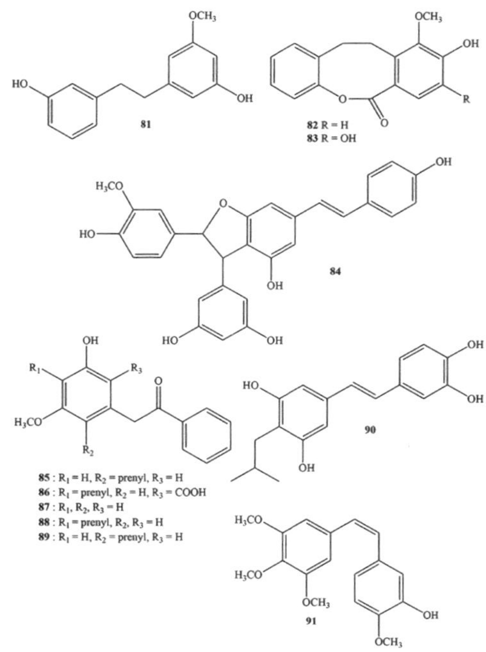

Figure 6. Structure of structure of stilbenoids **81–95**.

2.9. Lignan and their cytotoxic activities

Lignans based on (C6-C3)2 biphenolic skeleton and they consists on a large group of plant phenolic produced by the oxidative dimerization of two phenyl propanoid moieties. They are found in black berries, strawbwerries, raspberries, blue berries, broccoli, apricot, cabbage and many seeds. The antiestrogenic effects of lignans could help to reduce the risk of hormone-associated breast, uterine, ovarian and prostate cancers.

Pinoresinol, boehmenan and boehmenan H were found the most cytotoxic against human cancer cell lines Lu1, LNCaP and MCF-7 and a normal cell line HUVEC with ED_{50} values for Lu1 cell line was 0.8, 10.4 and 5.3 g mL^{-1}, for LNCaP cell line was 0.5, 9.5 and 7.7 g mL^{-1}, while the ED_{50} values for MCF-7 cell line was 1.7, 10.0 and 10.2 g mL^{-1}, ED_{50} values for HUVEC 1.1, 9.0 and 6.2 g mL^{-1} were observed for these lignans. The cytotoxic activity of pinoresinol against various cancer cell lines has been mentioned in the literature. It exhibited cytotoxicity against the KB cell line with an IC_{50} value of 2.2 μg mL^{-1} [90]. The cytotoxic activity of Sambucasinol A, B and C was evaluated by determining their inhibitory effects on human tumor cell lines A549, SK-OV-3, SK-MEL-2 and XF498 using the SRB bioassay. These compounds showed cytotoxicity against all cell lines, with IC_{50} values in the range of 11.07–19.62 μM [89]. The IC_{50} values for the pycnanthulignene A ranged from 0.20 μM (towards CCRF-CEM cells) for pycnanthulignene A to 195.12 μM (towards CEM/ADR5000 cells) for doxorubicin [41]. Phyllanthusmin C showed significant cytotoxicity against human leukemia HL-60 cells, human breast MCF-7 cells and human colon SW480 cell lines with IC_{50} values of 9.2, 19.2 and 20.5 μM, compared with the cisplatin with IC_{50} values of 1.7, 10.9 and 9.9 μM, respectively [92] (**Figure 8**).

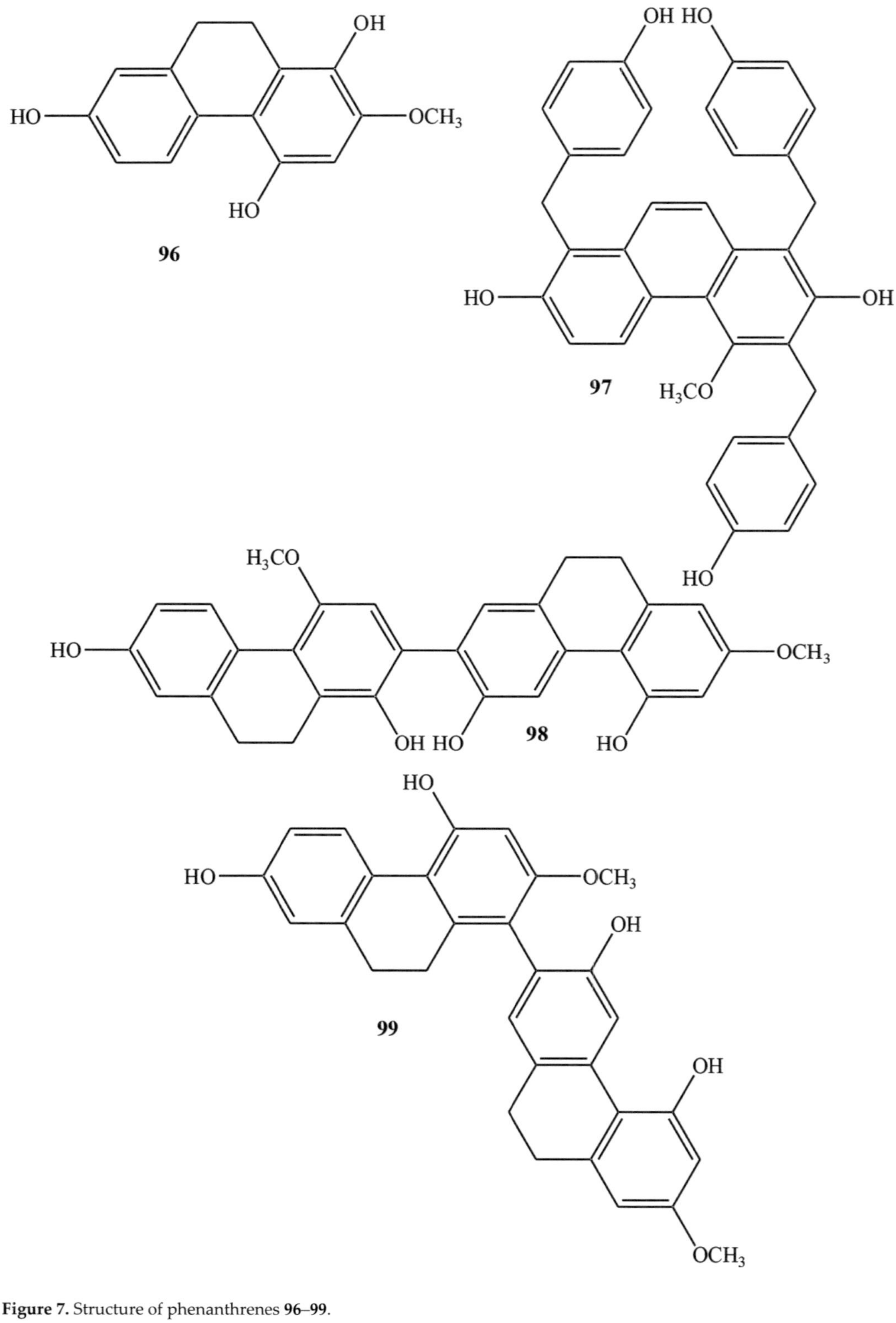

Figure 7. Structure of phenanthrenes **96–99**.

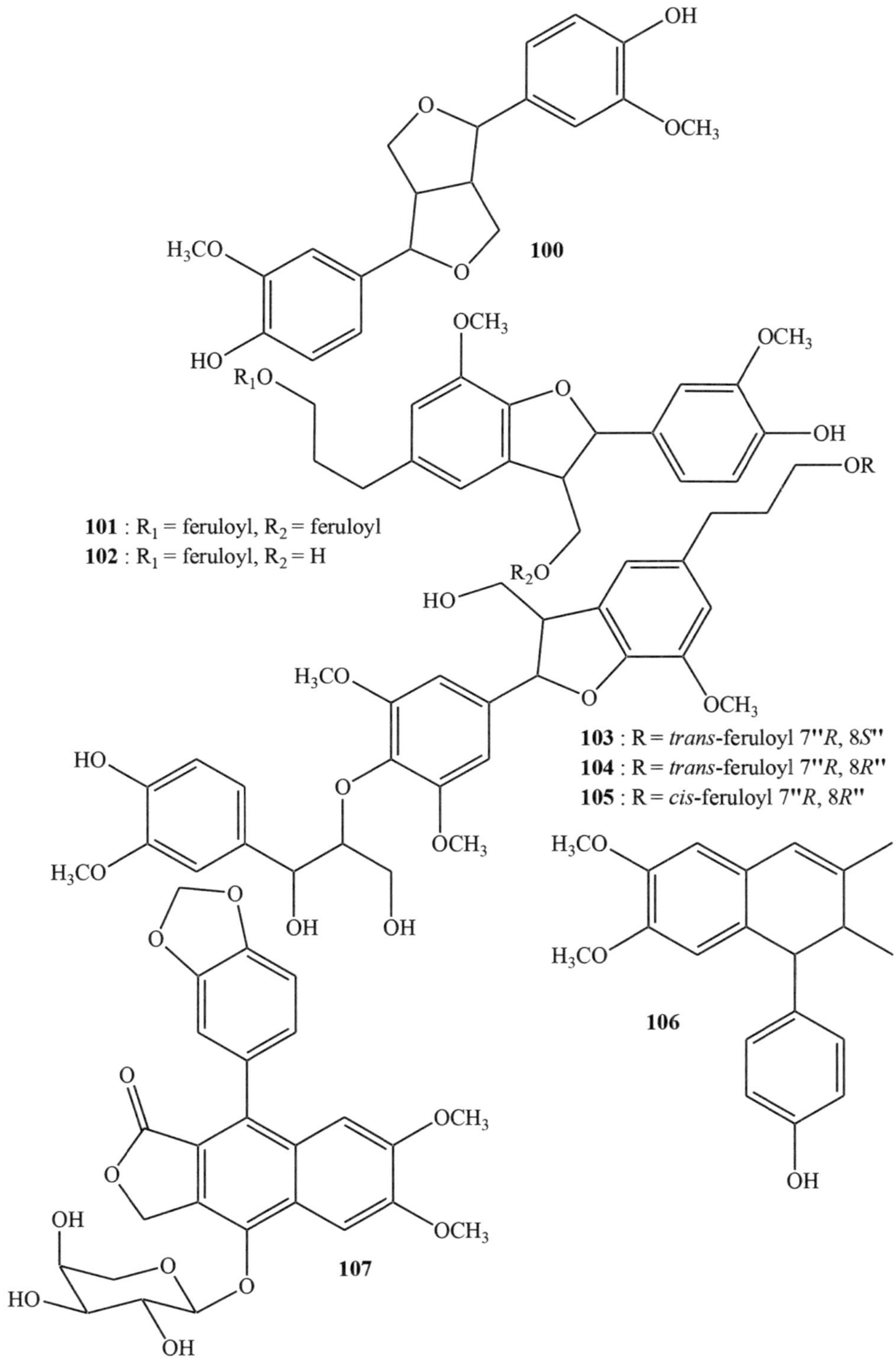

Figure 8. Structure of lignans **100–107**.

3. Conclusion

Phenolic plays an important role in human nutrition and health benefit and chemoprevention is one of the most realistic and promising approaches for the prevention of malignant disorders. The results of this chapter may help to identify the phenolic compounds from natural sources with optimized cytotoxic activity to be tested for the treatment of different types of tumors. Phenolic compounds are incorporated into them and the order of their incorporation efficiency is similar to its cytotoxic activity. The cytotoxic data of phenolic compounds compiled in this chapter and relationships presented here cannot only be useful in chemoprevention to choose the correct source of these compounds containing most active natural polyphenols, considering the individual genetic cancer risks and familial anamnesis but also in the selection of parent compounds to design and synthesize novel chemotherapy drugs starting from the valuable material given to us by the nature.

Acknowledgements

This book chapter was supported by a grant from the "Research Center of the Female Scientific and Medical Colleges", Deanship of Scientific Research, King Saud University.

Author details

Shagufta Perveen* and Areej Mohammad Al-Taweel

*Address all correspondence to: shagufta792000@yahoo.com

Department of Pharmacognosy, College of Pharmacy, King Saud University, Riyadh, Kingdom of Saudi Arabia

References

[1] Huang WY, Cai YZ. Natural phenolic compounds from medicinal herbs and dietary plants: potential use for cancer prevention. Nutrition and Cancer. 2010;**62**:1–20. doi:10.1080/01635580903191585

[2] World Health Organization, Cancer [Internet]. 2016. Available from: http://www.who.int/cancer/en/

[3] Enos TA, Keisuke Y, Kuniyoshi S, Ryuichiro K. Isoprenoid-substituted flavonoids from wood of *Artocarpus heterophyllus* on B16 melanoma cells: cytotoxicity and structural criteria. Fitoterapia. 2010;**81**:120–123. doi:10.1016/j.fitote.2009.08.001

[4] Jungwhoi L, Jae HK. Kaempferol inhibits pancreatic cancer cell growth and migration through the blockade of EGFR-related pathway in vitro. PLoS One. 2016. doi:10.1371/journal.pone.0155264.2016

[5] Victor K, Armelle TM, Maen Z, Bathelemy N, Gilbert DWFK, Simeon FK, Bonaventure TN, Thomas E. Cytotoxicity of two naturally occurring flavonoids (dorsmanin F and poinsettifolin B) towards multi-factorialdrug-resistant cancer cells. Phytomedicine. 2015;**22**:737–743. doi:10.1016/j.phymed.2015.04.007

[6] Victor K, Armelle TM, Maen Z, Bathelemy N, Gilbert DWFK, Simeon FK, Bonaventure TN, Thomas E. Cytotoxicity of three naturally occurring flavonoid derived compounds (artocarpesin, cycloartocarpesin and isobavachalcone) towards multi-factorialdrug-resistant cancer cells. Phytomedicine. 2015;**22**:1096–1102. doi:10.1016/j.phymed.2015.07.006

[7] Jinxiong C, Hongqiong L, Yuanxing D, Wei Z, Guangyu Y, Yadong G, Yongkuan C, Qiufen H. Three new flavonoids from the leaves of oriental tobacco and their cytotoxicity. Phytochemistry Letters. 2013;**6**:144–147. doi:10.1016/j.phytol.2012.12.001

[8] Anna KZ, Wanda KM, Malgorzata G, Katarzyna W, Edyta W, Miroslaw A. Isoxanthohumol- biologically active hop flavonoid. Fitoterapia. 2015;**103**:71–82. doi: 10.1016/j.fitote.2015.03.007

[9] Li-Ping P, He-Ping C, Mei-Ai C, Xiu-Li Z, Qing-Xiang G, Cheng-Shan Y, Chun-MingW. The antiangiogenic activity of Kushecarpin D, a novel flavonoid isolated from *Sophora flavescens* Ait. Life Sciences. 2013;**93**:791–797. doi:10.1016/j.lfs.2013.09.025

[10] Huanli X, Tao Y, Xiaohui L, Ye T, Xiaoliang C, Ru Y, Shuonan S, Xiukun L, Guanhua D. Luteolin synergizes the antitumor effects of 5-fluorouracil against human hepatocellular carcinoma cells through apoptosis induction and metabolism. Life Sciences. 2016;**144**:138–147. doi:10.1016/j.lfs.2015.12.002

[11] Han K, Meng W, Zhang JJ, Zhou Y, Wang YL, Su Y, Lin SC, Gan ZH, Sun YN, Min DL. Luteolin inhibited proliferation and induced apoptosis of prostate cancer cells through miR-301. Oncotargets and Therapy. 2016;**9**:3085–3094. doi:10.2147/OTT.S102862

[12] Sato Y, Sasaki N, Saito M, Endo N, Kugawa F, Ueno A. Luteolin attenuates doxorubicin-induced cytotoxicity to MCF-7 human breast cancer cells. Biological & Pharmaceutical Bulletin. 2015;**38**:703–709. doi:10.1248/bpb.b14-00780

[13] Eshvendar RK, Lakshmi NB, Rajaram MM, Athira KV, Ranadeep G, Chandana CB. Chemopreventive and therapeutic potential of chrysin in cancer: mechanistic perspectives. Toxicology Letters. 2015;**233**:214–225. doi:10.1016/j.toxlet.2015.01.008

[14] Daonian Z, Anhua W, Cheng C, Jinlan R. DICO, a novel nonaromatic B-ring flavonoid, induces G2/M cell cycle arrest and apoptosis in human hepatoma cells. Food and Chemical Toxicology. 2013;**57**:322–329. doi:10.1016/j.fct.2013.03.032

[15] Wea-Lung L, Deng-Yu L, Yean-Jang L, Nai-Fang C, Tsui-Hwa T. Antitumor progression potential of morusin suppressing STAT3 and NFkB in human hepatoma SK-Hep1 cells. Toxicology Letters. 2015;**232**:490–498. doi:10.1016/j.toxlet.2014.11.031

[16] Arung ET, Wicaksono BD, Handoko YA, Kusuma IW, Shimizu K, Yulia D, Sandra F. Cytotoxic effect of artocarpin on T47D cells. Journal of Natural Medicines. 2010;**64**:423–429. doi:10.1007/s11418-010-0425-6

[17] Cheng-Wei T, Wen-Sheng T, Liang-Tzung L, Chiang-Wen L, Feng-Lin Y, Chun-Ching L. Enhanced autophagic activity of artocarpin in human hepatocellular carcinoma cells through improving its solubility by a nanoparticle system. Phytomedicine. 2016;**23**:528–540. doi:10.1016/j.phymed.2016.02.010

[18] XueW, Zi-Jing S, Xin H, Run-Qi Z, Chun-Feng Z, Fei L, Chong-Zhi W, Chun-Su Y. Antitumor and immunomodulatory activity of genkwanin on colorectal cancer in the APCMin/+ mice. International Immunopharmacology. 2015;**29**:701–707. doi:10.1016/j.intimp.2015.09.006

[19] Harish CP, Ronald DB, Katherine MH, Jyoti A, Craig AE, Mohammad A, Farrukh A. Fisetin, a phytochemical, potentiates sorafenib-induced apoptosis and abrogates tumor growth in athymic nude mice implanted with BRAF-mutated melanoma cells. Oncotarget. 2015;**6**:28296–28311. doi:10.18632/oncotarget.5064

[20] Scoparo CT, Valdameri G, Worfel PR, Guterres FA, Martinez GR, Winnischofer SM, Di Pietro A, Rocha ME. Dual properties of hispidulin: antiproliferative effects on HepG2 cancer cells and selective inhibition of ABCG2 transport activity. Molecular and Cellular Biochemistry. 2015;**409**:123–133. doi:10.1007/s11010-015-2518-8

[21] Hui G, Qixiao J, Yantao H, Jianjun P, Chunbo W. Hispidulin potentiates the antitumor effect of sunitinib against human renal cell carcinoma in laboratory models. Cell Biochemistry and Biophysics. 2015;**71**:757–764. doi:10.1007/s12013-014-0260-6

[22] Li SZ, Li K, Zhang JH, Dong Z. The effect of quercetin on doxorubicin cytotoxicity in human breast cancer cells. Anticancer Agents in Medicinal Chemistry. 2013;**13**:352–355. doi:10.2174/1871520611313020020

[23] Akhilendra KM, Manjula V. Anticarcinogenic action of quercetin by downregulation of phosphatidylinositol 3-kinase (PI3K) and protein kinase C (PKC) via induction of p53 in hepatocellular carcinoma (HepG2) cell line. Molecular Biology Reports. 2015;**42**:1419–1429. doi:10.1007/s11033-015-3921-7

[24] Dandan W, Jixiang Z, Jing W, Jiao L, Fei L, Weiguo D. Hesperetin induces apoptosis of esophageal cancer cells via mitochondrial pathway mediated by the increased intracellular reactive oxygen species. Tumor Biology. 2016;**37**:3451–3459. doi:10.1007/s13277-015-4176-6

[25] Sanjeev S, Sanjay G. Apigenin: a promising molecule for cancer prevention. Pharmaceutical Research. 2010;**27**:962–978. doi:10.1007/s11095-010-0089-7

[26] Balbino LS, Mona NO, Paulo LCC, Bruno PSP, Alessandra BS, Tais A, Victor DAS, Maria FDC, Ramon SE, Marcienne T, Herve C, Marie-Pierre J, Vivaldo M, Silvia LC, Flavonoids suppress human glioblastoma cell growth by inhibiting cell metabolism, migration and by regulating extracellular matrix proteins and metalloproteinases expression. Chemico-Biological Interactions. 2015;**242**:123–138. doi:10.1016/j.cbi.2015.07.014.

[27] Yalda S, Maryam R, Leila H, Zohreh J, Quercetin-3-*O*-β-D-glucopyranoside, a dietary flavonoid, protects PC12 cells from H_2O_2-induced cytotoxicity through inhibition of reactive oxygen species, Food Chemistry. 2015;**167**:162–167. doi:10.1016/j.foodchem.2014.06.079

[28] Meixian X, Hanwen S, Guangwen S, Dingrong W, Feng H, Morgann L, Yali D, Jun L, Sinisa D, Gaungzhong Y, Chunhua S. Amplexicaule A exerts anti-tumor effects by inducing apoptosis in human breast cancer. Oncotarget. 2016;**7**:18521–18530. doi:10.18632/oncotarget.7848

[29] Yu S, Yan C, Yi Z, Fu-Qin G, Hao S, Xing-cong R, Yu C, Xu Feng, Jin-Ming Y. Triticuside A, a dietary flavonoid, inhibits proliferation of human breast cancer cells via inducing apoptosis. Nutrition and Cancer. 2013;**65**:891–899. doi:10.1080/01635581.2013.802001.

[30] Hongzhong L, Bing Y, Jing H, Tingxiu X, Xuedong Y, Jingyuan W, Fuling L, Li Z, Hongyuan L, Guosheng R. Naringin inhibits growth potential of human triple-negative breast cancer cells by targeting β-catenin signaling pathway. Toxicology Letters. 2013;**220**:219–228. doi:10.1016/j.toxlet.2013.05.006

[31] Jeong HK, Jae KL. Naringenin enhances NK cell lysis activity by increasing the expression of NKG2D ligands on Burkitt's lymphoma cells. Archives of Pharmacal Research. 2015;**38**:2042–2048. doi:10.1007/s12272-015-0624-5

[32] Balachandran C, Sangeetha B, Duraipandiyan V, Raj MK, Ignacimuthu S, Al-Dhabi NA, Balakrishna K, Parthasarathy K, Arulmozhi NM, Arasu MV. A flavonoid isolated from Streptomyces sp. (ERINLG-4) induces apoptosis in human lung cancer A549 cells through p53 and cytochrome c release caspase dependant pathway. Chemico-Biological Interactions. 2014;**224**:24–35. doi:10.1016/j.cbi.2014.09.019

[33] Mi JC, Sanghyun L, Yong P, Jisun L, Ki HK. Neuroprotective effects of phytosterols and flavonoids from *Cirsium setidens* and *Aster scaber* in human brain neuroblastoma SK-N-SH cells. Life Sciences. 2016;**148**:173–182. doi:10.1016/j.lfs.2016.02.035

[34] Panat NA, Singh BG, Maurya DK, Sandur SK, Ghaskadbi SS. Troxerutin, a natural flavonoid binds to DNA minor groove and enhances cancer cell killing in response to radiation, Chemico-Biological Interactions. 2016;**251**:34–44. doi:10.1016/j.cbi.2016.03.024

[35] Reneta G, Dimitrina Z, Silviya R, Nikolay D, Spiro K, Valentin L, Vanio M. Cytotoxic and hepatoprotective effects of *Bupleurum flavum* flavonoids on hepatocellular carcinoma HEP-G2 cells. British Journal of Pharmaceutical Research. 2016;**11**:1–8. doi:10.9734/BJPR/2016/25785

[36] Lokesh DN, Rit V, Jyotsana S, Spence F, Rhonda R, Sanjay A, Sharad SS. Anti-cancer effects of novel flavonoid vicenin-2 as a single agent and in synergistic combination with docetaxel in prostate cancer. Biochemical Pharmacology. 2011;**82**:1100–1109. doi:10.1016/j.bcp.2011.07.078

[37] Zhao Q, Zhao M, Parris AB, Xing Y, Yang X. Genistein targets the cancerous inhibitor of PP2A to induce growth inhibition and apoptosis in breast cancer cells. International Journal of Oncology, 2016;**49**:1203–1210. doi:10.3892/ijo.2016.3588

[38] Janet MP, Sankar NK, Raymond CB. Genistein inhibits human prostate cancer cell detachment, invasion and metastasis. The American Journal of Clinical Nutrition. 2014;**100**:431S–436S. doi:10.3945/ajcn.113.071290

[39] Lu Z, Zhou R, Kong Y, Wang J, Xia W, Guo J, Liu J, Sun H, Liu K, Yang J, Mi M, Xu H. S-equol, secondary metabolite of natural anticancer isoflavone daidzein, inhibits prostate cancer growth in vitro and in vivo, though activating the Akt/FOXO3a pathway. Current Cancer Drug Targets. 2016;**16**:455–465. doi:10.2174/156800961666615120 7105720

[40] Bhardwaj V, Tadinada SM, Jain A, Sehdev V, Daniels CK, Lai JC, Bhushan A. Biochanin A reduces pancreatic cancer survival and progression. Anticancer Drugs. 2014;**25**:296–302. doi:10.1097/CAD.0000000000000044.

[41] Victor K, Armelle TM, Eric CNN, Christophe CS, Maen Z, Augustin EN, Thomas E. Cytotoxicity of seven naturally occurring phenolic compounds towards multi-factorial drug-resistant cancer cells . Phytomedicine. 2016;**23**:856–863. doi:10.1016/j.phymed.2016.04.007

[42] Yan-Xi Z, Hong Z, Cheng P. Puerarin: a review of pharmacological effects. Phytotherapy Research. 2014;**28**:961–975. doi:10.1002/ptr.5083

[43] Sung-Gook C, Sang-Mi W, Seong-Gyu K. Butein suppresses breast cancer growth by reducing a production of intracellular reactive oxygen species. Journal of Experimental & Clinical Cancer Research. 2014;**33:**1–11. doi:10.1186/1756-9966-33-51

[44] Moon DO, Kim MO, Lee JD, Choi YH, Kim GY. Butein suppresses c-Myc-dependent transcription and Akt-dependent phosphorylation of hTERT in human leukemia cells. Cancer Letters. 2009;**286**:172–179. doi:10.1016/j.canlet.2009.05.028

[45] Yit CC, Das NP, Cytotoxic effect of butein on human colon adenocarcinoma cell proliferation, Cancer Letters. 1994;**82**:65–72. doi:10.1016/0304-3835(94)90147-3

[46] Cui Z, Song E, Hu DN, Chen M, Rosen R, McCormick SA. Butein induces apoptosis in human uveal melanoma cells through mitochondrial apoptosis pathway. Current Eye Research. 2012;**37**:730–9. doi:10.3109/02713683.2012.671436

[47] Xue B, Yaxin M, Guobin Z. Butein suppresses cervical cancer growth through the PI3K/AKT/mTOR pathway. Oncology Reports. 2015;**33**:3085–3092. doi:10.3892/or.2015.3922

[48] Fu P, Chun-Wang M, Qin-Mei Z, Jian-Ping C, Liang X. Cytotoxic evaluation against breast cancer cells of isoliquiritigenin analogues from *Spatholobus suberectus* and their synthetic derivatives. Journal of Natural Products. 2016;**79**:248–251. doi:10.1021/acs.jnatprod.5b00774

[49] Chun-Lin Y, Jian-Wen L, Dong-Zhi W, Yan-Hua L. Feng Qian. In vitro anti-tumor activity of 2′,4′-dihydroxy-6′-methoxy-3′,5′-dimethylchalcone against six established human cancer cell lines. Pharmacological Research. 2004;**50**:505–10. doi:10.1016/j.phrs.2004.05.004

[50] Huang HY, Niu JL, Zhao LM, Lu YH. Reversal effect of 2′,4′-dihydroxy-6′-methoxy-3′,5′-dimethylchalcone on multi-drug resistance in resistant human hepatocellular carcinoma cell line BEL-7402/5-FU. Phytomedicine. 2011;**18**:1086–1092. doi:10.1016/j.phymed.2011.04.001

[51] Victor K, Antoine HLN, Armelle TM, Benjamin W, Hippolyte KW, Pierre T, Thomas E. Cytotoxicity and modes of action of 4′-hydroxy-2′,6′-dimethoxychalcone and other flavonoids toward drug-sensitive and multidrug-resistant cancer cell lines. Phytomedicine. 2014;**21**:1651–57. doi:10.1016/j.phymed.2014.08.001

[52] Miranda CL, Stevens JF, Helmrich A, Henderson MC, Rodriguez RJ, Yang YH, Deinzer ML, Barnes DW, Buhler DR. Antiproliferative and cytotoxic effects of prenylated flavonoids from hops (*Humulus lupulus*) in human cancer cell lines. Food and Chemical Toxicology. 1999;**37**:271–285. doi:10.1016/S0278-6915(99)00019-8

[53] Tetsuro Y, Masato K, Etsu T, Masaya I, Hiroyuki O, Yasuo M, Junko H, Mitsunori S, Toyomasa K. Xanthohumol suppresses oestrogen-signalling in breast cancer through the inhibition of BIG3-PHB2 interactions. Scientific Reports. 2014;**7355**:1–9. doi:10.1038/srep07355

[54] Nadiah A, Muhammad NA, Swee KY, Kian LL, Wan YH, Mohd PA, Chai LH, Abdul RO, Jamil I, Noorjahan B. Flavokawain B induced cytotoxicity in two breast cancer cell lines, MCF-7 and MDA-MB231 and inhibited the metastatic potential of MDA-MB231 via the regulation of several tyrosine kinases *In vitro*. BMC Complementary and Alternative Medicine. 2016;**16**:1–14. doi:10.1186/s12906-016-1046-8

[55] Fang SC, Hsu CL, Yu YS, Yen GC. Cytotoxic effects of new geranyl chalcone derivatives isolated from the leaves of Artocarpus communis in SW 872 human liposarcoma cells. Journal of Agricultural and Food. 2008;**56**:8859–8868. doi:10.1021/jf8017436

[56] Sumiyoshi M, Taniguchi M, Baba K, Kimura Y. Antitumor and antimetastatic actions of xanthoangelol and 4-hydroxyderricin isolated from *Angelica keiskei* roots through the inhibited activation and differentiation of M2 macrophages. Phytomedicine. 2015;**22**:759–767. doi:10.1016/j.phymed.2015.05.005

[57] Keiichi T, Kou M, Noriya T, Reiko N, Satoru A, Yumiko K, Motohiko U, Daisuke H, Toshihiro A, Takashi S. Xanthoangelol, a major chalcone constituent of *Angelica keiskei*, induces apoptosis in neuroblastoma and leukemia cells. Biological and Pharmaceutical Bulletein. 2005;**28**:1404–1407. doi:10.1248/bpb.28.1404

[58] Victor K, Armelle TM, Bonaventure TN, Thomas E. Cytotoxicity of three naturally occurring flavonoid derived compounds (artocarpesin, cycloartocarpesin and isobavachalcone) towards multi-factorial drug-resistant cancer cells. Phytomedicine. 2015;**22**:1096–1102. doi:10.1016/j.phymed.2015.07.006

[59] Kuete V, Mbaveng AT, Zeino M, Ngameni B, Kapche GD, Kouam SF, Ngadjui BT, Efferth T. Cytotoxicity of two naturally occurring flavonoids (dorsmanin F and poinsettifolin B) towards multi-factorial drug-resistant cancer cells. Phytomedicine. 2015;**22**:737–743. doi:10.1016/j.phymed.2015.04.007

[60] Cazal CM, Choosang K, Severino VG, Fernandes JB, da Silva MF, Vieira PC, Nascimento MS, Almeida GM, Vasconcelos MH, Pakkong P, Pinto MM. Natural compounds with cell growth inhibitory activity in human tumor cell lines. Anticancer Agents in Medicinal Chemistry. 2013;**13**:1582–1589. doi:10.2174/18715206113666131125122857

[61] Ladislaus K. Mdee, Samuel O. Yeboah, Berhanu M. Abegaz, rhuschalcones II–VI, five new bichalcones from the root bark of *Rhuspyroides*. Journal of Natural Product. 2003;**66**:599–604. doi:10.1021/np020138q

[62] Yoo I, Yun B, Lee I, Ryoo I, Choung D, Han K. Three naphthalenes from root bark of Hibiscus syriacus. Phytochemistry. 1998;**47**:799–802. doi:10.1016/S0031-9422(97)00674-2

[63] Heng-Xing L, Huan-Qin D, Hai-An F, Xiao-Ping D, Abiodun HA, Li-Xin Z, Yong-Xian C. Bioactive compounds from Rumex plants. Phytochemistry Letters. 2010;**3**:181–184. doi:10.1016/j.phytol.2010.05.005

[64] Negera A, Fangfang P, Amra G, Jurgen G, Paul AF, Goran L, Kari R, Abiy Y, Mate E. Naphthalene derivatives from the roots of *Pentas parvifolia* and *Pentas bussei*. Journal of Natural Products. 2016;**79**:2181–2187. doi:10.1021/acs.jnatprod.6b00178

[65] Hui Z, Zengjun G, Nan W, Wenming X, Ling H, Nan L, Yanxia H. Two novel naphthalene glucosides and an anthraquinone isolated from *Rumex dentatus* and their antiproliferation activities in four cell lines. Molecules. 2012;**17**:843–850. doi:10.3390/molecules17010843

[66] Chandrasekar B, Veeramuthu D, Al-Dhabi NA, Stalin A, Balakrishna K, Ignacimuthu S, Tilton F. Isolation and characterization of Anthraquinone from *Streptomyces* sp. ERINLG-26 with anticancer activity against adenocarcinoma cell line COLO320. Applied Biochemistry and Microbiology. 2015;**51**:522–529. doi:10.1134/S0003683815050038

[67] Chandrasekar B, Veeramuthu D, Yuvaraj A, Balachandran S, Nobuhiko E, Naif AA, Savarimuthu I, Yoko I, Akinao O, Paramasivan TP. Isolation and characterization of 2-hydroxy-9,10-anthraquinone from *Streptomyces olivochromogenes* (ERINLG-261) with antimicrobial and antiproliferative properties. Brazilian Journal of Pharmacognosy. 2016;**26**:285–295. doi:10.1016/j.bjp.2015.12.003

[68] Balachandran C, Emi N, Arun Y, Yamamoto N, Duraipandiyan V, Inaguma Y, Okamoto A, Ignacimuthu S, Al-Dhabi NA, Perumal PT. In vitro antiproliferative activity of 2,3-dihydroxy-9,10-anthraquinone induced apoptosis against COLO320 cells through cytochrome c release caspase mediated pathway with PI3K/AKT and COX-2 inhibition. Chemico-Biological Interactions. 2016;**249**:23–35. doi:10.1016/j.cbi.2016.02.016

[69] Shi YQ, Fukai T, Sakagami H, Kuroda J, Miyaoka R, Tamura M, Yoshida N, Nomura T. Cytotoxic and DNA damage-inducing activities of low molecular weight phenols from rhubarb. Anticancer Research. 2001;**21**:2847–2853.

[70] Surat L, Wisanu M, Sorwaporn K. Antimalarial and cytotoxic phenolic compounds from *Cratoxylum maingayi* and *Cratoxylum cochinchinense*. Molecules. 2009;**14**:1389–1395. doi:10.3390/molecules14041389

[71] Sompong B, Chatchanok K, Chanita P, Akkharawit K, Kan C. Antibacterial and cytotoxic xanthones from the roots of *Cratoxylum formosum*. Phytochemistry. 2006;**67**:723–727. doi:10.1016/j.phytochem.2006.01.007

[72] Frank M, Niemann H, Bohler P, Stork B, Wesselborg S, Lin W, Proksch P. Phomoxanthone A-from mangrove forests to anticancer therapy. Current Medicinal Chemistry. 2015;**22**:3523–3532. doi:10.2174/0929867322666150716115300

[73] Manqin F, Samuel XQ, Yujuan X, Jijun W, Yulong C, Yuanshan Y, Gengsheng X. A new xanthone from the pericarp of *Garcinia mangostana*. Natural Product Communications. 2013;**8**:1733–1734.

[74] Victor K, Louis PS, Judith LNO, Hugues F, Benjamin W, Thomas E. Cytotoxicity and modes of action of three naturally occurring xanthones (8-hydroxycudraxanthone G, morusignin I and cudraxanthone I) against sensitive and multidrug-resistant cancer cell lines. Phytomedicine. 2014;**21**:315–322. doi:10.1016/j.phymed.2013.08.018

[75] Dya FD, Suresh A, Shigetoshi K, Hiroyuki M, Yasuhiro T. Heptaoxygenated xanthones as anti-austerity agents from *Securidaca Longepedunculata*. Bioorganic & Medicinal Chemistry. 2013;**21**:7663–7668. doi:10.1016/j.bmc.2013.10.027

[76] Asano J, Chiba K, Tada M, Yoshii T. Cytotoxic xanthones from *Garcinia hanburyi*. Phytochemistry. 1996;**41**:815–820. doi:10.1016/0031-9422(95)00682-6

[77] Shadid KA, Shaari K, Abas F, Israf DA, Hamzah AS, Syakroni N, Lajis NH. Cytotoxic caged-polyprenylated xanthonoids and a xanthone from *Garcinia cantleyana*. Phytochemistry. 2007;**68**:2537–2544. doi:10.1016/j.phytochem.2007.05.024

[78] Soleh K, Shu-Geng C, Xiao-Hua W, Jagadese JV, Tony S, Masdianto SH, Goh, Keng-Yeow Sim. Lateriflorone, a cytotoxic spiroxalactone with a novel skeleton, from *Garcinia lateriflora BL*. Tetrahedron Letters. 1999;**40**:157–160. doi:10.1016/S0040-4039(98)80046-4

[79] Zhang HZ, Kasibhatla S, Wang Y, Herich J, Guastella J, Tseng B, Drewe J, Cai SX. Discovery, characterization and SAR of gambogic acid as a potent apoptosis inducer by a HTS assay. Bioorganic and Medicinal Chemistry. 2004;**12**:309–317. doi:10.1016/j.bmc.2003.11.013

[80] Xiaohua W, Shugeng C, Sweehock G, Anne H, Benny KHT. Mitochondrial destabilisation and Caspase-3 activation are involved in the apoptosis of jurkat cells induced by gaudichaudione A, a cytotoxic xanthone. Planta Medica. 2002;**68**:198–203. doi:10.1055/s-2002-23142

[81] Zhengxiang X, Hong Z, Danqing X, Yuanzhi L, Wenwei F, Hongsheng T, Peng C, Ling Y, Hongxi X. Xanthones from the leaves of *Garcinia cowa* induce cell cycle arrest, apoptosis and autophagy in cancer cells. Molecules. 2015;**20**:11387–11399. doi:10.3390/molecules200611387

[82] Josiane AM, Ivania TAS, Thiago LA, Silvana MOS, Cleuza CS, Lucas URC, Ana LTGR, Joao EC, Debora BV, Celso VN, Debora BS, Vanessa K, Ricardo TF, Armando MP. Oncibauerins A and B, new flavanones from *Oncidium baueri* (Orchidaceae). Phytochemistry Letters. 2014;**9**:141–148. doi:10.1016/j.phytol.2014.06.004

[83] Qiao-Li L, Ling-Ling L, Xiaodong C, Nuo-Shu Z, Jin-Ao D. Bioactive cis-stilbenoids from the tubers of *Scirpus yagara*. Fitoterapia. 2013;**84**:170–73. doi:10.1016/j.fitote.2012.11.012

[84] Nen-Ling Z, Yue-Hui Z, Ri-Ming H, Man-Qin F, Zhi-Wei S, Jia-Zhong C, Ying-Jie H, Sheng-Xiang Q. Two new stilbenoids from *Cajanus cajan*. Zeitschrift fur Naturforschung B. 2012;**67b**:1314–18. doi:10.5560/ZNB.2012-0184

[85] Huang CP, Au LC, Chiou RY, Chung PC, Chen SY, Tang WC, Chang CL, Fang WH, Lin SB. Arachidin-1, a peanut stilbenoid, induces programmed cell death in human leukemia HL-60 cells. Journal of Agricultural and Food Chemistry. 2010;**58**:12123–29. doi:10.1021/jf102993j

[86] Jie M, Duowei W, Zhuo W, Wei T, Xianjing L, Jingjing D, YunW, Hongbao Y, Linjun Y, Yan C, Jinsong B, Zhen C, Yong Y. Combretastatin A-1 phosphate, a microtubule inhibitor, acts on both hepatocellular carcinoma cells and tumor-associated macrophages by inhibiting the Wnt/β-catenin pathway. Cancer Letters. 2016;**80**:134–143. doi:10.1016/j.canlet.2016.06.020

[87] Phanruethai P, Suwannee S, Chulabhorn M, Somsak R, Vilailak P. Cytotoxic and cancer chemopreventive properties of prenylated stilbenoids from *Macaranga siamensis*. Tetrahedron. 2015;**71**:5562–5571. doi:10.1016/j.tet.2015.06.058

[88] Ming HY, Le C, Mei HL, Xiang HZ, Ya BY, Zhong TD. Three new phenanthrenes from *Monomeria barbata* Lindl. Chinese Chemical Letters. 2010;**21**:325–328. doi:10.1016/j.cclet.2009.11.006

[89] Junju X, Hong Y, Chen Q, Yanli Z, Ying L, Yegao C. Two new biphenanthrenes with cytotoxic activity from *Bulbophyllum odoratissimum*. Fitoterapia. 2009;**80**:381–84. doi:10.1016/j.fitote.2009.05.007

[90] Young-Won C, William PJ, Ismail R, Soedarsono R, Leonardus BSK, Hee-Byung C, Norman RF, Geoffrey AC, Steven MS, John MC, Kinghorn AD. Cytotoxic lignans from the stems of *Helicteres hirsuta* collected in Indonesia. Phytotherapy Research. 2006;**20**:62–65. doi:10.1002/ptr.1806

[91] Won SS, Lalita S, Sun YK, Sang UC, Kang RL. Bioactive lignan constituents from the twigs of *Sambucus williamsii*. Bioorganic & Medicinal Chemistry Letters. 2016;**26**:1877–1880. doi:10.1016/j.bmcl.2016.03.023

[92] Shan Y, Jun-Jiang L, Jian-Qiang Z, Yan L, Dong W, Chong-Ren Y, Min X, Ying-Jun Z. New cytotoxic lignan glycosides from *Phyllanthus glaucus*. Natural Product Research. 2016;**30**:419–425. doi:10.1080/14786419.2015.1023198

Chapter 3

Phenolics in Foods: Extraction, Analysis and Measurements

Alfredo Aires

Additional information is available at the end of the chapter

http://dx.doi.org/10.5772/66889

Abstract

The increasing consumers demands to acquire healthier fruits and vegetables as well as the urgency in looking to natural compounds with antioxidant activity and enhanced antimicrobial activity against antibiotic-resistant pathogenic bacterial strains have encouraged a quick expansion of research studies about enhanced phenolic extraction and identification methods. Considering the importance of phenolics as natural compounds with antioxidant and antimicrobial activity, this chapter aims to present the most updated information about extraction methods, ranging from the traditional to the most advanced processes, as well as the access to the modern methods used in the identification and quantification of phenolics. The main goal of this chapter is to provide the reader with a broad view on the different protocols used to extract, identify and quantify phenolic compounds from different kinds of foods, including fruits and vegetables.

Keywords: phytochemicals, extraction, determination, colorimetric methods, HPLC, HPLC-MS

1. Introduction

The growing resistance of pathogenic bacterial isolates against the traditional chemical antibiotics as well as the resurgent of old disappeared diseases associated with the constant consumers' demanding of healthier, nutritious and safe food has led the researchers to focus on searching for new, safe and effective molecules. One class of such molecules is the class of polyphenols. Polyphenols are a ubiquitous class of compounds largely present in plants as their secondary metabolites, which are synthesized during their normal development [1] in response to several stressful biotic and abiotic factors [2, 3]. This class of compounds are a much diversified group derived from the amino acids phenylalanine and tyrosine and comprise simple phenols, hydroxybenzoic acids and cinnamic acid derivatives, flavonoids, coumarines, stilbenes and tannins, among others [4–6].

The results from the last decade's research have shown that polyphenols have important beneficial properties for human health, including antioxidative, antiaging, antibacterial and antimutagenic [7–11]. Moreover, the recent evidence of their interaction with proteins, DNA and other biological molecules has enhanced their exploitation for the production of new natural product-derived therapeutic agents. Despite these advantages, several limitations still persist, particularly those related with their extraction efficiency, which affects the large-scale use of some of these substances. The difficulties in screening, extracting, separation and purifying these compounds have increased the development of new and modern methods to address these limitations. In this context, the aim of this chapter is to present an updated review about sources, technologies and methods that have been developed until now to improve the extraction, detection, separation and full characterization of such beneficial compounds, with special emphasis to their possible application in the design of nutraceuticals and functional food products.

2. Foods as natural resources of phenolics

Polyphenols have been exhaustively studied in their different natural matrices such as fruits, vegetables, teas, algae and microalgae and more recently agro-food wastes (peels, seeds, pulps, stems and roots) [12–15]. In the three last decades, there has been a prolific publication of scientific studies showing that plant-derived foods and agro-food wastes from industrial transformation have huge quantities of polyphenols. In **Table 1** are summarized some recent studies, and as result from these and other studies, there is a diverse source of polyphenols in plant materials, but both type and amount seem to be highly influenced by their chemical nature, extraction methods, sample particle size, storage time and conditions, as well as by the presence other of interfering substances [25]. Also, their chemical structure and nature vary from simple to highly polymerized substances that include varying proportions of phenolic acids, phenylpropanoids, anthocyanins and tannins, among others [26–28]. Moreover, they might also exist in complex mixtures with carbohydrates, proteins and some quite insoluble high-molecular-weight phenolics [28]. Therefore, the phenolic extraction from plant materials is always a mixture of different steps, and many modifications of a particular method are often needed for the removal of unwanted non-phenolic substances such as waxes, fats, terpenes, pigments (chlorophylls and carotenoids). Solid-phase extraction (SPE) techniques, purification and fractionation based on acidity, are commonly used to remove unwanted non-phenolic substances or even other unwanted phenolics [29].

Although the recent advances in the technology had providing innovative approaches to obtain enriched polyphenol natural extracts, we must ware that their extraction efficiency will always be dependent of several factors in which the nature of samples and solvent, pH, temperature, light, length of extraction period, particle size, solvent/sample ratio and liquid-liquid or solid-liquid extraction process [25], among others, are the most critical.

Polyphenols	Source (some examples)
Phenolic acids	
Hydroxycinnamic acids	Cereals, coffee, cherries, citrus fruits and juices, peaches, plums, spinach, tomatoes, wheat flour, corn flour, rice flour, potato, olive mill wastewaters, winery sludge from red grapes, artichoke wastewaters, almonds
Hydroxybenzoic acids	Oilseeds, cereals, coffee, cowpeas, wheat flour, black currant, blackberry, raspberry, squash seeds and shell
Flavonoids	
Anthocyanins	Grapes, red wine, grape seeds, grape skins, winery by-products, fermented grape pomace, strawberries, back and red currants, raspberries, plums, red cabbage
Chalcones	Apples and apple juices,
Flavanols	Apples, grapes, leeks, tomatoes, curly kale, onions, lettuces, berries, beans, red grapes, black and green tea, red wine and red winery by-products, cider
Flavanones	Citrus fruits, citrus juices, orange peels and seeds wastes
Flavonols	Apples, apple peels, beans, leeks, lettuce, onions, tomatoes, olive leafs, broccoli inflorescences, chestnut, olives and olive fermented pomaces
Flavones	Spinach, citrus fruits, celery, pepper, capsicum pepper,
Isoflavones	Soybeans, soy flour, soy milk, soy processing waste
Stilbenes	Red grapes, grapes skins, grape seeds, red grape fermented pomaces
Xanthones	Mango fruits and mango peels fermented pomaces
Tannins	
Condensed tannins	Apples, grapes, peaches, pears, chestnut, hazelnuts, nuts
Hydrolyzable tannins	Pomegranate, raspberries

Table 1. Most common types of polyphenols found in foods and plant-derived products [14–24].

3. Methods used in extraction of polyphenols

It is widely accepted that the extraction step is one of the most important stage in isolation of polyphenols, but based in literature, there is no consensus about one single and effective standard extraction method. On contrary, there are several reported methods with very accurate results, and according to the literature in some cases, the solid-liquid extraction with different types of solvents is more adequate [30], and in others, the ultrasound-assisted extraction method (UAE) increases the extraction efficiency [31], while in others, this increment is higher when a microwave-assisted extraction (MAE) is used [32], and advanced methods such as pressurized fluid extraction (PFE), supercritical CO_2 extraction (SC-CO_2)

and enzyme-assisted extraction (EAE) are even better to enhance the content of polyphenols in the extracts [33–39]. Despite this diversity, all have the common fact that the extraction must be conducted carefully but exhaustively with simple, rapid and feasible procedures, and if possible open to automation [40]. In the next paragraphs, we present a summarized information of the most commonly used methods for the extraction of polyphenols in several pant and food matrices.

3.1. Classical solvent extraction

The classical solvent extraction of polyphenols usually includes extraction by maceration and percolation and by successive Soxhlet extraction [41–45].

The maceration, widely used in the past, is nowadays in underuse since other methods are more feasible. It is a simple procedure in which the powdered sample is soaked in an appropriate solvent in a closed container, normally under room temperature with constant or sporadic agitation [41, 46]. In the end of the extraction, the solid parts need to be separated from the solvent, which can be done by filtration, clarification and/or decantation [47]. This method is quite simple to hand but has the main disadvantage of time-consuming, requires a large volume of solvent [41, 42, 48, 49].

Similar to the maceration, the percolation method is characterized by placing the powdered sample in a closed container (normally cylindrical) in which the solvent is discharged from the top towards the bottom in a slow movement (drop wise). [41, 42, 50]. In this case, the filtration is not necessary because the percolator device has itself a filter which placed at the bottom, and we can only collect the final liquid. This method faces the same issues of maceration, which are time-consuming, large volumes of solvent, solubility of polyphenols, particle size of sample and contact time between solvent and sample.

In Soxhlet extraction [41, 42], the powdered samples are sealed in cellulose bags and placed in an extraction chamber located on top of a collecting flask beneath a reflux condenser, and after the addition of the solvent, the system is heated and the solvent condenses after reaching certain level of temperature [51–53]. A reflux occurs continuously. At the end, the liquid extract is collected to the flask positioned beneath the system [51–53]. The Soxhlet extraction is a continuous process with the advantage of being less time and less solvent consuming than the maceration or percolation methods [54]. However, some authors have stated that Soxhlet extraction must be handled carefully because the excess of temperature, always near to the boiling point, can destroy or modify some thermolabile polyphenols [44]. Others reported that Soxhlet extraction is used widely because of its convenience [41, 42, 44, 54]. Although these have variations, all these three methods have the common usage of organic solvents in a solid/liquid ratio. Solvents such as water, methanol, ethanol, acetone, n-hexane, chloroform, propanol and ethyl acetate have been most commonly used for the extraction of polyphenols (**Table 2**). The difference between solvents resides in their polarity (**Figure 1**) which affects their capacity in extract phytochemicals. The miscibility of organic solvents (**Figure 2**) with each other's or even other types of solvents is another fact to be considered in order to improve the polyphenol extraction yield as shown by several studies [59–63].

http://dx.doi.org/10.5772/66889

Solvent (alphabetical order)	Boiling point (°C)	Density (g/mL)	Solubility in H_2O (g/100g)	Polarity Index[1]	Eluent strength[2–4]	Dielectric constant at 20°C[1, 2, 4]
Acetone	56.2	0.786	Miscible	5.10	0.56	21.1
Acetonitrile	81.6	0.786	Miscible	6.20	0.65	36.64
Chloroform	61.2	1.498	0.8	4.40	0.40	4.81
Ethanol	78.5	0.789	Miscible	5.20	0.88	24.6
Ethyl acetate	77	0.894	8.7	4.30	0.58	6
Hexane	69	0.655	0.0014	0.06	0.01	1.89
Methanol	64.6	0.791	Miscible	6.60	0.95	32.6
2-Propanol	82.4	0.785	Miscible	4.30	0.82	18.3
Water	100	0.998	Miscible	10.20	>>1	78.54

[1]Data collected from http://macro.lsu.edu/HowTo/solvents.htm [55].
[2]Data collected from Speight (2005) [57].
[3]Data collected from Singh et al. (2014) [57].
[4]Data collected from Hakansson et al. (2016) [58].

Table 2. Important properties of some solvents commonly used in the extraction of polyphenols[1–4].

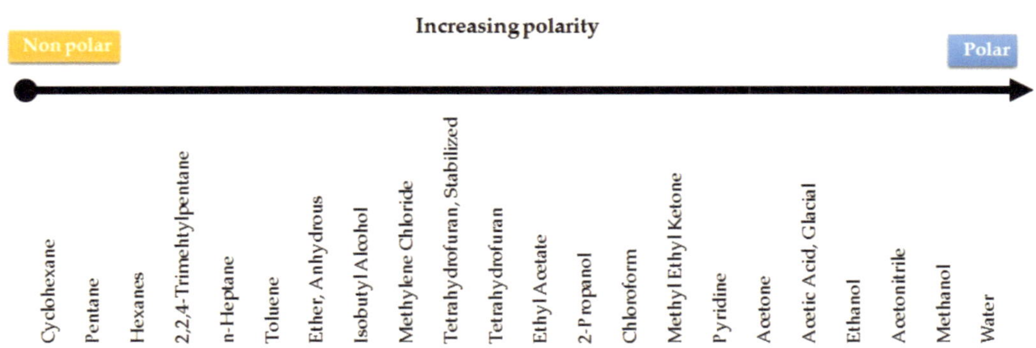

Figure 1. Polarity of the organic solvents most commonly used in phytochemicals extraction from natural sources. Adapted with permission from Refs. [55–56].

In general, organic solvents and their aqueous formulations are mostly used in the extraction of phytochemicals, but it is still no clear which solvent is most adequate for the extraction of polyphenols. For example, acetone showed to be very efficient in the extraction of polyphenols [59] from lychee (*Litchi chinenesis* Sonn.) flowers in comparison with methanol, ethanol or water. While in walnut (*Juglans regia* L.) green husks, the highest extraction yield of polyphenols (44.1%) was obtained when water was used as extraction solvent [60]. By other hand, in a recent study [61], it was found that aqueous and organic solvent have a higher extraction efficiency than absolute organic solvents. Similar situation was observed in *Phoradendron californicum* oak extracts [62], in which aqueous methanol was the solvent most efficient for the extraction of polyphenols.

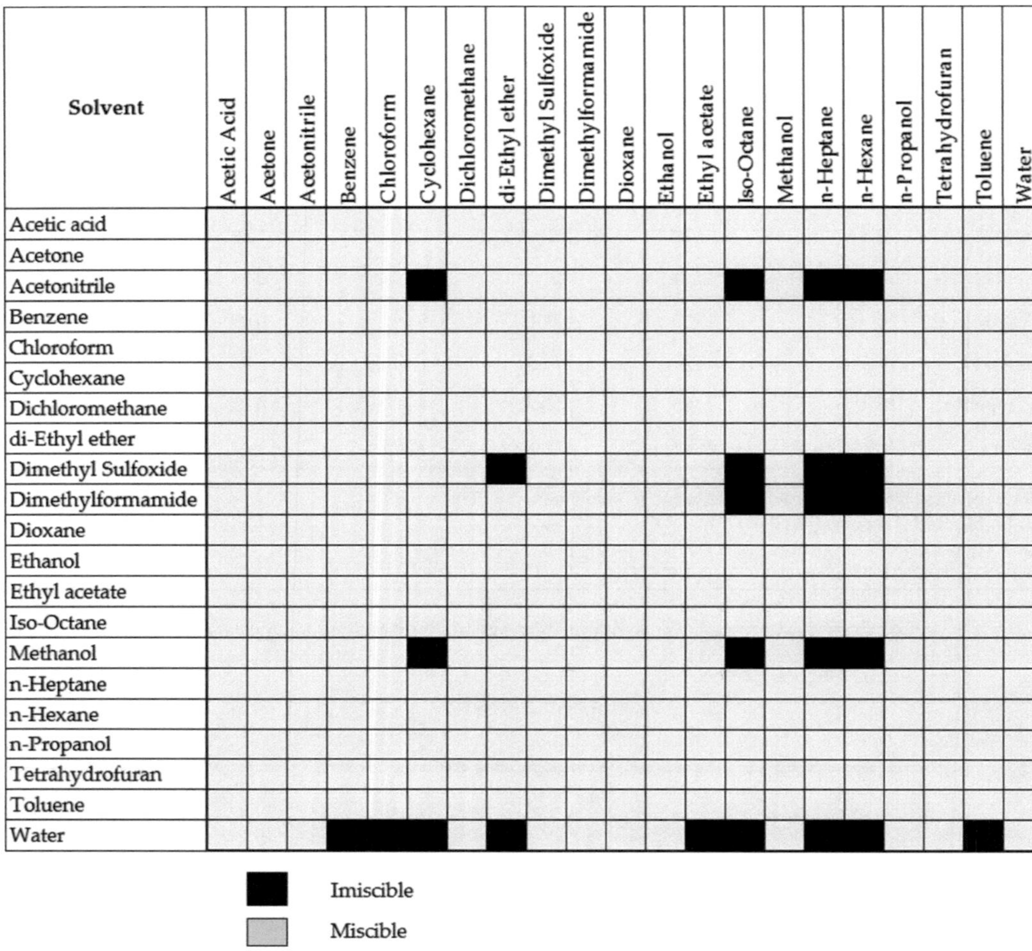

Figure 2. Miscibility of organic solvents used in the extraction of phytochemicals from natural sources. Adapted with permission from Refs. [56–58].

Based on the literature, there is no consensus about the best solvent to extract polyphenols. However, it has been widely accepted that higher polarity usually means better solubility of polyphenols into extraction solvents; however, differences in the structure of phenolic compounds may be critical for their solubility. Thus, the extraction of polyphenols and other phytochemicals must be prior tested and adapted to the solvent, because the diverse structures of polyphenols, such as multiple hydroxyl groups, conjugated or not with sugars, acids or alkyl groups, interfere in the extraction process. Therefore, it is very difficult to say what type of solvent is better to develop a standard method for all type of polyphenols, but the majority of the authors seem to agree that a good solvent system is the one that allows the maximization of polyphenols extracted without any modifications of their chemical nature. In this context, several factors must be considered when a specific solvent is selected, including (i) solvent power (selectivity); (ii) polarity; (iii) boiling temperature (should be low in order to facilitate removal of the solvent from the product); (iv) reactivity (the solvent should not react chemically with the extract neither should be decomposed quickly); (v) viscosity (low); (vi) stability (should be stable to heat, oxygen and light); (vii) safe in use (should be nonflammable and nontoxic for consumers and environment); (viii) if possible, suitability for reuse; and (ix) compatible with legislation for food applications.

3.2. Advanced methods of extraction

Classical extraction methods are dominated in many laboratory facilities mainly due to its simplicity and low economic cost. Nonetheless, many scientific reports have shown that maceration, percolation and Soxhlet extraction have low efficiency and several environmental issues due to the pollution caused in the environment when large volumes of organic solvents are used. Moreover, the classical extraction often requires a recovery step followed by evaporation to concentrate the extract, which makes it a high time-consuming process. To overcome these constraints, a number of methods have been developed in the last years such as microwave-assisted extraction (MAE); ultrasound-assisted extraction (UAE); supercritical CO_2 extraction (SC-CO_2); pressurized fluid extraction (PFE); enzyme-assisted extraction (EAE) or even combined approaches. From the hundreds of papers published until now, it seems that these novel extraction techniques can be an interesting choice for classical extraction methods, offering several advantages such as less extraction time length, less volume of solvents, less final toxic residues, higher extraction yields and better reproducibility.

Hundreds of works have been published (some of them listed in **Table 3**) [63–73] about the use of such methods to improve the extraction yield of polyphenols in different matrices. In the next paragraphs is presented a detailed description of each method.

3.2.1. The microwave-assisted extraction (MAE)

The MAE is a method that uses energy of microwave radiation to heat solvent in contact with the sample [74–77]. The heat produced increases the diffusivity of the solvent towards the powdered sample, extracting and diffusing the phytochemicals out of the matrix [77]. The disruption of hydrogen bonds, as a result of microwave-induced dipole rotation of molecules, enhances the penetration of the solvent into the matrix, allowing the dissolution of the components into the liquid matrices [78]. This method has the advantage of less consuming time and solvent volumes than the classical approach [77–79]. This method has been largely used for the extraction of monomeric polyphenols (short chains) such as phenolic acids and flavonoids [78–80], but it has been less used to extract polymeric polyphenols such tannins and anthocyanins, because polyphenols with a higher number of hydroxyl-type substituents (long-chains) and those sensitive to higher temperatures (e.g.,

Method	Botanic matrix	Results reported by the authors	References
MAE	Blueberries (*Vaccinium corymbosum* L.)	The usage of **MAE increased the yield of anthocyanins extracted**.	[63]
MAE	Grape seeds (*Vitis vinifera*)	The **MAE** was able **to extract maximum antioxidant phenolics** with lower solvent volume in a shorter time.	[64]
UAE	*Cassia auriculata* leaves	The **UAE** process **enhanced the phenolics extraction in less time**.	[65]
UAE	Mulberry pulp (*Morus nigra*)	The **UAE can be a reliable and economic** tool to extract both **anthocyanins** and **total polyphenols**	[66]
PLE	Mango (*Mangifera indica* L.)	**High yields** of **polyphenols** were obtained at flow rate.	[67]
PLE	Asparagus (*Asparagus officinalis* L)	**PLE revealed cheaper, faster and environmental friendly** and a **good alternative** for the extraction of natural compounds.	[68]
SC-CO_2	Pitanga leaves (*Eugenia uniflora* L.)	**SC-CO_2** allowed to obtain **extracts more concentrated in polyphenols**.	[69]
EAE	Pomegranate peels (*Punica granatum* L.)	The incorporation of **enzymes** in the maceration **improved the release of bound phenolics**.	[70]
EAE	Grape residues (*Vitis vinifera*)	**Celluclast®**, **Pectinex® Ultra®** and **Novoferm®** were used to release phenolic compounds from grape wastes. The pretreatment with **enzymes increased the yield of polyphenols extracted**.	[71]
Combined approaches	Broccoli inflorescences (*Brassica oleracea* L. var. italica)	A combined method of **EAE** + **UAE** was used to extract polyphenols. The combined methods enhanced the content of polyphenol and antioxidant activity	[72]
Combined approaches	Lemon balm (*Melissa officinalis L.*)	**EAE** + **PLE** were applied for the extraction of phytochemicals. The results showed that **EAE + PLE enhanced the total phenolic content and the antioxidant capacity**.	[73]

Table 3. Application of advanced methods in extraction of polyphenols in different foods and agro-waste matrices [63–73].

anthocyanins) may be degraded under MAE extraction conditions [79, 80]. The temperature used for extraction is proportional to the power (watts) and time and inversely proportional to the heat capacity of the solvent and the mass of sample [80]. Higher temperatures and small amounts of sample increase the rate of solvent diffusion and promote faster extraction kinetics [80].

Numerous phytochemical compounds, including polyphenols, have been extracted MAE system as shown in **Table 4**. It seems that MAE system provides higher polyphenols yield in less consuming time and solvents. However, there are some concerns when the MAE is used to extract polyphenols. Factors such as type of matrix, type and purity of solvent, the microwave application time, power, contact sample surface area and temperature can affect their efficiency. One of the most critical factors is the nature of solvent, which affects not only the solubility of the target components but also the efficiency of all physical process. The choice of solvent must take into account not only the affinity to the target phytochemicals but also the ability to absorb microwave energy [81]. For example, solvents like hexane or dichloromethane, which are transparent to microwaves, do not heat up under microwave [82, 83]; thus, they should not be used in this system. Others, such as ethanol, methanol, or even water, have good microwave absorbing capacity [83], and they get heated up faster; thus, the length of the time and microwave power must be adapted to the solvent to enhance the extraction process without any deleterious effect on thermolabile components.

3.2.2. The ultrasound-assisted extraction (UAE) method

The UAE is a very simple method that relies on the mechanical effect caused by the implosion of micro-sized bubbles, which cause a rapid tissue disruption allowing the release of compounds into the solvent [84]. This is a very simple method with relatively low cost, and it can be used on both small laboratory and large industrial scale [84, 85]. The use of UAE has been widely used in the last years in the extraction of polyphenols from different parts of plants such as leaves, stems, stalks, fruits, seeds [85–93]. In general, the experimental procedure involves the use of ultrasounds with frequencies ranging from 20 to 2000 kHz, which increases the permeability of cell walls and produces cavitation.

Several studies have reported that UAE allows a better and faster extraction of polyphenols with less degradation when compared with other extraction methods. For example, UAE shown to be highly efficient in the extraction of carnosic acid and rosmarinic compared to classical methods of extraction [94]. In a recent study [95], the maximum extraction yield of total polyphenols (13.2 mg/g dry weight) from spruce wood bark was obtained when UAE system was used. Also, an increment in anthocyanin content in purple sweet potato was observed when UAE was used [96]. All these studies have in common the same trend: under UAE, the rate speed dissolution of compounds into extraction solvent was always higher, and thus, the solvent volume used and need to extract phytochemicals was lower compared to the classical extraction methods. Based on these studies and others, it seems that UAE has the advantage of being less expensive due to lower solvent volume used, higher amount of samples tested and lower time needed to perform the extraction process. Also, they agree that the lower temperatures and shorter sonication periods (time) are better to enhance the extraction of polyphenols contributing also to the preservation of the thermolabile and unstable compounds. However, some studies [97, 98] reported that sonication for long periods (>40 min) with higher energy levels (above >20 kHz) could have a deleterious effect on phytochemicals due to the decrease of diffusion area and diffusion rate and increased diffusion distance, leading to a global decreased yield of total phenolic and flavonoid content. Moreover, under these conditions might occur the formation of free radicals and consequently undesirable changes in the drug molecules [97].

3.2.3. *Pressurized liquid extraction (PLE)*

The PLE method, also known as "accelerated solvent extraction (ASE)," is a very recent new technology for phytochemicals extraction including polyphenols, which associates high temperature and pressure [99]. In this method, high level of pressure (normally between 3.3 and 20.3 MPa) is combined with high level of temperatures (between 40 and 200°C) to improve the solubility and desorption of molecules, increasing their movement from matrix into solvents, and thus increasing the yield of polyphenols extracted [54]. According to Nieto et al. [99], the PLE method is an advanced technique that provides a faster extraction processes and requires a small amount of solvents when compared with the classical extraction approach. Moreover, it allows better the usage of water as extraction solvent, which is limited in the other previous methods. The use of water as an extraction solvent in PLE, as so-called subcritical water extraction (SWE), is always possible, particularly when elevated temperatures are used [100]. When temperatures around 200°C are used, a change in the dielectric water properties occurs, and then, the water behaves like a normal organic solvent, increasing their extraction efficiency [101]. The main advantages of PLE often reported by several researchers are cleanness of the extracts that PLE provides in comparison with classical maceration, Soxhlet, MAE and UAE, which results in reduced background noise during the subsequent analytical quantification, is especially important when the LC-MS analysis due to ion-suppression effects [102]. By opposition, the main limitations often reported are the low selectivity towards the analytes during extraction, and many interferents may be extracted during the extraction process, an exaggerated dilution of the analytes, especially when a large number of cycles are used, and the high requirements in instrumentation, which increases their costs [103–105]. However, these limitations in PLE are a well-known extraction technique and have been used for the extraction of polyphenols from several different matrices [106–111].

3.2.4. *The supercritical CO_2 extraction (SC-CO_2)*

The SC-CO_2 extraction is a process in which the CO_2 is used as supercritical fluid and probably is one of the most widely used fluid because it is nontoxic, nonflammable, inert cheap and easily available in high quantity with high grade of purity [112]. SC-CO_2 extraction is possible to use different combinations of temperature and pressure [112], making this method one of the most versatile for creating a multitude of end products. Due to the multitude of combinations, low temperatures (31.6°C, the critical point of carbon dioxide) and pressure (7.386 MPa) are needed, and the SC-CO_2 has been considered very popular in a lab-scale laboratorial facilities. Moreover, since low temperatures and pressure are used, there is a good preventing of thermal degradation of phytochemicals. The main advantage s of SC-CO_2 are [112–116] as follows: (i) more extraction capacity due to their higher diffusion coefficient and lower viscosity than the liquids, which increases a higher mass transfer from solid matrix towards solvents; (ii) it allows higher penetration of solvents into the matrices which increase the effectiveness and polyphenols extraction yield; (iii) it allows different combinations of pressure and temperature and thus allows a better adaptation of the extraction conditions to the different types of food and plant matrices, increasing the solubility of their different components in the supercritical fluids; (iv) it allows the CO_2 recycling at the end of the process, without any disgrace of chemical residue to environment at the end of the extraction and separation process.

3.2.5. Enzyme-assisted extraction (EAE)

The EAE is a recent method and is based on the capacity of the enzymes to degrade cell wall components into solvents, in general water, with high stability and high bioactivity [117]. In EAE, the enzymes added to food, plant matrices or agro-food wastes are capable to break and weaken the cell walls, increasing the exposure of their cellular components to extraction [71, 118], and thus increasing the capacity to extract polyphenols from the matrices. In fact, some phytochemicals are dispersed in plant cell cytoplasm, and even, some compounds are bound with the polysaccharide-lignin by hydrogen or hydrophobic chain, which are not accessible with a routine organic solvents [119, 120]. Thus, a previous treatment with enzymes can be the only choice, and an enzymatic pretreatment might be the unique and effective way to release bounded compounds from cells [121].

Cellulases, hemicelullases, pectinases and other enzymes may be used to hydrolyze efficiently the cell wall components, enabling the efficiency of extraction of phenolic compounds. Several papers have been published about the positive effect of EAE on increment of polyphenol extraction yield. In 2012, in a study with grape wastes [71], it was found a strong increment in the release of polyphenols when celluclast®, pectinex® and novoferm® enzymes were used. Similar trends were noted in other works [122, 123] which concluded that EAE should be regarded as an alternative method for improved extraction of insoluble-bound phenolics (linked to carbohydrates and proteins of cell wall matrices) from winemaking by-products. These and many other authors observed that the ability of enzymes to degrade cell walls and membranes enables the extraction efficiency of bioactive compounds, and in several situations, the EAE technology might be the unique way to extract effectively bioactive compounds from foods and agro-industrial by-products. In addition to these advantages, the EAE method has been recognized as one of the most eco-friendly methods, because it uses water as solvent instead of organic chemicals, often toxics [119], and is one of the modern extraction methods that are gaining more attention because of the need for eco-friendly extraction technologies.

3.2.6. Combined approaches

In some circumstances, it is possible to find different studies in which the extraction of phytochemical is done throughout combined methods. This occurs, particularly in situations in which a single extraction method is not as efficient as we would expect, and thus, a combination of extraction processes could be the unique effective method to obtain extracts with different polyphenols.

4. The identification and measurements of phenolics

There is a great diversity of studies about the development of new methods for polyphenol quantification. The high-performance liquid chromatography (HPLC) with or without mass spectrometry (MS) is one of the most commonly applied method to identify and quantify polyphenols. However, the classical spectrophotometric assay is still used, even if their results are limited.

4.1. The classical colorimetric methods

The classical spectrophotometry UV/Vis method [124], even with modifications, is still widely used to measure total phenolic content in plant materials. This method is based on the chemical reduction of polyphenols in an alkaline medium to form a blue chromophore complex (phosphomolybdic/phosphotungstic acid) that can be quantified by visible-light spectrophotometry (at 760 765 nm). Many studies have discussed the advantages and disadvantage of using routinely this method to quantify the level of polyphenols, and most of them seems to agree that although they are easy to perform, low cost, rapid and applicable routinely in the most laboratories, they are not accurate. In addition, the reagents used in the method do not react specifically with only polyphenols, and they react with any reducing substance like ascorbic acid, pigments, aromatic amines and sugars [125], and thus, these methods measure the total reducing capacity and not just the polyphenols compounds. Also, their reagents react with some nitrogen-containing compounds such as hydroxylamine and guanidine [126], thiols, many vitamins and some inorganic ions [127] Therefore, many researchers have chosen to use this method only as an indicative tool of total reduction capacity and not for a specific quantification of polyphenol compounds. However, these methods are still considered useful for a quick and prior screening of numerous samples, and for many applications, a simple measure of total amount of polyphenols is enough.

Similar to total polyphenols, total flavonoids can be measured by spectrophotometry methods, and the $AlCl_3$ method [128, 129] is the most vulgarized method used to determine the total flavonoid content. Vanillin and 4-(dimethylamino)-cinnamaldehyde (DMCA) assays are often used to determine the level of proanthocyanidins, in which the flavonoid catechin is used as standard [130, 131]. Like in total polyphenols, the vanillin or DMCA method can overestimate the amount of total flavonoids present in samples. The proanthocyanidins can also be determined by butanol-HCl [132] and bovine serum albumin (BSA) [133] methods. The butanol-HCl method is based on the cleavage of the flavonoid bonds by hot acid, followed by an auto-oxidation reaction which converts flavan-3-ols into anthocyanidins. The red extract formed has a maximum absorbance at around 550 nm. In the BSA method, the flavonoids complex is dissolved in an alkaline solution (sodium decyl sulphate-triethanolamine) followed by a reaction with ferric chloride solution to form a violet complex with a maximum absorbance at 510 nm.

Another spectrophotometric method widely used in the quantification of polyphenols is the UV/Vis spectrophotometry method to determine the anthocyanin content. The anthocyanins constitute one of the main class of polyphenols largely present in plant samples, particularly in red, blue and black color fruits such as grapes, blueberries, raspberries, redcurrants, blackcurrants, pomegranates and strawberries, among others. The quantification of anthocyanins is in general performed by the differential pH method [134] based on the property of the anthocyanin pigments to change the color with pH, in the wavelength ranging from 490 to 550 nm [134]. The anthocyanins suffer reversible structural modification with a change of pH, and this change allows to estimate spectrophotometrically the total amount of anthocyanins, even in the presence of degraded pigments and other interfering compounds.

All these spectrophotometric methods are considered simple and cheap, but only gives a general estimation about the content of each class of polyphenols but do not allow the quantification of polyphenols individually.

4.2. Chromatography

In the course of the last four decades, several chromatography methods were developed to overcome the main constraints of the classical spectrophotometry methods. The development of new technologies and software led to the appearance of improved methods capable of separation, identification and quantification of phytochemicals individually. These methods are generally based on the principle that a sample is composed of a mixture of components which are separated when the mixture passes through two phases: a mobile (liquid or gaseous) and a stationary (solid, liquid or gel). It is used for the qualitative and quantitative analysis, and the components are separated and analyzed according the properties of a given solution. The great diversity of combinations between the two phases makes possible an existence of several differentiated techniques.

4.2.1. Principles

The basic principles of chromatography are universal [135] and thus widely accepted by all researchers. The main principles are more or less the following ones: (i) chromatography is a physical and chemical method of separating, identification and quantification of different components of samples; (ii) the separation always dependent of the interaction between the components of the mixture with the mobile and stationary phases, and, thus a large combinations of the three are possible; (iii) the interaction of the matrices components with the mix of both phases is influenced by different intermolecular forces including ionic, dipole, nonpolar and effects of specific affinity or solubility; (iv) the mobile phase is generally named as eluent, and the absorbent material, named as stationary phase; (v) the analyte is the compound to be separated; (vi) a chromatogram is the visual output of the chromatograph; (vii) the instrument used for qualitative and quantitative detection of analytes after separation is named as detector; (viii) the separation of components present in the mixture occurs according to the different chemical affinity for the stationary phase, and it happens as the eluent advances on the stationary phase; (ix) the separation of compounds is slower when compounds have strong interaction with the stationary phase and faster when the components have weaker interaction with the stationary phase, and by this, the compounds will be separated from each other as they move over the support material; (x) the component to be analyzed must have solubility with the mobile phase, and different compounds have different retention time values; (xi) the identification and quantification of components in the mixtures are done by comparison with pure commercial standards of known commercial concentration, through analytical curves.

The chromatography can be classified according to several criteria [135, 136], but in general, the chromatography applied in separation, identification and quantification of phytochemicals is classified as:
(i) Gas chromatography (GC), when gas chromatography makes use of a pressurized gas cylinder and a carrier gas (e.g., helium), to carry the solute through the column. The most common detectors used in this type of chromatography are of thermal conductivity and flame ionization detectors. There are three types of GC as follows: (1) gas adsorption, (2) gas-liquid, and (3) capillary gas chromatography.
(ii) Liquid chromatography (LC), when a liquid adsorbent is used. This method is used in large-scale applications since adsorbents are relatively inexpensive. There is a liquid-liquid chromatography which is analogous to gas-liquid chromatography. The three types of modern LC are as follows: (1) reverse phase, (2) high performance and (3) size exclusion liquid chromatography, along with supercritical fluid chromatography.
(iii) Ion exchange chromatography (IEC), when charged molecules mobile phase passes through the column, until a binding site in the stationary phase appears and retains the molecules. There are two types of ion exchange chromatography: (1) cation exchange in which the stationary phase carries a negative charge, and (2) anion exchange in which the stationary phase carries a positive charge. The method is mainly used in the purification of biological materials.
(iv) Affinity chromatography (AC), which is a technique that involves the chemical modification of a given compound by attaching another compound with a specific affinity for the desired molecules. This method requires that the compounds to be analyzed must be inert and easily to modify, and otherwise, it can be very difficult to perform, and a large number of impurities can appear. Therefore, this type of technique is only used in advanced processes of purification.

4.2.2. The use of HPLC-DAD/UV-VIS and HPLC-MS

The high-performance liquid chromatography (HPLC), referred in the past as high-pressure liquid chromatography, like other chromatography methods is a technique used to separate, identify and quantify phytochemicals from plant mixtures and relies on pumps to pass a pressurized solvent containing the plant samples, foods or other matrices through a column filled with a solid adsorbent material (e.g., silica) [137]. The HPLC methods, however, differ from other liquid methods, particularly from "low pressures," because it uses high pressures (ranging from 50 to 350 bar), while the others normally use the force of gravity to pass the mobile phase through the column [137, 138]. Each component of the samples interacts differently with the adsorbent material of the column, causing a different flow rate for the different components in the mixture, thus leading to the separation of the components as they flow out the column. The columns used in the HPLC methods are made with smaller adsorbent particles size ranging from 2 to 50 μm [137, 138].

http://dx.doi.org/10.5772/66889

Although there is many variations in the HPLC equipment available in the market, the basic HPLC equipment includes a sampler (to carry the sample mixture into the mobile phase), pumps (to deliver mobile phase through the column, with a specific flow) and a detector (such as UV/Vis or photodiode array (PDA), which generates signal proportional to the amount of compound present in the sample mixture [138]. The signal detected allows the identification and quantification of sample components. Each compound detected has a specific retention time; however, due to the interaction strength of interactions between the analytes and stationary phases, the retention time can vary. Nowadays, modern HPLC equipment has a digital processor, which uses a software interface to control the instruments and provides data analysis. Other modern models are equipped with several pumps, which allow different combinations of various solvents at different ratios changing in time, creating a gradient in the mobile phase.

Nowadays, the classic HPLC evolved to HPLC coupled with a mass spectrometry detector (MS), called as LC-MS or HPLC-MS [138]. This new technique allows a more accurate identification which is based on the specific fragmentation of each separated molecule. This enhances the sensitivity and is oriented for the separation of chemicals with specific masses in a complex mixture. The separation of molecules or fragments occurs according to their mass-to-charge ratio in an analyzer by electromagnetic fields. The ions are detected by a qualitative and quantitative analysis, and the signal is processed into mass spectra. The HPLC-MS equipment is in general composed of three modules: (1) an ion source, which converts gas phase sample molecules into ions; (2) a mass analyzer, which sorts the ions by their masses by applying electromagnetic fields; and (3) a detector, which measures the value of the signal detected and provides data for the quantification of each ion present. In the last years, this new method has been strongly

Method	Purposes	Reference
HPLC-ESI-MSn,	Separation and identification of the main polyphenols in black currant (*Ribes nigrum* L.)	[139]
HPLC-MSn	Identification and quantification of phenolic compounds in hazelnut (*Corylus avellana* L.) kernels, oil and bagasse pellets	[140]
HPLC-DAD-ESI-MS/MS	Comparative fingerprint and quantification of polyphenol of aqueous extracts of *Cornus mas* and *Crataegus monogyna*.	[141]
HPLC-PDA–ESI-MS/MS	Identification of phenolic acids, flavonol glycosides in blueberry (*Vaccinium corymbosum* L.), blackberry (*Rubus fructicosus* L.), raspberry (*Rubus idaeus* L.) and cranberry (*Vaccinum vitis-idaea* L.)	[142]
HPLC-DAD-ESI-MS/MS	Characterization of chemical profile of mango (*Mangifera indica* L.)	[143]
HPLC-PDA–ESI-MS/MS	Determination of anthocyanins in cherry (*Prunus avium* L.) and cranberry (*Vaccinium oxycoccos* L.)	[144]
HPLC-DAD–ESI-MS/MS	Characterization of phenolic compounds from Turkish black tea (*Camellia sinensis* L. Kuntze)	[145]

Table 4. Some recent examples of the usage of HPLC-MS system for separation and quantification of polyphenols in different matrices.

implemented in academies in basic research, pharmaceutical and agro-chemical industries to study physical, chemical and biological properties of a great diversity of compounds, as well, and quality control of drugs, foods and natural products. In **Table 4** are presented some recent works [139–145] in which HPLC-MS was effectively used for polyphenol characterization of plants and food with very accurate results.

5. Conclusions

This chapter discusses the importance of polyphenols as well as the availability of different methods to extract them from its natural sources. The most widely used methods in the extraction of polyphenols are the classical ones which usually includes maceration, percolation and successive Soxhlet extractions. Although these methods are still in use, they involve long extraction times, huge quantities of solvent, higher accumulation of residues and very limited results. Therefore, new methods such as UAE, MA, PLE, S-CO_2 and EAE have been developed in the recent years with very feasible results. Also, the evolution of separation and identification techniques of polyphenols has evolved from a simple colorimetry method to the most advanced chromatography techniques. However, the growing demand for new bioactive molecules from natural sources enhances the continuous search for new and innovative methods to extract and separate new molecules, which never ends.

Author details

Alfredo Aires

Address all correspondence to: alfredoa@utad.pt

Centre for the Research and Technology for Agro-Environment and Biological Sciences (CITAB), University of Trás-os-Montes e Alto Douro (UTAD), Vila Real, Portugal

References

[1] Cheynier V, Comte G, Davies KM, Lattanzio V, Martens S: Plant phenolics: recent advances on their biosynthesis, genetics, and ecophysiology. Plant Physiology and Biochemistry. 2013, 72:1–20. doi:10.1016/j.plaphy.2013.05.009

[2] Lattanzio V: Phenolic compounds: introduction. In: Ramawat GK, Mérillon J-M (eds). Natural products: phytochemistry, botany and metabolism of alkaloids, phenolics and terpenes. Springer, Berlin Heidelberg, 2013. pp. 1543–80.

[3] Harborne JB: Introduction to ecological biochemistry 4 edition. Academic Press, London, United Kingdom, 2014.

[4] Harborne JB, Williams CA: Advances in flavonoid research since 1992. Phytochemistry. 2000, 55:481–504. doi:10.1016/S0031-9422(00)00235-1

[5] Han X, Shen T, Lou H: Dietary polyphenols and their biological significance. International Journal of Molecular Sciences. 2007, 8:950–88. doi:10.3390/i8090950

[6] Huang W-Y, Cai Y-Z, Zhang Y: Natural phenolic compounds from medicinal herbs and dietary plants: potential use for cancer prevention. Nutrition and Cancer. 2009, 62:1–20. doi:10.1080/01635580903191585

[7] Cao Y, Cao R: Angiogenesis inhibited by drinking tea. Nature. 1999, 398:381. doi:10.1038/18793

[8] Pandey KB, Rizvi SI: Plant polyphenols as dietary antioxidants in human health and disease. Oxidative Medicine and Cellular Longevity. 2009, 2:270–278. doi:10.4161/oxim.2.5.9498

[9] Del Rio D, Rodriguez-Mateos A, Spencer JPE, Tognolini M, Borges G, Crozier A: Dietary (poly)phenolics in human health: structures, bioavailability, and evidence of protective effects against chronic diseases. Antioxidants & Redox Signaling. 2012, 18:1818–92. doi:10.1089/ars.2012.4581

[10] Amararathna M, Johnston MR, Rupasinghe HPV: Plant polyphenols as chemopreventive agents for lung cancer. International Journal of Molecular Sciences. 2016, 17:1352. doi:10.3390/ijms17081352

[11] Gothai S, Ganesan P, Park S-Y, Fakurazi S, Choi D-K, Arulselvan P: Natural phyto-bioactive compounds for the treatment of type 2 diabetes: inflammation as a target. Nutrients. 2016, 8:461. doi:10.3390/nu8080461

[12] Hollman PCH, van Trijp JMP, Buysman MNCP, v.d. Gaag MS, Mengelers MJB, de Vries JHM, Katan MB: Relative bioavailability of the antioxidant flavonoid quercetin from various foods in man. FEBS Letters. 1997, 418:152–6. doi:10.1016/S0014-5793(97)01367-7

[13] Cai Y, Luo Q, Sun M, Corke H: Antioxidant activity and phenolic compounds of 112 traditional Chinese medicinal plants associated with anticancer. Life Sciences. 2004, 74:2157–84. doi:10.1016/j.lfs.2003.09.047

[14] Manach C, Scalbert A, Morand C, Rémésy C, Jiménez L: Polyphenols: food sources and bioavailability. The American Journal of Clinical Nutrition. 2004, 79:727–47. http://ajcn.nutrition.org/content/79/5/727.full.pdf+html

[15] Perez-Jimenez J, Neveu V, Vos F, Scalbert A: Identification of the 100 richest dietary sources of polyphenols: an application of the Phenol-Explorer database. European Journal of Clinical Nutrition.2010, 64:S112–S20. doi:10.1038/ejcn.2010.221

[16] Tsao R, Yang R, Young JC, Zhu H: Polyphenolic profiles in eight apple cultivars using high-performance liquid chromatography (HPLC). Journal of Agricultural and Food Chemistry. 2003, 51:6347–53. doi:10.1021/jf0346298

[17] Kawaii S, Tomono Y, Katase E, Ogawa K, Yano M: Quantitation of flavonoid constituents in citrus fruits. Journal of Agricultural and Food Chemistry. 1999, 47:3565–71. doi:10.1021/jf990153+

[18] Chun OK, Kim D-O, Moon HY, Kang HG, Lee CY: Contribution of individual polyphenolics to total antioxidant capacity of plums. Journal of Agricultural and Food Chemistry. 2003, 51:7240–5. doi:10.1021/jf0343579

[19] Lee KW, Kim YJ, Kim D-O, Lee HJ, Lee CY: Major phenolics in apple and their contribution to the total antioxidant capacity. Journal of Agricultural and Food Chemistry. 2003, 51:6516–20. doi:10.1021/jf034475w

[20] Naczk M, Shahidi F: Phenolics in cereals, fruits and vegetables: occurrence, extraction and analysis. Journal of Pharmaceutical and Biomedical Analysis. 2006, 41:1523–42. doi:10.1016/j.jpba.2006.04.002

[21] Anwar F, Sultana B, Rafique Asi M, Ali Shahid Chatha S: Antioxidant potential of extracts from different agro wastes: stabilization of corn oil. Grasas y Aceites. 2008, 59. doi:10.3989/gya.2008.v59.i3.510

[22] Teixeira A, Baenas N, Dominguez-Perles R, Barros A, Rosa E, Moreno DA, Garcia-Viguera C: Natural bioactive compounds from winery by-products as health promoters: a review. International Journal of Molecular Sciences. 2014, 15:15638–78. doi:10.3390/ijms150915638

[23] Almanasrah M, Roseiro LB, Bogel-Lukasik R, Carvalheiro F, Brazinha C, Crespo J, Kallioinen M, Mänttäri M, Duarte LC: Selective recovery of phenolic compounds and carbohydrates from carob kibbles using water-based extraction. Industrial Crops and Products. 2015, 70:443–50. doi:10.1016/j.indcrop.2015.02.051

[24] Fawole OA, Opara UL: Stability of total phenolic concentration and antioxidant capacity of extracts from pomegranate co-products subjected to in vitro digestion. BMC Complementary and Alternative Medicine. 2016, 16:1–10. doi:10.1186/s12906-016-1343-2

[25] Zhao F, Watanabe Y, Nozawa H, Daikonnya A, Kondo K, Kitanaka S: Prenylflavonoids and phloroglucinol derivatives from hops (*Humulus lupulus*). Journal of Natural Products. 2005, 68:43–9. doi:10.1021/np0499113

[26] D"Archivio M, Filesi C, Vari R, Scazzocchio B, Masella R: Bioavailability of the polyphenols: status and controversies. Int J Mol Sci. 2010, 11:1321–42. doi:10.3390/ijms11041321

[27] Tsao R: Chemistry and biochemistry of dietary polyphenols. Nutrients. 2010, 2:1231–46. doi:10.3390/nu2121231

[28] Jallali I, Megdiche W, M'Hamdi B, Oueslati S, Smaoui A, Abdelly C, Ksouri R: Changes in phenolic composition and antioxidant activities of the edible halophyte *Crithmum maritimum* L. with physiological stage and extraction method. Acta Physiologiae Plantarum. 2012, 34:1451–9. doi:10.1007/s11738-012-0943-9

[29] Sun B, Leandro MC, de Freitas V, Spranger MI: Fractionation of red wine polyphenols by solid-phase extraction and liquid chromatography. Journal of Chromatography A. 2006, 1128:27–38. doi:10.1016/j.chroma.2006.06.026

[30] Garcia-Salas P, Morales-Soto A, Segura-Carretero A, Fernández-Gutiérrez A: Phenolic-compound-extraction systems for fruit and vegetable samples. Molecules. 2010, 15:8813. doi:10.3390/molecules15128813

[31] Tsai CC, Chou CH, Liu YC, Hsieh CW: Ultrasound-assisted extraction of phenolic compounds from *Phyllanthus emblica* L. and evaluation of antioxidant activities. International Journal of Cosmetic Science. 2014, 36:471–6. doi:10.1111/ics.12143

[32] Chandrasekar V, Martín-González MFS, Hirst P, Ballard TS: Optimizing microwave-assisted extraction of phenolic antioxidants from red delicious and Jonathan apple pomace. Journal of Food Process Engineering. 2015, 38:571–82. doi:10.1111/jfpe.12187

[33] Pinelo M, Ruiz-Rodríguez A, Sineiro J, Señoráns FJ, Reglero G, Núñez MJ: Supercritical fluid and solid–liquid extraction of phenolic antioxidants from grape pomace: a comparative study. European Food Research and Technology. 2007, 226:199–205. doi:10.1007/s00217-006-0526-3

[34] Oliveira AL, Destandau E, Fougère L, Lafosse M: Isolation by pressurised fluid extraction (PFE) and identification using CPC and HPLC/ESI/MS of phenolic compounds from Brazilian cherry seeds (*Eugenia uniflora* L.). Food Chemistry. 2014, 145:522–9. doi:10.1016/j.foodchem.2013.08.065

[35] Marques LLM, Panizzon GP, Aguiar BAA, Simionato AS, Cardozo-Filho L, Andrade G, de Oliveira AG, Guedes TA, Mello JCPd: Guaraná (*Paullinia cupana*) seeds: selective supercritical extraction of phenolic compounds. Food Chemistry. 2016, 212:703–11. doi:10.1016/j.foodchem.2016.06.028

[36] Costa DC, Costa HS, Albuquerque TG, Ramos F, Castilho MC, Sanches-Silva A: Advances in phenolic compounds analysis of aromatic plants and their potential applications. Trends in Food Science & Technology. 2015, 45:336–54. doi:10.1016/j.tifs.2015.06.009

[37] Herrero M, Sánchez-Camargo AdP, Cifuentes A, Ibáñez E: Plants, seaweeds, microalgae and food by-products as natural sources of functional ingredients obtained using pressurized liquid extraction and supercritical fluid extraction. Trends in Analytical Chemistry. 2015, 71:26–38. doi:10.1016/j.trac.2015.01.018

[38] Sánchez-Camargo AdP, Montero L, Stiger-Pouvreau V, Tanniou A, Cifuentes A, Herrero M, Ibáñez E: Considerations on the use of enzyme-assisted extraction in combination with pressurized liquids to recover bioactive compounds from algae. Food Chemistry. 2016, 192:67–74. doi:10.1016/j.foodchem.2015.06.098

[39] Liu T, Sui X, Li L, Zhang J, Liang X, Li W, Zhang H, Fu S: Application of ionic liquids based enzyme-assisted extraction of chlorogenic acid from *Eucommia ulmoides* leaves. Analytica Chimica Acta. 2016, 903:91–9. doi:10.1016/j.aca.2015.11.029

[40] Dettmer K, Aronov PA, Hammock BD: Mass spectrometry-based metabolomics. Mass spectrometry reviews. 2007, 26:51–78.doi:10.1002/mas.20108

[41] Sticher O: Natural product isolation. Natural Product Reports. 2008, 25:517–54. doi:10.1039/B700306B

[42] Kaufmann B, Christen P: Recent extraction techniques for natural products: microwave-assisted extraction and pressurised solvent extraction. Phytochemical Analysis. 2002, 13:105–13. doi:10.1002/pca.631

[43] Santos-Buelga C, Gonzalez-Manzano S, Dueñas M, Gonzalez-Paramas AM: Extraction and isolation of phenolic compounds. In: Sarker DS, Nahar L (eds). Natural products isolation.. Humana Press, Totowa, NJ, 2012. pp. 427–64.

[44] Seidel V: Initial and bulk extraction of natural products isolation. In: Sarker DS, Nahar L (eds). Natural products isolation. Humana Press, Totowa, NJ, 2012. pp. 27–41.

[45] Bucar F, Wube A, Schmid M: Natural product isolation—how to get from biological material to pure compounds. Natural Product Reports. 2013, 30:525–45. doi:10.1039/C3NP20106F

[46] Olejar KJ, Fedrizzi B, Kilmartin PA: Influence of harvesting technique and maceration process on aroma and phenolic attributes of *Sauvignon blanc* wine. Food Chemistry. 2015, 183:181–9. doi:10.1016/j.foodchem.2015.03.040

[47] Ćujić N, Šavikin K, Janković T, Pljevljakušić D, Zdunić G, Ibrić S: Optimization of polyphenols extraction from dried chokeberry using maceration as traditional technique. Food Chemistry. 2016, 194:135–42. doi:10.1016/j.foodchem.2015.08.008

[48] Lima MdS, da Conceição Prudêncio Dutra M, Toaldo IM, Corrêa LC, Pereira GE, de Oliveira D, Bordignon-Luiz MT, Ninow JL: Phenolic compounds, organic acids and antioxidant activity of grape juices produced in industrial scale by different processes of maceration. Food Chemistry. 2015, 188:384–92. doi:10.1016/j.foodchem.2015.04.014

[49] Dai J, Mumper RJ: Plant phenolics: extraction, analysis and their antioxidant and anticancer properties. Molecules. 2010, 15:7313. doi:10.3390/molecules15107313

[50] Vongsak B, Sithisarn P, Mangmool S, Thongpraditchote S, Wongkrajang Y, Gritsanapan W: Maximizing total phenolics, total flavonoids contents and antioxidant activity of *Moringa oleifera* leaf extract by the appropriate extraction method. Industrial Crops and Products. 2013, 44:566–71.doi:10.1016/j.indcrop.2012.09.021

[51] Hayouni EA, Abedrabba M, Bouix M, Hamdi M: The effects of solvents and extraction method on the phenolic contents and biological activities in vitro of Tunisian *Quercus*

coccifera L. and *Juniperus* phoenicea L. fruit extracts. Food Chemistry. 2007, 105:1126–34. doi:10.1016/j.foodchem.2007.02.010

[52] Chemat F, Vian MA: Alternative solvents for natural products extraction. Springer, Berlin Heidelberg, 2014.

[53] Zhou K, Yu L: Effects of extraction solvent on wheat bran antioxidant activity estimation. LWT Food Science and Technology. 2004, 37:717–21. doi:10.1016/j.lwt.2004.02.008

[54] Azmir J, Zaidul ISM, Rahman MM, Sharif KM, Mohamed A, Sahena F, Jahurul MHA, Ghafoor K, Norulaini NAN, Omar AKM: Techniques for extraction of bioactive compounds from plant materials: a review. Journal of Food Engineering. 2013, 117:426–36. doi:10.1016/j.jfoodeng.2013.01.014

[55] Common solvent properties [Internet] 2016. Available from: http://macro.lsu.edu/HowTo/solvents.htm [Accessed: 2016-08-14].

[56] Speight JG: Lange's handbook of chemistry. McGraw-Hill, New York, 2005.

[57] Singh A, Nair G, Liplap P, Gariepy Y, Orsat V, Raghavan V: Effect of dielectric properties of a solvent-water mixture used in microwave-assisted extraction of antioxidants from potato peels. Antioxidants. 2014, 3:99. doi:10.3390/antiox3010099

[58] Hakansson E, Kaynak A, Kouzani A: Change in dielectric properties in the microwave frequency region of polypyrrole–coated textiles during aging. Materials. 2016, 9:609. doi:10.3390/ma9070609

[59] Liu S-C, Lin J-T, Wang C-K, Chen H-Y, Yang D-J: Antioxidant properties of various solvent extracts from lychee (*Litchi chinenesis* Sonn.) flowers. Food Chemistry. 2009, 114:577–81.doi:10.1016/j.foodchem.2008.09.088

[60] Fernández-Agulló A, Pereira E, Freire MS, Valentão P, Andrade PB, González-Álvarez J, Pereira JA: Influence of solvent on the antioxidant and antimicrobial properties of walnut (*Juglans regia* L.) green husk extracts. Industrial Crops and Products. 2013, 42:126–32. doi:10.1016/j.indcrop.2012.05.021

[61] Metrouh-Amir H, Duarte CMM, Maiza F: Solvent effect on total phenolic contents, antioxidant, and antibacterial activities of *Matricaria pubescens*. Industrial Crops and Products. 2015, 67:249–56.doi:10.1016/j.indcrop.2015.01.049

[62] Iloki-Assanga SB, Lewis-Luján LM, Lara-Espinoza CL, Gil-Salido AA, Fernandez-Angulo D, Rubio-Pino JL, Haines DD: Solvent effects on phytochemical constituent profiles and antioxidant activities, using four different extraction formulations for analysis of *Bucida buceras* L. and *Phoradendron californicum*. BMC Research Notes. 2015, 8:1–14. doi:10.1186/s13104-015-1388-1

[63] Routray W, Orsat V: MAE of phenolic compounds from blueberry leaves and comparison with other extraction methods. Industrial Crops and Products. 2014, 58:36–45. doi:10.1016/j.indcrop.2014.03.038

[64] Krishnaswamy K, Orsat V, Gariépy Y, Thangavel K: Optimization of microwave-assisted extraction of phenolic antioxidants from grape seeds (*Vitis vinifera*). Food and Bioprocess Technology. 2013, 6:441–55. doi:10.1007/s11947-012-0800-2

[65] Sharmila G, Nikitha VS, Ilaiyarasi S, Dhivya K, Rajasekar V, Kumar NM, Muthukumaran K, Muthukumaran C: Ultrasound assisted extraction of total phenolics from *Cassia auriculata* leaves and evaluation of its antioxidant activities. Industrial Crops and Products. 2016, 84:13–21.doi:10.1016/j.indcrop.2016.01.010

[66] Espada-Bellido E, Ferreiro-González M, Carrera C, Palma M, Barroso CG, Barbero GF: Optimization of the ultrasound-assisted extraction of anthocyanins and total phenolic compounds in mulberry (*Morus nigra*) pulp. Food Chemistry. 2017, 219:23–32. doi:10.1016/j.foodchem.2016.09.122

[67] Fernández-Ponce MT, Parjikolaei BR, Lari HN, Casas L, Mantell C, Martínez de la Ossa EJ: Pilot-plant scale extraction of phenolic compounds from mango leaves using different green techniques: kinetic and scale up study. Chemical Engineering Journal. 2016, 299:420–30.doi:10.1016/j.cej.2016.04.046

[68] Solana M, Boschiero I, Dall'Acqua S, Bertucco A: A comparison between supercritical fluid and pressurized liquid extraction methods for obtaining phenolic compounds from *Asparagus officinalis* L. The Journal of Supercritical Fluids. 2015, 100:201–8.doi:10.1016/j.supflu.2015.02.014

[69] Garmus TT, Paviani LC, Queiroga CL, Magalhães PM, Cabral FA: Extraction of phenolic compounds from pitanga (*Eugenia uniflora* L.) leaves by sequential extraction in fixed bed extractor using supercritical CO_2, ethanol and water as solvents. The Journal of Supercritical Fluids. 2014, 86:4–14.doi:10.1016/j.supflu.2013.11.014

[70] Mushtaq M, Sultana B, Anwar F, Adnan A, Rizvi SSH: Enzyme-assisted supercritical fluid extraction of phenolic antioxidants from pomegranate peel. The Journal of Supercritical Fluids. 2015, 104:122–31. doi:10.1016/j.supflu.2015.05.020

[71] Gómez-García R, Martínez-Ávila GCG, Aguilar CN: Enzyme-assisted extraction of antioxidative phenolics from grape (*Vitis vinifera* L.) residues. 3 Biotech. 2012, 2:297–300. doi:10.1007/s13205-012-0055-7

[72] Wu H, Zhu J, Yang L, Wang R, Wang C: Ultrasonic-assisted enzymatic extraction of phenolics from broccoli (*Brassica oleracea* L. var. italica) inflorescences and evaluation of antioxidant activity in vitro. Food Science and Technology International. 2015, 21:306–19. doi:10.1177/1082013214536174

[73] Miron TL, Herrero M, Ibáñez E: Enrichment of antioxidant compounds from lemon balm (*Melissa officinalis*) by pressurized liquid extraction and enzyme-assisted extraction. Journal of Chromatography A. 2013, 1288:1–9.doi:10.1016/j.chroma.2013.02.075

[74] Wang L, Weller CL: Recent advances in extraction of nutraceuticals from plants. Trends in Food Science & Technology. 2006, 17:300–12.doi:10.1016/j.tifs.2005.12.004

[75] Chemat F, Cravotto G: Microwave-assisted extraction for bioactive compounds: theory and practice. Springer, USA, 2012.

[76] Delazar A, Nahar L, Hamedeyazdan S, Sarker SD: Microwave-assisted extraction in natural products isolation. In: Sarker DS, Nahar L (eds). Natural products isolation. Humana Press, Totowa, NJ, 2012. pp. 89–115. doi:10.1007/978-1-61779-624-1_5

[77] Proestos C, Komaitis M: Application of microwave-assisted extraction to the fast extraction of plant phenolic compounds. LWT Food Science and Technology. 2008, 41:652–9. doi:10.1016/j.lwt.2007.04.013

[78] Pellati F, Prencipe FP, Bertelli D, Benvenuti S: An efficient chemical analysis of phenolic acids and flavonoids in raw propolis by microwave-assisted extraction combined with high-performance liquid chromatography using the fused-core technology. Journal of Pharmaceutical and Biomedical Analysis. 2013, 81–82:126–32.doi:10.1016/j.jpba.2013.04.003

[79] Pinela J, Prieto MA, Carvalho AM, Barreiro MF, Oliveira MBPP, Barros L, Ferreira ICFR: Microwave-assisted extraction of phenolic acids and flavonoids and production of antioxidant ingredients from tomato: a nutraceutical-oriented optimization study. Separation and Purification Technology. 2016, 164:114–24. doi:10.1016/j.seppur.2016.03.030

[80] Biesaga M, Pyrzyńska K: Stability of bioactive polyphenols from honey during different extraction methods. Food Chemistry. 2013, 136:46–54. doi:10.1016/j.foodchem.2012.07.095

[81] Zhang H-F, Yang X-H, Wang Y: Microwave assisted extraction of secondary metabolites from plants: current status and future directions. Trends in Food Science & Technology. 2011, 22:672–88. doi:10.1016/j.tifs.2011.07.003

[82] Hu X, Zhou Q: Comparisons of microwave-assisted extraction, simultaneous distillation-solvent extraction, soxhlet extraction and ultrasound probe for polycyclic musks in sediments: recovery, repeatability, matrix effects and bioavailability. Chromatographia. 2011, 74:489–95. doi:10.1007/s10337-011-2084-5

[83] Dudley GB, Richert R, Stiegman AE: On the existence of and mechanism for microwave-specific reaction rate enhancement. Chemical Science. 2015, 6:2144–52. doi:10.1039/C4SC03372H

[84] Vinatoru M: An overview of the ultrasonically assisted extraction of bioactive principles from herbs. Ultrasonics Sonochemistry. 2001, 8:303–13.doi:10.1016/S1350-4177(01)00071-2

[85] Chemat F, Zill e H, Khan MK: Applications of ultrasound in food technology: processing, preservation and extraction. Ultrasonics Sonochemistry. 2011, 18:813–35. doi:10.1016/j.ultsonch.2010.11.023

[86] Khan MK, Abert-Vian M, Fabiano-Tixier A-S, Dangles O, Chemat F: Ultrasound-assisted extraction of polyphenols (flavanone glycosides) from orange (*Citrus sinensis* L.) peel. Food Chemistry. 2010, 119:851–8. doi:10.1016/j.foodchem.2009.08.046

[87] Galvan d'Alessandro L, Kriaa K, Nikov I, Dimitrov K: Ultrasound assisted extraction of polyphenols from black chokeberry. Separation and Purification Technology. 2012, 93:42–7. doi:10.1016/j.seppur.2012.03.024

[88] Zou T-B, Xia E-Q, He T-P, Huang M-Y, Jia Q, Li H-W: Ultrasound-assisted extraction of mangiferin from mango (*Mangifera indica* L.) leaves using response surface methodology. Molecules. 2014, 19:1411. doi:10.3390/molecules19021411

[89] Sousa JN, Pedroso NB, Borges LL, Oliveira GAR, Paula JR, Conceição EC: Optimization of ultrasound-assisted extraction of polyphenols, tannins and epigallocatechin gallate from barks of *Stryphnodendron adstringens* (Mart.) Coville bark extracts. Pharmacognosy Magazine. 2014, 10:S318–S23.10.4103/0973-1296.133287

[90] Wong Paz JE, Muñiz Márquez DB, Martínez Ávila GCG, Belmares Cerda RE, Aguilar CN: Ultrasound-assisted extraction of polyphenols from native plants in the Mexican desert. Ultrasonics Sonochemistry. 2015, 22:474–81. doi:10.1016/j.ultsonch.2014.06.001

[91] Upadhyay R, Nachiappan G, Mishra HN: Ultrasound-assisted extraction of flavonoids and phenolic compounds from *Ocimum tenuiflorum* leaves. Food Science and Biotechnology. 2015, 24:1951–8. doi:10.1007/s10068-015-0257-y

[92] Wang W, Jung J, Tomasino E, Zhao Y: Optimization of solvent and ultrasound-assisted extraction for different anthocyanin rich fruit and their effects on anthocyanin compositions. LWT Food Science and Technology. 2016, 72:229–38. doi:10.1016/j.lwt.2016.04.041

[93] Makris D: Kinetics of ultrasound-assisted flavonoid extraction from agri-food solid wastes using water/glycerol mixtures. Resources. 2016, 5:7. doi:10.3390/resources5010007

[94] Zu G, Zhang R, Yang L, Ma C, Zu Y, Wang W, Zhao C: Ultrasound-assisted extraction of carnosic acid and rosmarinic acid using ionic liquid solution from *Rosmarinus officinalis*. International Journal of Molecular Sciences. 2012, 13:11027. doi:10.3390/ijms130911027

[95] Ghitescu R-E, Volf I, Carausu C, Bühlmann A-M, Gilca IA, Popa VI: Optimization of ultrasound-assisted extraction of polyphenols from spruce wood bark. Ultrasonics Sonochemistry. 2015, 22:535–41. doi:10.1016/j.ultsonch.2014.07.013

[96] Cai Z, Qu Z, Lan Y, Zhao S, Ma X, Wan Q, Jing P, Li P: Conventional, ultrasound-assisted, and accelerated-solvent extractions of anthocyanins from purple sweet potatoes. Food Chemistry. 2016, 197, Part A:266–72. doi:10.1016/j.foodchem.2015.10.110

[97] Wang L, Li D, Bao C, You J, Wang Z, Shi Y, Zhang H: Ultrasonic extraction and separation of anthraquinones from *Rheum palmatum* L. Ultrasonics Sonochemistry. 2008, 15:738–46. doi:10.1016/j.ultsonch.2007.12.008

[98] Annegowda HV, Anwar LN, Mordi MN, Ramanathan S, Mansor SM: Influence of sonication on the phenolic content and antioxidant activity of *Terminalia catappa* L. leaves. Pharmacognosy Research. 2010, 2:368–73. doi:10.4103/0974-8490.75457

[99] Nieto A, Borrull F, Pocurull E, Marcé RM: Pressurized liquid extraction: a useful technique to extract pharmaceuticals and personal-care products from sewage sludge. Trends in Analytical Chemistry. 2010, 29:752–64. doi:10.1016/j.trac.2010.03.014

[100] Teo CC, Tan SN, Yong JWH, Hew CS, Ong ES: Pressurized hot water extraction (PHWE). Journal of Chromatography A. 2010, 1217:2484–94. doi:10.1016/j.chroma.2009.12.050

[101] Plaza M, Turner C: Pressurized hot water extraction of bioactives. Trends in Analytical Chemistry. 2015, 71:39–54. doi:10.1016/j.trac.2015.02.022

[102] Sosa-Ferrera Z, Mahugo-Santana C, Santana-Rodríguez JJ: Analytical methodologies for the determination of endocrine disrupting compounds in biological and environmental samples. BioMed Research International. 2013, 2013:23. doi:10.1155/2013/674838

[103] Ramos L, Kristenson EM, Brinkman UAT: Current use of pressurised liquid extraction and subcritical water extraction in environmental analysis. Journal of Chromatography A. 2002, 975:3–29. doi:10.1016/S0021-9673(02)01336-5

[104] Ajila CM, Brar SK, Verma M, Tyagi RD, Godbout S, Valero JR: Extraction and analysis of polyphenols: recent trends. Critical Reviews in Biotechnology. 2011, 31:227–49. doi:10.3109/07388551.2010.513677

[105] Mustafa A, Turner C: Pressurized liquid extraction as a green approach in food and herbal plants extraction: a review. Analytica Chimica Acta. 2011, 703:8–18. doi:10.1016/j.aca.2011.07.018

[106] Runnqvist H, Bak SA, Hansen M, Styrishave B, Halling-Sørensen B, Björklund E: Determination of pharmaceuticals in environmental and biological matrices using pressurised liquid extraction—are we developing sound extraction methods? Journal of Chromatography A. 2010, 1217:2447–70. doi:10.1016/j.chroma.2010.02.046

[107] Erdogan S, Ates B, Durmaz G, Yilmaz I, Seckin T: Pressurized liquid extraction of phenolic compounds from *Anatolia propolis* and their radical scavenging capacities. Food and Chemical Toxicology. 2011, 49:1592–7. doi:10.1016/j.fct.2011.04.006

[108] Liazid A, Barbero G, Azaroual L, Palma M, Barroso C: Stability of anthocyanins from red grape skins under pressurized liquid extraction and ultrasound-assisted extraction conditions. Molecules. 2014, 19:21034: doi:10.3390/molecules191221034

[109] Watanabe J, Oki T, Takebayashi J, Takano-Ishikawa Y: Extraction efficiency of hydrophilic and lipophilic antioxidants from lyophilized foods using pressurized liquid extraction and manual extraction. Journal of Food Science. 2014, 79:C1665–C71. doi:10.1111/1750-3841.12570

[110] Santos J, Oliveira MBPP, Ibáñez E, Herrero M: Phenolic profile evolution of different ready-to-eat baby-leaf vegetables during storage. Journal of Chromatography A. 2014, 1327:118–31. doi:10.1016/j.chroma.2013.12.085

[111] Otero-Pareja M, Casas L, Fernández-Ponce M, Mantell C, Ossa E: Green extraction of antioxidants from different varieties of red grape pomace. Molecules. 2015, 20:9686. doi:10.3390/molecules20069686

[112] Morgan ED: Natural products | supercritical fluid chromatography. Reference Module in Chemistry, Molecular Sciences and Chemical Engineering. Elsevier, Netherlands. 2013.

[113] Reverchon E, De Marco I: Supercritical fluid extraction and fractionation of natural matter. The Journal of Supercritical Fluids. 2006, 38:146–66. doi:10.1016/j.supflu.2006.03.020

[114] Seabra IJ, Braga MEM, de Sousa HC: Statistical mixture design investigation of CO_2–ethanol–H_2O pressurized solvent extractions from tara seed coat. The Journal of Supercritical Fluids. 2012, 64:9–18. doi:10.1016/j.supflu.2012.01.005

[115] Cavalcanti RN, Navarro-Díaz HJ, Santos DT, Rostagno MA, Meireles MAA: Supercritical carbon dioxide extraction of polyphenols from pomegranate (*Punica granatum* L.) leaves. Chemical Composition, Economic Evaluation and Chemometric Approach. 2012. 2012, 1. doi:10.5539/jfr.v1n3p282

[116] De Zordi N, Cortesi A, Kikic I, Moneghini M, Solinas D, Innocenti G, Portolan A, Baratto G, Dall'Acqua S: The supercritical carbon dioxide extraction of polyphenols from propolis: a central composite design approach. The Journal of Supercritical Fluids. 2014, 95:491–8. doi:10.1016/j.supflu.2014.10.006

[117] Puri M, Sharma D, Barrow CJ: Enzyme-assisted extraction of bioactives from plants. Trends in Biotechnology. 30:37–44. doi:10.1016/j.tibtech.2011.06.014

[118] Swer TL, Chauhan K, Paul PK, Mukhim C: Evaluation of enzyme treatment conditions on extraction of anthocyanins from *Prunus nepalensis* L. International Journal of Biological Macromolecules. 2016, 92:867–71. doi:10.1016/j.ijbiomac.2016.07.105

[119] Singh RP, Chidambara Murthy KN, Jayaprakasha GK: Studies on the antioxidant activity of pomegranate (*Punica granatum*) peel and seed extracts using in vitro models. Journal of Agricultural and Food Chemistry. 2002, 50:81–6. doi:10.1021/jf010865b

[120] Ambigaipalan P, de Camargo AC, Shahidi F: Phenolic compounds of pomegranate byproducts (outer skin, mesocarp, divider membrane) and their antioxidant activities. Journal of Agricultural and Food Chemistry. 2016, 64:6584–604. doi:10.1021/acs.jafc.6b02950

[121] Sumczynski D, Kotásková E, Družbíková H, Mlček J: Determination of contents and antioxidant activity of free and bound phenolics compounds and in vitro digestibility of commercial black and red rice (*Oryza sativa* L.) varieties. Food Chemistry. 2016, 211:339–46. doi:10.1016/j.foodchem.2016.05.081

[122] Montella R, Coïsson JD, Travaglia F, Locatelli M, Bordiga M, Meyrand M, Barile D, Arlorio M: Identification and characterisation of water and alkali soluble oligosaccharides

from hazelnut skin (*Corylus avellana* L.). Food Chemistry. 2013, 140:717–25. doi:10.1016/j.foodchem.2013.01.061

[123] Papillo VA, Vitaglione P, Graziani G, Gokmen V, Fogliano V: Release of antioxidant capacity from five plant foods during a multistep enzymatic digestion protocol. Journal of Agricultural and Food Chemistry. 2014, 62:4119–26. doi:10.1021/jf500695a

[124] Singleton VL, Rossi JA: Colorimetry of total phenolics with phosphomolybdic-phosphotungstic acid reagents. American Journal of Enology and Viticulture. 1965, 16:144–58 http://www.ajevonline.org/content/16/3/144

[125] Blainski A, Lopes G, de Mello J: Application and analysis of the Folin Ciocalteu method for the determination of the total phenolic content from *Limonium Brasiliense* L. Molecules. 2013, 18:6852. doi:10.3390/molecules18066852

[126] Ikawa M, Schaper TD, Dollard CA, Sasner JJ: Utilization of Folin−Ciocalteu phenol reagent for the detection of certain nitrogen compounds. Journal of Agricultural and Food Chemistry. 2003, 51:1811–5. doi:10.1021/jf021099r

[127] Everette JD, Bryant QM, Green AM, Abbey YA, Wangila GW, Walker RB: Thorough study of reactivity of various compound classes toward the Folin−Ciocalteu reagent. Journal of Agricultural and Food Chemistry. 2010, 58:8139–44. doi:10.1021/jf1005935

[128] Chun OK, Smith N, Sakagawa A, Lee CY: Antioxidant properties of raw and processed cabbages. International Journal of Food Sciences and Nutrition. 2004, 55:191–9. doi:10.1080/09637480410001725148

[129] Silva L, Pezzini B, Soares L: Spectrophotometric determination of the total flavonoid content in *Ocimum basilicum* L. (Lamiaceae) leaves. Pharmacognosy Magazine. 2015, 11:96–101. doi:10.4103/0973-1296.149721

[130] Tai A, Sawano T, Yazama F, Ito H: Evaluation of antioxidant activity of vanillin by using multiple antioxidant assays. Biochimica et Biophysica Acta. 2011, 1810:170–7. doi:10.1016/j.bbagen.2010.11.004

[131] McMurrough I, McDowell J: Chromatographic separation and automated analysis of flavanols. Analytical Biochemistry. 1978, 91:92–100. doi:10.1016/0003-2697(78)90819-9

[132] Porter LJ, Hrstich LN, Chan BG: The conversion of procyanidins and prodelphinidins to cyanidin and delphinidin. Phytochemistry. 1985, 25:223–30. doi:10.1016/S0031-9422(00)94533-3

[133] Hagerman AE, Butler LG: Protein precipitation method for the quantitative determination of tannins. Journal of Agricultural and Food Chemistry. 1978, 26:809–12. doi:10.1021/jf60218a027

[134] Smith MAL, Marley KA, Seigler D, Singletary KW, Meline B: Bioactive properties of wild blueberry fruits. Journal of Food Science. 2000, 65:352–6. doi:10.1111/j.1365-2621.2000.tb16006.x

[135] Weston A, Brown PR: High performance liquid chromatography & capillary electrophoresis: principles and practices. Elsevier Science, Cambridge, Massachusetts, United States 1997.

[136] Rouessac F, Rouessac A: Chemical analysis: modern instrumentation methods and techniques. Wiley, Chichester, United Kingdom, 2013.

[137] Corradini D: Handbook of HPLC, second edition. CRC Press, Cambridge, Massachusetts, United States. 2016.

[138] Niessen WMA, Tinke AP: Liquid chromatography-mass spectrometry general principles and instrumentation. Journal of Chromatography A. 1995, 703:37–57. doi:10.1016/0021-9673(94)01198-N

[139] Vagiri M, Ekholm A, Andersson SC, Johansson E, Rumpunen K: An optimized method for analysis of phenolic compounds in buds, leaves, and fruits of black currant (*Ribes nigrum* L.). Journal of Agricultural and Food Chemistry. 2012, 60:10501–10. doi:10.1021/jf303398z

[140] Slatnar A, Mikulic-Petkovsek M, Stampar F, Veberic R, Solar A: HPLC-MSn identification and quantification of phenolic compounds in hazelnut kernels, oil and bagasse pellets. Food Research International. 2014, 64:783–9. doi:10.1016/j.foodres.2014.08.009

[141] Badalica-Petrescu M, Dragan S, Ranga F, Fetea F, Socaciu C: Comparative HPLC-DAD-ESI(+)MS fingerprint and quantification of phenolic and flavonoid composition of aqueous leaf extracts of *Cornus mas* and *Crataegus monogyna*. Relation to Their Cardiotonic Potential. 2014. 2014, 42. doi:10.10.15835/nbha4219270

[142] Diaconeasa ZA, Florica R, Rugin DA, Lucian C, Socaciu C: HPLC/PDA¨CESI/MS identification of phenolic acids, flavonol glycosides and antioxidant potential in blueberry, blackberry, raspberries and cranberries. Journal of Food and Nutrition Research. 2014, 2:781–5. doi:10.12691/jfnr-2-11-4

[143] Oliveira BG, Costa HB, Ventura JA, Kondratyuk TP, Barroso MES, Correia RM, Pimentel EF, Pinto FE, Endringer DC, Romão W: Chemical profile of mango (*Mangifera indica* L.) using electrospray ionisation mass spectrometry (ESI-MS). Food Chemistry. 2016, 204:37–45. doi.10.1016/j.foodchem.2016.02.117

[144] Karaaslan NM, Yaman M: Determination of anthocyanins in cherry and cranberry by high-performance liquid chromatography–electrospray ionization–mass spectrometry. European Food Research and Technology. 2016, 242:127–35. doi:10.1007/s00217-015-2524-9

[145] Kelebek H: LC-DAD–ESI-MS/MS characterization of phenolic constituents in Turkish black tea: effect of infusion time and temperature. Food Chemistry. 2016, 204:227–38. doi:10.1016/j.foodchem.2016.02.132

Chapter 4

Synthesis and Characterization of Phenolic Lipids

Mohamed Hussein Hamdy Roby

Additional information is available at the end of the chapter

http://dx.doi.org/10.5772/66891

Abstract

Omega-3 polyunsaturated fatty acids (ω3 PUFAs) from fish oils promote well-established health and antiaging benefits that justify their use as functional ingredients in dietary supplements, healthy foods, and nutraceutical products. Dietary supply is needed because human metabolism exhibits limited to synthesize ω3 PUFAs. However, the practical use of such lipids as food ingredients is often limited due to their high susceptibility to oxidation, which is responsible for the undesirable off-flavor and odor of rancid oils, associated with the loss of nutritional value. Produced phenolic lipids were a good solution for all these problems. These phenolic lipids are preferentially produced by enzymatic bioprocesses that exhibit high selectivity toward polyfunctional substrates and mild reaction conditions compared with chemical synthesis pathways. This chapter presents the acylation of phenolic compounds and lipids using enzyme under various operating conditions. In conclusion, the acylation of lipids with natural phenolic compounds resulted in the formation of a lipophilic ester that should be able to stabilize the oil, fats, and emulsions against oxidation. Acylation of lipids to phenolic compounds that have antioxidant properties thus protects the lipids from oxidation, and the phenolic lipid derivatives carry the combined health beneficial properties of lipids and the phenolic molecules.

Keywords: phenolic compounds, omega-3, phenolic lipids, enzymatic esterification, oxidative stability, neuroprotection, fish oil

1. Introduction

In the recent years, interests for natural substances with beneficial activity to human have sharply risen. In fact, there is a significant increase in nutraceuticals and pharmaceutical products, based on natural compounds. The main interest has been observed for natural substances with strong antioxidant activity because oxidative stress induced by multiple factors is the main cause of many pathological conditions such as inflammation, cancer, coronary heart disease and even skin aging. Also, there has been a significant consumer interest in health enhancing the role of specific foods or physiologically active food components.

Unsaturated lipids have been widely recognized for their role in the maintenance of human health. These lipids, especially those from the omega-3 (ω-3) series, have been linked to inhibitory effects on atherosclerosis cardiovascular and Alzheimer's diseases [1–3]. However, the use of such lipids remains strongly limited due to their high susceptibility to autoxidation. To overcome this difficulty, a lot of researches have been carried out focusing on the development and the use of antioxidants that could delay or even prevent omega-3 lipid oxidative degradation. In this context, natural plant phenols were perceived by many researchers as potential substitutes for controversial synthetic antioxidants; however, the major drawback of these compounds is their low solubility in matrices that strongly restrains their use in food applications [4, 5].

The hydrophilic nature of phenolic compound reduces their effectiveness in oil-based formulae and emulsions [6]. The synthesis of more lipophilic derivatives, especially esters, could help to increase their lipophilicity and then their interactions with lipidic phases that need to be stabilized. To achieve this goal, acylation with fatty acids appears as a promising way (lipophilization) that could extend the scope of application of phenolic antioxidants in lipid-rich food matrices. When applied to polyunsaturated lipids, this approach is expected to provide stable ingredients with high nutritional value and high antioxidant potential. Additional effects could be an increased bioavailability of phenols as well as cumulative and even synergistic biological activities [7, 8].

Many studies reported the enzymatic synthesis of phenolic lipids based on the ability of lipases to catalyze the acylation of phenolic compounds with either fatty acids or triacylglycerols (TAGs) [9–13]. Main advantages of enzyme-catalyzed processes include the use of mild reaction conditions that limit substrate degradation and high selectivity that avoids the production of undesirable compounds and facilitates further purification protocols [14].

2. Fatty acids and phenolic compounds

Dietary fat is an essential component for digestion, absorption, and transport of fat-soluble vitamins and phytochemicals, such as carotenoids and lycopenes. Dietary fat contributes approximately 34% of the energy in the human diet. Because fat is a main source of energy (9 kcal/g), humans are able to obtain adequate energy with a reasonable daily composition of fat-containing food item products.

2.1. Fatty acids

Fatty acids are classified as saturated fatty acid (SFA), monounsaturated fatty acid (MUFA) and polyunsaturated fatty acid (PUFA). The essential fatty acids (EFAs) refer to those polyunsaturated fatty acids (PUFAs) that must be provided in our food because these EFAs cannot be synthesized in our body, and they are necessary for a good health. The main two families of EFAs are omega-3 (ω-3) and omega-6 (ω-6). ω-3 and ω-6 structures are based on the position of the double bond from the methyl (omega) terminal of the aliphatic carbon chain [1, 15]. The parent fatty acid of the ω-6 series is linoleic acid (18:2n-6) and the parent fatty acid of the ω-3 series is linolenic acid (18:3n-3.). ω-3 includes alpha-linolenic acid (ALA), docosahexaenoic acid (DHA), and eicosapentaenoic acid (EPA) (**Figure 1**).

Human body can synthesized omega-3 and omega-6 from linoleic acid and linolenic acid, respectively, through a series of desaturation (addition of a double bond) and elongation (addition of two carbon atoms) reactions [16]. Unlike linolenic and linoleic acid, oleic acid (18:1n-9) is consumed in substantial amounts in the Western diet and is not an essential fatty acid. There is a little eicosatrienoic acid (ETA, 20:3n-9) in cell membranes, however, probably because of the overwhelming competition from dietary linoleic acid for the relevant desaturase and elongase enzymes. The pathways for desaturation and elongation of ω-3 and ω-6 fatty acids are given in **Figure 2**.

2.2. Health benefits of omega-3 fatty acids

The ω-3 fatty acids provide a wide range of benefits from general improvements in health to protect against inflammation and disease. Several studies have indicated that the consumption of ω-3 fatty acids provides benefits in reducing the risk of cardiovascular diseases [1, 2].

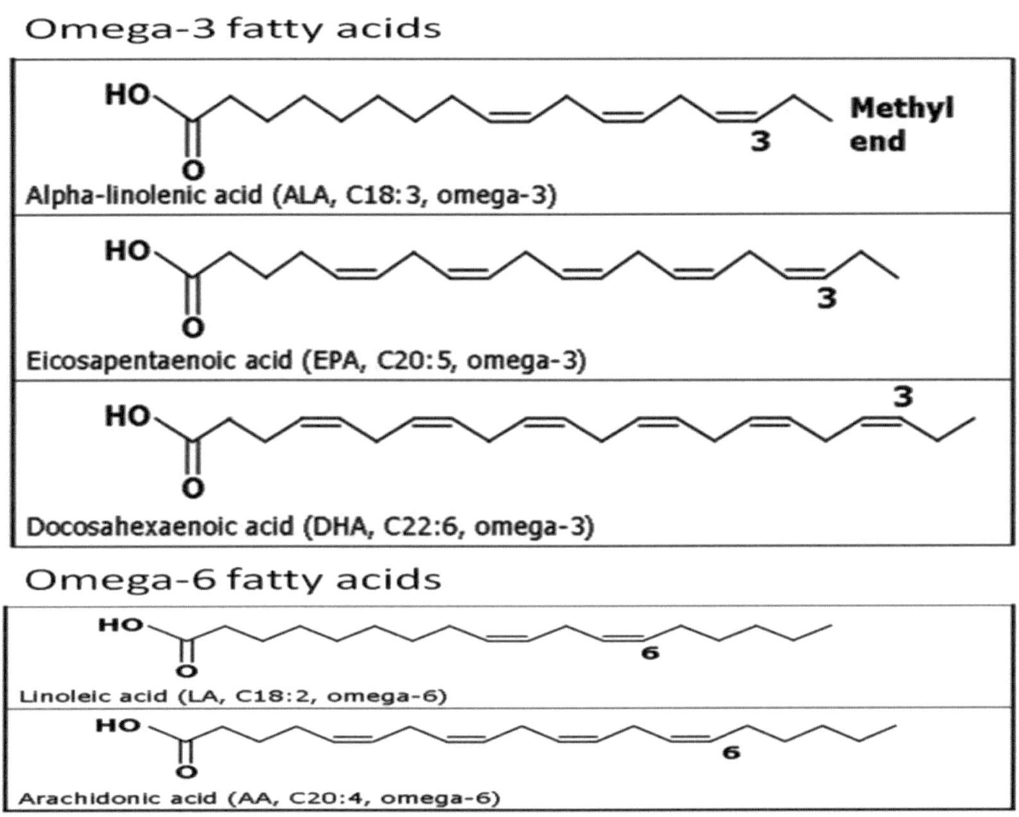

Figure 1. Omega-3 and omega-6 fatty acids.

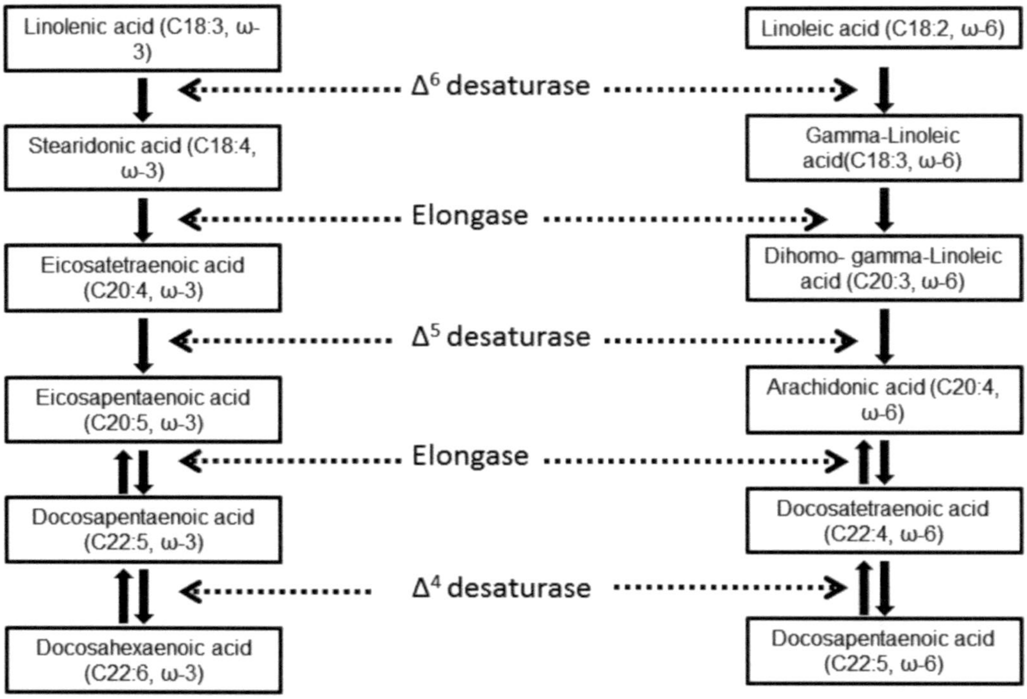

Figure 2. Desaturation and elongation pathway of ω-3 and ω-6 fatty acids.

DHA and EPA have been used in a number of small clinical trials to understand their efficacy and shown to possess immunomodulatory properties depending on their localization in different cell types. DHA is selectively incorporated into retinal cell membranes and postsynaptic neuronal cell membranes, suggesting that it plays important roles in vision and nervous system function [17–19]. DHA content in the brain may be particularly important, since animal studies have shown that depletion of DHA in the brain can be resulted in learning deficits. It is not clear how DHA affects brain function, but changes in DHA content of neuronal cell membranes could alter the function of ion channels or membrane-associated receptors, as well as the availability of neurotransmitters [20, 21]. Increasing ω-3 fatty acid intake enhances the DHA content of cell membranes, resulting in higher proportions of DHA in the body (**Figure 3**).

The ω-3 fatty acids are reported to associate with the brain development; also, it is important for the vision and the functions of the reproductive system. This may be due to the fact that DHA is a component of brain nerve synapses, in the eye's retina, in the testes, and in sperms and plays a vital role in the development and functions of these organs and systems [20]. The nervous system contains approximately 35% PUFAs as its lipid content; most of which are long-chain (LC) PUFAs. In addition, higher prenatal intake of DHA has been shown to be associated with improved visual, cognitive, and motor development in offspring. Children given ω-3 PUFA-supplemented formula demonstrated enhanced visual and mental capabilities [19],

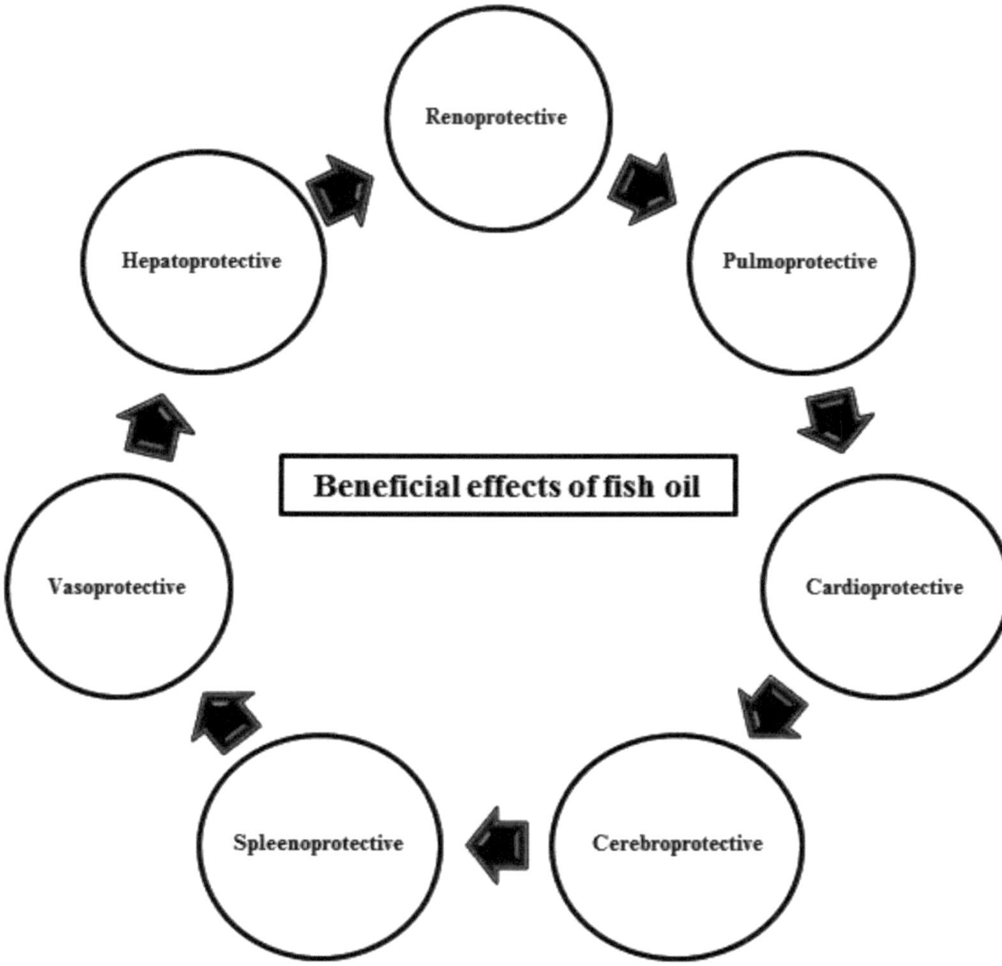

Figure 3. Beneficial effects of omega-3 oil for human body.

while in human adults, clinical studies have suggested a low intake or inadequate. The ω-3 fatty acids possess antithrombotic properties, which in combination with their anti-inflammatory effect is likely to positively aid cardiovascular disease treatment. DHA and EPA also appear to possess anticancer and antiapoptotic effects. Additionally, these PUFAs suppress gene expression of lipogenic genes in the liver and trigger adipose fatty acid oxidation, suggesting a potential role against obesity [15, 22].

2.3. Phenolic compounds

Phenolic compound is chemically defined as a substance that contains an aromatic ring containing one or more hydroxyl substitute including functional derivatives [23]. In general, phenolic compounds are present in a wide variety of food plants as esters or glycosides conjugated with other compounds, such as flavonoids, alcohols, hydroxyl fatty acids, sterols and glucosides. Phenolic compounds found in foods may be categorized accordingly to three groups, simple phenols and phenolic acids, hydroxycinnamic acid derivatives and flavonoids. The simple phenols include the monophenols, such as p-cresol found in berry fruits (e.g., raspberry, blackberry) and diphenols, such as hydroquinone found commonly in vanilla [5, 24].

Phenolic compounds play a major role in the protection against oxidation processes. The antioxidant properties of phenolic compounds can act as free radical scavengers, hydrogen donators, metal chelators and singlet oxygen quenchers [25, 26].

2.3.1. Nutritional and antioxidant properties

Phenolic compounds are natural antioxidants that are present in food or in the body, to delay or stop the oxidation of that substance. The main advantages of these natural antioxidant are (1) they are readily acceptable by the consumers; (2) they are considered to be safe; (3) no safety tests are required by legislation; and (4) this natural antioxidant is identical to the food which people have taken over a hundred years or have been mixing with food. Phenolic compounds are associated with nutritional and organoleptic qualities of foods from plant origin [24, 26]. Phenolic compounds at low concentration protect foods from autoxidation, but at high concentration, they can cause undesirable discoloration as a result of their interaction with the carbohydrate or protein components.

Among naturally found phenolic compounds, phenolic acids are of high interest due to their potential biological properties [27, 28]. Many phenolic acids are known to be potent antioxidants through their radical scavenging activity, and due to their chemical structure, the reactivity of phenolic acids increases as the number of hydroxyl and methoxyl groups increases [29]. The consumption of fruits, vegetables, and soft drinks such as tea and coffee, which contain phenolic compounds, has been linked to lower risk of some diseases, such as cancer and CVD [30, 31]. However, the use of phenolic acids as natural antioxidants in foods and nutraceutical supplements has the limitation of low solubility in oil-based media. Nevertheless, lipase-catalyzed reactions of lipids with phenolic acids could produce structured lipids with phenolic moieties, which would have health benefits and improved solubility characteristics [32–35].

2.4. Synthesis of phenolic lipid (PL) compounds

Phenolic lipids (PL) are types of fats and oils modified to improved nutritional or physical properties by incorporating phenol compound on the glycerol backbone. Phenolic lipids play an important role as antioxidant and biological active compounds, but their contents in the nature are minor, and the procedures for separation and purification are not easy, very expensive and take a long time, which makes their applications in the food or cosmetic industry very inconvenient. Consequently, the synthesis of PL has attracted more attention in recent years because it is a good way to improve the hydrophobic nature of phenolic compounds, which could be achieved by chemical or enzymatic synthesis.

2.4.1. *Chemical synthesis of phenolic lipids*

Chemical synthesis is a traditional method that is used for PL preparation. Synthesis of PL through chemical synthesis could be done by using Friedel-Crafts acylation reaction or Fisher acid catalysis esterification. These processes are generally carried out at relatively high temperatures and pressures under anhydrous conditions, using rather unspecific alkali metal or alkali catalysts. Some related works have been provided in this topic, one of them is the work of Qianchun et al. [36] about the chemical synthesis of phytosterol esters of polyunsaturated fatty acids (PUFAs), which could be used in different formulations of functional foods. Direct esterification of phytosterols with PUFA was catalyzed by sodium bisulfate to produce sterol esters of PUFA without organic solvent. The modeling of sodium bisulfate with superfluous fatty acids as solvents to synthesize phytosterol esters of PUFA was successfully performed with degree of esterification up to 96% and less oxidative products in the reaction process [36].

The chemical esterification of flavonoids with some fatty acids was provided by [37] and its product exhibited lipophilic, antiradical and antioxidant properties. In works reported by Zhong and Shahidi [38, 39] on epigallocatechin gallate (EGCG), the predominant catechin in tea was structurally modified by esterification with fatty acids, including stearic acid (SA), docosapentaenoic acid (DPA), eicosapentaenoic acid (EPA) and docosahexaenoic acid (DHA). The esterification of EGCG with these fatty acids using acylating agents, namely, the acyl chlorides, resulted in yields of 65.9, 42.7 and 30.7 for SA, EPA and DHA, respectively [39]. This esterification leads to produce various compounds that have anti-inflammatory effect and also shows higher inhibition effect against hydroxyl and peroxyl radical-induced DNA scission [38]. Phenolic lipid (PL) chemical synthesis meets some partial needs; this pattern possesses a low degree of regioselectivity and is generally accompanied by drastic reaction conditions, many intermediary stages and purification steps to remove by-products and catalyst residues. The main drawbacks to chemical transesterification are (1) non-selectivity leading to random distribution of FAs, (2) isomerization of sensitive PUFAs by the alkali catalyst, (3) production of fatty acid soaps and unwanted by-products and (4) requiring substantial post-treatment and downstream processes, especially when food applications are concerned.

2.4.2. *Enzymatic synthesis of phenolic lipids (PLs)*

The application of enzymes is widely in different fields such as pharmaceutical, cosmetic and food industry. Enzymatic synthesis of PL from fats and oils is receiving a lot of attention as a method for their modification because of the advantages of milder reaction conditions, minimization of side reactions and by-product formation, a selective specificity, a wider variety of pure synthetic substrates, fewer purification steps and a more environmentally friendly process [40]. Even if enzymes may be more expensive than chemical reagents, the enzyme-catalyzed acylation is a well-mastered technique for synthesis of selectively modification of PL at present. A high degree of conversion to the desired products could be achieved under the optimal reaction conditions. The enzymatic processes can be used in the production of fats and oils containing beneficial fatty acids and phenolic compounds. Some reviews have given a comprehensive understanding and shown a whole outline on the enzymatic synthesis of PL [41–46].

In particular, enzymes appear to be very effective for the synthesis of molecules involving the grafting of a lipophilic moiety or a hydrophilic one. This review will be described and discussed in some of the recent works in the field of enzyme-assisted acylation of fatty acids with phenolic compounds in order to modify the hydrophilic/lipophilic properties of the initial molecules to obtain new products with multifunctional properties combining, for example, antimicrobial, antioxidant and emulsifying properties. The enzymatic synthesis of phenolic lipids has been reported previously in Refs. [10, 45, 47–49]. A lot of enzymes can be used in the synthesis of PL and selectivity is the most important characteristics of enzymes used in phenolic lipid synthesis. Lipase is the most enzymes used in this type of process because of high selectivity, lower overall reaction time and fewer side reactions when compared with chemical methods [50]. An example of a synthesis reaction catalyzed by the lipase is shown in **Figure 4**. This overwhelming interest is based largely on consumers' desire to maintain overall well-being with minimal effort and an industries' ability to respond to this need. Furthermore, with the consumption of manufactured foods continually on the rise, there is a distinct advantage to providing more healthful choices for consumers. The concept of a natural phenolic lipid composed of a long-chain aliphatic and phenolic moiety readily fits this mold, particularly since the inclusion of unsaturated lipids into these compounds could result in additional nutritional benefits. Lipases constitute the most important group of biocatalysts for biotechnological applications.

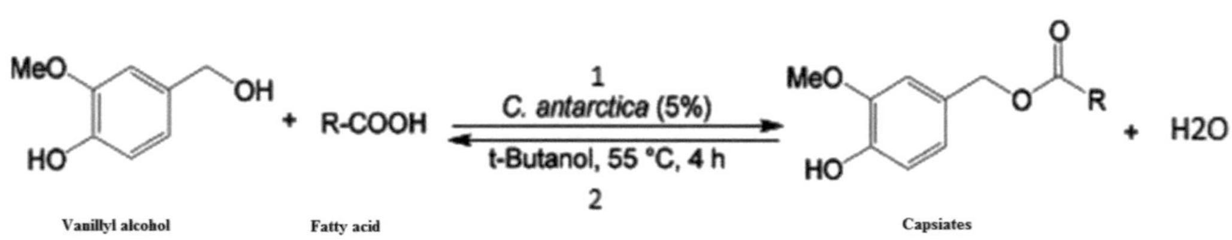

Figure 4. Enzymatic synthesis reaction capsiates (fatty vanillyl alcohol acid ester) catalyzed by lipase Novozym 435® [51].

2.5. Lipases

2.5.1. Definition, sources, and applications

Lipase enzymes are defined as glycerol ester hydrolyses that can hydrolyze tri-, di-, and monoacylglycerols [52, 53]. Lipases are soluble in water as a result of their protein nature, but it could act on lipids, which are water insoluble, at the interface between oil and water [53, 54] and catalyze esterification and transesterification in addition to the hydrolytic activity on TAG [55–57]. Lipases are originated from a wide variety of sources including animals, plants and microorganisms. Animal lipases include pregastric esterase, pancreatic lipase, and lingual lipases [53]. Plants such as wheat germ and castor beans also contain lipases [58, 59]. Finally, microbial sources including yeast (*Candida* and *Geotrichum*), molds (*Rhizopus, Aspergillus*), and bacteria (*Bacillus, Pseudomonas*) [60]. Lipases are widely used because of their ready availability, low cost of production, utility in food, biotechnology, and pharmacology [61]. Novel biotechnological applications have been successfully established using lipases for the synthesis phenolic lipids, the production of pharmaceuticals, agrochemicals, and flavor compounds [52, 62, 63]. Moreover, the use of lipases in the food industry is increasing due to the need for the production of esters, biodegradable polyesters, and specific FAs [64].

2.5.2. Mechanism of action

Lipase-catalyzed reactions have been gained a lot of interest over the last years; the major reason for this is that lipase can promote either ester formation or ester hydrolysis. Moreover, lipase can control the acylation and deacylation to produce specific fatty acids and triacylglycerols (i.e., phenolic lipids). Lipase-catalyzed reactions can be classified into three groups which are hydrolysis, esterification, and transesterification [65].

2.5.2.1. Hydrolysis

Hydrolysis of lipids by lipases refers to the splitting of fat into its constituent acids and alcohols in the presence of water. Lipase-catalyzed hydrolysis can be used for the preparation of fatty acids from oils, especially for the selective hydrolysis and concentrations of PUFAs from edible oils [10]. Furthermore, lipase catalyzed hydrolysis reactions only in the presence of amount of water. This is due to the fact that water molecules participate in the breaking of covalent bond in the substrate as well as subsequent incorporation of their elements into these bonds to form reaction products [66].

Different products are determined during the extent of hydrolysis reaction as shown in **Figure 5**. Mixtures of monoacylglycerols, diacylglycerols and free fatty acids are produced; the more complete the hydrolysis, the higher the concentration of free fatty acids in the final reaction medium. In the end of lipase- hydrolysis reactions glycerol esters-enriched in w-3 fatty acids were produced from fish oil. Reactions are ideal for removal of fatty acids from unstable oils, including conjugated or highly unsaturated fatty acids, which effectively reduce unwanted oxidation reactions [67]. Lipase catalyzed hydrolysis reactions produce glycerol esters enriched in ω-3 fatty acids from fish oil [68, 69]. Because natural fish oils do not contain more than about one-third of their fatty acids from the ω-3 family, hydrolysis reactions are particularly helpful for the purpose of concentration.

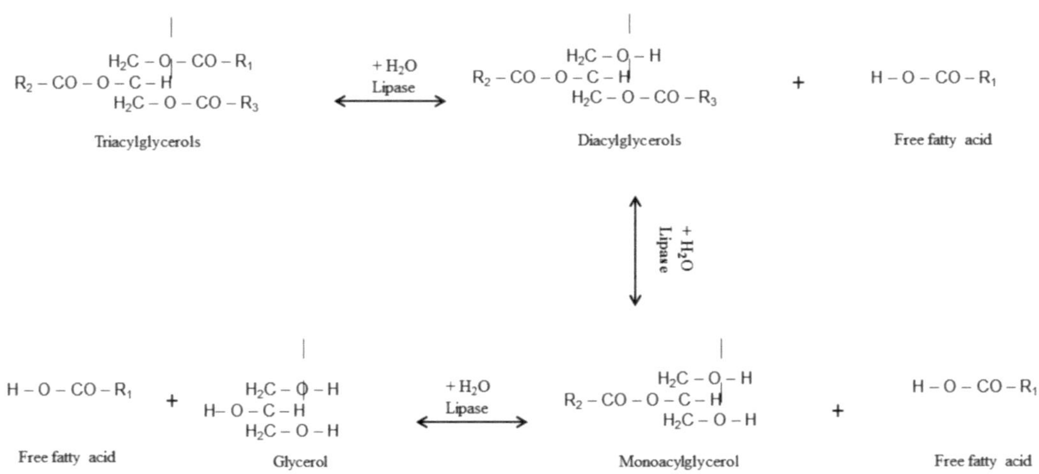

Figure 5. Enzymatic hydrolysis of triacylglycerol molecule. Reverse reaction corresponds to synthesis by esterification. R1, R2 and R3 are different acyl groups.

2.5.2.2. Esterification

Esterification is the reverse reaction of hydrolysis and is used to synthesize selected products under appropriate reaction conditions [70]. The products of an esterification reaction are usually an ester and water. The water content of esterification reaction system strongly effects on lipase activity. Low water content shifts the equilibrium of the reaction to favor the synthesis of lipids. So that additional techniques were used to drive synthesis reaction including removal of water that formed during the process by evaporation under reduced pressure [71] or by adding molecular sieves to adsorb the water. Direct enzymatic esterification of some primary alcohols and selected carboxylic acids was catalyzed by the *Candida antarctica* and *Rhizomucor miehei* lipases. The reactions were performed in solvent-free medium with the removal of water [72].

2.5.2.3. Transesterification

Transesterification is a process of acyl exchange between two molecules. This process normally takes place between an ester and alcohol (alcoholysis), an ester and an acid (acidolysis), or an ester with another ester (interesterification), and no water is involved in the reaction. Acidolysis is one of the most frequently used reactions to incorporate novel fatty acids into TAG in several researches [13, 73, 74]. Interesterification involving hydrolysis and esterification, firstly hydrolysis of the TAG molecule, then followed by re-synthesis of the liberated fatty acids onto the glycerol molecule. Interesterification is another main strategy to incorporate PUFAs into TAGs. The literature reported extensive research work on the interesterification reaction [75, 76]. Lipase-catalyzed alcoholysis, acidolysis, and interesterification reactions are described clearly in **Figure 6**.

2.5.3. Selectivity and specificity of lipase

It is beneficial to have knowledge about lipase selectivity/specificity to guide research to the best choice of lipase for particular fatty acid or for synthesis of PL containing ester of a specific fatty acid. Specificity generally refers to the ability of enzyme to differentiate between several substrates. Lipases can be divided according to their specificity into three groups: (i) nonspecific lipases, (ii) acyl-group specific lipases, and (iii) positional specific lipases. Nonspecific lipase can catalyze the release of FA from any position on the glycerol molecule. Acyl-group specific lipases catalyze the release of a particular type of FA from the TAG molecules, while positional lipases attack sn-1,3 positions on the TAG molecule. The use of positional specific lipases has led to the production of useful TAG mixtures whose composition could not be produced by simple chemical transesterification. In recent years, positional specific lipases have been intensively used in research purposes and food industry sectors [77–79].

2.6. Enzyme reactions in organic solvent media (OSM)

Enzymes in organic solvents have manifested good selectivity and stability; however, catalytic activities in this environment are generally lower than in aqueous solutions. This could be partly explained by the fact that in low water environments, enzymes are less flexible. On

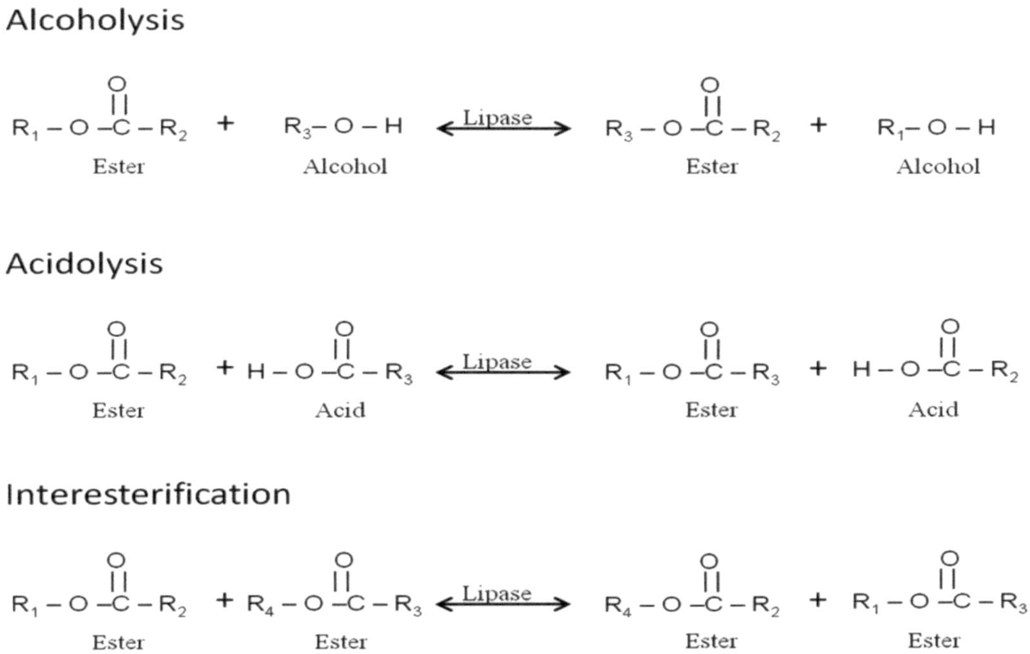

Figure 6. Lipase-catalyzed transesterification reactions. R1, R2, and R3 are different acyl groups.

the other hand, the activities of enzymes also depend on the type of organic solvent, since some are known to inactivate or denature biocatalysts. Meanwhile, the advantages of using organic solvent media are increased solubility of hydrophobic compounds that permits for greater interactions between substrates and enzymes as well as advantageous, partitioning of substrates and products; specifically, this is because partitioning of products away from the enzyme can decrease the possibility of inhibition due to excess product around the biocatalyst [51, 72].

When enzymes are placed in OSM, they exhibit novel characteristics such as altered chemo- and stereoselectivity, enhanced stability, increased rigidity, insolubility, and high thermal stability [80]. It has also been reported that the thermal stability of lipases can be improved in organic solvent systems since the lack of water prevents the unfolding of the lipase at high temperatures [81]. The activity of lipase in OSM depends on the nature and concentration of the substrate and source of the enzyme. Moreover, the organic solvent used can dramatically affect the activity of the lipase. Lipases are more active in n-hexane, n-heptane, and isooctane as compared to other solvents, such as toluene, ethyl acetate, and acetonitrile [82, 83]. It has been reported that the hydrophobicity of the solvent can affect the degree of acyl migration during transesterification using a 1,3-specific lipase [84]. Since the choice of organic solvents based on minimization of acyl migration may conflict with maximization of transesterification, acyl migration is usually minimized by reducing reaction times [85]. With increasing concern for the environment, synthesis of PL in solvent-free systems [86–88] and ionic liquids systems [89] has been extensively studied.

The enzymatic synthesis of vanillyl-PUFA esters from fish oil and vanillyl alcohol in acetone solvent medium was studied by [10]. Lipase-catalyzed esterification of vanillyl alcohol with different fatty acids was carried out by [51] to the synthesis of capsiate analogs. Equimolar concentration of vanillyl alcohol and fatty acid was solubilized in *tert*-butanol and esterified using *C. antartica* lipase (Novozym 435) at 55°C for 4 h.

2.7. Enzyme reactions in solvent-free medium (SFM)

Enzymatic catalysis in solvent-free medium (SFM) has attracted considerable interest in the recent years [90]. It used as an efficient approach to the synthesis of natural products, pharmaceuticals and food ingredients. Under nonaqueous conditions, the industrial utility of enzymes can be improved, recovery of product and enzyme is eased, and the ability to catalyze reactions that are not favorable in aqueous solutions [91]. However, it would be technically beneficial if the enzymatic reactions were performed in mixtures of substrates in the absence of solvents. Lipase-catalyzed PL has been extensively studied in systems using organic solvents; however, if such a process is intended to be used in the food industry, it is preferred to develop solvent-free systems. The downside of organic solvents is that they are expensive, toxic and flammable and their use involves higher investment costs to meet safety requirements [80]. On the other hand, solvent-free systems, which are a simple mixture of reactants and the biocatalyst, present the advantages of using nearly nonaqueous organic solvents while offering greater safety, reduction in solvent extraction costs, increased reactant concentrations and consequently higher volumetric productivity defined as kg product per unit of reactor volume [53, 80].

Phenolic lipids have been received increasing attention in the food area, since they are a good way for providing nutraceutical FA to consumers. Hong et al. [47] studied the esterification of vanillyl alcohol with conjugated linoleic acid under vacuum in solvent-free system. Further studies on the enzymatic synthesis of structured phenolic lipids in SFM have also been conducted by [34, 44, 92]. In these studies, phenolic acids were esterified with fatty acids resulted in the formation of more lipophilic constituents that can be used as a nutraceutical product. In addition, feruloylated mono- and diacylglycerols were synthesized in SFM using *C. antarctica* lipase, and the yield was 96% [92].

Lipase-catalyzed synthesis in SFM has a number of advantages as compared to that in OSM, including the use of a smaller reaction volume, maximization of substrate concentration and with no additional solvent recovery. In addition, downstream processing is easier as fewer purification steps are required providing significant cost savings, as well as toxic organic solvents are completely avoided (clean conversions), and an increase in the volumetric productivity can be achieved [80]. However, there are some problems with the use of SFM, mainly, the high viscosity of the medium as well as the production of high amounts of glycerol, free FAs as by-products. These by-products affect the reaction equilibrium and limit the mass transfer rate [93]. Thus, the development of a bioprocess for the lipase-catalyzed synthesis in SFM is of major interest but with great challenge.

2.8. Parameters affecting the enzyme activity and conversion yield of phenolic lipids

Grafting of phenolic compound substrates with lipids is the major difficulty to overcome in such lipase-catalyzed reactions. Several parameters must be considered in order to achieve the reaction in satisfactory kinetics and yields and to overcome the fact that the two substrates greatly differ in terms of polarity and solvent affinity.

2.8.1. Effect of solvent

The interesting strategy is to carry out the synthesis reaction without using solvent. However, when it is not possible, the choice of an adequate solvent is important. The type of organic solvent employed can dramatically affect the reaction kinetics and catalytic efficiency of lipases [94]. Two factors must be considered when solvent is selected; solvent affects the enzyme activity and solvent effect on the equilibrium position of the desired reaction. Polarity of the solvents is an important characteristic which determine the effect of solvents on enzymatic catalysis reactions. Log P value, the partition coefficient between water and octanol, is used as the indicator of solvent polarity. Laane et al. [95] reported that solvents with log $P < 2$ are not suitable for enzyme-catalyzed systems, since they strip off the essential water from the enzyme and therefore inactivate them. However, solvents with log P values in the range of 2–4 were weak water distorters, in which enzymes display medium activity and solvents with log $P > 4$ are ideal media for enzyme-catalyzed systems since they do not distort the essential water from the enzyme. Therefore, intermediate polarity media are often chosen. Other factors that must be taken into account in determining the most appropriate solvent for given reaction include solubility of reactants, solvent inertness, density, viscosity, surface tension, toxicity, flammability, waste disposal, and cost [96]. A good contact between the substrates must be obtained, and the selected solvent must be solubilizing them at least partially.

Various authors have tried to find original strategies to improve enzyme activity in organic solvent [71, 93, 94]. The effect of solvent concentration on the conversion yield of phenolic lipids synthesized from flaxseed oil and phenolic acids was demonstrated by [41]. Solvent concentration of 7% was the best concentration with 61.1% of conversion yield.

2.8.2. Lipase conditioning

Another important parameter in the synthesis reactions of phenolic lipids (PLs) is concerning with the enzyme itself and especially its conditioning. Various techniques for lipase conditioning have greatly improved during the last 10 years in the field of enzyme immobilization, chemical modification, or molecular engineering [97, 98]. Lipases are used after immobilization on a support. Different carrier materials are employed, and the resulting immobilized enzyme usually exhibits an improved thermostability compared to its free form. Moreover, the use of immobilized enzymes allows an easy removal and recovery of the biocatalyst once the reaction is over [99]. Lipase from *Candida rugosa* was immobilized onto montmorillonite via two techniques, i.e., adsorption and covalent-binding montmorillonite [100].

2.8.3. Influence of water activity

Water content refers to the total amount of water present in the reaction system. Controlling of water activity is very important in lipid modification processes. Water content in the reaction system is a determining factor in whether the reaction equilibrium will progress toward hydrolysis or ester synthesis [101]. While ester synthesis depends on low water content, too low water activity prevents all reaction from occurring. The monolayer of water on the surface of enzyme is required to maintain the three-dimensional structure of the enzyme, which is essential to enzymatic activity [102]. This layer acts as a buffer between the enzyme surface and the bulk reaction medium. However, too much water can cause hydrolysis of the TAG [14]. The activity of lipases at different water activities is dependent on the source of the enzyme and the type of solvent and immobilization support used [103]. Lipases from molds have shown to be more tolerant to low water activity than bacterial lipases. The optimal water content for most interesterification reactions by different lipases has been reported to be in the range of 0.04 to 11% (w/v) [104].

However, the amount of water in the system should be minimized in order to decrease the by-products. Lipases tend to retain the greatest degree of original activity, when immobilized on hydrophobic supports. Therefore, when the immobilized lipase contacts with oil in water emulsion, the oil phase tends to associate with and permeate the support, which can be assumed that an ordered hydrophobic network of lipid molecules will surround the support. Any water that reaches the enzyme for participation in the reaction must diffuse from the bulk emulsion. Thus, to avoid diffusional limitations, the oil phase must be well saturated with water [105].

Zhao et al. [106] investigated the effect of different reaction parameters on the enzymatic acidolysis of lard with capric acid catalyzed by Lipozyme TL-IM. They achieved the highest incorporation of capric acid (35.56 mol%) without added water. The amount of incorporation was almost constant up to 10% added water but decreased significantly above this amount. The current research work shows that Lipozyme TL-IM-catalyzed interesterification can easily be moved to the industrial sector for commercial exploitation. Both stirred tank reactors [107] and PBR [108, 109] can be used for the production of plastic fats, and the control of water activity in the system presents no particular difficulty, as is often the case in other lipase-application systems, in which the lipase activity was not affected by the reduction of water content in the system [107, 110].

2.8.4. Molecular sieve

In order to promote the synthesis of phenolic lipids by shifting the reaction toward synthesis rather than hydrolysis, a reduction of water content in the reaction mixture can be accomplished through the addition of molecular sieve pellets as dehydrating agents. Li et al. [111] reported that the addition of molecular sieves increased the rate and conversion yield; this is due to the effect of the molecular sieves to sequester the water layer from the enzyme molecule which is essential for the water-enzyme interaction. Mellou et al. [112] found that the conversion yield of rutin during esterification reaction with oleic acid catalyzed by immobilized *C. antarctica* lipase B in different solvents was varied from 37 to 71% under the use of molecular sieves (100 mg/ml). However, Karboune et al. [113] observed 28 and 35% decrease in the maximum conversion yield upon the addition of 10 mg/ml of molecular sieves to the lipase-catalyzed biosynthesis of cinnamoylated lipids. This could be explained by the fact that molecular sieves promote the lipase-catalyzed synthesis reactions by dehydrating; however, excess of molecular sieves will capture the necessary water of enzyme, which may inhibit the enzyme activity.

2.8.5. Substrate composition and concentration (molar ratio)

Chemical structures of the phenolic compounds have an effect on the conversion yield of the end products. Different studies presented the effect of chemical structure of phenolic compounds which are hydroxylated or methoxylated derivatives of cinnamic, phenyl acetic and benzoic acids on the conversion yield [34, 35, 42, 44]. The presence of a hydroxyl group in the sn-2 position has a negative inductive effect. Thus, TAG is hydrolyzed at a faster rate as compared to DAGs, which are hydrolyzed faster than MAGs. Substrate conformation can also affect the reaction rate, since the hydrophobic tunnel in the lipase accepts aliphatic chains and aromatic rings easier than branched structures. Moreover, oxidation of substrates, such as PUFAs, could cause inhibition and decrease in lipase activity due to the production of hydroperoxides and their consequent breakdown to free radicals.

Substrate concentration has an effect on the rate of enzyme hydrolysis and transesterification. So, the selection of a suitable substrate molar ratio in terms of reaction efficiency (incorporation level of acyl donors per unit time) and productivity (product quantity per unit time) in a reaction system is very important. The choice of the proper substrate molar ratio is also related to the downstream processing expenses and associated difficulties of separating FFAs or acyl donors by evaporation and/or distillation. Previous studies have shown that high substrate molar ratio would require a shorter reaction time, move the reaction equilibrium to the product formation, and improve the acyl incorporation [114, 115]. Yang et al. [114] reported the positive effect of substrate molar ratio on the interesterification reaction between EPA and DHA ethyl esters and tripalmitin. They indicated that the optimization results suggested a molar ratio of 6 along with an enzyme load of 20% (Lipozyme TL-IM) and a 17.9 h reaction time would provide the highest incorporation. However, due to the downstream purification expenses, they decided to select the optimal conditions to be a molar ratio of 5 along with a 20% enzyme load and 20 h reaction time. Lee et al. [115] investigated the synthesis of 1,3-dioleoyl-2-palmitoyl glycerol-rich HMFS from tripalmitin-rich fraction and ethyl oleate by lipase-catalyzed interesterification. Similarly, these authors reported an increase of OPO content (25.7%) with an increase of substrate molar ratio up to a ratio of 1:6 of tripalmitin-rich fraction to ethyl oleate.

The study of Sabally et al. [32] investigated the enzymatic transesterification of selected PAs with TAGs, including trilinolein and trilinolenin in organic solvent media (OSM), and reported that the affinity of Novozym 435 was found to be greater for DHCA than that for ferulic acid; these authors suggested that the presence of both the methoxyl substituent and the double bond on the side of the aromatic ring of the ferulic acid could explain its lower affinity for the transesterification reactions with TAG.

Karboune [42] study the effect of PA structure on the bioconversion yield (BY) of phenolic lipids (PLs) obtained by acidolysis of FSO with selected PAs, including hydroxylated and/or methoxylated derivatives of cinnamic, phenyl acetic and benzoic acids in OSM, using Novozym 435 as biocatalyst. The overall findings showed that the BY of PL was dependent on the structural characteristics of PAs. The highest BY was obtained with cinnamic acid (74%). In addition, Karboune et al. [42] concluded that the presence of p-OH groups on the benzene cycle of cinnamic acid derivatives may have an inhibitory effect on the lipase activity, since the BY decreased to 45 and 11%, respectively, when *p*-coumaric and caffeic acids were used as substrates. The inhibitory effect of p-OH substituent was most likely due to their electronic donating effect rather than to their steric hindrance in the enzyme-active site as the inhibition was much less significant (56%) in the presence of a double bond on the side chain conjugated with the aromatic ring of DHPA.

2.8.6. Reaction temperature

Temperature changes effect on different parameters including enzyme stability, affinity, and preponderance of the competing reactions [71]. Temperature normally affects lipase activity, and high temperatures usually increase the initial transesterification rate. However, high reaction temperatures deactivate the enzyme due to its protein nature [35]. The optimal temperature used in transesterification reactions is mainly based on considering properties of feedstock, such as melting behavior at different temperatures as well as the reaction system that is with or without solvent. In a solvent-free system, the temperature is maintained high enough to keep the substrates in liquid state [40].

The optimal temperature for the most immobilized lipases ranges from 30–60°C, while it tends to be lower for free lipases. Heat stability of lipase also depends on whether a substrate is present. This is because substrates remove excess water from the immediate vicinity of the enzyme, hence limiting its overall conformational mobility. Ishihara et al. [116] studied the effect of temperature on vanillyl alcohol acylation with nonanoic acid to give vanillyl nonanoate in *n*-hexane solvent medium. The authors found that the optimum temperature for enzymatic acylation was 70°C. Higher temperatures than 70°C lead to decrease the conversion yield due to the deactivation of enzyme at high temperature. The effect of temperature on the synthesis of capsiate analog by lipase-catalyzed esterification of vanillyl alcohol and conjugated linoleic acid (CLA) was presented [47]. The range of temperature tested was from 30 to 60°C. The results demonstrated that the yield increased when the temperature increased from 30 to 50°C. However, when temperature increased to 60°C, there is no increase effect on the yield.

2.8.7. Enzyme concentration

Normally, as the enzyme concentration increases, the reaction equilibrium will be shifted quickly toward the synthesis [117]. However, for economic reasons, it is important to reduce the enzyme loading and the reaction time. In addition, the presence of high enzyme concentration in the reaction medium may increase the probability of its collision with the substrate subsequently enhancing the reaction rate [118]; however, after reaching certain enzyme concentration, the conversion yield was constant. Carrin et al. [117] reported that during the Lipozyme TL-IM-catalyzed acidolysis of sunflower oil with palmitic acid and stearic acid mixture, the extent of palmitic and stearic acids incorporation was enhanced by increasing the amount of enzyme in the reaction; however, when the enzyme concentration was greater than 8% by weight of substrates, there was no significant increase in the esterification yield. The effects of lipase concentration on the synthesis of capsiate analog were depicted in the work of [47].

2.8.8. Agitation speed

In a heterogeneous enzymatic system, it is important to ensure that the rate of substrate diffusion will not limit the rate of the synthesis reaction. The increase in agitation speed may decrease the boundary liquid layer surrounding the porous support, leading to lower diffusion limitations. Lue et al. [102] reported an increase of the enzymatic activity from 108.6 to 156.5 nmol/g/min, when the agitation speed of the system was increased from 0 to 200 rpm. The increase in the enzymatic activity indicated that external diffusion limitations of substrates did occur within the range of agitation applied. Kumari et al. [118] reported that carrying the reaction at the optimum agitation speed can limit the external mass transfer limitations, in the case of immobilized enzymes, where the reactants need to diffuse from the bulk oil to the external surface of the enzyme particles and from there, subsequently to the interior pores of the catalyst. In addition [44] investigated the effect of agitation speed on the conversion yield of phenolic lipids synthesized from flaxseed oil and DHCA; the results have shown that the conversion yield increased significantly from 39 to 62.5% when the agitation speed was increased from 50 to 150 rpm, before it was decreased to 44.8% at agitation speed of 250 rpm. The low conversion yield could be attributed to insufficient agitation rate, a condition in which a hydrophilic layer of glycerol may be formed around the enzyme, limiting hence the mass transfer rate of the oil to the surface of the lipase.

2.8.9. Carbon chain length

The effect of carbon chain length of fatty alcohols on the reaction rate was examined by [119]; the esterification of C4–C18 straight-chain fatty alcohol with dihydrocaffeic acid (DHCA), as a model of phenolic acid, was systematically evaluated. The results indicated that the conversion of DHCA was significantly affected by the number of carbon chain of fatty alcohols. Conversion yield of 95% was achieved within 3 days when hexanol was used as an acyl acceptor, while only 56 and 44% conversions were achieved when 1-butanol and octadecanol were employed, respectively. The conversions of ferulic and caffeic acids under the same conditions were much lower than was that of DHCA. In another by [120], various alkyl cinnamates were formed in high to moderate yield by lipase-catalyzed esterification of cinnamic acid and its analogs with fatty alcohols in vacuo at moderate temperatures in the absence of drying agents and solvents.

Several carboxylic acids of different chain lengths from acetic, propionic, butyric, caproic, and caprylic acids were tested via an enzymatic esterification reaction to produce hexyl ester in *n*-hexane and supercritical carbon dioxide ($SCCO_2$). The reactions were carried out at 40°C, and the amount of enzyme used was 13.8 g/mol alcohol. Substrates were added at equimolar concentrations, with sufficient stirring to avoid external diffusion control. The results in both solvents show that the reaction rate increases with the chain length of the acid, but the final yields were similar.

2.9. Analysis and characterization of phenolic lipids

The structural analyses of phenolic lipids have been carried out using a wide range of various techniques. These mainly include thin-layer chromatography (TLC), high-performance liquid chromatography (HPLC), gas-liquid chromatography (GLC), and liquid chromatography-mass spectrometry (LC-MS). Thin-layer chromatography has been used for initial qualitative analyses of substrates by employing a wide range of organic solvent mixtures. Products from the esterification reactions are characterized and analyzed by TLC using silica gel G-25 plates [10, 47]. The elution solvents used depend on the nature of synthesized compounds. In the study of [10], the elution solvent used was chloroform/methanol mixture (80:20, v/v) and pure chloroform; the plates were visualized under UV light (254 nm), meanwhile, in the work of Hong et al. [47], the elution solvent was n-hexane/diethyl ether/formic acid (160:40:5.5, v/v/v), and the plate was visualized with 0.2% (w/v) 2,7 dichlorofluorescein in methanol solution under UV light.

High-performance liquid chromatography (HPLC) has often been used over other instrumentations and has shown scientifically to be the overall preferred method of choice for quantification and separation of phenolic lipids following synthesis reactions. Phenolic lipids were separated on C18 reverse-phase column using a gradient elution system with UV detection at 280 nm [10]. Gas-liquid chromatography (GLC) analysis has been conducted for determining the fatty acid composition of the synthesized phenolic lipids. REF has reported on the GC analysis of phenolic lipid esters through the use of a CP-Sil CB-MS column linked to an FID detector.

Recent research on phenolic lipids has also made using liquid chromatography-mass spectrometry (LC-MS) that is considered being one of the most powerful techniques used for the characterization of biomolecules due to its high sensitivity and specificity. Generally, its application is oriented toward the specific detection and potential identification of chemicals in the presence of other chemicals (in a complex mixture). LC-MS has been used for the structural characterization of lipids and phenolic lipids [10, 116].

Many lipid systems have been studied by Fourier transform infrared spectroscopy (FTIR) in order to determine several aspects including the degree and the form of unsaturation of the acyl groups as well as their length [121]. The infrared region of the electromagnetic spectrum extends from 14,000 to 50 cm^{-1} and is divided into three areas: the far infrared from 400 to 50 cm^{-1}, the mid infrared region from 4,000 to 400 cm^{-1}, and the near infrared from 14,000 to 4,000 cm^{-1} [122].

2.10. Application of phenolic lipids

Phenolic lipids, compounds which have been known for a century, are more recently being extensively studied not only from the biological but also from the chemical point of view. Phenolic lipids used as novel antioxidants that synthesized enzymatically. These natural antioxidants increased the antioxidant capacity and the oxidative stability of the edible oils [123]. These products can be used as nutraceuticals for their nutritional value and antioxidant capacity as well as natural ingredients for their physicochemical characteristics [39]. Enzymatic esterification of omega-3 PUFAs with vanillyl alcohol leads to protect these compounds from oxidation, and the PUFA-phenolic derivatives prepared confer the combined health beneficial properties of PUFA and the phenolic molecules [10]. Studies of Zhong and Shahidi [38, 39] indicated that antioxidant activity of esters produced from the esterification of EGCG with

PUFA (EPA and DHA) was superior to that of parent compound in retarded of the oxidation of bulk oil and emulsion. The results suggest that these lipophilic derivatives of EGCG could be considered for use in food preservation and health promotion [38].

Recently, most of the observed activities of phenolic lipids were rather nonspecific and resulted from their amphiphilic and phenolic nature. Further investigation on various aspects of biology may open new opportunities to exploit their properties, as, for example, chemo-preventive and antitumor agents, and to develop pharmaceuticals based on phenolic lipid compounds.

Author details

Mohamed Hussein Hamdy Roby

Address all correspondence to: mhr00@fayoum.edu.eg

Faculty of Agriculture, Food Science and Technology Department, Fayoum University, Fayoum, Egypt

References

[1] Gogus, U. and C. Smith, n-3 Omega fatty acids: a review of current knowledge. International Journal of Food Science & Technology, 2010. **45**(3): pp. 417–436.

[2] Ruxton, C.H.S., et al., The health benefits of omega-3 polyunsaturated fatty acids: a review of the evidence. Journal of Human Nutrition and Dietetics, 2004. **17**(5): pp. 449-459.

[3] Prostek, A., et al., The influence of EPA and DHA on markers of inflammation in 3T3-L1 cells at different stages of cellular maturation. Lipids in Health and Disease, 2014. **13:3**.

[4] Pinchuk, I., et al., Evaluation of antioxidants: scope, limitations and relevance of assays. Chemistry and Physics of Lipids, 2012. **165**(6): pp. 638-647.

[5] Acosta-Estrada, B.A., J.A. Gutierrez-Uribe, and S.O. Serna-Saldivar, Bound phenolics in foods, a review. Food Chemistry, 2014. **152**: pp. 46–55.

[6] Laguerre, M., et al., How to boost antioxidants by lipophilization? Biochimie, 2013. **95**(1): pp. 20–26.

[7] Shahidi, F. and Y. Zhong, Novel antioxidants in food quality preservation and health promotion. European Journal of Lipid Science and Technology, 2010. **112**(9): pp. 930–940.

[8] Stasiuk, M. and A. Kozubek, Biological activity of phenolic lipids. Cellular and Molecular Life Sciences, 2010. **67**(6): pp. 841–860.

[9] Bouallagui, Z., et al., Hydroxytyrosol acyl esters: biosynthesis and activities. Applied Biochemistry and Biotechnology, 2011. **163**(5): pp. 592–599.

[10] Mbatia, B., et al., Enzymatic synthesis of lipophilic rutin and vanillyl esters from fish byproducts. Journal of Agricultural and Food Chemistry, 2011. **59**(13): pp. 7021–7027.

[11] Stergiou, P.Y., et al., Advances in lipase-catalyzed esterification reactions. Biotechnology Advances, 2013. **31**(8): pp. 1846–1859.

[12] Liu, L.Y., C. Jin, and Y. Zhang, Lipophilic phenolic compounds (Lipo-PCs): emerging antioxidants applied in lipid systems. RSC Advances, 2014. **4**(6): pp. 2879–2891.

[13] Wang, J.K. and F. Shahidi, Acidolysis of p-coumaric acid with omega-3 oils and antioxidant activity of phenolipid products in in vitro and biological model systems. Journal of Agricultural and Food Chemistry, 2014. **62**(2): pp. 454–461.

[14] Christopher, L.P., K. Hemanathan, and V.P. Zambare, Enzymatic biodiesel: challenges and opportunities. Applied Energy, 2014. **119**: pp. 497–520.

[15] Murray, M., Omega-3 Polyunsaturated fatty acids and their metabolites as inhibitors of mammalian tumorigenesis. Phytochemistry Reviews, 2014. **13**(1): pp. 139–156.

[16] Domenichiello, A.F., et al., Whole body synthesis rates of DHA from alpha-linolenic acid are greater than brain DHA accretion and uptake rates in adult rats. Journal of Lipid Research, 2014. **55**(1): pp. 62–74.

[17] Kuratko, C.N., et al., The relationship of docosahexaenoic acid (DHA) with learning and behavior in healthy children: a review. Nutrients, 2013. **5**(7): pp. 2777–2810.

[18] Igarashi, M., et al., Kinetics of eicosapentaenoic acid in brain, heart and liver of conscious rats fed a high n-3 PUFA containing diet. Prostaglandins, Leukotrienes and Essential Fatty Acids (PLEFA), 2013. **89**(6): pp. 403–412.

[19] Swanson, D., R. Block, and S.A. Mousa, Omega-3 fatty acids EPA and DHA: health benefits throughout life. Advances in Nutrition, 2012. **3**(1): pp. 1–7.

[20] Cunnane, S.C., et al., Docosahexaenoic acid homeostasis, brain aging and Alzheimer's disease: can we reconcile the evidence? Prostaglandins, Leukotrienes, and Essential Fatty Acids, 2013. **88**(1): pp. 61–70.

[21] Arab-Tehrany, E., et al., Beneficial effects and oxidative stability of omega-3 long-chain polyunsaturated fatty acids. Trends in Food Science & Technology, 2012. **25**(1): pp. 24–33.

[22] Wu, F.J., et al., The protective effect of eicosapentaenoic acid-enriched phospholipids from sea cucumber *Cucumaria frondosa* on oxidative stress in PC12 cells and SAMP8 mice. Neurochemistry International, 2014. **64**: pp. 9–17.

[23] Rice-Evans, C.A., N.J. Miller, and G. Paganga, Structure-antioxidant activity relationships of flavonoids and phenolic acids. Free Radical Biology and Medicine, 1996. **20**(7): pp. 933–956.

[24] Velderrain-Rodriguez, G.R., et al., Phenolic compounds: their journey after intake. Food & Function, 2014. **5**(2): pp. 189–197.

[25] Carocho, M. and I. Ferreira, The role of phenolic compounds in the fight against cancer - a review. Anti-Cancer Agents in Medicinal Chemistry, 2013. **13**(8): pp. 1236–1258.

[26] Gulcin, I. and S. Beydemir, Phenolic compounds as antioxidants: carbonic anhydrase isoenzymes inhibitors. Mini-Reviews in Medicinal Chemistry, 2013. **13**(3): pp. 408–430.

[27] Torres de Pinedo, A.T., P., Penalver and J.C., Morales, Synthesis and evaluation of new phenolic-based antioxidants: structure-activity relationship. Food Chemistry, 2007. **103**(1): pp. 55–61.

[28] Torres de Pinedo, A.T., et al., Synthesis of new phenolic fatty acid esters and their evaluation as lipophilic antioxidants in an oil matrix. Food Chemistry, 2007. **105**(2): pp. 657–665.

[29] Silva, F.A.M., et al., Phenolic acids and derivatives: studies on the relationship among structure, radical scavenging activity, and physicochemical parameters. Journal of Agricultural and Food Chemistry, 2000. **48**(6): pp. 2122–2126.

[30] Yang, C.S. and J.M. Landau, Effects of tea consumption on nutrition and health. The Journal of Nutrition, 2000. **130**(10): pp. 2409–2412.

[31] Kuriyama, S., The relation between green tea consumption and cardiovascular disease as evidenced by epidemiological studies. The Journal of Nutrition, 2008. **138**(8): pp. 1548S–1553S.

[32] Sabally, K., et al., Lipase-catalyzed transesterification of trilinolein or trilinolenin with selected phenolic acids. Journal of the American Oil Chemists' Society, 2006. **83**(2): pp. 101–107.

[33] Sabally, K., et al., Lipase-catalyzed synthesis of phenolic lipids from fish liver oil and dihydrocaffeic acid. Biocatalysis and Biotransformation, 2007. **25**(2–4): pp. 211–218.

[34] Sorour, N., et al., Enzymatic synthesis of phenolic lipids in solvent-free medium using flaxseed oil and 3,4-dihydroxyphenyl acetic acid. Process Biochemistry, 2012. **47**(12): pp. 1813–1819.

[35] Stamatis, H., V. Sereti, and F.N. Kolisis, Enzymatic synthesis of hydrophilic and hydrophobic derivatives of natural phenolic acids in organic media. Journal of Molecular Catalysis B: Enzymatic, 2001. **11**(4–6): pp. 323–328.

[36] Qianchun, D., et al., Chemical synthesis of phytosterol esters of polyunsaturated fatty acids with ideal oxidative stability. European Journal of Lipid Science and Technology, 2011. **113**(4): pp. 441–449.

[37] Mainini, F., et al., Synthesis, molecular characterization and preliminary antioxidant activity evaluation of quercetin fatty esters. Journal of the American Oil Chemists' Society, 2013. **90**(11): pp. 1751–1759.

[38] Zhong, Y. and F. Shahidi, Lipophilised epigallocatechin gallate (EGCG) derivatives and their antioxidant potential in food and biological systems. Food Chemistry, 2012. **131**(1): pp. 22–30.

[39] Zhong, Y. and F. Shahidi, Lipophilized epigallocatechin gallate (EGCG) derivatives as novel antioxidants. Journal of Agricultural and Food Chemistry, 2011. **59**(12): pp. 6526–6533.

[40] Torres de Pinedo, A.T., et al., Efficient lipase-catalyzed synthesis of new lipid antioxidants based on a catechol structure. Tetrahedron, 2005. **61**(32): pp. 7654–7660.

[41] Sabally, K., et al., Lipase-catalyzed transesterification of dihydrocaffeic acid with flaxseed oil for the synthesis of phenolic lipids. Journal of Biotechnology, 2006. **127**(1): pp. 167–176.

[42] Karboune, S., R., St-Louis and S., Kermasha, Enzymatic synthesis of structured phenolic lipids by acidolysis of flaxseed oil with selected phenolic acids. Journal of Molecular Catalysis B: Enzymatic, 2008. **52–53**: pp. 96–105.

[43] Zheng, Y., et al., Enzymatic synthesis of novel feruloylated lipids and their evaluation as antioxidants. Journal of the American Oil Chemists' Society, 2010. **87**(3): pp. 305–311.

[44] Sorour, N., et al., Lipase-catalyzed synthesis of structured phenolic lipids in solvent-free system using flaxseed oil and selected phenolic acids as substrates. Journal of Biotechnology, 2012. **158**(3): pp. 128–136.

[45] Tan, Z., et al., Enzymatic synthesis of phytosteryl docosahexaneates and evaluation of their anti-atherogenic effects in apo-E deficient mice. Food Chemistry, 2012. **134**(4): pp. 2097–2104.

[46] Aziz, S. and S. Kermasha, Assessment of the antioxidant capacity and oxidative stability of esterified phenolic lipids in selected edible oils. Journal of Food Science, 2014. pp. **H730 – H737**.

[47] Hong, S.I., et al., Lipase-catalyzed synthesis of capsiate analog using vanillyl alcohol and conjugated linoleic acid under vacuum system. Process Biochemistry, 2012. **47**(12): pp. 2317–2322.

[48] Speranza, P. and G.A. Macedo, Lipase-mediated production of specific lipids with improved biological and physicochemical properties. Process Biochemistry, 2012. **47**(12): pp. 1699–1706.

[49] Tan, Z. and F. Shahidi, Phytosteryl sinapates and vanillates: chemoenzymatic synthesis and antioxidant capacity assessment. Food Chemistry, 2013. **138**(2–3): pp. 1438–1447.

[50] Li, C., et al., Chemical and enzymatic synthesis of a library of 2-phenethyl esters and their sensory attributes. Food Chemistry, 2014. **154**: pp. 205–210.

[51] Reddy, K.K., et al., Evaluation of the antioxidant activity of capsiate analogues in polar, nonpolar, and micellar media. Journal of Agricultural and Food Chemistry, 2011. **59**(2): pp. 564–569.

[52] Houde, A., A. Kademi, and D. Leblanc, Lipases and their industrial applications. Applied Biochemistry and Biotechnology, 2004. **118**(1–3): pp. 155–170.

[53] Sandoval, G., et al., Lipases: an overview, in Lipases and Phospholipases. 2012, Humana Press, Totowa, NJ. pp. 3–30.

[54] Jaeger, K.-E. and T. Eggert, Lipases for biotechnology. Current Opinion in Biotechnology, 2002. **13**(4): pp. 390–397.

[55] Leung, D.Y.C., X. Wu, and M.K.H. Leung, A review on biodiesel production using catalyzed transesterification. Applied Energy, 2010. **87**(4): pp. 1083–1095.

[56] Gog, A., et al., Biodiesel production using enzymatic transesterification – current state and perspectives. Renewable Energy, 2012. **39**(1): pp. 10–16.

[57] Rodrigues, R.C. and M.A.Z.C. Ayub, Effects of the combined use of *Thermomyces lanuginosus* and *Rhizomucor miehei* lipases for the transesterification and hydrolysis of soybean oil. Process Biochemistry, 2011. **46**(3): pp. 682–688.

[58] Barros, M., L.F. Fleuri, and G.A. Macedo, Seed lipases: sources, applications and properties – a review. Brazilian Journal of Chemical Engineering, 2010. **27**: pp. 15–29.

[59] Villeneuve, P., Plant lipases and their applications in oils and fats modification. European Journal of Lipid Science and Technology, 2003. **105**(6): pp. 308–317.

[60] Sandoval, G., et al., Lipases production by solid-state fermentation: the case of *Rhizopus homothallicus* in perlite, in Lipases and Phospholipases, G. Sandoval, Editor. 2012, Humana Press, Totowa, NJ. pp. 227–237.

[61] Sandoval, G., et al., Lipases as biocatalysts for the synthesis of structured lipids, in Lipases and Phospholipases. 2012, Humana Press, Totowa, NJ. pp. 403–433.

[62] Sandoval, G., A. Baldessari, and L. Iglesias, Lipases in green chemistry: acylation and alcoholysis on steroids and nucleosides, in Lipases and Phospholipases. 2012, Humana Press, Totowa, NJ. pp. 457–469

[63] Dhake, K.P., D.D. Thakare, and B.M. Bhanage, Lipase: a potential biocatalyst for the synthesis of valuable flavour and fragrance ester compounds. Flavour and Fragrance Journal, 2013. **28**(2): pp. 71–83.

[64] Jala, R.C.R., et al., Lipases as biocatalysts for the synthesis of structured lipids, Methods in Molecular Biology. 2012. pp. 403–433.

[65] Kamal, M.Z., et al., Lipase in aqueous-polar organic solvents: activity, structure, and stability. Protein Science, 2013. **22**(7): pp. 904–915.

[66] Hampson, J.W. and T.A. Foglia, Effect of moisture content on immobilized lipase-catalyzed triacylglycerol hydrolysis under supercritical carbon dioxide flow in a tubular fixed-bed reactor. Journal of the American Oil Chemists' Society, 1999. **76**(7): pp. 777–781.

[67] Bispo, P., et al., Preparation of triacylglycerols rich in omega-3 fatty acids from sardine oil using a *Rhizomucor miehei* lipase: focus in the EPA/DHA ratio. Applied Biochemistry and Biotechnology, 2014. **172**(4): pp. 1866–1881.

[68] Cerdàn, L.E., et al., Synthesis of polyunsaturated fatty acid-enriched triglycerides by lipase-catalyzed esterification. Journal of the American Oil Chemists' Society, 1998. **75**(10): pp. 1329–1337.

[69] Miranda, K., et al., Optimization of structured diacylglycerols production containing omega-3 fatty acids via enzyme-catalysed glycerolysis of fish oil. European Food Research and Technology, 2013. **236**(3): pp. 435–440.

[70] Zniszczol, A. and K.Z. Walczak, Lipases aided esterification of (2,2-dimethyl-1,3-dioxolan-4-yl) methanol. Letters in Organic Chemistry, 2014. **11**(1): pp. 6–12.

[71] Herbst, D., S. Peper, and B. Niemeyer, Enzyme catalysis in organic solvents: influence of water content, solvent composition and temperature on *Candida rugosa* lipase catalyzed transesterification. Journal of Biotechnology, 2012. **162**(4): pp. 398–403.

[72] Sanchez, D.A., G.M. Tonetto, and M.L. Ferreira, Enzymatic synthesis of 1,3-dicaproyglycerol by esterification of glycerol with capric acid in an organic solvent system. Journal of Molecular Catalysis B-Enzymatic, 2014. **100**: pp. 7–18.

[73] Kocak, D., et al., Acidolysis reaction of terebinth fruit oil with palmitic and caprylic acids to produce low caloric spreadable structured lipid. Journal of the American Oil Chemists' Society, 2013. **90**(7): pp. 999–1009.

[74] Ray, J., et al., Kinetic study of the acidolysis of high oleic sunflower oil with stearic-palmitic acid mixtures catalysed by immobilised *Rhizopus oryzae* lipase. Biochemical Engineering Journal, 2013. **73**: pp. 17–28.

[75] Tarnowska, K., et al., Interesterification of goose fat and rapeseed oil mixture using *Candida rugosa* lipase immobilized in alginate beads. Rivista Italiana Delle Sostanze Grasse, 2013. **90**(2): pp. 95–105.

[76] Ilyasoglu, H., Production of human fat milk analogue containing α-linolenic acid by solvent-free enzymatic interesterification. LWT--Food Science and Technology, 2013. **54**(1): pp. 179–185.

[77] Suo, J.J., et al., The derivation of a chiral substituent code for secondary alcohols and its application to the prediction of enantioselectivity. Journal of Molecular Graphics & Modelling, 2013. **43**: pp. 11–20.

[78] Hermansyah, H., et al., Rigorous kinetic model considering positional specificity of lipase for enzymatic stepwise hydrolysis of triolein in biphasic oil-water system. Bioprocess and Biosystems Engineering, 2010. **33**(7): pp. 787–796.

[79] Mbatia, B., et al., Enzymatic enrichment of omega-3 polyunsaturated fatty acids in Nile perch (*Lates niloticus*) viscera oil. European Journal of Lipid Science and Technology, 2010. **112**(9): pp. 977–984.

[80] Dossat, V.R., D. Combes, and A. Marty, Efficient lipase catalysed production of a lubricant and surfactant formulation using a continuous solvent-free process. Journal of Biotechnology, 2002. **97**(2): pp. 117–124.

[81] Rahman, R.N.Z.R.A., et al., High-yield purification of an organic solvent-tolerant lipase from *Pseudomonas* sp. strain S5. Analytical Biochemistry, 2005. **341**(2): pp. 267–274.

[82] Namal Senanayake, S.P.J. and F. Shahidi, Enzyme-catalyzed synthesis of structured lipids via acidolysis of seal (*Phoca groenlandica*) blubber oil with capric acid. Food Research International, 2002. **35**(8): pp. 745–752.

[83] Lima, V.M.G., et al., Activity and stability of a crude lipase from *Penicillium aurantiogriseum* in aqueous media and organic solvents. Biochemical Engineering Journal, 2004. **18**(1): pp. 65–71.

[84] Kim, I.-H., et al., Lipase-catalyzed acidolysis of perilla oil with caprylic acid to produce structured lipids. Journal of the American Oil Chemists' Society, 2002. **79**(4): pp. 363–367.

[85] Jennings, B. and C. Akoh, Enzymatic modification of triacylglycerols of high eicosapentaenoic and docosahexaenoic acids content to produce structured lipids. Journal of the American Oil Chemists™ Society, 1999. **76**(10): pp. 1133–1137.

[86] Zheng, Y., et al., Dual response surface-optimized process for feruloylated diacylglycerols by selective lipase-catalyzed transesterification in solvent free system. Bioresource Technology, 2009. **100**(12): pp. 2896–2901.

[87] Sun, S., et al., A novel, two consecutive enzyme synthesis of feruloylated monoacyl- and diacyl-glycerols in a solvent-free system. Biotechnology Letters, 2007. **29**(12): pp. 1947–1950.

[88] Chaibakhsh, N., et al., Optimized lipase-catalyzed synthesis of adipate ester in a solvent-free system. Journal of Industrial Microbiology & Biotechnology, 2009. **36**(9): pp. 1149–1155.

[89] Guo, Z. and X. Xu, Lipase-catalyzed glycerolysis of fats and oils in ionic liquids: a further study on the reaction system. Green Chemistry, 2006. **8**(1): pp. 54–62.

[90] Feltes, M.M.C., et al., Enzymatic production of monoacylglycerols (MAG) and diacylglycerols (DAG) from fish oil in a solvent-free system. Journal of the American Oil Chemists' Society, 2012. **89**(6): pp. 1057–1065.

[91] Jin, J.N., S.H. Lee, and S.B. Lee, Enzymatic production of enantiopure ketoprofen in a solvent-free two-phase system. Journal of Molecular Catalysis B: Enzymatic, 2003. **26**(3–6): pp. 209–216.

[92] Sun, S., et al., Solvent-free enzymatic transesterification of ethyl ferulate and monostearin: optimized by response surface methodology. Journal of Biotechnology, 2012. **164**(2): pp. 340–345.

[93] Zheng, Y., et al., Enzymatic synthesis and characterization of novel feruloylated lipids in selected organic media. Journal of Molecular Catalysis B-Enzymatic, 2009. **58**(1–4): pp. 65-71.

[94] Fu, B. and P.T. Vasudevan, Effect of solvent - co-solvent mixtures on lipase-catalyzed transesterification of canola oil. Energy & Fuels, 2010. **24**(9): pp. 4646–4651.

[95] Laane, C., et al., Rules for optimization of biocatalysis in organic solvents. Biotechnology and Bioengineering, 2009. **102**(1): pp. 1–8.

[96] Akoh, C.C., Structured lipids, in Food Lipids: Chemistry, Nutrition and Biotechnology, C.C. Akoh, Editor. 2002, Marcel Dekker: New York. pp. 877–908.

[97] Mateo, C., et al., Improvement of enzyme activity, stability and selectivity via immobilization techniques. Enzyme and Microbial Technology, 2007. **40**(6): pp. 1451–1463.

[98] Plou, F.J., et al., Enzymatic acylation of di- and trisaccharides with fatty acids: choosing the appropriate enzyme, support and solvent. Journal of Biotechnology, 2002. **96**(1): pp. 55–66.

[99] Horchani, H., et al., Enzymatic synthesis of eugenol benzoate by immobilized *Staphylococcus aureus* lipase: optimization using response surface methodology and determination of antioxidant activity. Bioresource Technology, 2010. **101**(8): pp. 2809–2817.

[100] Reshmi, R. and S. Sugunan, Superior activities of lipase immobilized on pure and hydrophobic clay supports: characterization and catalytic activity studies. Journal of Molecular Catalysis B-Enzymatic, 2013. **97**: pp. 36–44.

[101] Farnet, A.M., et al., The importance of water availability in the reaction equilibrium of hydrolases in forest litters from a Mediterranean area: a study on lipases. European Journal of Soil Science, 2013. **64**(5): pp. 661–666.

[102] Lue, B.-M., et al., Lipase-catalyzed esterification of cinnamic acid and oleyl alcohol in organic solvent media. Journal of Chemical Technology & Biotechnology, 2005. **80**(4): pp. 462–468.

[103] Petersson, A.E., P. Adlercreutz, and B. Mattiasson, A water activity control system for enzymatic reactions in organic media. Biotechnology and Bioengineering, 2007. **97**(2): pp. 235–241.

[104] Xiong, J., et al., Lipase-catalyzed transesterification synthesis of geranyl acetate in organic solvents and its kinetics. Food Science and Technology Research, 2014. **20**(2): pp. 207–216.

[105] Chandel, C., A. Kumar, and S.S. Kanwar, Enzymatic synthesis of butyl ferulate by silica-immobilized lipase in a non-aqueous medium. Journal of Biomaterials & Nanobiotechnology 2011. **2**(4): pp. 400–408.

[106] Zhao, H., et al., Lipase catalyzed acidolysis of lard with capric acid in organic solvent. Journal of Food Engineering, 2007. **78**(1): pp. 41–46.

[107] Zhang, H., et al., Production of margarine fats by enzymatic interesterification with silica-granulated *Thermomyces lanuginosa* lipase in a large-scale study. Journal of the American Oil Chemists' Society, 2001. **78**(1): pp. 57–64.

[108] Xu, X. and C.C. Akoh, Enzymatic production of Betapol and other specialty fats, in Lipid Biotechnology, T.M. Kuo and H. Gardner, Editors. 2002, Marcel Dekker: New York. pp. 461–478.

[109] Ronne, T.H., et al., Enzymatic interesterification of butterfat with rapeseed oil in a continuous packed bed reactor. Journal of Agricultural and Food Chemistry, 2005. **53**(14): pp. 5617–5624.

[110] Alim, M.A., et al., Enzymatic transesterification of fractionated rice bran oil with conjugated linoleic acid: optimization by response surface methodology. LWT - Food Science and Technology, 2008. **41**(5): pp. 764–770.

[111] Li, Z., et al., Lipase-catalyzed esterification of conjugated linoleic acid with l-carnitine in solvent-free system and acetonitrile. Bioprocess and Biosystems Engineering, 2007. **30**(5): pp. 331–336.

[112] Mellou, F., et al., Enzymatic esterification of flavonoids with unsaturated fatty acids: effect of the novel esters on vascular endothelial growth factor release from K562 cells. Process Biochemistry, 2006. **41**(9): pp. 2029–2034.

[113] Karboune, S., et al., Lipase-catalyzed biosynthesis of cinnamoylated lipids in a selected organic solvent medium. Journal of Biotechnology, 2005. **119**(3): pp. 281–290.

[114] Yang, T., M.-B. Fruekilde, and X. Xu, Applications of immobilized *Thermomyces lanuginosa* lipase in interesterification. Journal of the American Oil Chemists' Society, 2003. **80**(9): pp. 881–887.

[115] Lee, J.H., et al., Optimized synthesis of 1,3-dioleoyl-2-palmitoylglycerol-rich triacylglycerol via interesterification catalyzed by a lipase from *Thermomyces lanuginosus*. New Biotechnology, 2010. **27**(1): pp. 38–45.

[116] Ishihara, K., et al., One-procedure synthesis of capsiate from capsaicin by lipase-catalyzed dynamic transacylation. World Journal of Microbiology and Biotechnology, 2010. **26**(7): pp. 1337–1340.

[117] Carrin, M.E. and G.H. Crapiste, Enzymatic acidolysis of sunflower oil with a palmitic–stearic acid mixture. Journal of Food Engineering, 2008. **84**(2): pp. 243–249.

[118] Kumari, A., et al., Enzymatic transesterification of Jatropha oil. Biotechnology for Biofuels, 2009. **2**(1): pp. 1.

[119] Yang, Z., Z. Guo, and X. Xu, Enzymatic lipophilisation of phenolic acids through esterification with fatty alcohols in organic solvents. Food Chemistry, 2012. **132**(3): pp. 1311–1315.

[120] Vosmann, K., P. Weitkamp, and N. Weber, Solvent-free lipase-catalyzed preparation of long-chain alkyl phenylpropanoates and phenylpropyl alkanoates. Journal of Agricultural and Food Chemistry, 2006. **54**(8): pp. 2969–2976.

[121] Mitra, S. and P.C. Bhowmik, Spectroscopic study of the sorption of isoxaflutole and its diketonitrile metabolite in dissimilar soils. Bulletin of Environmental Contamination and Toxicology, 2000. **64**(4): pp. 518–525.

[122] Guillén, M.D. and N. Cabo, Infrared spectroscopy in the study of edible oils and fats. Journal of the Science of Food and Agriculture, 1997. **75**(1): pp. 1–11.

[123] Roby, M. H., Allouche, A., Dahdou, L., De Castro, V. C., da Silva, P. H. A., Targino, B. N., Humeau, C. Enzymatic production of bioactive docosahexaenoic acid phenolic ester. Food Chemistry, 2015. 171: pp. 397–404.

Chapter 5

Anthocyanin Pigments: Importance, Sample Preparation and Extraction

Julia Martín, María José Navas,
Ana María Jiménez-Moreno and Agustín G. Asuero

Additional information is available at the end of the chapter

http://dx.doi.org/10.5772/66892

Abstract

Anthocyanins are naturally occuring pigments belonging to the group of flavonoids, a subclass of the polyphenol family. They are common components of the human diet, as they are present in many foods, fruits and vegetables, especially in berries and red wine. There were more studies conducted on effect of processing and storage on changes and stability of colors of anthocyanins in foods such as fruits and also for their use as natural colorants. Besides, the interest on anthocyanins is still growing also owing to their strong antioxidant activity against many chronic diseases, numerous studies about their medicinal, therapeutical and nutritional value were also conducted. There are pieces of evidence regarding the positive association of their intake with healthy biological effects. They act as antioxidants both in the foodstuffs in which they are found and in the organism that take in foods rich in anthocyanins. Many efforts have been carried out to develop new analytical techniques for identification and quantification of anthocyanins in plant materials, as well as their effects *in vivo* and *in vitro*. With this in mind, an overview to general considerations concerning (i) polyphenol and flavonoid history; (ii) chemical structure, color and intake of anthocyanins and (iii) sample preparation and extraction methods are presented in this chapter.

Keywords: anthocyanins, pigments, sample preparation, extraction

1. Introduction

The progress in the last years in the interdisciplinary fields of chemistry, separation science, biology and pharmacy has boosted the natural product chemistry research [1–4], providing a valuable information about many classes of naturally occurring dietary phytochemicals. Among these phytochemicals, polyphenols are the worldwide redox-active secondary metabolites of a phenolic nature [5–12]. The importance of secondary metabolites and their crucial role in many important functional aspects of plant life was recognized for the first time in the second-half of the ninetieth century by Julius Sachs (1873) [13, 14]. Polyphenols are natural compounds occurring in plants [15–18], including foods such as fruits, vegetables, cereals, tea, coffee and wine.

The study mainly focuses on organoleptical properties of polyphenols [19] and their physiological importance to plants [20]. Later, polyphenols are found to be recognized by their nutritional value, since they may help reduce the risk of chronic diseases [1, 21–24] and, in general, have a positive effect on health, because of their free radical scavenging capacity [25–27], which, among other biological effects, increases antioxidant activity and prevents cellular oxidation.

The research on phenolic compounds is mainly focused on anthocyanins [28–29], natural pigments and common components of the human diet (foods, fruits and vegetables, especially in berries and in red wine), As they provide for much of the red to blue pigmentation of flowers and fruits and have physiological functions in vegetative tissues. Their biosynthetic pathway has been the subject of much research and the associated biosynthetic and regulatory genes are well defined. Besides considerable interest in coloring properties of anthocyanins, they have also attracted attention due to their antioxidant activity [30–34] and their property is closely related, to a large extent, with their chemical structure. The pH-dependent ground-state chemistry of anthocyanins is extremely rich. In the past 20 years, the health benefits of anthocyanins have become the subject of intensive research.

Analytical chemistry plays an importance role in this context [35–37] which determines the identity and quantities of anthocyanins in natural products, as well as their effects *in vivo* and *in vitro*.

This chapter intends to reflect the interdisciplinary nature of the research that is currently carried out in anthocyanin pigments through an update of the state-of-art of a series of previously published reviews on this field in the year 2012 [28, 29, 38, 39]. First, general considerations concerning polyphenols with emphasis on their role as secondary metabolites are made. Flavonoid classification, structure, biological activities, databases, intake and dietary sources are also contemplated. Second, aspects of anthocyanin concerning its early history and chemical structure, color and intake are dealt. It should be noted that anthocyanins are readily distinguished from other flavonoids as they undergo rearrangements in response to pH. The antioxidant activity of anthocyanins is depending on their chemical structure. Finally, special attention is paid to analytical methodologies involved in the isolation, determination and characterization of bioactive polyphenols in plants, fruits and vegetables, herbal drugs, medicinal plants and wines, including sample-handling strategies, a feature of analysis often ignored. The use of nonthermal technologies in the assisted extraction of anthocyanins will be covered in future reports.

2. Polyphenols

Plant phenols are among the most abundant and widely represented class of existing plant natural products [40] thanks to the continuous evolution of new genes brought about by gene duplication and mutation and subsequent recruitment and adaptation to specific functions.

The amino acids phenylalanine and tyrosine (derived from the shikimic acid pathway) are the most common origin of polyphenols [41, 42]. Chemically, polyphenols belong to four main classes (**Figure 1**): flavonoids, phenolic acids (hydroxy derivatives of benzoic acid and cinnamic acid, i.e., p-hydroxybenzoic, protocatechuic, vanillic and syringic acids) and their esters (chlorogenic, caftaric, coutaric and fertaric acids), stilbenes (resveratrol, pterostilbene, piceatannol), characterized by a double bond (1,2-diarylethene) connecting the phenolic rings and lignans (pinoresinol, podophyllotoxin, steganacin), characterized by a 1,4-diarylbutane structure, i.e., having 2-phenylpropane units. Flavonoids and phenolic acids account for 60 and 30%, respectively, of the total dietary polyphenols [43].

When the phenolic molecules are not attached to sugar moieties are known as the aglycone form; while those molecules conjugated with one or more sugar residues are called glycosides. Most phenolic compounds are found in nature associated with mono- or polysaccharides or functional derivatives [44] such as esters or methyl esters, varying widely in their hydroxylation pattern and can be glycosylated or acylated.

Despite the fact that most of the literature on phenolic compounds focuses mainly on those found in fruits, vegetables, wines and teas. However, many phenolic compounds present in fruits and vegetables (phenolic acids and flavonoids) are also found in cereals [45, 46].

For decades, this family of compounds has attracted the attention [47] since three British scientists open the door to understand separation, structural elucidation and taxonomical distribution of phenolic compounds. As mentioned above, traditionally, the interest was focused on organoleptical properties of polyphenol such as color, astringency, bitterness, astringency and a range of other tactile or "mouth feel" characteristics [48, 49], as well as their physiological role in plants in the reproduction, pathogenesis and symbiosis. In last decades, polyphenols

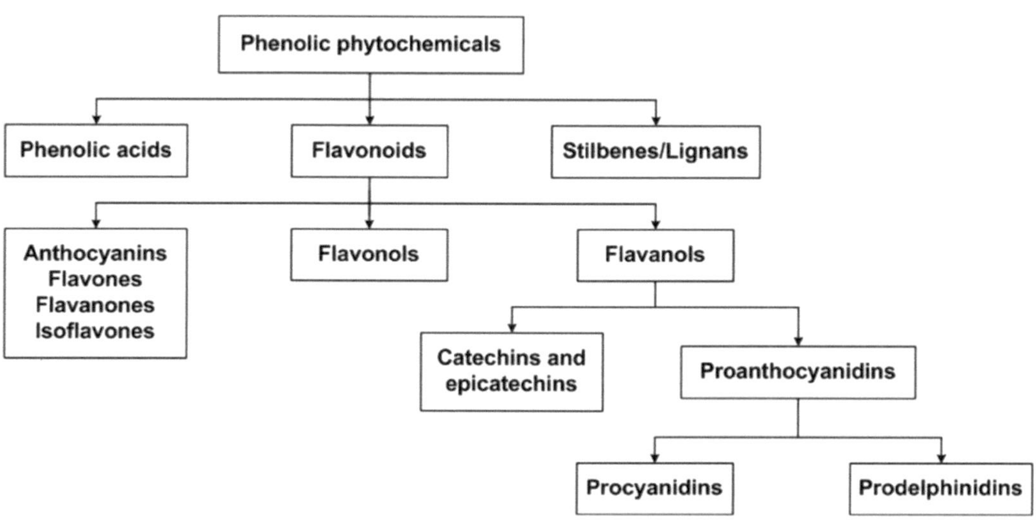

Figure 1. Types of phytochemicals [38].

are increasingly recognized due their nutritional value, since they may help reduce the risk of chronic diseases [50–53]. The capacity of phenolic compounds to trap free radicals depends upon their structure, in particular of the hydrogen atoms of the aromatic group that can be transferred to the free radicals [54, 55] and of the capacity of the aromatic compound to cope with the uncoupling of electrons as a result of the surrounding displacement of the electron-π system [27]. The polyphenols are still gaining attention.

Although the percentage of absorbed natural polyphenols is usually quite low [56], researchers have seen a large quantity of metabolites of polyphenols in the form of simple phenolic acids in the blood. The amount and form in which plant phenolic substances are administered influence greatly the physiological effects connected with their consumption [57]. About 1 g of polyphenols per day is commonly ingested with foods, being the most abundant antioxidant in the diet (about 10 times higher than the intake of vitamin C and 100 times that of vitamin E) [28, 29, 58]. The daily intake of polyphenols is difficult to estimate and depend on several factors. In the literature there are about 1000 peer-reviewed publications [28] concerning the polyphenol content in food. Recently, the construction and application of a database with polyphenols content in foods has facilitated this task [59]. The Institute Nationale de la Recherche Agronomique [60] has developed a new Phenol-Explorer database [59] covering over 60,000 foods useful to epidemiologists, food scientists and food manufacturers. The content of polyphenols of the 100 richest dietary sources can range from 15,000 mg per 100 g in cloves to 10 mg per 100 mL in rose wine [61]. EuroFir [62] is another database to build national food composition in different countries within the framework of the European Food Information Resources Network (i.e., Spanish [63] and Irish [64] databases).

3. Flavonoids

The flavonoids constitute the largest group of polyphenols (of low molecular weight) and are considered to be responsible for the color and taste of many fruits and vegetables [65, 66]. Since first identified in the mid-1800s, more than 9000 flavonoid structures have been described, with formulas, references and biological information [67, 68] and the list is constantly increase. These include over 600 different naturally occurring anthocyanins [69] that are widely distributed among at least 27 families, 73 genera and innumerable species.

Many of these compounds are yellow in color, as the Latin root suggests. They are present particularly in the epidermis of leaves and the skin of fruits [9, 70]. Flavonoids can accumulate in vacuolar compartments, or be secreted, for example, as part of root exudates. Most attracting is the accumulation of flavonoids on the surface of leaves and flowers [70]. Flavonoids play roles in many facets of plant physiology [71] and their influence on the transport of the plant hormone auxin is one of their most important roles.

Around 5000 of the flavonoids studied have antioxidant activity. Because the number of phytochemicals already identified is only a small part of those that exist in nature, there is considerable interest in new methods [37, 72–78] of separation, isolation and characterization of polyphenol structures from foods.

Concerning the chemical structures of flavonoids in which two aromatic rings are present that linked by three carbons in an oxygenated heterocycle, i.e., a flavan (2-phenyl-benzo-γ pyran) nucleus consisting of two benzene rings combined with an oxygen-containing pyran ring, the parent compound bearing a tricyclic (C6-C3-C6) skeleton. The heterocyclic benzopyran ring is known as the C ring, the fused aromatic ring as the A ring and the phenyl constituent as the B ring. The structural differences in each flavonoid family result from variations in the number/substitution pattern of the hydroxyl and methoxy groups [28, 79, 80], as well as different glycosylation patterns and the presence of a C2-C3 double bond in the heterocycle pyran ring.

Compounds are classified according to differences in their heterocycle (C ring): flavonols, flavones (catechins), flavanones, chalcones, dihydrochalcones and dihidroflavonols, anthocyanins and isoflavonoids (isoflavones) [29, 55], varying in the oxidation state (degree of saturation) of the heterocyclic central pyran ring. When unsaturation is present, the geometry of the molecule is planar, as in the case of anthocyanins, flavones and flavonols.

Flavonoids usually occur as glycosides in plants, reflecting a biological strategy increasing their polarity and necessary for storage in the plant cell vacuoles and decreasing their reactivity to interact with macromolecules [81–83]. While flavan-3-ols (catechins and theaflavins) are present in either a free form or as gallic acid esters (e.g., in tea). The glycosidic linkages appear to be important for the absorption of flavonoids [44].

Williams and Grayer [84] have stated: "Flavonoids continue to capture the interest of scientists from many different disciplines because of their structural diversity, biological and ecological significance (e.g. the colored pigments in many flower petals) and health promoting and anticancer properties." Recent advances in genomics, proteomics and metabolomics provide new approaches in the field of flavonoids in plant: protection against oxidative diseases, ability to modulate the activity of various enzymes and interactions with specific receptors are among the most significant health benefits [67, 85].

Flavonoids of dietary significance are present in edible plants in widely varying combinations [86, 87]. Unlike traditional vitamins, flavonoids are not essential for short-term wellbeing. The daily intake is almost at the same level as the sum of other antioxidants, including carotene, vitamin C and vitamin E, can range from several hundred mg up to 1–2 g [88] and although flavonoids are not essential for short-term may exhibit potential health benefits at modest long-term intake [85, 89, 90]. **Table 1** [91–98] shows the dietary sources of flavonoids. Great differences in flavonoid intake and food sources were observed between a large Mediterranean cohort and non-Mediterranean populations (U.S. and Finland as non-Mediterranean countries) [99]. The mean intake for a Spanish population was 313 mg/day [99]. Estimated per capita daily flavonoid intake is 182 and 177 mg for the UK and Ireland, respectively [100].

Flavonoid subclass	Prominent food flavonoids	Typical rich food sources
Anthocyanins	Cyanidin Delphinidin	Bilberries, black and red currants, blueberries, cherries, chokeberries, elderberries, grapes, strawberries, pomegranate
Chalcones	Cinnamon Methylhydroxychalcone	Apples, pears, strawberries, tomatoes, cinnamon
Flavanols	Catechin Epigallocatechin Epigallicatechingallate	Apples, blueberries, grapes, onions, lettuces, red wine, tea, chocolate, apricots, sour cherries, grape juice, mint
Flavanonols	Taxifolin or dihydroquercetin Aromaderin or dihydrokaempherol	Grapes, red onion, açai palm
Flavanones	Hesperetin Naringenin Eriodiclyol	Citrus fruits and juices, peppermint
Flavonols	Quercetine Kaempferol Myricetin	Apples, bean, blueberries, buck wheat, cranberries, endive, leeks, broccoli, lettuces, onions, olive, pepper, tomatoes
Flavones	Apigenin Luteolin	Citrus fruits, celery, parsley, spinachs, rutin, olives, artichoke
Isoflavones	Genistein Daidzein Glyatein	Soybeans, grape seed/skin, chick peas, black beans, green peas
Xanthones	Mangostin	Mango, mangosteen, bark of pear, apples, cherries

Table 1. Dietary sources of flavonoids [28, 91–98].

4. Anthocyanins

Anthocyanins are the largest group of phenolic pigments and the most important group of water-soluble pigments in plants [68, 69, 101–104], responsible for the red, purple and blue colors found in many fruits, vegetables, cereal grains and flowers, being odorless and nearly flavorless and contributing to taste as a moderately astringent sensation. Anthocyanins are almost universally found in higher plants (occurring in about 30 families), but in general anthocyanins seem to be absent [103] in the liverworts, algae and other lower plants, although some of them have been identified in mosses and ferns. **Figure 2** shows a picture of plant species rich in anthocyanins [105–107].

Anthocyanins are found mainly in the skin, except for certain types of red fruit [94], in which they also occur in the flesh (cherries and strawberries).

Anthocyanin biosynthesis was one of the first branches of the general propanoid metabolism [41, 108, 109], for which biosynthetic enzymes and corresponding transcription factors were identified, given the ease of visualization and control of mutants and genetic imbalances.

Anthocyanins have characteristic physicochemical properties that confer them its unique color and stability [30, 32, 110–112]. They are highly reactive molecules and thus sensitive to

Figure 2. Natural sources of anthocyanin pigments [105–107].

degradation reactions. Oxygen, temperature, light, enzymes and pH are among the factors that may affect anthocyanins chemistry and, consequently, their stability and color [113]. In the following, aspects of anthocyanins concerning its chemical structure, color, antioxidant activity and intake are dealt.

4.1. Chemical structure

Chemically, anthocyanins are glycosylated polyhydroxy or polymethoxy derivatives of 2-phenylbenzopyrilium [68, 103, 114], usually with molecular weights ranging from 400 to 1200 (medium-size biomolecules) and containing two benzyl rings (A and B). Anthocyanins usually contain a single glucoside unit, but many anthocyanins contain two, three, or more sugars attached at multiple positions [79], or occurring as oligosaccharide side chains. Intensity and type of the color of anthocyanins is affected by the number of hydroxyl and methoxyl groups [68, 115]: if more hydroxyl groups are present then the color goes toward a more bluish shade; and redness is increased if more methoxyl groups are present. The major anthocyanins are shown [116, 117] in **Figure 3** and **Table 2**.

Anthocyanins mainly exist in glycosidic forms in fruits and with the exception of blueberries, fruits usually contain anthocyanin derived from only one or two of the aglycone bases. Grapes offer a richer anthocyanin profile than many other fruits [118] (red grapes may contain a mixture of more than 20 pigments [85, 119]. Various berries and black currants are the anthocyanin-richest fruits [120–122]. The eggplant is only one common vegetable [123] that contains a high level of anthocyanins.

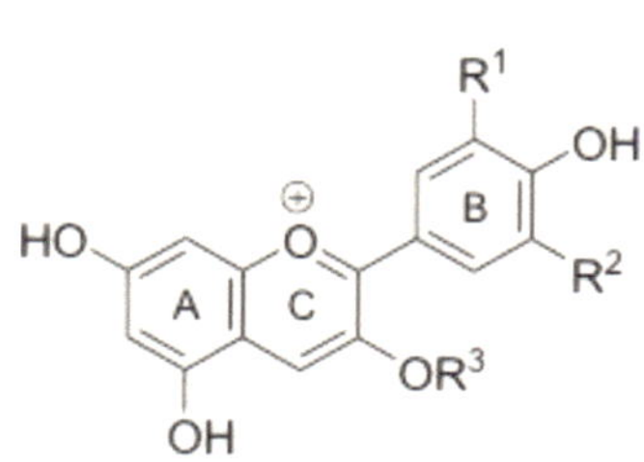

Figure 3. Structure of the major anthocyanins-3-O-glucoside present in fruits [29, 116, 117].

Anthocyanidin	Abbrev.	R_1	R_2	λ_{max} (nm)*		Some of the produced colors
				R_3 = H	R_3 = gluc	
Delphinidin	Dp	OH	OH	546	541	Purple, mauve and blue
Petunidin	Pt	OH	OCH_3	543	540	Purple
Malvidin	Mv	OCH_3	OCH_3	542	538	Purple
Cyanidin	Cy	OH	H	535	530	Magenta and crimson
Peonidin	Pn	OCH_3	H	532	528	Magenta
Pelargonidin	Pg	H	H	520	516	Orange salmon
Source: Ref. [29].						

Table 2. Major anthocyanins-3-O-glucoside present in fruits [29].

The anthocyanins are all amphoteric [32, 34] forming salts with either acids or bases. In addition, anthocyanins occur in plants as salts (indicated by the positive charges on the heterocyclic ring) and their color in plant cells depends mainly upon their mode of combination. The conjugated bonds in their structures (light-conjugated double bonds carrying a positive charge), which absorb light at about 500 nm, are the basis of the bright red, blue and purple color of fruits and vegetables [124] as well as the autumn foliage of deciduous trees. Every color except green has been observed (either natural or synthetic), depending on aspects such as a kind of substituent present in the B-ring, the local pH, the state of aggregation of the anthocyanins, complexation by organic molecules, or, as in the case of blue color [125], complexation by metal cations.

In spite of the increasingly large number of structures, they are derived from only about 30 different anthocyanidins [69, 126], most of them are based on cyanidin (31%), delphinidin (22%), or pelargonidin (18%). The other common anthocyanidins (peonidin, malvidin and petunidin), which contain methoxy group(s) on their B-ring (**Figure 4**), represent together 21% of the isolated anthocyanins. One new methylated anthocyanidin, 7-O-methylcyanidin, five new desoxyanthocyanidins and a novel type of anthocyanidin called pyroanthocyanidin [11, 127] have also been reported. In spite of the structural diversity of anthocyanins, the three nonmethylated anthocyanidins are the most widespread in nature [128], which are present in 80% of pigmented leaves, 69% of fruits and 50% of flowers.

4.2. Antioxidant activity

The relationship between diet and health has been known since ancient times and recent studies demonstrated the relevance of many food components in modulating health [1]. Due to anthocyanin's positive charge (**Figure 3**), the number and arrangement of aromatic hydroxyl groups, the extent of structural conjugation and the presence of electron-donating and electron-withdrawing substituents in the ring structure made anthocyanins very effective donors of hydrogen to highly reactive free radicals (such as superoxide (O_2^-), singlet oxygen ($^1O_2$), peroxide (RCOO·), hydrogen peroxide (H_2O_2) and hydroxyl radical (OH·) and reactive nitrogen species in a terminator reaction) and, thereby preventing further radical formation

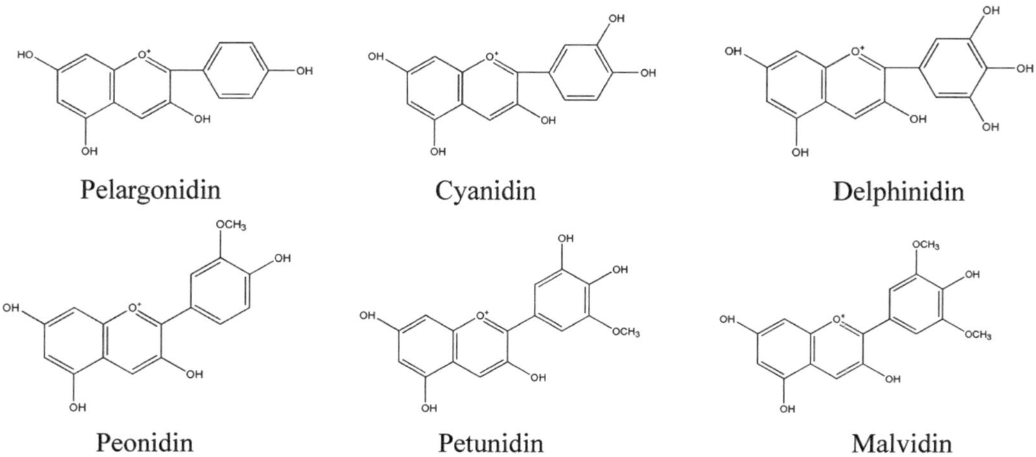

Figure 4. Structure of the main anthocyanins.

[55, 91, 129]. This effect protect cells from oxidative damage, which leads to aging and various diseases [102, 130–133], such as cancer, neurological and cardiovascular diseases, inflammation, diabetes and bacterial infections. The antioxidant capacity of phenolic compounds is also attributed to chelate metal ions involving in the production of free radicals [134], thereby reducing metal-induced peroxidation. Anthocyanin bioavailability has been the subject [135–138] of recent reviews.

Laboratory-based evidence was provided (the potential health benefits of anthocyanins [139]. Consumption of diets rich in natural bioactive components (i.e., fruits and vegetables) as an alternative to pharmaceutical medication has been a subject [104, 140, 141] of considerable interest to researchers. In recent investigations carried out in population-based anthocyanins have been linked to a decrease in the incidence of several diseases, such as diabetes mellitus, cancer and cardiovascular diseases. *In vivo* studies have reported evidence regarding the positive association of their intake with healthy biological effects. However, much work remains to achieve definitive conclusions [139, 142] and the need for additional basic and applied research in this area is evident.

4.3. Color

Anthocyanins are the pigment compounds responsible for pale yellow, orange, red, magenta, violet and blue colors [143]. Carotenoids and betalains confer yellow and red colors [144], although only the families of Caryophyllales (except for Caryophyllaceae and Molluginaceae) produce betalains. Up to now, no plants producing both anthocyanins and betalains [109] have been discovered. Anthocyanins, carotenoids and other pigments contribute to the UV patterns that are visible to insects and serve [145] to signal flowers that are attractive to pollinators. **Figure 5** shows a schematic representation of the biochemical process involving anthocyanins in color plants [146].

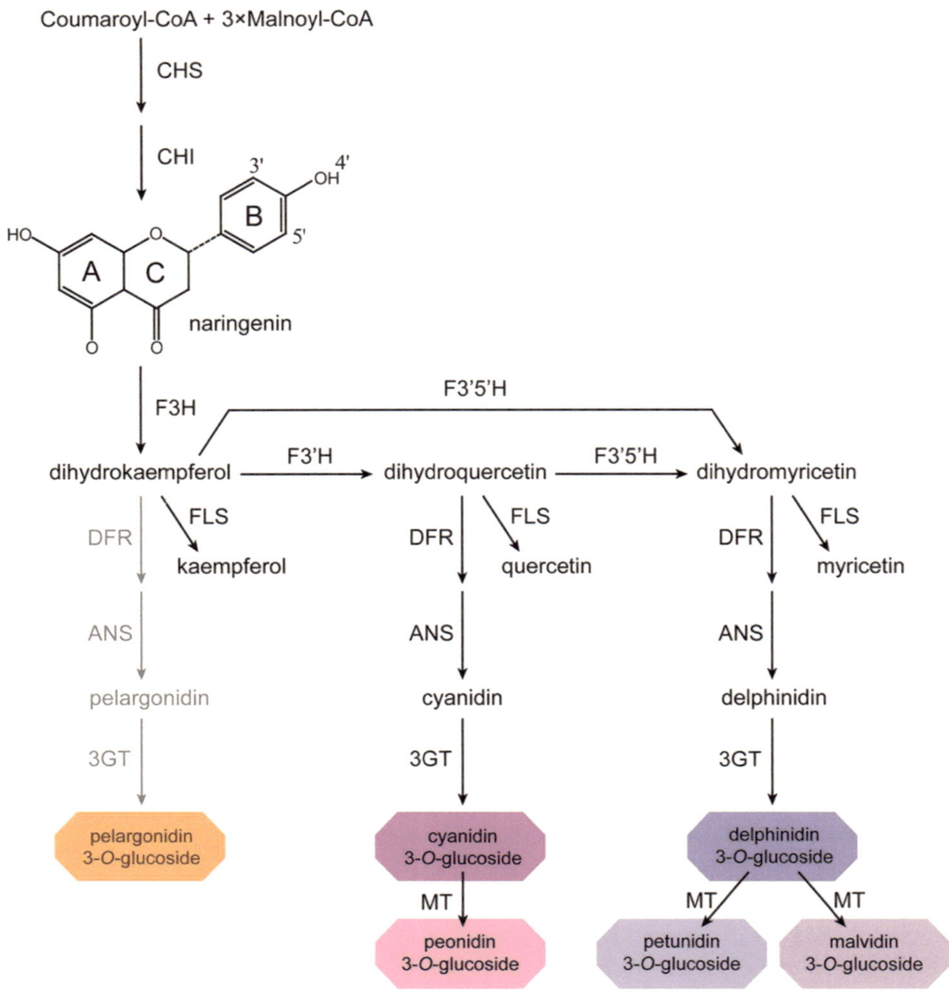

Figure 5. Schematic of the general flavonoid biosynthetic pathway relevant to flower color (ANS: anthocyanidin synthase; CHI: chalcone isomerase; CHS: chalcone synthase; DFR: dihydroflavonol 4-reductase; F3H: flavonoid-3′ hydroxylase; F3′5H: flavonoid-3′5′hydroxylase FLS: flavonole synthase; 3GT: flavonoid 3-O-glucosyltransferase; MT: Malonyl transferase) [146].

In food chemistry, anthocyanins have been studied [68, 147, 148] in relation to changes and stability of colors in foods such as fruits during processing and storage and also for their use as natural colorants. Indeed, many types of anthocyanin food colorants have been developed and are now available to customize the appearance of foods. In horticulture, color conversion of flower pigments has become possible by new findings of anthocyanin research. Creation of flowers in new colors enriches our life; for example, the creation of blue roses [109] is a noteworthy achievement. Genetic engineering is the key technology for converting flower color and it became possible after the discovery of genes involved in anthocyanin biosynthesis and elucidation of their expression mechanisms.

Additionally, color may act as a "fingerprint" of a food product, being related to its flavor and at the same time [149] an estimate of its overall quality. In this sense, special attention is paid to the application of anthocyanin analysis in classification of wine [150]. Anthocyanins can be used as markers to classify wines according to the grape variety [49, 151], although this requires a complex separation with very high chromatographic efficiency, together with advanced statistical methods, especially when dealing with aged red wines, because of the formation [152] of pyranoanthocyanins (formed through the reaction of anthocyanins with small molecules).

4.4. Copigmentation

Copigmentation is almost always a variation toward blueness. This phenomenon induced by the presence of substances [153, 154] that are themselves colorless or only slightly colored in wine has received considerable attention [155, 156]. The basic role of copigments is to protect [125] the colored flavylium cation from the nucleophilic attack of the water molecule. The copigmentation complexes are easily disrupted by dilution returning to the pH-dependent equilibria among the structural forms of anthocyanins. It has been suggested that in acylated anthocyanins, the acyl groups interact with the basic anthocyanin structure, thus avoiding the formation of the hydrated species. This is the basis of characterizing [157] the color due to copigmentation. Copigmentation of flavonoids other than anthocyanins is also possible [158], but it is either a rare or an understudied phenomenon.

4.5. Intake

Anthocyanins are widely ingested by humans, mainly due to consumption [159] of red fruits (like berries and red grapes), vegetables such as red cabbage, red wines, cereals and purple corn. Accurate estimation of anthocyanin contents in foods and daily intake is critical in food science, nutrition and other related research fields. The type and concentration of anthocyanins differ widely among different fruits and vegetables. Intake levels of anthocyanins varies widely with region, season and among individuals with different social, cultural and educational backgrounds. High intake levels of anthocyanins can be achieved with the regular consumption of fruits (blueberries, blackberries, raspberries, strawberries, red grapes and saskatoon berries). Depending on nutritional habits, the daily intake has been estimated in the range [58] from several milligrams to hundreds of milligrams per person, while the consumption of other phytonutrients such as carotenoids, vitamin E and vitamin C are estimated at 5, 12 and 90 mg/day, respectively.

Regular consumers of red wine are likely to have [149] significantly higher intakes. A glass of red wine provides around 115 mg of polyphenols, contributing toward a total intake [58] of phenolic compounds of 1171 mg/person/day. In the United States, an average daily intake of anthocyanins has been estimated [119] at 215 mg during the summer and 180 mg during the winter. Wu et al. [160] estimated that the mean daily intake of anthocyanins is 12.5 mg/person in the United States; such a huge difference of the total anthocyanin daily intake estimation must result from different food intake data. The influence of methodological differences in the assessment, as well as nutritional, social and cultural differences of the investigated populations, may also explain the wide range of anthocyanin consumption estimated by different authors. Anthocyanin intake in the German young shows differences between girls and boys [161], decreasing from young childhood to adolescence.

Up to now, anthocyanins have not been detected in processed food such as canned food, bread, or cereals. Also, although prepared baby food containing blueberries, rich in anthocyanins, are expected to find these compounds, analyses have hardly detected them [162]. In young infants [163] the anthocyanidin intake was found to be zero.

Due to antioxidant and other potential beneficial properties, grapes, various berries, red cabbage and other anthocyanin-rich foods are becoming more popular. Berry extracts are also being commercialized as nutraceuticals and dietary supplements [164] to meet consumer demands.

Currently, there is no recommended intake level of anthocyanins for optimal health or to avoid adverse effects; however, future research and continued consumer interest will undoubtedly present opportunities for pursing dietary guidance recommendations.

5. Sample preparation and extraction of anthocyanins

The presence of phytochemical bioactive compounds in food and dietary supplements poses difficult problems in connection with the optimization of their extraction process and determination. Aspects related to Ref. [165] the complexity of sample matrices, the presence of varying forms of bioactive substances and interaction with other components need to be solved. A number of factors including pH, metal ions, complex formation, light, temperature, enzymes, sugars, oxygen and ascorbic acid exert influence [166] on the stability of anthocyanins. The role of analytical chemistry is vital in this context, e.g., [34, 76, 91, 167–171] promoting advances in separation science.

Most phenolic compounds are made up of only C, H and O, differing in some cases even by only one atom and in many others by constitutional or stereochemical isomers. The identification of species proves thus be difficult because of subtle structural changes, being necessary for this purpose to apply [38] complementary techniques. Anthocyanins may be forming part of complexes, may be present in matrices of a complex nature and may appear in distinct equilibrium forms [30, 32, 34], depending of the pH of the medium. Acid dissociation, rate constant and tautomerization constants are of great importance in the analysis of bioactive compounds and in the interpretation of their mechanisms of action.

Time-consuming processes are involved [172] in the isolation, purification and determination of the structures of anthocyanins, which must be accomplished with care. However, there is no universal sample pretreatment technique applicable to all kind of samples. The primary steps required, sampling, sample preservation and sample preparation, are not always properly documented [173] in the analytical literature.

5.1. Sample preparation

A number of strategies are used [37, 174, 175] for the characterization of phenolic samples in plant materials. In any case extraction techniques and semipreparative isolation methods are usually applied prior to the separation and quantification steps. After the first operations of drying and of powdering the plant material, it follows [176]: (i) a previous extraction step of the plant materials as well as a preliminary consequent purification step; (ii) fractionation of the mixture in order to isolate pure pigments and (iii) the final characterization and identification of pure anthocyanins compounds (**Table 3**).

In order to avoid sample oxidation, thermal degradation, chemical and biochemical changes under mild extraction conditions [177] are recommended and drying, lyophilized, or frozen samples should be used. The deterioration processes of the compounds may be avoided by the

Sample pretreatment	Air-drying, freeze drying
	Milling
	Grinding
	Homogenization
	Filtration
	Centrifugation
Extraction	Direct
	Solid extraction (SE)
	Liquid-liquid extraction (LL)
	Soxhlet extraction
	Microwave-assisted extraction (MAE)
	Ultrasound-assisted extraction (UAE)
	Pressurized fluid extraction (PFE)
	- pressurized liquid extraction (PLE/ASE)
	- subcritical water extraction (SWE)
	- supercritical fluid extraction (SPE)
	Enzyme-assisted extraction (EAE)
	Solid phase microextraction (SPME)
	Membrane extraction (ME)
	High hydrostatic pressure (HHP)
	Electric fields (EF)
Purification	Solid phase extraction (SPE)
	Column chromatography (CC)
	Countercurrent chromatography (CCC)
Analysis	Spectrophotometric assays
	Gas chromatography (GC)
	Liquid chromatography techniques (LC)
	Mass spectrometry (MS)

	Nuclear magnetic resonance (NMR)
	Capillary electrophoresis (CE)
	Thin layer chromatography (TLC)
	Voltammetry
	Others
	Hyphenated techniques
	GC-MS, LC-MS, LC-DAD-ES-MS/MS,
	CE-MS, LC-NMR, others

Table 3. Strategies for preparation and characterization of anthocyanin samples from plant materials [38].

addition of antioxidants compounds, by using inert atmospheres or working in the absence of light. However, no definitive procedure for storage and collection has been established.

To determine either the target analytes (in their various conjugated forms) or the aglycones is an important question [175, 178] to answer. When dealing with plant, food products and biological matrices, the instant conjugates are usually search, whereas in the other instances it is necessary to carry out a preliminary hydrolysis, e.g., an enzymatic or chemical (acidic or alkaline) treatment. Intentional hydrolysis for obtaining the aglycones of some flavonoid or derivatization of some fatty acids to esters is sometimes intentionally incorporated [79] to the extraction process.

Extraction represents an important phase [34, 38, 39, 79, 169] in the isolation, identification and utilization of anthocyanins. Anthocyanins are usually recovered by mean of a solvent extraction procedure. Parameters such as solvent-extraction kind and its composition, liquid-to-solid ratio, extraction time and temperature require [179] proper optimization. The flavylium cation form of anthocyanins is red and stable in a highly acidic medium. Thus, the extraction solution should be enough (slightly) acid to maintain the flavylium cation form [180], but not so much as to cause partial hydrolysis of the acyl moieties in acylated anthocyanins. Protocols of extraction and analysis of plant materials and biological fluids are, however, difficult to accomplish because of [181] the structural diversity of anthocyanins and their susceptibility to heat, pH, metal complexes and copigmentation.

In the same way as flavonoids, in general, anthocyanins possess aromatic rings that contains polar substituent groups (—OH, —C=O, or —OCH_3) and glucosyl residues, which altogether [28, 29] constitute a polar molecule. Flavonoid glycosides are of a more polar nature, whereas aglicones are extracted either with alcohols or alcohol-water mixtures. Cold acidified solvents (polar organic solvents, water) under mild conditions [167, 182] are used for the extraction of anthocyanins. The organic solvent usually used is methanol. However, solvents such as acetone, ethanol, or acetonitrile may be used. These solvents system denature the cell membranes [93] also dissolving and stabilizing the anthocyanins. Acetic acid at about 7% or trifluoroacetic acid at about 3% are usually used; the organic solvent content [183] varying from 50 to 100%

in the mixture. When a mineral acid is used it may assist [174] to the loss of the attached acyl group. Sulfurous water also allows [181] the reduction of organic solvent and cost extraction.

Phytochemical recovery of a good antioxidant from various sources may be achieved by using solvent extraction. The conventional solvent extraction procedure suffers from the drawback of requiring subsequent extraction and cleanup prior analysis. In addition, health and safety risks are associated with the use of large amounts of organic solvents, being on the other hand environmentally unfriendly. A modern trend toward [184]: (1) the use of samples smaller in size, volume, or organic solvent content; (2) an extraction with increasing selectivity or specificity; (3) improved recoveries and reproducibility; (4) greater automation facilities. A variety of modern techniques have been developed for this purpose, including solid phase extraction (SPE), countercurrent chromatography (CCC), microwave-assisted extraction (MAE), supercritical fluid extraction (SFE), pressurized hot water extraction (PHWE) and high hydrostatic pressure extraction (HHP), among others. Selected applications of sample preparation techniques on anthocyanin compounds are listed in **Table 4** [185–213], an extension of applications previously published by the authors in [39]. The applications of other novel nonthermal techniques will be the subject of further study. **Figure 6** shows a schematic representation of a highly separation and purification methodology based on a macroporus polymeric adsorbent for the determination of anthocyanins in bilberry [198].

Solid phase extraction				
Type of matrix/analyte	**Extraction/cleanup technique**	**Sorbent**	**Final analysis**	**References**
Petals of the oboshibana plant (*Commelina communis*)/anthocyanidin 3,5-diglucosides and their related chemicals	Sonication with 10% TFA aqueous solution/SPE	Discovery DPA-6S and DSC-SCX	UPLC-DAD	[185]
Urine, serum and feces/ anthocyanin metabolites	Addition of a preservative (10% w/v ascorbate in 0.5 mM EDTA) prior to SPE	Strata-X SPE cartridges	LC-MS/MS	[186]
Grapes/anthocyanins	SPE	Vinylbenzene-based cartridges	UPLC-DAD	[187]
Red wines/anthocyanins and derived pigments	SPE	Zip-Tip® pipette tips filled with C18 stationary phase	ToF-MS	[188]
Chinese black rice wine/ phenolic constituents	Acidification at pH 2 with HCl/SPE	Oasis HLB	LC-MS/MS	[189]
Wine lees/anthocyanidins, proanthocyanidins and anthocyanins	Liquid phase: SPE with filtration previous. Solid residue: MAE with a mixture 60:40 (v/v) ethanol-water	HySpere C8 EC cartridges (end-capped silica-based octyl phase, particle size 10 μm, 10 × 2 mm i.d.)	LC-MS/MS	[190]
Hybrid grape/monomeric, nonanthocyanins, condensed tannins and anthocyanins	Acidification with HCl 0.01 M and centrifugation/SPE	Oasis HLB	LC-DAD	[191]

Countercurrent chromatographic methods			
Type of matrix/analyte	**Solvent system**	**Elution mode/comment**	**References**
Crude mulberry/ cyanidin-3-glucoside and cyanidin-3-rutinoside	MTBE/n-butanol/ acetonitrile/0.01% TFA (1:3:1:5, v/v)	The upper phase was used as the stationary phase and the lower phase as the mobile phase	[192]
Blue honeysuckle fruits/ cyanidin 3-glucoside	MTBE/n-butanol/acetonitrile/ water/TFA (2:2:1:5:0.01, v/v). HSCCC	Head–tail elution mode with the upper organic phase as stationary phase	[193]
Mulberry fruit/anthocyanins	MTBE, 1-butanol, acetonitrile, water and TFA (10:30:10:50:0.05; %, v/v)	Stationary phase: upper organic phase. Mobile phase: lower aqueous phase. The elution was in "head to tail" mode	[194]
Petals of *Chaenomeles sinensis*/ Anthocyanidins	n-butanol/MTBE/ acetonitrile/0.1% aqueous TFA (0.715:1.0:0.134:1.592, v/v/v/v). HSCCC	The lower phase in the solvent separator was pumped into the mobile phase bottle, the pumps for pumping MTBE, n-butanol, acetonitrile and 0.1% TFA aqueous were set to a certain flow rate, the solvents were mixed and separated into two layers; and after rinsing the solvent separator, the upper phase flowed into the stationary phase bottle	[195]

Adsorbents and eluting agents			
Solution (target compounds)	**Adsorbent**	**Eluting and/or regenerating agent**	**References**
Purple-fleshed potato/ anthocyanins	Amberlite XAD-7HP	75 vol.% acidic (7 vol.% acetic acid) aqueous ethanol	[196]
Blackberries/anthocyanins	Polyamide resin	Deionized water and ethanol (0.1% v/v HCl, pH = 3)	[197]
Bilberry based/anthocyanins	Copolymerization of divinylbenzene and ethylene glycol dimethyl acrylate	Ethanol	[198]
Blueberries/anthocyanins and polyphenols	FPX66 resin	3 bed volumes of 95% ethanol	[199]
Jamun (*Syzygium cumini* L.)/ anthocyanins	Amberlite XAD7HP	Aqueous acidified ethanol (above 40%, v/v)	[200]
Aronia melanocarpa berries/ antioxidant phenolics	XAD 7HP resin	Ethanol-water mixtures	[201]
Strawberry/aroma compounds and Anthocyanins	Cross-linked acrylonitrile-co-divinylbenzene (AN/DVB)	Methanol	[202]
Muscadine (*Vitis rotundifolia*) juice pomace/anthocyanins	FPX-66 resin	Three bed volumes of aqueous ethanol (70%)	[203]
Black carrot/anthocyanins	T-10-coded	1% acetic acid solution and methanol	[204]

Pressurized fluid extraction

Matrix/compounds	Solvent	T (°C)	Pressure/ cycles	Extraction time	Technique	References
Blackberry (*Rubus fruticosus* L.)/anthocyanins	Ethanol:water 50% v/v	100	7.5 Mpa	55 min	UPLC-QToF-MS and UPLC-UV-Vis	[205]
Red grape pomace/ anthocyanins	Ethanol:water 50% v/v	120	90 bar	90 min	LC-Uv-vis	[206]
Solanum stenotomun peel/ antioxidants	Ethanol in water acidified to pH 2.6	80	100 bar	3 h	LC-Uv-vis	[207]
Jabuticaba skins/ anthocyanins	Ethanol	553 K	5 Mpa	31 min	Uv-vis	[208]
Purple-fleshed sweet potato genotypes/ anthocyanins	Acetic acid:methanol:water mixture of 7:75:18% (v/v)	100	1500 psi	20 min	LC-Uv-vis	[209]

Microwave assisted extraction

Sample/analyte	Comment	Result	Technique	References
Blackberry/ anthocyanins	Ethanol was used as solvent. Microwave power of 469 W, a solvent concentration of 52%, a liquid–solid ratio of 25 g mL^{-1} and a time of 4 min	MAE was found to be the most effective method for improving the yield and antioxidant capacity among the other methods tested	Vis	[210]
Blueberry leaves/ anthocyanins	80 mL of solvent consisting of 30% ethanol and 1.5 M citric acid combination in a ratio of 97:3 (v/v)	Time extraction and microwave power level were observed to be significant factors affecting the extraction of phenolic compounds. MAE was the best extraction method among MAE, ultrasonic extraction and 24-h room temperature extraction	Folin-Ciocalteu reagent (spectrophotometric method)	[211]
Black currant marc/ anthocyanins	Maximum yields of anthocyanins were achieved at pH2 with an extraction time of 10 min with a microwave power of 700 W.	A significant reduction of extraction time was achieved using MAE and the final anthocyanin concentration in the solvent phase of MAE increased by 20 % compared to the conventional extraction	LC-DAD	[212]
Mulberry/ anthocyanins	59.6% acidified methanol, 425 W power, 25 (v/w) liquid-to-solid ratio and 132 s time	In comparison with conventional extraction, MAE is more rapid and efficient for extracting anthocyanins from mulberry	LC-MS	[213]

Source: Ampliation from Ref. [39].

Table 4. Selected applications of extraction techniques applied to anthocyanins.

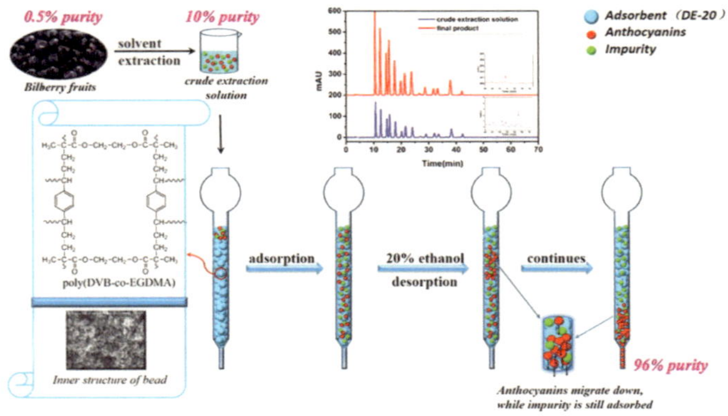

Figure 6. Schematic representation of a highly separation and purification of anthocyanins from bilberry based on macroporus polymeric adsorbent [198].

6. Identification and quantification: a primer

The identification of anthocyanins has a critical role in taxonomic [214] and adulteration [215] studies, besides anthocyanins might replay synthetic days. HPLC, especially in the reversed phase, is the most widely used separation technique. Due to pH-dependent interconversions among various molecular forms of the anthocyanins, a highly acidic mobile phase (pH < 2) is required to ensure that they are maintained predominantly in the flavylium cationic form for maximum chromatographic efficiency. However, even at low pH [31, 32, 35, 110] some interconversion between the anthocyanin flavylium cationic and carbinol pseudobasic forms occurs.

Regarding chromatographic detection techniques for the study of anthocyanins we have [168, 180, 216, 217] diode array detection (DAD) and MS or tandem mass spectrometry (MS/MS) among the most widely used. Spectroscopy is the main technique used due to its simplicity and low cost providing very useful qualitative and quantitative information (anthocyanins have a specific and intense absorbance band in the range of 520–560 nm) [218–220], however the difficulty in obtaining reference compounds and the spectral similarities of the anthocyanins represent important drawbacks.

Various MS instruments, as well as the advances in nuclear magnetic resonance (NMR) have given a fresh impetus to anthocyanin analysis [78, 83, 221]. MS/MS is particularly suited for structure elucidation and compound identification [217, 222, 223] since information pertaining to the aglycone moiety, type and number of sugars and other substituents can be obtained and many of the previously proposed reaction mechanisms for the formation of polymeric anthocyanins and other new pigments have been verified. NMR identification of anthocyanin compounds [78] offers new promising approaches for analysis of complex phenolic mixtures. NMR is based primarily on the analysis of ^{1}H NMR spectra but important structural information can also be provided by ^{13}C NMR [170] and, especially for compounds that have many quaternary carbons, by combining homo and heteronuclear 2D and 3D techniques. However, the relatively high capital costs are still an impediment [218] to their routine use in enforcement laboratories, a fact that must be taken into consideration.

The almost universal distribution of anthocyanins in flowering plants makes them also suitable for chemotaxonomic considerations [224] both at the family and genus level. Differential anthocyanins profiles may be used [164] for the detection and adulteration in specific commodities of berry fruit products. In the last few years, special attention has been paid [225] to the isolation and characterization of compounds that may delay the onset of aging, as occurs with some berry phenolics. The extremely low levels of anthocyanins usually present in biological samples [57, 168, 181, 186] (blood plasma and body tissues) possess further challenges to their identification and quantification, together with the lack of commercially available anthocyanin standards.

7. Final comments

In last decades, polyphenol chemistry has experienced an explosion of knowledge, being anthocyanins one of the most widely studied groups [27, 226, 227], due to its great potential for practical applications in various fields, contributing in addition this to obtain a better understanding of the chemistry of life.

Anthocyanins occur in all plant tissues including leaves, stems, roots, flowers and fruits imparting color. Anthocyanins are responsible for the red, purple and dark blue colors of many fruits and berries [68, 228–230]. Anthocyanins have antioxidant activity [55, 129, 231–233] preventing radical formation. These nontoxic natural pigments have received considerable attention from such as food, pharmaceutical and nutritional industries due to their potential applications in color-processed food and medicines [31, 33, 147, 234] which may replace synthetic dyes.

It was only a few decades ago that anthocyanins were regarded as highly degradable compounds and the research studies mainly were focused on their chemical structures, color stability, use as food constituents and changes in foods during storage. Anthocyanins are now recognized as food constituents with potential health benefits [102, 131] and research related to these properties has markedly progressed at the molecular level. Anthocyanins will continue to attract researchers across various disciplines, including those involved in the creation of new flower varieties with novel colors. Research on the health benefits of anthocyanins will provide information [172] on underlying molecular mechanisms and absorption and metabolism. Moreover, once these benefits are proven in humans, development of foods and dietary supplements in a capsule form [164, 235, 236] can be accelerated to promote the proven functions, i.e. berry extracts are being commercialized as nutraceuticals and as dietary supplements to fulfill consumer demands.

The development of analytical techniques to determine the identity and quantities of anthocyanins in natural products, as well as their effects *in vivo* and *in vitro*, is challenging. Up to date, there is no universal extraction procedure suitable for extraction of all plant phenolics. The choice of an extraction method should maximize pigment recovery [237] with a minimal amount of adjuncts and minimal degradation or alteration of the natural state.

Solvent extraction involving the use of acidic solvents has been the most commonly used method [36, 79] for the recovery of diverse compounds found in flavonoids, including anthocyanins. The traditional solid-liquid or liquid-liquid extraction offers good recovery. Nevertheless, they are often described as laborious, time and solvent consuming and prone to errors. However, in recent years there are trends toward other environmentally and economically friendlier extraction techniques [38, 39, 110, 167, 184, 206, 238] using a smaller amount of (nontoxic) solvents and sample sizes, reducing working time and increasing selectivity, specificity, recovery and potential of automation. MAE, SFE, PLE, or PHWE are among the greener techniques that have experienced a large increase in recent years to extract anthocyanins from plant material and other samples.

Abbreviations

CCC	Countercurrent chromatography
DAD	Diode array detection
EDTA	Ethylenediaminetetraacetic acid
HHP	High hydrostatic pressure extraction
HPLC	High-performance liquid chromatography
HSCCC	High-speed countercurrent chromatography
MAE	Microwave-assisted extraction
MS	Mass spectrometry
MS/MS	Triple quadrupole mass spectrometer
MTBE	Methyl tert-butyl ether
NMR	Nuclear magnetic resonance
PHWE	Pressurized hot water extraction
PLE	Pressurized liquid extraction
QToF	Quadrupole-time of flight
SFE	Supercritical fluid extraction
SPE	Solid phase extraction
TFA	Trifluoroacetic acid
UPLC	Ultra performance liquid chromatography
Uv-Vis	Ultraviolet-visible spectroscopy

http://dx.doi.org/10.5772/66892

Author details

Julia Martín, María José Navas, Ana María Jiménez-Moreno and Agustín G. Asuero*

*Address all correspondence to: asuero@us.es

Department of Analytical Chemistry, Faculty of Pharmacy, The University of Seville, Seville, Spain

References

[1] Zhang H, Tsao R. Dietary polyphenols, oxidative stress and antioxidant and anti-inflammatory effects. Curr. Opin. Food Sci. 2016;**8**:33–42.

[2] Bohn T, McDougall GJ, Alegría A, Alminger M, Arrigoni E, Aura A-M, Brito C, Cilla A, El SN, Karakaya S, Martínez-Cuesta MC, Santos CN. Mind the gap-deficits in our knowledge of aspects impacting the bioavailability of phytochemicals and their metabolites. A position paper focusing on carotenoids and polyphenols. Mol. Nutr. Food Res. 2015;**59**:1307–1323.

[3] Ziyatdinova GK, Budnikov HC. Natural phenolic antioxidants in bioanalytical chemistry: state of the art and prospects of development. Russ. Chem. Rev. 2015;**84**(2):194–224.

[4] Li RH. Health-promoting components of fruits and vegetables in the diet. Adv. Nutr. 2013;**4**(3):384S–392S.

[5] Ferrazzano GF, Amato I, Ingenito A, Zarrelli A, Pinto G, Pollio A. Plant polyphenols and their anti-carcinogenic properties: a review. Molecules. 2011;**16**(2):1486–1507.

[6] Burton-Freeman B. Postprandial metabolic events and fruit derived phenolics: a review. Br. J. Nutr. 2010;**104**(S3):S1–S14.

[7] Chong MF-F, Macdonald R, Lovergrove JA. Fruit polyphenols and CVD risk: a review of human intervention studies. Br. J. Nutr. 2010;**104**(S3):S28–S39.

[8] Crozier A, del Rio D, Clifford MN. Bioavailability of dietary flavonoids and phenolic compounds. Mol. Aspects Med. 2010;**31**(6):446–467.

[9] Crozier A, Jaganath IB, Clifford MN. Dietary phenolics: chemistry, bioavailability and effects on health. Nat. Prod. Rep. 2009;**26**(8):1001–1043.

[10] Martin KR, Appel CL. Polyphenols as dietary supplements. A double-edged sword. Nutr. Diet. Suppl. 2010;**2**:1–12.

[11] Pereira DM, Valentao P, Pereira JA, Andrade PB. Phenolics: from chemistry to biology. Molecules. 2009;**14**(6):2202–2211.

[12] Whiting DA. Natural phenolic compounds 1900-2000: a bird's eye view of a century's chemistry. Nat. Prod. Rep. 2001;**18**(6):583–606.

[13] Neumann KH, Kumar A, Imani J. Plant cell and tissue culture – a tool in biotechnology: basic and application. In Principles and Practice. Chapter 10, Berlin: Springer-Verlag, 2009. pp. 181–182.

[14] Hartmann T. From waste products to ecochemicals: fifty years research of plant secondary metabolism. Phytochem. 2007;**68**(22–24):2831–2846.

[15] Shahidi F, Ambigaipalan P. Phenolic and polyphenolics in foods, beverages and spices: antioxidant activity and health effects – a review. J. Funct. Foods 2015;**18**:820–827.

[16] Lima GPP, Vianello F, Correa CR, da Silva-Campos RA, Borguini, MG. Polyphenols in fruits and vegetables and its effect on human health. Food Nutr. Sci. 2014;**5**:1065–1082.

[17] Cicerale S, Lucas L, Keast R. Biological activities of phenolic compounds present in virgin olive oil. Int. J. Molec. Sci. 2010;**11**(2):458–479.

[18] González-Gallego J, García-Mediavilla V, Sánchez-Campos S, Tuñon MJ. Fruit polyphenols, immunity and inflammation. Br. J. Nutr. 2010;**104**(S3):S15–S27.

[19] Cheynier V. Polyphenols in foods are more complex than often thought. Am. J. Clin. Nutr. 2005;**81**:223S–229S.

[20] Grotewold E. The genetics and biochemistry of floral pigments. Annu. Rev. Plant Biol. 2006;**57**:761–780.

[21] Joseph SV, Edirisinghe I, Burton-Freeman B. Fruit polyphenols: a review of anti-inflammatory effects in humans. Crit. Rev. Food Sci. Nutr. 2016;**56**(3):419–444.

[22] Khurana S, Venkataraman K, Hollingsworth A, Piche M, Tai TC. Polyphenols: benefits to the cardiovascular system in health and in aging. Nutrients. 2013;**5**:3779–3827.

[23] Peluso I, Miglio C, Morabito G, Ioannone F, Serafini M. Flavonoids and immune function in human: a systematic review. Crit. Rev. Food Sci. Nutr. 2015;**55**(3):383–395.

[24] Umeno A, Horie M, Murotomi K, Nakajima Y, Yoshida Y. Antioxidative and antidiabetic effects of natural polyphenols and isoflavones. Molecules. 2016;**21**:708–723.

[25] Halake K, Birajdar M, Lee J. Structural implications of polyphenolic antioxidants. J. Ind. Eng. Chem. 2016;**35**:1–7.

[26] Martin SN, Barros L, Ferreira ICR. In vivo antioxidant activity of phenolic compounds: facts and gaps. Trends Food Sci. Technol. 2016;**28**:1–12.

[27] Quideau S, Deffieux D, Douat-Casassus C, Pouységu L. Plant polyphenols: chemical properties, biological activities and synthesis. Angew. Chem. Int. Ed. 2011;**50**(3):586–621.

[28] Martín Bueno J, Ramos-Escudero F, Sáez-Plaza P, Muñoz AM, Navas MJ, Asuero AG. Analysis and antioxidant capacity of anthocyanin pigments. Part I: general considerations concerning polyphenols and flavonoids. Crit. Rev. Anal. Chem. 2012;**42**:102–125.

[29] Martín Bueno J, Sáez-Plaza P, Ramos-Escudero F, Jímenez AM, Fett R, Asuero AG. Analysis and antioxidant capacity of anthocyanin pigments. Part II: chemical structure, color and intake of anthocyanins. Crit. Rev. Anal. Chem. 2012;**42**:126–151.

[30] Pina F, Oliveira J, de Freitas V. Anthocyanins and derivatives are more than flavylium cations. Tetrahedron. 2015;**71**:3107–3114.

[31] Pina F. Chemical applications of anthocyanins and related compounds. A source of bio-inspiration. J. Agric. Food Chem. 2014;**62**(39):6885–6897.

[32] Pina F, Melo MJ, Laia CAT, Parola J, Lima JC. Chemistry and applications of flavylium compounds: a hanfull of colours. Chem. Soc. Rev. 2012;**41**:896–908.

[33] Santos-Buelga C, Mateus N, de Freytas V. Anthocyanins, plant pigments and beyond. J. Agric. Food Chem. 2014;**62**:6879–6884.

[34] Silva VO, Freitas AA, Maçanita AL, Quina FH. Chemistry and photochemistry of natural plant pigments: the anthocyanins. J. Phys. Org. Chem. 2016;**29**:594–599. DOI: 10.1002/poc.3534.

[35] Alberts P, Stander MA, de Villiers A. Advanced ultra high pressure liquid chromatography-tandem mass spectrometric method for the screening of red wine anthocyanins and derived pigments. J. Chromatogr. A. 2012;**1235**:92–102.

[36] Giusti MM, Jing P. Analysis of anthocyanins, In Food Colorants: Chemical and Functional Properties, Socaciu, C. (Ed.), Chapter 6. Boca Raton, FL: Taylor and Francis, 2008. pp. 479–506.

[37] Gleichenghagen M, Schieber A. Current challenges in polyphenol analytical chemistry. Curr. Opin. Food Sci. 2016;**7**:43–49.

[38] Navas MJ, Jiménez-Moreno AM, Martín Bueno J, Sáez-Plaza P, Asuero AG. Analysis and antioxidant capacity of anthocyanin pigments. Part III: an introduction to sample preparation and extraction. Crit. Rev. Anal. Chem. 2012;**42**:284–312.

[39] Navas MJ, Jiménez-Moreno AM, Martín Bueno J, Sáez-Plaza P, Asuero AG. Analysis and antioxidant capacity of anthocyanin pigments. Part IV: extraction of anthocyanins. Crit. Rev. Anal. Chem. 2012;**42**:313–342.

[40] Boudet AM. Evolution and current status of research in phenolic compounds. Phytochem. 2007;**68**(22–24):2722–2735.

[41] Vogt T. Phenylpropanoid biosynthesis. Mol. Plant. 2010;**3**(1):2–20.

[42] Pourcel L, Routaboul JM, Cheynier V, Lepiniec L, Debeaujon I. Flavonoid oxidation in plants: from biochemical properties to physiological functions. Trends Plant Sci. 2007;**12**(1):29–36.

[43] Nichenametla SN, Taruscio TG, Barney DL, Exon JH. A review of the effects and mechanisms of polyphenolic in cancer. Crit. Rev. Food Sci. Nutr. 2006;**46**(2):161–183.

[44] Kren V. Glycoside vs aglycon: the role of glycosidic residue in biological activity. In Glycoscience. Fraser-Reid B, Tasuka K, Thiem J. (Eds.), Berlin: Springer-Verlag, 2008.

[45] Masisi K, Beta T, Moghadasian MH. Antioxidant properties of diverse cereal grains: a review on in vitro and in vivo studies. Food Chem. 2016;**196**:90–97.

[46] Dykes L, Rooney LW. Phenolic compounds in cereal grains and their health benefits. Cereal Foods World. 2007;**52**(3):105–111.

[47] Robins RJ, Gibbons S, Marston A. The fiftieth anniversary meeting of the phytochemical society of Europe: Churchill College Cambridge. 11–14 April 2007: Highlights in the Evolution of Phytochemistry. Phytochem. 2007;**68**(22–24):2699–2704.

[48] Mateus N. The chemistry of wine flavors – polyphenols. (in Portuguese) Revista Real Academia Gallega de Ciencias. 2009;**28**:5–22.

[49] Zimman A, Waterhouse AL, Kennedy JA. Short history of red wine color. ACS Symposium Series No. 1886, Chapter 1. Washington, DC: ACS, 2004. pp. 1–6.

[50] Valdes L, Cuervo A, Salazar N, Ruas-Madiedo P, Gueimonde M, Gonzalez S. The relationship between phenolic compounds from diet and microbiota: impact on human health. Food Funct. 2015;**6**:2424–2439.

[51] Ciriminna R, Meneguzzo F, Fidalgo A, Ilharco LM, Pagliaro M. Extraction, benefits and valorization of olive polyphenols. Eur. J. Lipid Sci. Technol. 2016;**118**:503–511.

[52] Rigacci S, Stefani M. Nutraceutical properties of olive oil polyphenols. An itinerary from cultured cells through animal models to humans. Int. J. Mol. Sci. 2016;**17**:843–871.

[53] Sarkar D, Shelty K. Metabolic stimulation of plant phenolics for food preservation and health. Ann. Rev. Food Sci. Technol. 2014;**5**:395–413.

[54] Croft KD. Dietary polyphenols: antioxidants or not?. Arch. Biochem. Biophys. 2016;**595**:120–124.

[55] Leopoldini M, Russo N, Toscano M. The molecular basis of working mechanism of natural polyphenol antioxidants. Food Chem. 2011;**125**(2):288–306.

[56] Forester SC, Waterhouse AL. Metabolites are key to understanding health effects of wine polyphenolics. J. Nutr. 2009;**139**(9):1824S–1832S.

[57] Vacek J, Ulrichova J, Klejdus B, Simanek V. Analytical methods and strategies in the study of plant polyphenols in clinical samples. Anal. Methods. 2010;**2**(6):604–613.

[58] Scalbert A, Manach C, Morand C, Rémésy C, Jiménez L. Dietary polyphenols and the prevention of diseases. Crit. Rev. Food Sci. Nutr. 2005;**45**(4):287–306.

[59] Neveu V, Perez-Jimenez J, Vos F, Crespy V, du Chaffaut L, Du Mennen L, Knox C, Eisner R, Cruz J, Wishart D, Scalbert A. Phenol-Explorer: an online comprehensive data base on polyphenol contents in foods. Database: J. Biol. Databases Curat (Oxford). 2010; doi:10.1093/database/bap024.

[60] Scalbert A. INRA, Phenol-Explorer: Database on Polyphenol Content in Foods; http://www.phenol-explorer.eu

[61] Perez Jimenez J, Neveu V, Vos F, Scalbert A. Identification of the 100 richest dietary sources of polyphenols: an application of the Phenol-Explorer Database. Eur. J. Clin. Nutr. 2010;**64**(Suppl. 3):S112–S120.

[62] Finglas PM, Denny AR. EuroFIR approach to develop tools to assure the quality of the data compilation process. Food Chem. 2009;**113**(3):759–767

[63] Burgos MAM, Martinez-Victoria I, Milá R, Farrán A, Farre R, Ros G, Yago MD, Audi N, Santana C, Millán MBL, López SR, Mañas M, Martinez-Victoria E. Building a unified Spanish food database according to EuroFIR specifications. Food Chem. 2009; **113**(3):784–788.

[64] Black CJ, Ireland J, Moller A, Roe M, Walton J, Flynn A, Finglas PM, Kiely M. Development of an on-line Irish food composition database for nutrients. J. Food Compos. Anal. 2011;**24**(7):1017–1023.

[65] Hoensch HP, Oertel R. The value of flavonoids for the human nutrition: short review and perspectives. Clin. Nutr. Exp. 2015;**3**:8–14.

[66] Grotewold, E. (Ed.) The Science of Flavonoids. New York: Springer, 2006.

[67] Fraga CG. (Ed.) Plant Phenolics and Human Health: Biochemistry, Nutrition and Pharmacology. New York: Wiley, 2010.

[68] He J, Giusti M. Anthocyanins: natural colorants with health-promoting properties. Ann. Rev. Food Sci. Technol. 2010;**1**(1):163–187.

[69] Andersen OM, Jordheim M. The anthocyanins. In Flavonoids: Chemistry, Biochemistry and Applications. Andersen OM, Markham KR. (Eds.), Boca Raton, FL: Taylor and Francis, 2006. pp. 471–551.

[70] Onylagha JC, Grotewold E. The biology and structural distribution of surface flavonoids. Recent Res. Develop. Plant Sci. 2004;**2**:53–71.

[71] Buer CS, Imin N, Djordjevic A. Flavonoids: new roles for old molecules. J. Integr. Plant Biol. 2010;**52**(1):96–111.

[72] Casassa LF, Haberlson JF. Extraction, evolution and sensory impact of phenolic compounds during red wine maceration. Ann. Rev. Food Sci. Technol. 2014;**5**:83–109.

[73] de Villiers A, Venter P, Pasch H. Recent advances and trends in the liquid-chromatography-mass spectrometry analysis of flavonoids. J. Chromatogr. A. 2016;**1430**:16–78.

[74] Khoddami A, Wilkes, MA, Roberts TH. Techniques for analysis of plant phenolic compounds. Molecules. 2013;**18**:2328–2375.

[75] Lorrain B, Ky I, Pechamat L, Teissedre P-L. Evolution of analysis of polyphenols from grapes, wines and extracts. Molecules. 2013;**18**:1076–1100.

[76] Motilva M-J, Serra A, Macià A. Analysis of food polyphenols by ultra high-performance liquid chromatography coupled to mass spectrometry: an overview. J. Chromatogr. A. 2013;**1292**:66–82.

[77] Szultka M, Papaj K, Rusin A, Szeja W, Buszewski B. Determination of flavonoids and their metabolites by chromatographic techniques. Trends Anal. Chem. 2013;**47**:47–67.

[78] Ziaullah HP, Rupasinghe V. Application of NMR spectroscopy in plant polyphenols associated with human health. Chapter 1. In Applications of NMR Spectroscopy. Bentham Science Publishers. Ltd. Published by Elsevier Inc. Sharjah, United Arab Emirates.Vol. 2, 2015. pp. 3–92.

[79] Andersen OM, Markham KR. Flavonoids, Chemistry, Biochemistry and Applications. Boca Raton, FL: Taylor and Francis, 2006.

[80] Kumar S, Pandey AK. Chemistry and biological activities of flavonoids: an overview. Sci. World J. 2013; **2013**. http://dx.doi.org/10.1155/2013/162750.

[81] Xiao J, Capanoglu E, Jassbi AR, Miron A. Advance on the flavonoid C-glycosides and health benefits. Crit. Rev. Food Sci. Nutr. 2016;**56**:S29–S45.

[82] Agati G, Azzarello E, Pollastri S, Tattini M. Flavonoids as antioxidants in plants: location and functional significance. Plant Sci. 2012;**196**:67–76.

[83] Corradini E, Foglia, P, Giansanti, P, Gubbiotti, R, Samperi, R, Lagana, A. Flavonoids: chemical properties and analytical methodologies of identification and quantitation in foods and plants. Nat. Prod. Res. 2011;**25**(5):469–495.

[84] Williams CA, Grayer RJ. Anthocyanins and other flavonoids. Nat. Prod. Rep. 2004;**21**(4):539–573.

[85] Jaganatah IB, Crozier A. Dietary Flavonoids and phenolic compounds, In Plant Phenolics and Human Health: Biochemistry, Nutrition and Pharmacology. Fraga CG. (Ed.) New York: Wiley, 2010. pp. 1–50.

[86] Bhagwat S, Haytowitz DB, Wasswa-Kintu SI, Holden JM. USDA develop is a database for flavonoids to assess dietary intakes. Procedia Food Sci. 2013;**2**:81–96.

[87] Biesaga M, Pyrzynska K. Analytical procedures for determination of quercetin and its glycosides in plant material. Crit. Rev. Anal. Chem. 2009;**39**(2)95–107.

[88] Pan M-H, Lai C-S, Ho C-T. Anti-inflammatory activity of natural dietary flavonoids. Food Funct. 2010;**1**(1):15–31.

[89] Ahmad A, Kaleem M, Ahmed Z, Shafia H. Therapeutic potential of flavonoids and their mechanism of action against microbial and viral infections. A review. Food Res. Int. 2015;**77**:221–235.

[90] Chen L, Teng HT, Xie Z, Cao H, Cheang WS, Skalicka-Woniak K, Georgiev MI, Xiao J. Modifications of dietary flavonoids towards improved bioactivity: an update on structure-activity relationship. Crit. Rev. Food Sci. Technol. 2016; dx.doi.org/10.1080/10408398.2016.1196334

[91] Ignat I, Volf I, Popa VI. A critical review of methods for characterisation of polyphenolic compounds in fruits and vegetables. Food Chem. 2011;**126**(4):1821–1835.

[92] Han X, Shen T, Lou H. Dietary polyphenols and their biological significance. Int. J. Mol. Sci. 2007;**8**(9):950–988.

[93] Naczk M, Shahidi F. Phenolics in cereals, fruits and vegetables: occurrence, extraction and analysis. J. Pharm. Biomed. Anal. 2006;**41**(5):1523–1542.

[94] Manach C, Donovan JL. Pharmacokinetics and metabolism of dietary flavonoids in humans. Free Radical Res. 2004;**38**(8):771–785.

[95] Beecher GR. Overview of dietary flavonoids: nomenclature, occurrence and intake. J. Nutr. 2003;**133**(10):3248S–3254S.

[96] Aherne SA, O'Brien NM. Dietary flavonols: chemistry, food content and metabolism. Nutrition. 2002;**18**(1):75–81.

[97] Duthie GG, Duthie SJ, Kyle JAM. Plant polyphenols in cancer and heart disease: implications as nutritional antioxidants. Nutr. Res. Rev. 2000;**13**(1):79–106.

[98] Rice-Evans C, Spencer JP, Schroeter H, Rechner AR. Bioavailability of flavonoids and potential bioactive forms in vivo. Drug Metabol. Drug Interact. 2000;**17**(1–4):291–310.

[99] Zamora-Ros R, Andres-Lacueva C, Lamuela-Reventos RM, Berenguer T, Jakszyn P, Barricate A, Ardanaz E, Amiano R, Dorronsoro M, Larrañaga N, Martinez C, Sanchez MJ, Navarro C, Chiriaque MD, Tormo MJ, Quiros JR, Gonzalez CA. Estimation of dietary sources and flavonoids intake in a Spanish adult population (EPIC-Spain). J. Am. Dietetic Assoc. 2010;**110**(3):390–398.

[100] Beking K, Vieira A. An assessment of dietary flavonoids intake in the UK and Ireland. Int. J. Food Sci. Nutr. 2011;**62**(1):17–19.

[101] Fang J. Classification of fruits based on anthocyanin types and relevance to their health effects. Nutrition. 2015;**31**:1301–1306.

[102] Wallace TC, Giusti MM. Anthocyanins in Health and Disease. New York, NY: CRC Press, 2014.

[103] Gould K, Davies K, Winefields C. (Eds.) Anthocyanins, Biosynthesis, Functions and Applications. New York: Springer-Verlag, 2009.

[104] Motohashi N, Sakagami H. Functionability of anthocyanins as alternative medicine. Topics Heterocyclic Chem. 2008;**15**:1–48.

[105] Borghesi E, González-Miret ML, Escudero-Gilete ML, Malorgio F, Heredia FJ, Meléndez-Martínez AJ. Effects of salinity stress on carotenoids, anthocyanins and color of diverse tomato genotypes. J. Agric. Food Chem. 2011;**59**:11676–11682.

[106] Borghesi E, Ferrante A, Gordillo B, Rodríguez-Pulido FJ, Cocetta G, Trivellini A, Mensuali-Sodi A, Malorgio F, Heredia FJ. Comparative physiology during ripening in tomato rich-anthocyanins fruits. Plant. Growth. Regul. 2016;**80**:207–214. doi 10.1007/s10725-016-0158-y.

[107] Fernández-Lara R, Gordillo B, Rodríguez-Pulido FJ, González-Miret ML, del Villar-Martínez AA, Dávila-Ortiz G, Heredia FJ. Assessment of the differences in the phenolic composition and color characteristics of new strawberry (Fragaria x ananassa Duch.) cultivars by HPLC–MS and imaging tristimulus colorimetry. Food Res. Int. 2015;**76**:645–653.

[108] He F, Mu L, Yan G-L, Liang N-N, Pan Q-H, Wang J, Reeves MJ, Duan C-Q. Biosynthesis of anthocyanins and their regulation in colored grapes. Molecules. 2010;**15**(12):9057–9091.

[109] Tanaka Y, Brugliera F, Kalc G, Senior M, Dyson B, Nakamura N, Katsumoto Y, Chandler S. Flower color modification by engineering of the flavonoid biosynthetic pathway: practical perspectives. Biosci. Biotechnol. Biochem. 2010;**74**(9):1760–1769.

[110] Silva S, Costa Em, Calhau C, Morais Rm, Pintado Me. Anthocyanin extraction from plant tissues: a review. Crit. Rev. Food Sci. Nutr. 2015; doi:10.1080/10408398.2015.1087963.

[111] Petrov V, Slavcheva S, Stanimirov S, Pina F. Origin of the metastable statility in flavylium multistate systems. J. Phys. Chem. A. 2015;**119**(12):2908–2918.

[112] Pina F. Anthocyanins and related compounds. Detecting the change of regime between rate control by hydration or by tautomerization. Dyes Pigments. 2014;**102**:308–314.

[113] Prior RL, Wu X. Anthocyanins: structural characteristics that result in unique metabolic patterns and biological activities. Free Radical Res. 2006;**40**(10):1014–1028.

[114] Zhao C-L, Yu Y-Q, Chen Z-J, Wen G-S, Wei F-G, Zheng Q, Wang C_D, Xiao X-L. Stability-increasing effects of anthocyanin glycosyl acylation. Food Chem. 2017;**214**:119–128.

[115] Heredia FJ, Francia-Aricha EM, Rivas-Gonzalo JC, Vicario IM, Santos-Buelga C. Chromatic characterization of anthocyanins from red grapes I. pH effect. Food Chem. 1998;**63**(4):491–498.

[116] Kahkonen MP, Heinonen M. Antioxidant activity of anthocyanins and their aglycons. J. Agric. Food Chem. 2003;**51**(3):628–633.

[117] Giusti MM, Rodríguez-Saona LE, Griffin D, Wrolstad RE. Electrospray and tandem mass spectroscopy as tools for anthocyanin characterization. J. Food Agric. Chem. 1999;**47**(11):4657–4664.

[118] McCallum JL, Yang R, Young JC, Strommer JN, Tsao R. Improved high performance liquid chromatographic separation of anthocyanin compounds from grapes using a novel mixed-mode ion exchange reverse-phase column. J. Chromatogr. A. 2007;**1148**(1):38–45.

[119] Clifford MN. Anthocyanins – nature, occurrence and dietary burden. J. Sci. Food Agric. 2000;**80**(7):1063–1072.

[120] Manganaris GA, Goulas V, Vicente AR, Terry LA. Berry antioxidants: small fruits providing large benefits. J. Sci. Food Agric. 2014;**94**(5):815–833.

[121] Jimenez-Garcia S, Guevara-Gonzalez RG, Miranda-Lopez R, Feregrino-Perez A, Torres-Pacheco I, Vazauez-Cruz MA. Functional properties and quality characteristics of bioactive compounds in berries: biochemistry, biotechnology and genomics. Food Res. Int. 2013;**54**(1):1195–1207.

[122] Norberto S, Silva S, Meireles M, Faria A, Pintado M, Calhau, C. Blueberry anthocyanins in health promotion: a metabolic overview. J. Funct. Foods. 2013;**5**(4):1518–1528.

[123] Horbowicz M, Kosson R, Grzesink A, Debski H. Anthocyanin of fruits and vegetables – their occurrence, analysis and role in human nutrition. Veg. Crops Res. Bullet. 2008;**68**:5–22.

[124] Wang LS, Stoner GD. Anthocyanins and their role in cancer prevention. Cancer Lett. 2008;**269**(2):281–290.

[125] Quina FH, Moreira PF, Vautier-Giongo C, Rettori D, Rodrigues RF, Freitas AA, Silva PF, Maçanita AL. Photochemistry of anthocyanins and their biological role in plant tissues. Pure Appl. Chem. 2009;**81**(9):1687–1694.

[126] Tsao R. Chemistry and biochemistry of dietary polyphenols. Nutrients. 2010;**2**:1231–1246.

[127] Andersen OM. Recent advances in the field of anthocyanins—main focus on structures. In Recent Advances in Polyphenol Research. Daayf F, Lattanzio V. (Eds.), Blackwell Publishing: Chichester, UK, Vol. 1, 2001. pp. 167–201.

[128] Ghosh D, Konishi T. Anthocyanin and anthocyanin-rich extracts: role in diabetes and eye function. Asia Pac. J. Clin. Nutr. 2007;**16**(2):200–208.

[129] Ali HM, Almagribi W, Al-Rashidi MN. Antiradical and reluctant activities of anthocyanidins and anthocyanins, structure-activity relationship and synthesis. Food Chem. 2016;**194**:1275–1282.

[130] Cassidy A, Bertoia M, Chiuve S, Flint A, Forman J, Rimm EB. Habitual intake of anthocyanins and flavanones and risk of cardiovascular disease in men. Am. J. Clin. Nutr. 2016;**104**:587–594.

[131] Liobikas J, Skemiene K, Trumbeckaite S, Borutaite V. Anthocyanins in cardioprotection: a path through mitochondria. Pharmacol. Res. 2016; **113**:808–815.

[132] Faria A, Fernandes I, Norberto S, Mateus N, Clhau C. Interplay between anthocyanins and gut microbiota. J. Agric. Food Chem. 2014;**62**:6898–6902.

[133] Chahar MK, Sharma N, Dobhal MP, Yoshi YC. Flavonoids: a versatile source of anticancer drugs. Pharmacogn. Rev. 2011;**5**(9):1–12.

[134] Liu Z-Q. Chemical methods to evaluate antioxidant ability. Chem. Rev. 2010;**110**(10): 5675–5691.

[135] Lila MA, Burton-Freeman B, Grace M, Kalt W. Unraveling anthocyanin bioavailability for human health. Annu. Rev Food Sci. Technol. 2016;**7**:375–393.

[136] Fang J. Bioavailability of anthocyanins. Drug Metab. Rev. 2014;**46**(4):508–520.

[137] Fernandes I, Faria A, Calhau C, de Freitas V, Mateus N. Bioavailability of anthocyanins and derivatives. J. Funct. Foods. 2014;**7**:54–66.

[138] Yang M, Si K, Wo S, Ok C. Food matrix affecting anthocyanin bioavailability: review. Curr. Med. Chem. 2011;**18**(2):291–300.

[139] Wallace TC. Anthocyanins in cardiovascular disease. Adv. Nutr. 2011;**2**:1–7.

[140] Rajasekaran A, Sivagnanam G, Xavier R. Nutraceuticals as therapeutic agents: a review. Res. J. Pharm. Technol. 2008;**1**(4):328–340.

[141] Wu X, Prior R. Anthocyanins: analysis and distribution in selected medicinal plants. Chem. Lett. 2008;**12**(1):9–22.

[142] Wallace TC, Giusti MM. Anthocyanins. Adv. Nutr. 2015;**6**:20–22.

[143] Mlodzinski E. Survey of plant pigments: molecular and environmental determinants of plant colors. Acta Biol. Crac. Ser. Bot. 2009;**51**(1):7–16.

[144] Tanaka Y, Ohmiya A. Seeing is believing: engineering anthocyanin and carotenoid biosynthetic pathway. Curr. Opin. Biotechnol. 2008;**19**(2):190–197.

[145] Mortensen A. Caratenoids and other pigments as natural colorants. Pure Appl. Chem. 2006;**78**(8):1477–1491.

[146] Sun W, Liang L, Meng X, Li Y, Gao F, Liu X, Wang S, Gao X, Wang L. Biochemical and molecular characterization of a flavonoid 3-O-glycosyltransferase responsible for anthocyanins and flavonols biosynthesis in Freesia hybrida. Front. Plant Sci. 2016;**7**:1–14.

[147] Wrolstad RE, Culver CA. Alternatives to those artificial FD&C food colorants. Annu. Rev. Food Sci. Technol. 2012;**3**:59–77.

[148] Ngamwonglumlert L, Devahastin S, Chiewchan N. Natural colorants: pigment stability and extraction yield enhancement via utilization of appropriate pretreatment and extraction methods. Crit. Rev. Anal. Chem. 2015; DOI: 10.1080/10408398.2015.1109498

[149] Mateus N, de Freitas V. Anthocyanins as food colorants. In Anthocyanins, Biosynthesis, Functions and Applications. Gould K, Davies K, Winifield C. (Eds.), Chapter 9, New York: Springer, 2009. pp. 283–304.

[150] Hartmanova L, Ranc V, Papouskova B, Bednar P, Havlicek V, Lemr K. Fast profiling of anthocyanins in wine by desorption nano-electrospray ionization mass spectrometry. J. Chromatogr. A. 2010;**1217**(25):4223–4228.

[151] Pisano PL, Silva, MF, Olivieri AC. Anthocyanins as markers for the classification of Argentinean wines according to botanical and geographical origin. Chemometric modelling of liquid chromatography-mass spectrometry data. Food Chem. 2015;**175**:174–180.

[152] Rentzsch M, Schwarz M, Winterhalter P. Pyranoanthocyanins – an overview on structures, occurrence and pathways of formation. Trends Food Sci. Technol. 2007;**18**(10):526–534.

[153] Trouillas P, Sancho-Garcia, JC, de Freitas V, Gierschner J, Otyepka M, Dangles O. Stabilizing and modulating color by copigmentation: insights from theory and experiment. Chem. Rev. 2016;**116**(9):4937–4982.

[154] Lambert SG, Asenstorfer RE, Williamson NM, Iland PG, Graham P. Copigmentation between malvidin-3-glucoside and some wine constituents and its importance to colour expression in red wine. Food Chem. 2011;**125**(1):106–115.

[155] Bimpilas A, Panagopoulou M, Tsimogiannis D, Oreopoulou V. Anthocyanin copigmentation and color of wine: the effect of naturally obtained hydroxycinnamic acids as cofactors. Food Chem. 2016;**197**(Pt A):39–46.

[156] Heras-Roger J, Diaz-Romero C, Darias-Martin J. What gives a wine its strong red color? Main correlations affecting copigmentation. J. Agric. Food Chem. 2016;**64**(34):6567–6574.

[157] González-Manzano S, Santos-Buelga C, Dueñas M, Rivas-Gonzalo JC, Escribano-Bailón T. Colour implications of self-association processes of wine anthocyanins. Eur. Food Res. Technol. 2008;**226**(3):483–490.

[158] Davies K. Plant pigments and their manipulation. Ann. Plant Rev. Vol. 14. Boca Raton, FL: CRC Press, 2004.

[159] Galvano F, La Fauci L, Lazzarino G, Fogliano V, Ritieni A, Ciappellano S, Battistini NC, Tavazzi B, Galvano G. Cyanidins: metabolism and biological properties. J. Nutr. Biochem. 2004;**15**(1):2–11.

[160] Wu X, Beecher GR, Holden JM, Haytowitz DB, Gebhardt SE, Prior RL. Concentration of anthocyanins in common foods in the United States and estimation of normal consumption. J. Agric. Food Chem. 2006;**54**(11):4069–4075.

[161] Drossard C, Bolzenius K, Kunz C, Kersting M. Anthocyanins in the diet of children and adolescents: intake, sources and trends. Eur. J. Nutr. 2013;**52**(2):667–76.

[162] United States Department of Agriculture. Agricultural Research Service.http://www.ars.usda.gov/

[163] Drossard C, Fröhling B, Dietrich H, Kersting M. Anthocyanin analysis in banana fruit. A mistake. Am. J. Clin. Nutr. 2011;**93**:865–866.

[164] Jing P, Giusti MM. Contribution of berry anthocyanins to their chemopreventive properties. In Berries and Cancer Protection. Stoner GD, Seeram, NP. (Eds.), Chapter 1. New York: Springer, 2011. pp. 1–38.

[165] Luthria DL, Natarajan SS. Influence of sample preparation of the assay of isoflavones. Planta Medica. 2009;**75**(7):704–710.

[166] Cavalcanti RN, Santos DT, Meireles MAA. Non-thermal stabilization mechanisms of anthocyanins in model and food systems: an overview. Food Res. Int. 2011;**44**(2):409–509.

[167] Baiano A, Del Nobile MA. Antioxidant compounds from vegetable matrices: biosynthesis, occurrence and extraction systems. Crit. Rev. Food Sci. Nutr. 2016;**56**(12):2053–2068.

[168] Jing P, Giusti MM. Analysis of anthocyanins in biological samples. In Anthocyanins in Health and Disease. Wallace TC, Giusti MM. (Eds.), Chapter 4. Boca Raton, FL: Taylor and Francis, 2014. pp. 115–139.

[169] Fontana AR, Antoniolli A, Bottini R. Grape pomace as sustainable source of bioactive compounds: extraction, characterization and biotechnological applications of phenolics. J. Agric. Food Chem. 2013;**51**:8987–9003.

[170] Kirby CW, Wu T, Tsao R, McCallum JL. Isolation and structural characterization of unusual pyranoanthocyanins and related anthocyanins from Staghorn sumac (*Rhus typhina* L.) via UPLC-ESI-MS, (1)H, (13)C and 2D NMR spectroscopy. Phytochem. 2013;**94**:284–293.

[171] Li D, Wang P, Luo Y, Zhao M, Chen F. Health benefits of anthocyanins and molecular mechanisms: update from recent decade. Crit. Rev. Food Sci. Nutr. 2015; DOI: 10.1080/10408398.2015.1030064.

[172] Jordheim M. Isolation, identification and properties of pyranoanthocyanins and anthocyanin forms. PhD Thesis. Department of Chemistry, University of Bergen, 2007.

[173] Luthria DL. Significance of sample preparation in developing analytical methodologies for accurate estimation of bioactive compounds in functional foods. J. Sci. Food Agric. 2006;**86**(14):2266–2272.

[174] Dai J, Mumper RJ. Plant phenolics: extraction, analysis and their antioxidant and anticancer properties. Molecules. 2010;**15**(10):7313–7352.

[175] Rijke E de, Out P, Niessen WMA, Ariese F, Gooier C, Brinkman UA. Analytical separation and detection methods for flavonoids. J. Chromatogr. A. 2006;**1112**(1–2):31–63.

[176] Strack D, Heilemann J, Wray V, Dirks H. Structures and accumulation patterns of soluble and insoluble phenolics from Norway spruce needles. Phytochem. 1989;**28**(8):2071–2078.

[177] Tura D, Robards K. Sample handling strategies for the determination of biophenols in food and plants. J. Chromatogr. A. 2002;**975**(1):71–93.

[178] Rijke E de. Trace-level Determination of Flavonoids and Their Conjugates. Application to Plants of the Leguminosae Family. Amsterdam: Academisch Proefschrift, Wrije University, 2005.

[179] Zou TB, Wang M, Gan RY, Ling WH. Optimization of ultrasound-assisted extraction of anthocyanins from mulberry, using response surface methodology. Int. J. Mol. Sci. 2011;**12**(5):3006–3017.

[180] Welch CR, Wu Q, Simon JE. Recent advances in anthocyanins analysis and characterization. Curr. Anal. Chem. 2008;**4**(2):75–101.

[181] Mazza G, Cacace JE, Kay DC. Methods of analysis for anthocyanins in plants and biological fluids. J. AOAC Int. 2004;**87**(1):129–145.

[182] Shi J, Nawaz H, Pohorli J, Mittal G, Kakuda Y, Kiang Y. Extraction of polyphenolics from plant material for functional foods-engineering and technology. Food Rev. Int. 2005;**21**(1):139–166.

[183] Nicoue EE, Savards S, Belkacemi K. Anthocyanins in wild blueberries of Quebec: extraction and identification. J. Agric. Food Chem. 2007;**55**(14):5625–5635.

[184] Liu EH, Qi LW, Cao J, Li, P. Li, CY, Peng YB. Advances of modern chromatographic and electrophoretic methods in separation and analysis of flavonoids. Molecules. 2008;**13**(10):2521–2544.

[185] Asada T, Koi Y, Arakawa R, Zhu F, Sadaoka M, Tamura H. Isolation techniques for anthocyanidin 3,5-diglucosides and their related chemicals using supramolecules technique and two solid-phase extraction cartridges. J. Chromatogr. A. 2014;**1351**:21–28.

[186] de Ferrars RM, Czank C, Saha S, Needs PW, Zhang Q, Saki Raheem K, Botting NP, Kroon PA, Kay CD. Methods for isolating, identifying and quantifying anthocyanin metabolites in clinical samples. Anal. Chem. 2014;**86**:10052–10058.

[187] Ferreiro-González M, Carrera C, Ruiz-Rodríguez A, Barbero GF, Ayuso J, Palma M, Barroso CG. A new solid phase extraction for the determination of anthocyanins in grapes. Molecules. 2014;**19**:21398–21410.

[188] Ivanova Petropulos V, Dörnyei A, Stefova M, Stafilov T, Vojnoski B, Márk L, Hermosín-Gutiérrez I, Kilár K. Application of a novel small-scale sample cleanup procedure prior to MALDI-TOF-MS for rapid pigment fingerprinting of red wines. Food Anal. Methods. 2014;**7**:820–827.

[189] Wang Y, Liu Y, Xiao C, Liu L, Hao M, Wang J, Liu X. simultaneous determination of 15 phenolic constituents of Chinese black rice wine by HPLC-MS/MS with SPE. J. Food Sci. 2014;**79**:1100–1105.

[190] Delgado de la Torre MP, Ferreiro-Vera C, Priego-Capote F, Luque de Castro MD. Anthocyanidins, proanthocyanidins and anthocyanins profiling in wine lees by solid-phase extraction–liquid chromatography coupled to electrospray ionization tand-em mass spectrometry with data-dependent methods. J. Agric. Food Chem. 2013;**61**: 12539–12548.

[191] Manns DC, Mansfield AK. A core-shell column approach to a comprehensive high-performance liquid chromatography phenolic analysis of *Vitis vinifera* L. and interspecific hybrid grape juices, wines and other matrices following either solid phase extraction or direct injection. J. Chromatogr. A. 2012;**1251**:111–121.

[192] Choi S-J, Choi J, Uk Lee C, Hee Yoon S, Kyung Bae S, Chin Y-W, Kim J, Dong Yoon K. Rapid separation of cyanidin-3-glucoside and cyanidin-3-rutinoside from crude mulberry extract using high-performance countercurrent chromatography and establishment of a volumetric scale-up process. J. Sep. Sci. 2015;**38**:1828–1836.

[193] Chen L, Xin X, Lan R, Yuan Q, Wang X, Li Y. Isolation of cyanidin 3-glucoside from blue honeysuckle fruits by high-speed counter-current chromatography. Food Chem. 2014;**152**:386–390.

[194] Sheng F, Wang Y, Zhao X, Tian N, Hu H, Li P. Separation and identification of anthocyanin extracted from mulberry fruit and the pigment binding properties toward human serum albumin. J. Agric. Food Chem. 2014;**62**:6813–6819.

[195] Li S, Guo L, Liu C, Fu Z, Zhang Y. Combination of supercritical fluid extraction with counter-current chromatography to isolate anthocyanidins from the petals of *Chaenomeles sinensis* based on mathematical calculations. J. Sep. Sci. 2013;**36**: 3517–3526.

[196] Heinone J, Farahmandazad H, Vuorinen A, Kallio H, Yang B, Sainio T. Extraction and purification of anthocyanins from purple-fleshed potato. Food Bioprod. Process. 2016;**99**:136–146.

[197] Wen Y, Chen H, Zhou X, Deng Q, Zhao C, Gong X. A polyamide resin based method for adsorption of anthocyanins from blackberries. New J. Chem. 2016;**40**:3773–3780.

[198] Yao L, Zhang N, Wang C, Wang C. Highly selective separation and purification of anthocyanins from bilberry based on a macroporous polymeric adsorbent. J. Agric. Food Chem. 2015;**63**:3543–3550.

[199] Buran TJ, Sandhu AK, Li Z, Rock CR, Yang WW, Gu L. Adsorption/desorption characteristics and separation of anthocyanins and polyphenols from blueberries using macroporous adsorbent resins. J. Food Eng. 2014;**128**:167–173.

[200] Jampani C, Naik A, Raghavarao KSMS. Purification of anthocyanins from jamun (*Syzygium cumini* L.) employing adsorption. Sep. Purific. Technol. 2014;**125**:170–178.

[201] D'Alessandro LG, Vauchel P, Przybylski R, Chataigné G, Nikov I, Dimitrov K. Integrated process extraction-adsorption for selective recovery of antioxidant phenolics from *Aronia melanocarpa* berries. Sep. Purific. Technol. 2013;**120**:92–101.

[202] Ozgen S, Sarıoglu K. Synthesis and characterization of acrylonitrile-co- divinylbenzene (AN/DVB) polymeric resins for the isolation of aroma compounds and anthocyanins from strawberry. Food Bioprocess. Technol. 2013;**6**:2884–2894.

[203] Sandhu AK, Gu L. Adsorption/desorption characteristics and separation of anthocyanins from muscadine (*Vitis rotundifolia*) juice pomace by use of macroporous adsorbent resins. J. Agric. Food Chem. 2013;**61**:1441–1448.

[204] Sarıoglu K, Erdem T, Ozgen S. Synthesis and characterization of lipophilic polymeric adsorbent resins for recovery of anthocyanins from black carrot. Food Anal. Methods. 2013;**6**:1621–1628.

[205] Da Fonseca Machado AP, Pasquel-Reátegui JL, Fernández Barbero G, Martínez J. Pressurized liquid extraction of bioactive compounds from blackberry (*Rubus fruticosus* L.) residues: a comparison with conventional methods. Food Res. Int. 2015;**77**:675–683.

[206] Otero-Pareja MJ, Casas L, Fernández-Ponce MT, Mantell C, Martínez de la Ossa EJ. Green extraction of antioxidants from different varieties of red grape pomace. Molecules. 2015;**20**:9686–9702.

[207] Casas Cardoso L, Mantell Serrano C, Torrez Quintero E, Pereyra López C, Medrano Antezana R, Martínez de la Ossa EJ. High pressure extraction of antioxidants from Solanum stenotomun peel. Molecules. 2013;**18**:3137–3151.

[208] Santos DT, Veggi PC, Meireles MA. Optimization and economic evaluation of pressurized liquid extraction of phenolic compounds from jabuticaba skins. J. Food Eng. 2012;**108**:444–452.

[209] Truong VD, Hu Z, Thompson RL, Yencho GC, Pecota KV. Pressurized liquid extraction and quantification of anthocyanins in purple-fleshed sweet potato genotypes. J. Food Compos. Anal. 2012;**26**:96–103.

[210] Wen Y, Chen H, Zhou X, Deng Q, Zhao Y, Zhao C, Gong X. Optimization of the microwave-assisted extraction and antioxidant activities of anthocyanins from blackberry using a response surface methodology. RSC Adv. 2015;**5**:19686–19695.

[211] Routray W, Orsat V. MAE of phenolic compounds from blueberry leaves and comparison with other extraction methods. Ind. Crop. Prod. 2014;**58**:36–45.

[212] Pap N, Beszédes S, Pongrácz E, Myllykoski L, Gábor M, Gyimes E, Hodúr C, Keiski RL. Microwave-assisted extraction of anthocyanins from black currant marc. Food Bioprocess Tech. 2013;**6**:2666–2674.

[213] Zou T, Wang D, Guo H, Zhu Y, Luo X, Liu F, Ling W. Optimization of microwave-assisted extraction of anthocyanins from mulberry and identification of anthocyanins in extract using HPLC-ESI-MS. J. Food Sci. 2012;**71**:46–50.

[214] Picariello G, Ferranti P, Garro G, Manganiello G, Chianese L, Coppola R, Addeo F. Profiling of anthocyanins for the taxonomic assessment of ancient purebred *V. Vinifera* red grape varieties. Food Chem. 2014;**146**:15–22.

[215] Gardana C, Ciappellano S, Marinoni L, Fachechi C, Simonetti P. Bilberry adulteration: identification and chemical profiling of anthocaynins by different analytical methods. J. Agric. Food Chem. 2014;**45**:10998–11004.

[216] Barnes JS, Nguyen HP, Shen S, Schug KA. General method for the extraction of blueberryanthocyanins and identification using high performance liquid chromatography-electrospray ionization-ion trap-time of flight-mass spectrometry. J. Chromatogr. A 2009;**16**(23):4728–4735.

[217] Jin H, Liu Y, Yang F, Wang J, Fu D, Zhang X, Peng X, Liang X. Characterization of anthocyanins in wild Lycium tuthenium Murray by HPLC-DAD/QTOF-MS/MS. Anal. Meth. 2015;**7**:4947–4956.

[218] Scotter MJ. Emerging and persistent issues with artificial food colour: natural colour additives as alternatives to synthetic colour in food and drink. Qual. Assurance Saf. Crops Foods 2011;**3**(1):28–39.

[219] Scotter MJ. Methods for the determination of European union-permitted added natural colours in foods: a review. Food Addit. Colorants Part A. 2011;**28**:527–596.

[220] Scotter, M. Review and evaluation of available methods of extraction and analysis for approved natural colours in food and drink; Report Number FD10/02. London: Food Standard Agency, March 2010.

[221] Ferrari E, Foca G, Vignali M, Tassi L, Ulrich A. Adulteration of the anthocyanin content of red wines: perspectives for authentication by Fourier transform-near infrared and ^{1}HNMR spectroscopies. Anal. Chim. Acta. 2011;**701**(2):139–151.

[222] Flamini R. Recent applications of mass spectrometry in the study of grape and wine polyphenols. ISRN Spectroscopy 2013;**2013**:1–45.

[223] Li SY, He F, Zhu BQ, Xing RR, Reeves MJ, Duan CQ. A systematic analysis strategy for the accurate detection of anthocyanin pigments in red wines. Rapid Commm. Mass Spectrom. 2016;**30**(13):1619–1626.

[224] Song W, Wang HJ, Bucheli P, Zhang PF, Wei DZ, Lu YH. Phytochemical profiles of different mulberry (*Morus* sp.) species from China. J. Agric. Food Chem. 2009;**57**(19):9133–9140.

[225] Paredes-López O, Cervantes-Ceja ML, Vigna-Perez M, Hernandez-Perez T. Berries: improving human health and healthy aging and promoting quality life. A review. Plant Foods Hum. Nutr. 2010;**65**(3):299–308.

[226] Benvenuti S, Bortolotti E, Maggini R. Antioxidant power, anthocyanin content and organoleptic performance of edible flowers. Sci. Hort. 2016;**199**(16):170–177.

[227] Smeriglio A, Bareca D, Belloccoco E, Trombetta D. Chemistry, pharmacology and health benefits of anthocyanins. Phytother. Res. 2016;**30**(8):1256–1286.

[228] Li X, Ma H, Huang H, Li D, Yao S. Natural anthocyanins from phytoresources and their chemical researches. Nat. Prod. Res. 2013;**27**(4–5):456–469.

[229] Iwashina T. Contribution to flower colors of flavonoids including anthocyanins: a review. Nat Prod. Commun. 2015;**10**(3):529–544.

[230] Kuskoski EM, Vega JM, Rios JJ, Fett R, Troncoso AM, Asuero AG. Characterization of anthocyanins from the fruits of baguaçu (*Eugenia umbelliflora* Berg). J. Agric. Food Chem. 2003;**51**(18):5450–5454.

[231] Kuskoski EM, Asuero AG, Morales MT, Fett R. Wild tropical fruits and frozen fruit pulps: antioxidant activity, polyphenols and anthocyanins. (in Portuguese) Ciência Rural. 2006;**36**(4):1283–1287.

[232] Kuskoski EM, Asuero AG, García-Parrilla MC, Troncoso AM, Fett R. Actividad antioxidante de pigmentos antociánicos. Cienc. Tecnol. Campinas. 2005;**24**(4):691–693.

[233] Kuskoski EM, Asuero AG, Troncoso AM, Mancini-Filho J, Fett , R. Application of different chemical methods to determine antioxidant activity in fruit pulp. (in Spanish) Food Sci. Technol. 2005;**25**(4):726–732.

[234] Veitch, NC, Grayer RJ. Flavonoids and their glycosides, including anthocyanins. Nat. Prod. Rep. 2011;**28**(10):1626–1695.

[235] Yousuf B, Gul K, Wani AA, Singh P. Health benefits of anthocyanins and their encapsulation for potential use in food systems; a review. Crit. Rev. Food Sci. Nutr. 2016;**56**(13):2223–2230.

[236] Robert P, Fredes C. The encapsulation of anthocyanins from berry-type fruits. Trends in foods. Molecules. 2015;**20**(4):5875–5888.

[237] Rodriguez-Saona L, Wrolstad RE. Extraction, isolation and purification of anthocyanins. Curr. Protoc. Food Anal. Chem. 2001;F1.1–F1.11.

[238] Bucar F, Wube A, Schmid M. Natural product, isolation. How to get from biological material to pure compounds. Nat. Prod. Rep. 2013;**30**:525–545.

Chapter 6

Flavonoid Phenolics in Red Winemaking

L. Federico Casassa

Additional information is available at the end of the chapter

http://dx.doi.org/10.5772/67452

Abstract

This chapter reviews the chemical diversity of flavonoid phenolics in grapes (*Vitis vinifera* L.) with impact on the sensory properties of red wines. Anthocyanins, flavan-3-ols, tannins, and polymeric pigments are discussed from a chemical, technological, and sensory perspective. Anthocyanins, responsible for the color of red wines, reach a peak of extraction after 4 or 5 days of maceration, followed by a decrease in their concentration as maceration progresses. Flavan-3-ols and oligomeric tannins from skins are responsible for bitterness and extracted within the first days of maceration, whereas extraction of seed-derived tannins requires longer maceration times. Matrix effects, including the presence of anthocyanins, polysaccharides, and other cell-wall components affect the rate of retention of tannins into wine. Polymeric pigments, bearing astringent and bitter properties different from those of intact tannins, are formed from covalent reactions between anthocyanins and tannins, putatively accounting for the changes in mouthfeel and textural properties of red wines during maceration and aging. Different maceration techniques applied during red wine production affect the rate, quantity, and the chemical composition of wine phenolics. Understanding of the factors that modulate phenolic retention into wine should allow the winemaker to adjust maceration variables to meet stylistic and/or commercial specifications.

Keywords: flavonoid, phenolic, anthocyanins, flavan-3-ols, tannins, polymeric pigments, maceration, sensory analysis

1. Introduction

The term "phenolics," however overarching, generally bears a positive connotation for grape growers and winemakers alike. In spite of the use (and abuse) of the concept that touts phenolics as naturally occurring, health-promoting compounds in plant-derived food and beverages, it is in wine, like in perhaps no other beverage, where this term has been so widely discussed both in the popular press and in academia [1–4]. So this begs the question: why all the hype surrounding phenolics in wine? Perhaps the first reason stems from the fact the most phenolics bear color. Color in food and beverages have always captivated human beings. Louis Pasteur, the prominent French chemist and microbiologist, used the term "wine color" a whopping 119 times in its seminal treatise *"Etudes sur le vin"* [5]. Pigments were among the first organic compounds studied in wine, perhaps as an unconscious acknowledgement to the fact that it is color, through the sense of vision, the first attribute human beings appraise when approaching a glass of wine. The term *"oenin"* to characterize the grape anthocyanin malvidin-3-glucoside, one of the main pigments found in grapes and wines, was first used in 1915 [6]. The amphoteric nature of anthocyanins and their pH-dependent colored forms was recognized as early as 1806: "The colouring matter of the Alicant raisin is the same as that of the red fruits and common red wines; it has the singular property of becoming red by the acids; although blue by nature, it becomes green with the alkalis…" [7]. Tannins, which contribute indirectly to wine color, were also discussed by Pasteur [5] and later in 1895, by E. Manceau, then a scientists at Möet et Chandon (France) who published a method to study tannins in, expectedly, Champagne wines [8].

Another possible explanation for the early interest in the study of phenolics in wine is the fact that specific phenolic compounds were early recognized as determinants of the flavor and mouthfeel properties of red wines. The tactile sensation of astringency and the taste sensation of bitterness in red wines were recognized, again, by Pasteur [5], but no link to phenolics was made at the time. Later in 1958, E.C. Bate-Smith, a prominent British phytochemist, stated that tannins in wines were responsible for the "liquoring properties and body" of wines and were "intimately" concerned with the perception of quality [9]. Interest on the sensory aspects of phenolic compounds in wines quickly sparkled a series of studies on how to maximize the extraction of phenolics into wine during winemaking.

Indeed, Eugene W. Hilgard, a German-born UC-Berkeley professor, conducted perhaps the earliest studies on the effect of different processing techniques during the fermentation of red grapes in California (USA). His findings, though made between 1885 and 1890, were accurate and were confirmed decades later using much more advanced analytical tools. Hilgard noted that, for example, during red winemaking "maceration[1] of the wine on the pomace after fermentation increases tannins but adds nothing to color" [10]. Hilgard also noted that "it is quite certain that, according to the method of fermentation used, the extraction of the pomace and the consequent tint of the wine may seriously differ" [11]. This chapter expands on these later thoughts, i.e., the factors that underpin the extraction and retention of phenolic compounds into wine, along with the chemical and sensory implications they brought about to the finished wines. Before discussing how phenolics are extracted during red winemaking, it is

[1]Maceration is a crucial step during red winemaking whereby the fermentation solids (skins, seeds, lees, and eventually stems) are kept in contact with the fermenting must/wine. It is during maceration that phenolic compounds, free aroma, and aroma precursors are extracted into wine. Winemakers also refer to maceration as "skin contact time" or "maceration time."

pertaining to present a succinct classification and occurrence of phenolic compounds in grapes and wines as well as an overview of the factors that underpin phenolic reactivity in wines.

2. Classification, occurrence, and general reactivity

The term "polyphenol," sometimes also referred to as phenol or *flavonoid*, encompasses about 4000 bioactive compounds of plant and fungal origin which have more than one aromatic phenol ring within the structure as opposed to *nonflavonoids* or simple phenols [12, 13]. Phenolic compounds are very reactive in wines due to the high electron density of the aromatic ring (s), which is enhanced by the presence of one or more electron-donating hydroxyl groups [12, 14]. As a result, the electron density at the -*ortho* and -*para* positions activates the carbons on those sites with partial negative charges, thus enabling the structure to undergo electrophilic substitutions at these positions (**Figure 1**). The reactivity of the phenol ring is further increased by the weak acidic character of the hydroxyl group that can donate a proton (H^+) due to the partial migration of electrons from the oxygen in the hydroxyl to the aromatic ring [14]. In addition, the pK_a of the most acidic phenolic hydrogen is above 7, which explains why polyphenols are easily deprotonated forming phenolate ions at physiological pH. Phenolate ions can donate electrons directly to oxygen, thereby forming reactive semiquinones [15]. However, at pH values between 3 and 4, such as the ones normally encountered in red wines, phenols are largely protonated and as such cannot directly donate electrons to oxygen [15–17].

Phenols can also readily participate in noncovalent interactions with various components of the wine matrix. The hydroxyl groups can act both as H-bond donors or acceptors. The phenol group(s) are also amphiphilic agents, as the hydroxyl groups are hydrophilic but the aromatic ring is hydrophobic, allowing the structure to become engaged in hydrophobic interactions as well [12]. Another distinctive property of phenolics is that, upon variation in the matrix

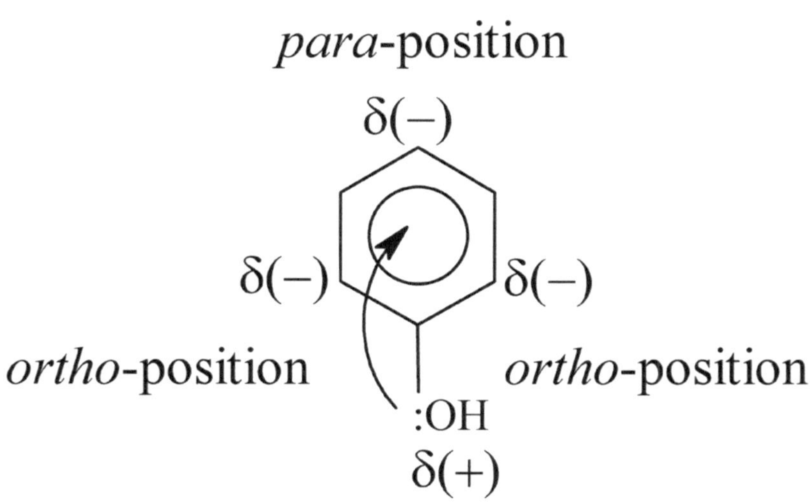

Figure 1. Activated sites of the phenol ring. Redrawn from Ref. [14].

conditions such as pH [14] or oxidation-reduction potential [18], phenolics can act both as electrophiles (i.e., electron-loving molecules) and nucleophiles, thus readily reacting with electron-rich or electron-deficient compounds, respectively (**Figure 2**).

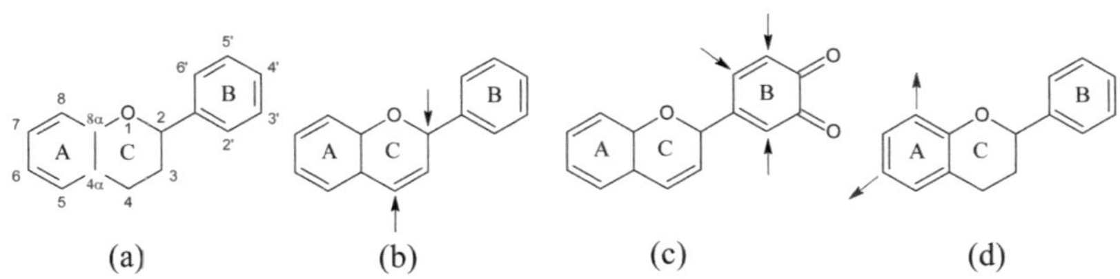

Figure 2. (a) Basic *flavonoid* ring structure and numbering, (b and c) sites for electrophilic substitutions, and (d) sites for nucleophilic substitutions. Redrawn from Refs. [13, 25].

For the purpose of this chapter, occurrence, reactivity, chemical structure, and sensory properties of phenolic compounds are discussed exclusively in the context of *Vitis vinifera* L. grapes and wines. Also, only the most important polyphenols of the *flavonoid* group, namely anthocyanins, flavan-3-ols and tannins, as well as their reaction products, will be discussed. Phenolics belonging to the *nonflavonoid* class include benzoic (e.g., gallic, hydroxibenzoic, protocatechuic, vanillic, and syringic acids) and cinnamic acid derivatives (*p*-coumaric, cafeic, ferulic, and sinapic acids). Also, the hydroxylated stilbenes are included in the *nonflavonoid* class, of which *trans*- and *cis*-resveratrol (3,5,4-trihydroxystilbene), as well as their glucose derivatives (*trans*- and *cis*-piceids), have all been identified in grapes and wines [19, 20]. Although quantitatively much less important than the *flavonoid* class, *nonflavonoids* phenolics play a direct role in both chemical and coupled-enzymatic oxidation reactions in white and red wines [21, 22]. Cinnamic acids can act as copigments,[2] inducing changes in color through the phenomenon of copigmentation[3] [23] and can also impact wine aroma when metabolized by yeast of the genera *Brettanomyces/Dekkera* to generate volatile ethyl-phenols [24].

Phenolic *flavonoids* possess a three-ring system, composed of 15 carbon atoms in the form C6-C3-C6 [12, 13]. The central C ring contains oxygen forming a pyran ring that can adopt various oxidation states, and it is fused to two aromatics rings, termed A and B (**Figure 2a**). In the *flavonoid* family, the B-ring is fused to the pyran ring at position 2. The A ring is derived from the phloroglucinol structure and is the most conserved portion of the C6-C3-C6 backbone. Furthermore, different substitutions in the B ring define different compounds within a family. The members of the *flavonoid* class found in grapes and wines all have the same substitution pattern of hydroxyl groups at carbons 5 and 7 of the A ring. On the other hand, differences in the oxidation state and substitution on the C ring define the different *flavonoid* classes [13]. Thus,

[2]Copigments are typically noncolored phenolic and/or nonphenolic compounds able to engage in copigmentation reactions with anthocyanins.

[3]As defined by Boulton [34], the phenomenon of copigmentation is due to molecular associations between pigments and other (usually non-colored) organic molecules in solution. These associations cause the pigments to exhibit far greater color than would be expected from their concentration. When anthocyanins engage in copigmentation, both hyperchromic (increase of absorbance) and bathochromic (shift of absorbance toward blue hues) are normally observed.

anthocyanins represent the highest oxidation state of the C ring, and on the other extreme, flavan-3-ols represent the most reduced state of the said ring.

From a chemical and sensory standpoint, the three most relevant phenolic classes within the *flavonoid* family are flavan-3-ols (the "building blocks" of tannins), anthocyanins, and tannins. A heterogeneous family of reaction products resulting from the reaction between anthocyanins and tannins, the so-called polymeric pigments, are not originally present on grapes but formed during winemaking operations and wine aging (Section 3.4). This chapter will focus on the chemical and sensory aspects of flavan-3-ols, anthocyanins, tannins, and polymeric pigments in red wines, providing along the way an overview of their extraction patterns during red winemaking.

3. Chemical, sensory aspects, and extraction of phenolic flavonoids during red winemaking

3.1. Anthocyanins

3.1.1. Occurrence, general chemistry, and sensory aspects

Anthocyanins occur as vacuolar components in the berry skin tissue (and in the mesocarp of the so-called teinturier varieties) and are present as monomers of six glycosylated forms, namely malvidin, cyanidin, petunidin, peonidin, delphinidin, and pelargonidin [26, 27]. Glycosylation occurs at the C3 position and renders the molecule water-soluble [28]. Acylation in turn occurs in the C6 position of the glucose moiety by esterification with an aromatic (*p*-coumaric, caffeic, ferulic, and sinapic acids) or an aliphatic acid (acetic, malic, malonic, oxalic, and succinic acids) (**Figure 3**). The acylation of the sugar might promote the chemical stability of the anthocyanin molecule [29], possibility through stacking of the acyl groups with the pyrilium ring of the flavylium cation, thereby reducing the susceptibility to the nucleophilic attack of water [30].

Anthocyanins are red pigments responsible for the color of red wines and owe their spectral properties to the resonant structure given by a 10-electron system, partially delocalized between the pyran C and A rings, as well as to the extended conjugated system of unsaturated bonds in the structure [13, 31]. As a result, the maximum absorbance (λ_{max}) of these compounds varies between 475 and 540 nm depending upon the aglycone (known as anthocyanidin) and substitution patterns of the C ring [26]. The λ_{max} between 475 and 540 nm confers anthocyanins blue to red color hues. *O*-methylation in aglycones such as malvidin, petunidin, and peonidin causes a slight reddening effect (**Figure 4**) and reduces the reactivity of the nearby phenolic hydroxyl groups, thereby increasing the stability of the molecule [32]. On the other hand, the increase in the number of free hydroxyl groups in the B ring increases blueness (i.e., bathochromic shift), which in turn renders the structure more polar but less stable. Acylation has also been suggested to produce a bathochromic shift, giving more purple tones, possibly as a result of intramolecular copigmentation reactions [30]. These and other features can be observed in **Figure 4** for purified anthocyanins isolated from Cabernet Sauvignon grapes after preparative HPLC fractionation of anthocyanins and confirmation by mass spectroscopy.

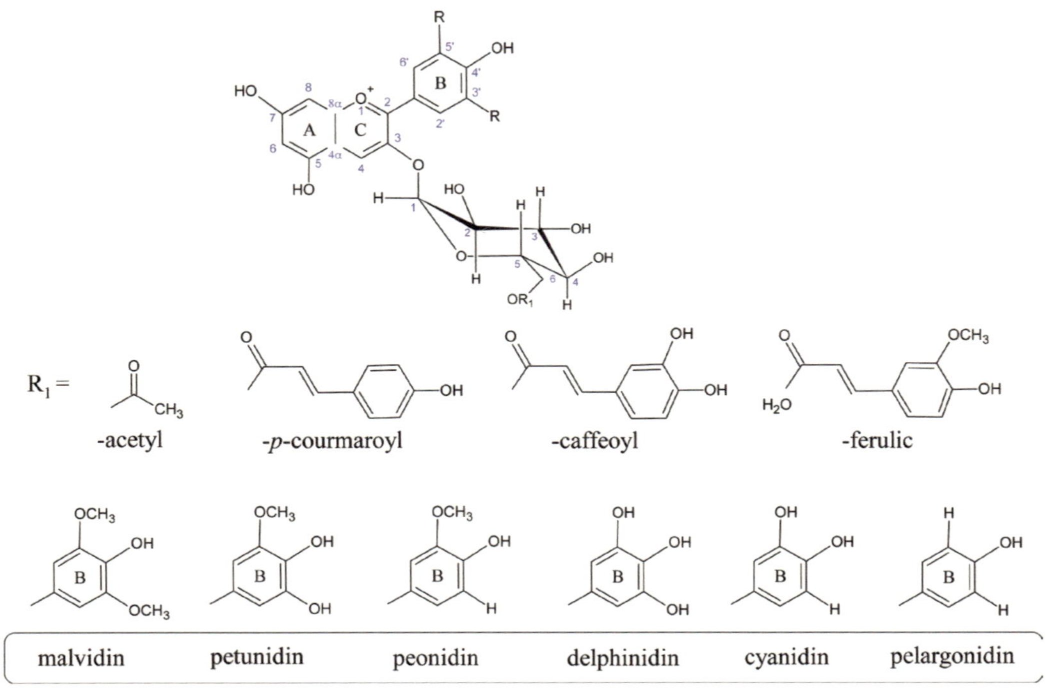

Figure 3. Basic anthocyanin structure, acyl groups, and the main six anthocyanindins found in *V. vinifera* grapes and wines.

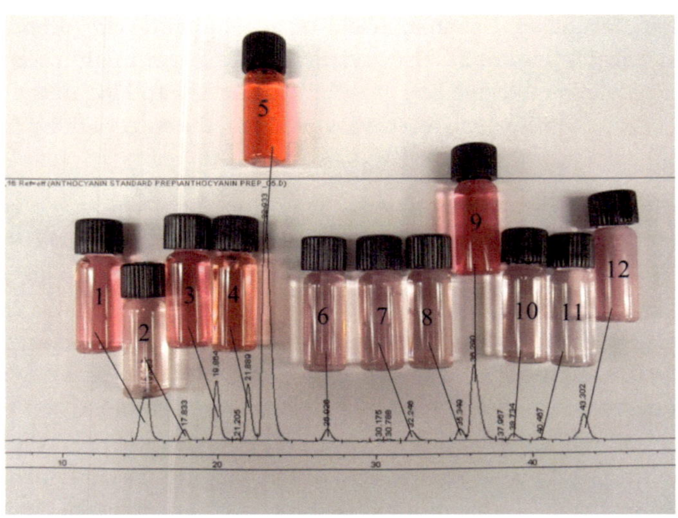

Figure 4. Preparative HPLC chromatogram showing separation of anthocyanin monoglucosides and acyl-glucosides of Cabernet Sauvignon grapes from Washington State (USA). Anthocyanin fractions are shown approximately above of each major peak and were collected directly from the elutant at pH ~ 1.8. Peak assignment: (1) delphinidin-3-*O*-glucoside, (2) cyanidin-3-*O*-glucoside, (3) petunidin-3-*O*-glucoside, (4) peonidin-3-*O*-glucoside, (5) malvidin-3-*O*-glucoside, (6) delphinidin-3-*O*-acetyl-glucoside, (7) petunidin-3-*O*-aceyl-glucoside, (8) peonidin-3-*O*-acetyl-glucoside, (9) malvidin-3-*O*-acetyl-glucoside, (10) malvidin-3-(6-*O*-caffeoyl)-glucoside, (11) petunidin-3-(6-*O*-coumaroyl)-glucoside, (12) peonidin-3-(6-*O*-coumaroyl)-glucoside (shoulder), and malvidin-3-(6-*O*-coumaroyl)-glucoside.

The flavylium cation of the anthocyanins is responsible for the chromatic properties of young red wines, with a molar extinction coefficient (ε) of 29,500 M^{-1} cm^{-1} for malvidin-3-glucoside in 0.1% HCl methanolic solution [33]. As a result of this, some of the color observed in aged red wines can still be attributable to the flavylium cation. However, upon crushing and during winemaking, anthocyanins undergo a variety of electrophilic and nucleophilic substitutions giving rise to cycloaddition and condensation products, and oligomeric and polymeric pigments (Section 3.4 and **Figure 13**). These transformations invariably have an impact on wine color, and it is usually during maceration and postmaceration when the most noticeable changes in wine color take place (**Figure 5**). For example, at the beginning of maceration, the intense purple-red color and hyperchromic shift (i.e., increase in absorbance) observed in red wines is the result of intra- and intermolecular and self-association copigmentation reactions [34]. Subsequently, with the disruption of copigmentation mediated by increasing ethanol levels, perceived color decreases and shifts toward more red tones [35]. As maceration winds up and aging ensues, the incorporation of anthocyanins into vitisins A, B, vinyl-catechol derivatives, and xanthylium salts, to name some possible reactions that lead to the formation of the so-called pyranoanthocyanins (**Figure 13**), causes a shift in color from deep-red to orange or brick-orange hues [36]. It is also expected that the incorporation of anthocyanins into

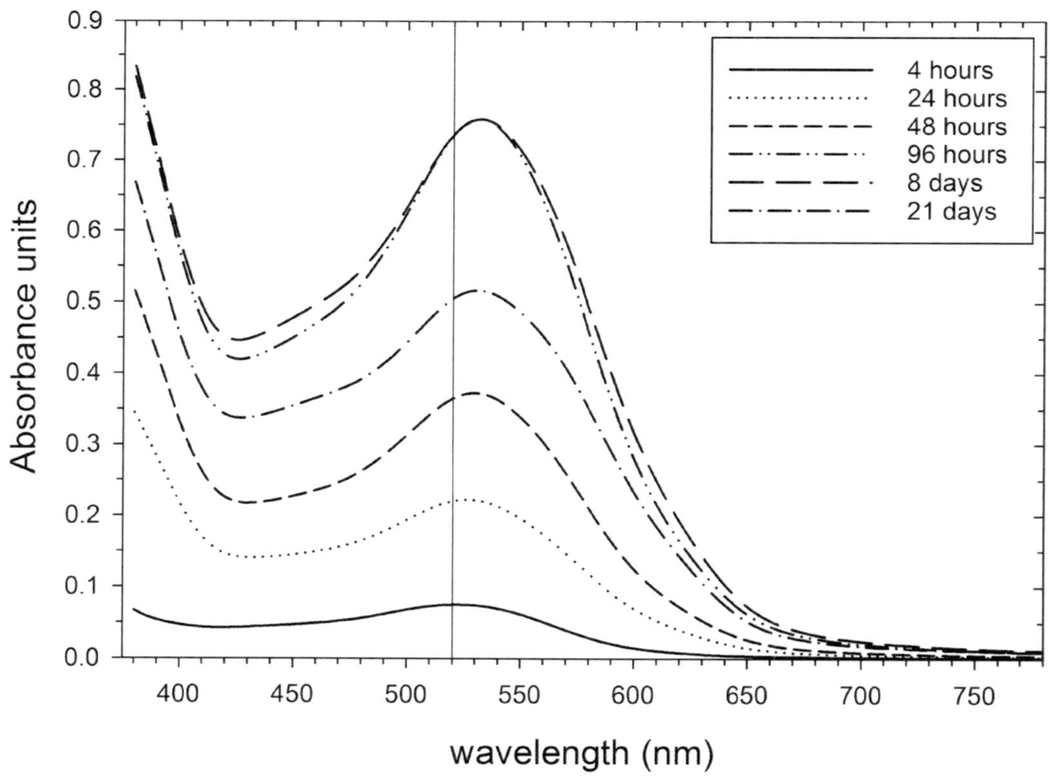

Figure 5. Evolution of the full-length visible spectrum of a Malbec wine produced with different maceration times, ranging from 4 hours to 21 days. The vertical line indicates the 520-nm wavelength, which is approximately the wavelength of maximum absorbance for most anthocyanins. Notice the drop in the absorbance values of the full spectrum in the wine produced with 21 days maceration relative to the wine produced with 8 days of maceration.

oligomeric and polymeric tannins (to form the so-called polymeric pigments) should cause a reduction in the molar extinction coefficient of the flavylium form, thus leading to a decrease in color saturation as winemaking progressed [37]. In the case of pyranoanthocyanins, these do not necessarily appear to have lower molar extinction coefficients relative to the native anthocyanins [38], and, in fact, their molar extinction coefficient is much more stable toward pH swings [39].

Isolated anthocyanins are tasteless or indistinctly flavored [40]; however, upon reaction with oligomeric or polymeric tannins during winemaking, oligomeric and polymeric pigments are formed (**Figure 14**) and these can in turn modulate astringency (Section 3.4).

3.1.2. Extraction during winemaking

The diffusion of anthocyanins into the must requires the breakdown of two biological barriers, namely the cell wall, including the degradation of the pectic substances in the middle lamella, and the tonoplast of the vacuoles of the skin subepidermal cells [41]. Normal operations during crushing ensure the breakdown of cellular walls, and native enzymatic reactions allow for the degradation of pectic substances and polysaccharides in the middle lamella. The diffusion process is favored by the water-soluble nature of anthocyanins, resulting in maximum extraction rates and a peak of extraction within the first 3 to 7 days of maceration [35, 42–48] as observed in **Figure 6**. The rate of extraction of the different anthocyanins seems to be similar [45, 47] and proportional to their original concentration in grapes. However, some studies have found that wines have a relative higher amount of malvidin-3-glucoside than that originally present in the grapes [49, 50]. Establishment of unequivocal extraction and retention patterns of different anthocyanin forms may be complicated by the fact that acylated anthocyanins may undergo acid hydrolysis upon extraction into the fermenting must/wine [50], thereby releasing the monoglucoside forms and/or by the fact that, for example, acylated derivatives are preferentially adsorbed by wine lees [51].

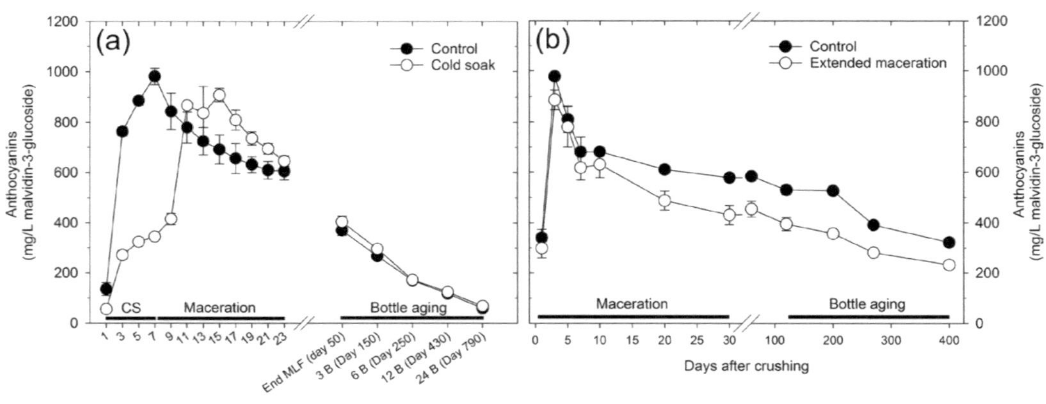

Figure 6. Overview of the extraction of and evolution of anthocyanins during (a) prefermentative cold soak (CS), maceration, end of malolactic fermentation (end MLF), and up to 24 months of bottle ageing (B) of Malbec wines processed with a maceration length of 15 days (control) and with cold soak for 7 days + 15 days of maceration (cold soak), and (b) maceration and bottle aging of Cabernet Sauvignon wines processed with a maceration length of 10 days (control) and 30 days (extended maceration). Adapted from Refs. [43, 55].

The early extraction of anthocyanins may also influence the solubility and retention of oligomeric and polymeric tannins through the formation of polymeric pigments [52–54]. Following the peak of anthocyanin extraction, a variable drop in concentration, which can be as high as 60% from peak concentration, is typically observed [42, 43, 55] (**Figure 6a**). Loss of anthocyanins during maceration in fermenting must and wines has been attributed to a variety of factors, including ionic adsorption by the negatively charged yeast cell walls and yeast lees during postmaceration [56, 57], adsorption onto bitartrate crystals and particulate matter [35], incorporation into small and large polymeric pigments [43, 58], formation of pyranoanthocyanins [56, 59], and oxidative cleavage of the heterocyclic C-ring leading to direct anthocyanin degradation [60]. A decrease in copigmentation as a result of an increasing concentration of ethanol in the fermenting must (which increases the hydrophobic character of the medium thereby disrupting the copigmentation complex) also contributes to both the loss of anthocyanins and a decrease in wine color [34, 35, 61]. At the end of maceration, the levels of anthocyanins recovered in the wine, relative to the grape initial content, have been reported to be around 40% [26].

Counter to what intuition may suggest, there appears to be a negative relationship between maceration length and anthocyanins retained in the resulting wine [43, 47, 58, 62, 63] (**Figure 6b**). Moreover, analysis of anthocyanins recovered in the pomace hardly increased the recovery yield, suggesting that a major proportion of anthocyanins were converted to other species or were irreversibly adsorbed on the solid material during maceration [26, 47]. This discrepancy was originally attributed to the enhanced formation of polymeric pigments in wines undergoing extended maceration. One study found that Merlot wines produced with extended maceration were lower in anthocyanins but higher in polymeric pigments relative to control wines [58]. However, the formation of polymeric pigments during extended maceration cannot fully explain the observed anthocyanin loss [47]. This suggests that, in addition to anthocyanin losses by adsorption, degradation reactions may be at play during extended maceration. Ultimately, the extraction patterns during winemaking and aging and thus the final concentration of anthocyanins are both modulated by the variety and the maceration technique the winemaker decides to put in place, which can be as simple as delaying the onset of alcoholic fermentation (e.g., the technique known as cold soak[4]) or increasing maceration time (e.g., extended maceration) (**Figure 6a** and **b**, respectively).

3.2. Flavan-3-ols

3.2.1. Occurrence, general chemistry, and sensory aspects

In *V. vinifera* grapes and wines, flavan-3-ols occur both in seeds and skins as five monomers: (+)-catechin, (-)-epicatechin, (+)-gallocatechin, (+)-epigallocatechin, and (-)-epicatechin-3-*O*-gallate [64–66] (**Figure 7**). The glycosyl derivatives of four of these monomers have also been

---

[4]Cold soak, also known as prefermentative cold soak, is a winemaking technique in which the onset of alcoholic fermentation is delayed by keeping the must (and thus the fermentation solids) at temperatures ranging from 5 to 15°C by means of a cooling system in jacketed fermenters or by the use of solid CO_2 (dry ice). During cold soak, the extraction of water-soluble compounds is sought, as opposed to the extraction that takes place in the presence of ethanol during alcoholic fermentation. The duration of cold soak is defined by the winemaker and it can be as short as 2 days and as long as 14 days.

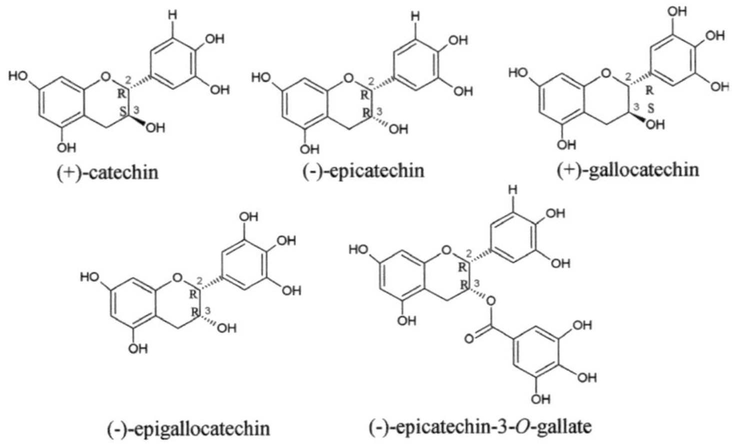

Figure 7. Monomeric flavan-3-ols found in *V. vinifera* grapes and wines.

recently reported in Merlot grapes and wines [67]. Flavan-3-ols occur in several isomeric forms. The carbons at the C2 and C3 positions of the flavan-3-ol backbone are two asymmetric centers, such as the five monomeric flavan-3-ols are grouped into two diastereomers pairs, with configurations 2*R*:3*S* for (+)-catechin and (+)-gallocatechin and 2*R*:3*R* for (-)-epicatechin, (-)-epigallocatechin, and (−)-epicatechin-3-*O*-gallate [68]. These different isomeric configurations, in turn, have an impact on bioavailability [69], antioxidant and radical scavenging properties [70], and, ultimately, on sensory properties [71, 72], as further discussed later.

In seeds, flavan-3-ols are located in thin-walled cells between the external cuticle and the inner lignified layers. The (sometime observed) seed browning during berry ripening is thought to be the result of both monomeric flavan-3-ols and tannins undergoing oxidation [73]. Seeds contain only (+)-catechin, (−)-epicatechin, and (-)-epicatechin-3-*O*-gallate [64, 74]. In the skins, flavan-3-ols occur in the subepidermal cell as shapeless or spherical inclusions free in the vacuoles but also associated with the tonoplast [75]. Skins contains (+)-catechin, (-)-epicatechin, (-)-epicatechin-3-*O*-gallate, and, additionally, (-)-epigallocatechin [76], as well as trace amounts of (+)-gallocatechin and (-)-epigallocatechin gallate [66, 77].

Quantitative differences also occur within the berry tissues. Seeds concentrate the vast majority of flavan-3-ols of the berry [66, 75–79]. For example, flavan-3-ol concentrations of about 179 mg/kg fresh weight (FW) have been found in Cabernet Sauvignon seeds whereas skins of the same variety only have 4.8 mg/kg FW [80]; similar results are reported for other varieties [81, 82]. In wines, the content of monomeric flavan-3-ols varies from 29 to 41 mg/L in Tempranillo and Graciano wines [78], 107 to 176 mg/L in Pinot Noir wines [83], 182 mg/L in Tannat wines [84], and up to 288 mg/L in Cabernet Sauvignon wines [43].

Due to the reduced state of the C ring of the flavan-3-ol structure, and thus favorable one-electron donation properties, flavan-3-ols can react with several wine electrophiles. The condensation of monomeric flavan-3-ols with anthocyanins either by a direct covalent reaction between them [85, 86] or mediated by acetaldehyde [87, 88] is one of the main reactions with impact on color during wine aging and is further addressed in Section 3.4.

Flavan-3-ols are colorless and do not absorb light in the visible spectrum, but have a peak of UV absorbance between 270 and 280 nm [89, 90]. Catechin and epicatechin are both susceptible to enzymatic [91] and nonenzymatic oxidation [92, 93], with both mechanisms resulting in a change in spectral properties from colorless to yellow hues [89, 94]. The end products of these reactions are quinones. Quinones are powerful electrophiles that can readily react with wine nucleophiles such as sulfur compounds (e.g., volatile thiols, common in Sauvignon Blanc wines) and hydrogen sulfide, thus decreasing their volatility and their influence on wine aroma [95, 96].

Flavan-3-ols have defined taste attributes. Flavan-3-ols have a marked bitter taste, and their influence on the development of the bitterness sensation was recognized as early as 1966, when Rossi and Singleton isolated an ether-soluble fraction from grape seeds containing (+)-catechin, (-)-epicatechin, and (-)-epicatechin gallate [97]. Addition of this fraction at 200 mg/L to a white wine showed no contribution to astringency but significantly increased bitterness. Later, it was found that the chiral difference between the two main wine flavan-3-ols produces a significant difference in temporal perception of bitterness: (-)-epicatechin is significantly more bitter and had a significantly longer duration of bitterness than (+)-catechin [71, 98, 99]. The more planar conformation of the C ring of (-)-epicatechin compared with the less planar (+)-catechin may facilitate the diffusion of the molecule to the gustative receptor; in addition, the higher lipophilicity of epicatechin (relative to catechin) could also explain its higher comparative bitterness [71, 100].

Although it is assumed that both (+)-catechin and (-)-epicatechin cannot precipitate proteins, protein-induced precipitation of these flavan-3-ols has been confirmed by peptide models of salivary proline-rich-proteins (PRPs) whereby PRPs did indeed interact with flavan-3-ols having masses below 500 Da [101, 102]. Further confirmation was reported in model wine studies whereby a time-intensity sensory procedure found both (+)-catechin and epicatechin to elicit astringency [98, 103]. Also, (-)-epicatechin-3-*O*-gallate and (+)-catechin were found to precipitate PRPs when the molar ratio of flavan-3-ols to protein exceeded 27 [104]. In Cabernet Sauvignon wines, the perception of astringency was at least partially explained by the simultaneous occurrence of a comparatively higher concentration of flavan-3-ols and tannins [105]. Astringency of monomeric flavan-3-ols may be the result of cooperative precipitation of, or cooperative binding with, proteins due to the presence of one 1,2-dihydroxy or 1,2,3-trihydroxy groups [103, 106], and this may be enhanced by the presence of tannins [105]. These studies highlight the potential role of flavan-3-ol monomers on astringency perception in red wines.

3.2.2. Extraction during winemaking

Release of flavan-3-ols from skins occurs early during winemaking, within the first 2 or 3 days of skin contact [107–110]. For example, levels of (+)-catechin and (-)-epicatechin after 5 days of skin contact represented between 80 and 85% of the maximum content attained later at pressing in Grenache wines [60]. In Tempranillo and Grenache-Graciano blends, the release of flavan-3-ols occurred early, between days 3 and 5 of maceration, and remained unchanged during a postmaceration period of 1 week followed alcoholic fermentation [109]. In this same study, (-)-epigallocatechin, only located on the skins, was extracted rapidly, suggesting that the extraction of flavan-3-ols and small oligomers from skins occurs during the first days of maceration [109].

Release of flavan-3-ols from seeds requires longer maceration times. For example, the maximum extraction of flavan-3-ols from seeds occurred after 2 to 3 weeks of maceration in model wines, and under those extended conditions, the seeds contributed almost 90% of the total flavan-3-ol content of the final wines [108]. Also, in model wines containing only seeds of the variety Monastrel, the levels of (+)-catechin and (-)-epicatechin increased from 5 mg/L at day 2 to 27 mg/L at day 10 [111]. This represents a slower rate of extraction relative to that of flavan-3-ols from skins. Moreover, longer maceration times, favored by the winemaking technique known as extended maceration, enhance the overall extraction of flavan-3-ols and, more specifically, that of epicatechin-3-*O*-gallate from seeds [43] (**Figure 8b**). Other studies have also reported that the percentage contribution of galloylated subunits increases along with maceration time [109, 111–113], and this, together with the content of epigallocatechin (exclusive of the skins), has been used as a surrogate to estimate the percentage contribution of seeds (and skins) to the overall wine's flavan-3-ol and tannin content [114].

Since extraction of flavan-3-ol from seeds occurs toward the end of a regular maceration period (e.g., 15 days), the extraction of both flavan-3-ols and tannins from seeds was assumed to be mediated by the dissolutive effect of ethanol on the lipidic outer coat of the seeds, which will typically occur toward the end of the fermentation process [115]. From this perspective, higher ethanol levels should selectively favor the extraction of seed phenolics in general and that of flavan-3-ols in particular [48, 110]. However, model wine experiments have shown that extraction of both flavan-3-ols and tannins can occur in the absence of ethanol, and that the role of ethanol may be to solely increase the rate of extraction [111]. Current evidence supports the notion of maceration length, and not alcohol, as being one of the main drivers of seed tannin extraction, at least under typical final ethanol concentrations (13–14.5%, v/v) [47, 116]. Thus, an alternative explanation for the late release and extraction into wine of seed flavan-3-ols and tannins is that extraction from the seeds may only occur after the seeds had attained maximal hydration. In studying seed extraction in model wine solutions, and noticing a lag phase

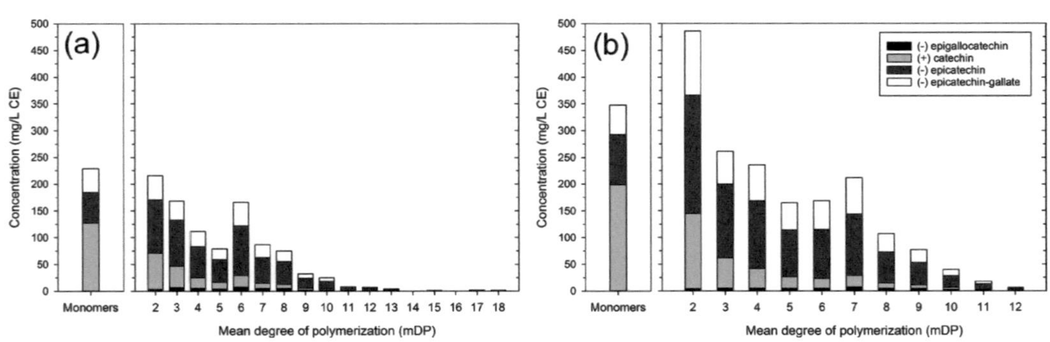

Figure 8. Monomeric flavan-3-ol concentration and composition and tannin size distribution by concentration and composition in Cabernet Sauvignon wines grouped as a function of the maceration length treatment. (a) Control wines (10 days of maceration) and (b) extended maceration wines (30 days of maceration). CE: catechin equivalents. Adapted from [43].

http://dx.doi.org/10.5772/67452

during the time course of maceration, Singleton and Draper noted that "…The lag may be an expression of the fact that most cells do not surrender their constituents for extraction while they are still living. Swelling and osmotic-pressure phenomena may also be involved…" [117]. If this is accepted, then the onset of seed tannin and flavan-3-ol extraction will occur once seeds have reached full hydration, whereby the leakiness of the parenchyma cells outside the true seed coat would allow the prompt release of flavan-3-ols [111, 118].

3.3. Tannins

3.3.1. Occurrence, general chemistry, and sensory aspects.

Tannins, also known as proanthocyanidins, and/or *condensed* tannins,[5] encompass oligomeric (degree of polymerization ≤ 2 and < 5) and polymeric flavonoids (degree of polymerization ≥ 5) made up by the five monomeric flavan-3-ols shown in **Figure 7** [12, 119]. As defined by Bate-Smith and Swain (1962), tannins are "water soluble compounds having molecular weights between 500 and 3000 and, besides giving the usual phenolic reactions, they have special properties such as the ability to precipitate alkaloids, gelatin and other proteins" [120]. Haslam [12] further noted that tannins could also form complexes with polysaccharides and that molecular weights as high as 20,000 can be found in nature. These annotations certainly pertain to grapes and wines. Wine tannins readily interact with native or yeast-derived polysaccharides [121, 122], and this interaction can modulate sensory properties such as astringency [121, 123] (**Figure 9**). Furthermore, based on the degree of polymerization, molecular weights as high as 5000, 5200, and 22,000 have been reported in seed, wine, and skin extracts, respectively, from *V. vinifera* [124], highlighting the polymeric nature of grape and wine tannins.

Oligomeric and polymeric tannins are defined by the connectivity and nature of the interflavanic bond. This refers to a covalent connection between two flavan-3-ols, also referred as "subunits"; the *average number* of constitutive subunits in the oligomer or polymer (assuming a normal distribution of tannin sizes as a function of molecular weight) is referred to as degree of polymerization. A technique based on HPLC analysis, known as phloroglucinolysis, allows for the calculation of the so-called mean degree of polymerization (mDP) [125].

Two types of interflavanic bonds have been observed. Interflavanic bonds connecting either the carbon at position 4 in the extension subunit and the carbon at position 8 in the terminal subunit (C4 → C8), or the carbon at position 4 in the extension subunit and the carbon at position 6 in the terminal subunit (C4 → C6), are collectively known as B-type linkages [12]. The flavan-3-ol subunits can also be connected by two single bonds, one between C4 and C8, and another between the hydroxyl groups at C7 and C2, forming an ether bond known as A-type linkage [126, 127]. The occurrence of these three types of bonds is not uniformly distributed within grape and wine tannins. The C4 → C8 bonds are found in skins, seeds, and wines [68]; however, C4 → C6 and A-type linkages are formed as a result of rearrangements and intra and/or intermolecular oxidation reactions during winemaking [128].

[5] The relevant literature reviewed here uses several terms to refer to red wine tannins, including "tannin," "condensed tannin," and "proanthocyanidin." The term "tannin" will be used throughout this review.

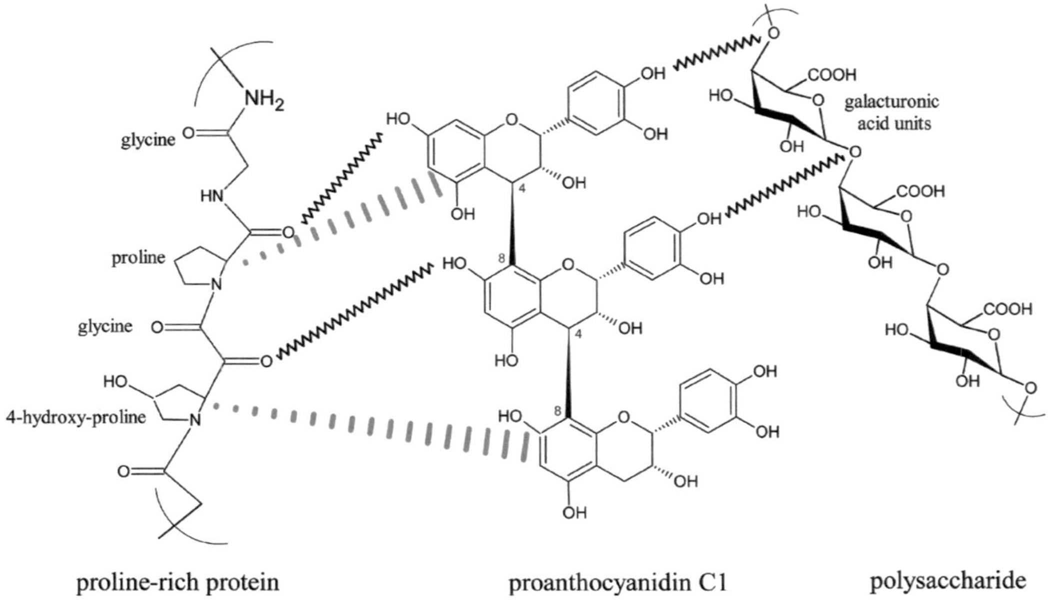

Figure 9. Interaction between proline-rich proteins, a tannin trimer, and a polysaccharide fragment. Narrow black saw-like lines are suggested sites of hydrogen bonding. Wide gray lines are suggested regions of hydrophobic interactions. Redrawn and modified from Ref. [123].

The three connectivities described (C4 → C8, C4 → C6, and C7 → C2) can, in turn, display the usual two stereochemistries of monomeric flavan-3-ols, with the *R* stereochemistry referred to as α and the *S* stereochemistry as β [129]. As a result of the enantiomeric and conformational properties of the flavan-3-ols subunits, the number of possible structures increases in a factorial fashion as the number of subunits in the oligomer or polymer increases. For example, considering only the C4 → C8 linkages, about 10^5 unique tannin structures have been estimated to exist.

Oligomeric tannins include tannin dimers, trimers and tetramers of different connectivities. B-type dimers isolated from grapes and wines are composed of (+)-catechin and (-)-epicatechin and include the dimers B1 (epicatechin-(4β→8)-catechin), B2 (epicatechin-(4β→8)-epicatechin), B3 (catechin-(4α→8)-catechin), and B4 (catechin-(4α→8)-epicatechin) (**Figure 10**). Moreover, dimers having a C4 → C6 connectivity (B5 to B8 series) and epigallocatechin have also been observed [130]. Likewise, the α and β dimers B1, B2, B3, and B4 of epicatechin-3-*O*-gallate and either catechin or epicatechin have also been identified in seeds, skins, and wines [131–133]. Dimers, trimers, tetramers, and pentamers containing A-type bonds were recently found in grape seeds of both white and red *V. vinifera* varieties [127]. Tannin trimers found in grapes and wines consist of three flavan-3-ol units linked by two C4 → C8 interflavan bonds (C-type) or one C4 → C6 interflavan bond in the terminal or extension subunits, and one C4 → C6 interflavan bond in the remaining subunit (T-type) [130, 134, 135]. A nonexhaustive account of the most common connectivities in dimers and trimers observed in *V. vinifera* grapes and wines is shown in **Figure 10**.

http://dx.doi.org/10.5772/67452

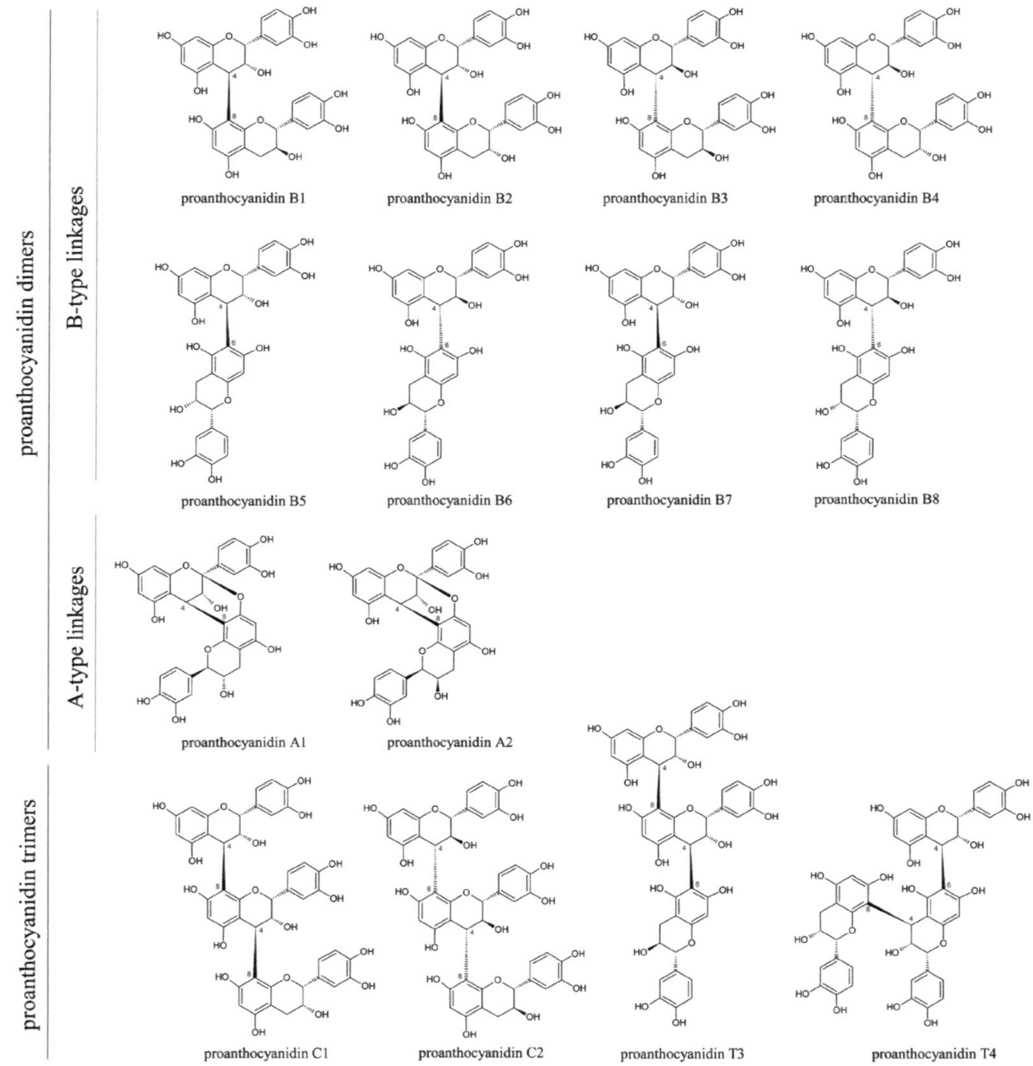

Figure 10. Overview of some of the most abundant tannin dimers and trimers observed in *V. vinifera* grapes and wines. Only tannins based on (+)-catechin and (-)-epicatechin are shown; however, dimers and trimers containing (-)-epigallo-catechin and (-)-epicatechin-3-*O*-gallate also occur in *V. vinifera*.

Polymeric tannins have an mDP ≥ 5 subunits. Early studies by Czochanska and colleagues on tannins from 22 plant sources with molecular weights ranging from 1500 to 5000 elucidated the structure and stereochemistry of the interflavanic bond in these polymers by ^{13}C NMR and confirmed the presence of C4 → C8, C4 → C6, and A-type (C7 → C2) linkages [136]. Therefore, the observations described previously for oligomers also apply to polymeric tannins.

The term "proanthocyanidin," commonly used in the literature to refer to tannins, stems from the ability of these compounds to release anthocyanins upon cleavage of the interflavanic bond under heat and acidic conditions. Based on the last reaction, tannins are classified as *procyanidins*, when they release cyanidin (characteristic of tannins containing (+)-catechin, (-)-epicatechin, and (-)-epicatechin-3-*O*-gallate), and *prodelphinidins*, when they release delphinidin (as is the case of tannins containing (-)-epigallocatechin) [137]. The spectral features mentioned for monomeric flavan-3-ols also apply for oligomeric and polymeric tannins.

Oligomeric and polymeric tannins occur both in skins and seeds, although their quantitative and qualitative composition, as well as their molecular weight and distribution, differ within these tissues and also in the resulting wines. On a whole berry basis, up to 80% of the total extractable tannin pool of the berry is located in the seeds [12]. In seeds at physiological maturity, seed tannins measured by protein precipitation vary from 2.15 to 4 mg/g FW for varieties such as Cabernet Sauvignon, Merlot, and Syrah [43, 47, 138–140]. Conversely, skin tannins vary between 0.35 and 1 mg/g FW [43, 47, 139, 140]. In wines, total tannins (by protein precipitation) measured in 1325 commercial red wines ranged from 30 to 1895 mg/L, with a mean concentration of 544 mg/L [141].

Seed-derived polymeric tannins are co-located with the flavan-3-ols in the cells below the external cuticle [142]. Seed tannins associated with cell walls reportedly have a higher mDP than their cellular counterparts [143]. The mDP of seed tannins of different grape varieties vary between 2 and 22 [73, 78, 134, 143, 144–146], which represents a 11-fold variation. Seed tannins are composed of (+)-catechin, (-)-epicatechin, and epicatechin-3-*O*-gallate, with monomeric flavan-3-ols, dimers (a portion of them bearing A-type linkages), and trimers being the predominant species [74, 130, 134]. The percentage of galloylation of seed tannins has been estimated to be around 30% [26]. In the skins, tannins occur as vacuolar components in subepidermal, thick-walled cells, but they are also found in the tonoplast and cell walls [75, 147]. Skin tannins associated with cell walls have a higher mDP than those found as free cytoplasmic components; however, tannins found in the cytoplasm are more abundant [147]. Skin tannins typically have the highest molecular weight within the berry, varying from 6 up to 85 subunits in varieties such as Cabernet Franc, Cabernet Sauvignon, Graciano, Merlot, Pinot Noir, Syrah, Carmenère, and Tempranillo [78, 124, 134, 144–148]. This represents a 14-fold variation. Skin tannins are composed primarily of (-)-epicatechin, (-)-epigallocatechin, and (+)-catechin. However, trace amounts of epicatechin-3-*O*-gallate have also been found [147], giving a percentage of galloylation estimated to typically be <5% [26].

Wine tannins are composed of all the five monomeric flavan-3-ol subunits, with (-)-epicatechin, followed by (+)-catechin and (-)-epigallocatehin being the major constituents [78]. Moreover, the proportion of each subunit is affected by both grape variety and winemaking technique, and, like with flavan-3-ols, a relationship between an increased percentage of galloylated subunits and maceration length has been observed [43, 113, 135]. For wines, the reported mDP values vary between 2 and 17 subunits [47, 124, 148–151]. However, oligomeric tannins with an mDP < 5 are the major constituents of wine tannins [43, 109, 124, 152] (**Figure 8a** and **Figure 8b**). Taking into consideration only the reported mDP values and subunit composition of seeds, skins, and wines, it seems as though tannin composition and distribution in wine resembles more of that of the seeds than that of the skins and that this feature is amplified as maceration time increases (**Figure 8b**).

As noted previously, upon crushing and during winemaking, tannins not only undergo modifications in their original molecular weight, but they can also undergo further reactions and re-arrangements with anthocyanin, to form the so-called polymeric pigments (Section 3.4), as well as with other flavan-3-ols and/or tannins dimers and trimers [153].

The ability of wine tannins to interact with proteins provides the physiological basis of the sensation of astringency, of paramount importance in red wines. Astringency is a tactile (not a taste) sensation with a marked temporal aspect that appears as a pucker feeling (typically in the upper lip) and/or feeling of dryness in the palate, which arises from a sudden loss of lubrication of the oral epithelium [154–157]. Astringency is not confined to a particular region of the mouth but is a diffuse surface phenomenon, which typically develops between 15 and 20 s [158, 159].

The development of astringency necessitates the presence of both tannins and proteins capable of interact with each other. The wine provides the tannins. The proteins are provided by the subject's saliva. The saliva of humans and other mammals contains proline-rich proteins (PRPs) [160]. Salivary PRPs constitute 70% of the proteins in saliva and are made up of three types: the acidic, glycosylated, and basic proteins, which comprise roughly 30, 23, and 17% of unstimulated saliva, respectively [161]. While acidic- and glycosylated proline-rich proteins have specific biological roles, basic PRPs display high affinity for tannins [162, 163]. Basic PRPs have a relatively preserved sequence across a wide range of species of mammals, composed of a 19-aminoacid sequence dominated by proline, glutamine, and glycine [164]. This sequence is repeated with variations between 5 and 15 times to result in a protein of ~150 amino acids in length that is extended and random-coiled in confirmation [164, 165]. The flexibility of the protein is further improved by the existence of several low-energy conformations [166, 167]. The biological role of astringency in mammals has been intensively debated but there is now evidence that astringency may be an evolutionary defense mechanism against dietary anti-feeding factors [168]. Tannins can have a variety of harmful effects on animals, including sequestration of iron and inhibition of digestive enzymes [160]. The role of PRPs may be to bind to the tannins and precipitate them, thereby preventing harmful effects in the gastrointestinal tract [169].

The interaction between PRPs and tannins results from hydrogen bonding between the tertiary amide or carbonyl groups of the proline subunits and the hydroxyl groups of the tannin [167, 170]. However, hydrophobic interactions whereby the hydrophobic face of the aromatic ring of the phenol interacts with the pyrrolidine ring of the protein may also cooperatively aid in the complexation process [163, 170, 171] (**Figure 9**). Likewise, the galloyl ring in epicatechin-3-*O*-gallate and tannins containing this flavan-3-ol provides supplementary aromatic surfaces that may engage in hydrophobic complexation with the proline ring as well [163, 172]. From this perspective, increase in the percentage of galloylation in a given tannin should result in enhanced perceived astringency.

The development of astringency follows a three-stage process, consistent with the in-mouth temporal development of this sensation [173, 174]. In the first stage, the binding of multidentate tannins to several sites on the protein causes the randomly coiled protein to coil around the phenol, compacting the protein. In the second stage, the tannin fractions of the protein-tannin complexes cross-link forming polyphenol bridges and creating protein dimers.

During the third stage, the dimers aggregate to form larger complexes that eventually precipitate out of solution. In wines, the spectrum of subtle differences in astringency sensations was compiled in the "red wine mouthfeel wheel" in which astringency is categorized in 7 subqualities and 30 subattributes [175], which highlights the complex nature of this sensation in red wines.

Perhaps the most determining factor on astringency development in red wines is, simply, tannin concentration [105, 176–178]. However, additional effects include wine pH and ethanol content [179, 180], viscosity [181, 182], and the presence of sugars [183, 184] and polysaccharides [121, 185]. Also, the tannin stereochemistry, composition, and connectivity affect the perception of astringency. For example, the dimer (+)-catechin-(+)-catechin linked via a C4 → C8 interflavan bond has lower astringency than its C4 → C6 counterpart or than the C4 → C8 (+)-catechin-(-)-epicatechin dimer [103]. Second to tannin concentration, tannin size (molecular weight) also modulates the development of astringency [186–189]. Precipitation of salivary PRPs is enhanced by the presence of high molecular weight tannins that have some structural flexibility, as in the case of tannins containing freely rotating interflavan bonds and gallolyl groups, due to a larger number of available binding sites for interaction with the proline residues [102]. Larger tannins can also engage in self-association, thereby promoting complex aggregation. However, the relationship between polymer size and perceived astringency does not progress linearly. For example, at equimolecular concentrations, Vidal et al. found that a wine-like solution containing tannins with an mDP of 70 is only two times more astringent than one with an mDP of 3 [188]. Empirical evidence that astringency elicited by grape tannins is mostly concentration-driven can be found by comparatively assessing astringency in seeds (overall high tannin concentration, with tannins of low molecular weight) with that of skins (relative lower tannin concentration than seeds but with tannins of high molecular weight). Seeds typically elicit a much higher astringency sensation than skins in spite of being composed of low molecular weight tannins and this may be due to the fact that seed tannins are typically at much higher concentrations than skin tannins on a berry fresh weight basis.

Lastly, astringency perception in red wines is also influenced by factors extrinsic to the nature of the tannin and PRP's interactions, including salivary flow rate [159], sensitivity to the bitter agent 6-*n*-propylthiouracil (PROP), known as PROP status [190], and frequency of exposure [183, 191]. Sensitivity to PROP seems to have equivocal effects on astringency development; however, individuals with low salivary flow rate (1.92 g/min) perceive astringency later and with higher intensity than individuals with high salivary flow rate (3.73 g/min) [159].

Some oligomeric and polymeric tannins can also interact with gustatory receptors, and they can thus elicit a mild bitter response [26]. Time-intensity sensory studies have shown that bitterness perception decreases from flavan-3-ols monomers to trimers [103]. More generally, bitterness decreases as the polymer size increases, probably because of the difficulty of large polymers to diffuse inside the taste bud pores [26].

3.3.2. Extraction during winemaking

Tannin dimers and trimers follow extraction kinetics similar to those reported for flavan-3-ols, and as such these oligomers can be extracted in the absence of ethanol, for example during prefermentative cold soak [111]. In Tempranillo wines after a 2-day cold soak, the concentration of tannin dimers, trimers, and tetramers increased from 22, 23, and 0 mg/L, respectively, to 27, 30, and 6 mg/L, respectively, after postfermentative maceration for 1 week [109]. In another study, the levels of the B2 dimer and the C trimer increased from about 6 mg/L at day 2 to about 22 mg/L (B2 dimer) and to 11 mg/L (C trimer) after 20 days of maceration in the white variety Viura [108]. These results suggests that as the oligomers increases in size, their extraction and retention into wine progresses more slowly.

Extraction of tannins (mDP ≥ 5) into wine during winemaking has been followed using different analytical approaches, ranging from protein precipitation [47, 58] to acid-catalyzed depolymerization followed by HPLC analysis [111, 146]. Expectedly, the comparison of extraction patterns established by both methods is tenuous at best, as protein precipitation and HPLC greatly differ in the amount, composition, and molecular weight of the tannins that each method quantify. Moreover, extraction patterns of skin and seed tannins are not necessarily similar and they certainly diverge from each other as maceration progresses; as such it is often difficult to unequivocally ascertain the source of tannin extraction during the time course of maceration. Using acid-catalyzed depolymerization and HPLC, the extraction of skin tannins into wine was found to follow a Boltzmann sigmoid model [146]. In this model, a lag phase of initially slow extraction is observed due to the period of time required for the tannins to diffuse out of the berry cells and into the fermentor. The extent of this lag phase may be modulated by the degree of berry crushing or the ethanol concentration. This lag phase is followed by a plateau concentration, which is reached when tannin concentration is at its apparent maximum. Likewise, the extraction of tannins from seeds has also been modeled using a Boltzmann sigmoid extraction pattern, but this model has only proved valid for the first 6 days of contact with the fermentation solids [111]. In model wine experiments containing only seeds with varying ethanol levels, these same authors observed an initial slow extraction of tannins, which was attributed to the period of time required for these tannins to diffuse out of the seed cells and into the solution. As with skins, this lag period was followed by a plateau, the value of which increased with ethanol content, suggesting an initial effect of the ethanol in the degradation of the outer protective layers of the seeds. However, from day 6 to day 10, tannin extraction into wine increased linearly. The authors attributed this observation to the hydration of seed cells and subsequent cellular leakiness, leading to a somewhat abrupt release of seed tannins.

In Merlot wines, tannins measured by protein precipitation increased almost linearly during the first 7 days of maceration. Moreover, extended maceration for 20 days increased tannin extraction from a mean of 469 mg/L in control wines to 985 mg/L in the extended maceration wines [58]. Also, in Merlot wines, protein precipitable tannins were followed at 2-day intervals during a 30-day maceration period. Tannin extraction occurred quickly and almost linearly from day 2 to day 10, then reached a plateau between day 10 and 22, and was followed by an almost linear increase up to the peak of extraction [47] (**Figure 11a**). While the observed plateau between day 10 and 20 is consistent with seed hydration, the ensuing linear increase is consistent with the extraction of seed tannins [111]. Indeed, after 30 days of maceration, the proportion of seed-derived tannins in finished wines was estimated to be about 73% in Cabernet Sauvignon [43] and between 73 and 80% in Merlot wines [47, 116]. It has also been suggested that postfermentation extraction of tannins could be the result of a desorption mechanism whereby ethanol may disrupt the noncovalent interactions of the previously extracted tannins that were bound to cell-wall material [192]. It is unclear if this linear increase is the result of one or a

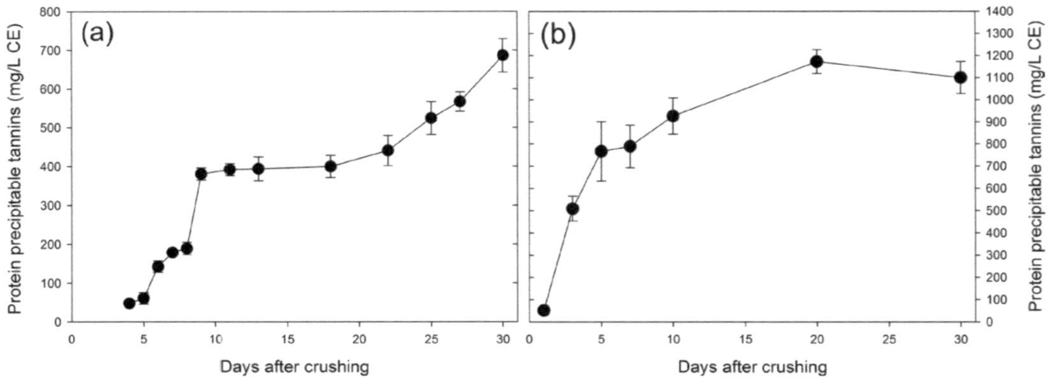

Figure 11. Extraction patterns of protein precipitable tannins during a 30-day maceration period in (a) Merlot and (b) Cabernet Sauvignon wines. CE: catechin equivalents. Adapted from Refs. [43, 47].

combination of both mechanisms, but in the range of 11–14.4% ethanol this supposed "desorptive" effect seems to be inexistent [47, 116]. It is also unclear if the extraction patterns shown in **Figure 11** will apply to other varieties with vastly different seed and skin phenolic composition, such as Pinot Noir, Sangiovese, Tempranillo, or Nebbiolo.

The extraction into wine and the ensuing fate of oligomeric and polymeric tannins during postmaceration and aging is also governed by the matrix composition, including the presence (or lack thereof) of anthocyanins and the presence of other compounds known to bind with wine tannins, primarily mannoproteins from yeast origin, polysaccharides, and other cell-wall components. Indeed, wine tannins can also react via noncovalent interactions with cell-wall material present during fermentation. These noncovalent interactions include hydrogen bonding and hydrophobic interactions. For example, the hydrogen-bonding-mediated interaction between tannins and polysaccharides is illustrated in **Figure 12**. As tannins increase in molecular weight, each additional monomer increases the number of sites that can form hydrogen bonds between the tannin and cell-wall components [192–195]. An increase in polymerization also increases the hydrophobic character of the tannin. The presence of galloylated units, more common on seed than in skins tannins, enhances the tannin-polysaccharide interaction as well, as highly galloylated tannins may be encapsulated by hydrophobic pockets and pores in the polysaccharide network [195, 196]. These findings had led to the hypothesis that the failure to recover high molecular weight tannins in wine is the result of tannin-cell-wall interactions occurring during winemaking and that the binding capacity of the cell walls is influenced by both tannins and polysaccharide structure and composition [192]. A series of studies conducted by Bindon and colleagues had confirmed this hypothesis. Bindon et al. found a significant relationship between the tannin molecular mass and the proportion of tannins adsorbed by skin cell-wall polysaccharides, with the end result being that high molecular weight tannins (>15,000 g/mol) are not extractable and/or removed from the wine by interaction with cell-wall components [194, 197–199]. In practical terms, the proportion of bound (and thus "missing") tannins during Cabernet Sauvignon winemaking have been estimated to range from 17 to 29% of those originally measured in the fruit [200, 201].

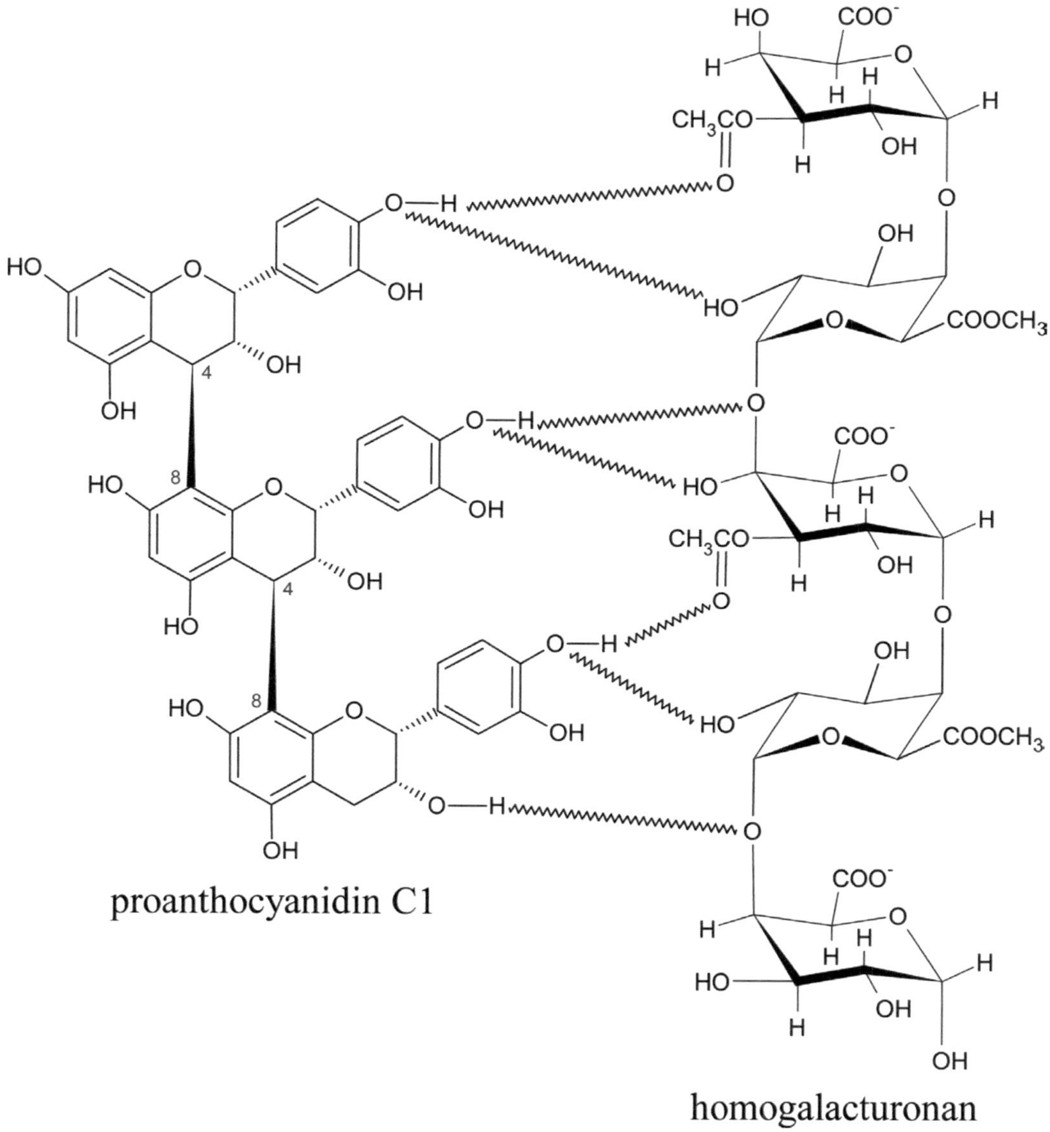

Figure 12. Hydrogen-bonding-mediated interaction between the hydroxyl groups of the tannin trimer C1 and the oxygen atoms and glycosidic linkages of the polysaccharide homogalacturonan. Redrawn and modified from Ref. [192].

3.4. Polymeric pigments and other anthocyanin-derived pigments

3.4.1. Occurrence, general chemistry, and sensory aspects

An overview of the variety of reactions anthocyanins can undergo during maceration and aging is presented in **Figure 13**. Upon crushing (and releasing from vacuoles) and during aging, anthocyanins can readily react with a variety of electrophiles and nucleophiles, including other anthocyanins [202], flavan-3-ols via direct condensation [203–206], or acetaldehyde-mediated [204, 205, 207, 208], dimeric or trimeric tannins [86, 209], lactic acid [204], glyceraldehyde [210], acetaldehyde [211, 212], pyruvic acid [212, 213], glyoxylic acid [214, 215], and a number of other aldehydes, including furfural, 5 hydroxymethylfurfural, isovaleraldehyde, benzaldehyde, propionaldehyde, isobutyraldehyde, formaldehyde, and 2-methylbutyraledehyde [215, 216]. Aldehydes are formed either from the metabolism of *Saccharomyces cerevisiae* during alcoholic fermentation [212, 217] or from the metal-catalyzed oxidation of

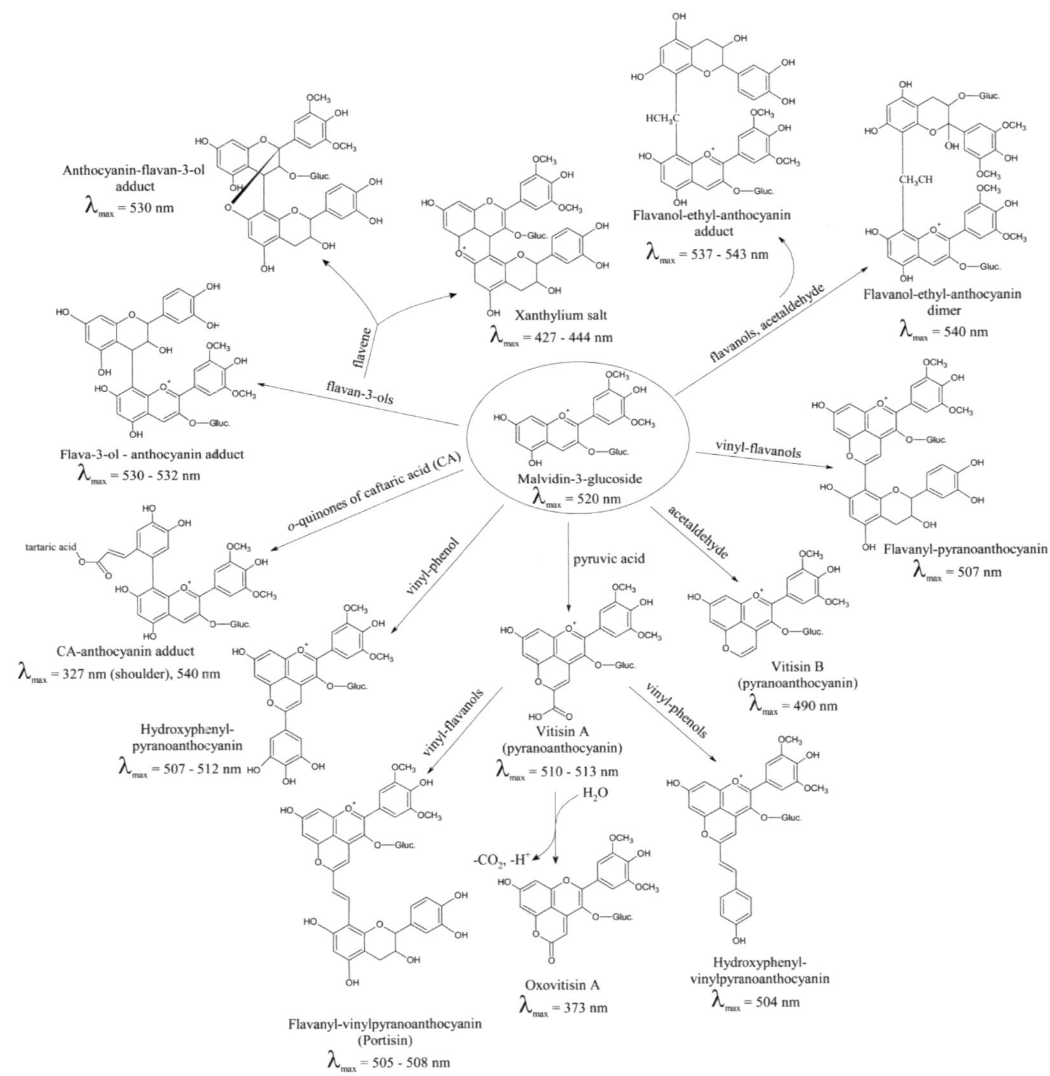

Figure 13. Main chemical pathways involving anthocyanins during winemaking and aging of red wines. Malvidin-3-glucoside is shown as the starting anthocyanin, but other anthocyanins may also participate in these reactions. Redrawn and adapted from Ref. [219] and complemented with data from Refs. [26, 204, 220, 221].

several wine substrates, especially ethanol (to acetaldehyde), most particularly thorough the Fenton Reaction [17, 218]. All these anthocyanin-derived products have spectral properties different from that of the native anthocyanins (**Figure 13**). Ultimately, these reactions together with the formation of oligomeric and polymeric pigments account for the evolution of wine color, from deep purple, due to monomeric anthocyanins and self-association copigmentation reactions in young wines, to orange, brick-red tones as the wine ages.

Polymeric pigments encompass a variety of winemaking artifacts formed by the covalent association, either direct or mediated by an aldehyde (e.g., acetaldehyde, glyceraldehyde, and glyoxilic acids) between anthocyanins and tannins. Relative to intact anthocyanins, polymeric pigments share two fundamental features: they are (1) partially resistant to bisulfite bleaching and (2) more resilient to pH changes [222, 223]. This is because polymerization protects by steric hindrance of the chromophore of the anthocyanin from the attack of water and other nucleophiles, oxidation, or the bleaching effect of SO_2 [222, 224]. The presence of anthocyanins during maceration increases the solubility and retention of tannins [52–54], an observation classically attributed to the formation of polymeric pigments. Singleton hypothesized that the glucose moiety in the anthocyanin and the polarity of the flavylium cation may decrease the precipitability of the resulting polymeric pigment [225]. In a subsequent experiment in which white wines were produced with different portions of added tannins and anthocyanins, the retention of tannins and formation of polymeric pigments increased in the presence of added anthocyanins [53]. However, this later study also found that the stoichiometric addition of anthocyanins relative to tannins approached an ideal proportion, where excessive anthocyanins did not increase pigmented polymer formation. This finding suggests that the proportion of anthocyanins and tannins during maceration can condition anthocyanin and tannin stability and the subsequent formation of polymeric pigments.

Polymeric pigments modulate wine color, long-term color stability, and possibly astringency changes during winemaking and aging. From the perspective of color, the UV-visible spectra of anthocyanin-flavan-3-ol adducts (i.e., dimers) resulting from condensation with aldehydes is bathochromically shifted (i.e., bluish color) compared to those of their precursors, with a shift in absorbance of 10 nm for linear substituents and of 20 nm for branched substituents [26, 207, 216] (**Figure 13**). However, the chromatic properties of polymeric pigments are less clear. Indeed, a numerical value for the molar extinction coefficient of polymeric pigments at wine pH is not yet available, but (indirect) experimental evidence suggests that it should be comparatively lower than that of the intact anthocyanins [37, 43, 226]. This, together with the disappearance and/or transformation of anthocyanins, may in turn explain why the color of red wine not only changes in hue but also decreases in saturation during aging.

Evidence for the existence of polymeric pigments in wine is abundant [37, 43, 143, 227–229], and some possible structures are shown in **Figure 14**. In Pinot Noir wines, pigmented polymers were isolated and characterized as mixtures of dimers to octamers in which the anthocyanin moiety was linked to the flavan-3-ol by B-type and A-type linkages [227]. In a separate study, ultrafiltration and gel adsorption chromatography combined with ^{1}H, ^{13}C, and 2D-NMR were used to characterize a high molecular weight tannin polymer (>5 kDa) isolated from a Bordeaux red wine [229]. The structural backbone of this polymer consisted of a tannin chain with (-)-epicatechin,

(+)-catechin, and (-)-epicatechin-3-*O*-gallate as extension and terminal subunits. The presence of acetaldehyde bridges was also observed in the A ring of some subunits as well as that of pyranoanthocyanins linked to the backbone via C4 → C6 or C4 → C8 linkages. Interestingly, polysaccharides were also found to be present within the structure, although these were not covalently linked to the tannin backbone. Organic and phenolic acids as well as aminoacids were also found to be part of the polymeric fraction structure. This appears to be the first report to completely elucidate the heterogeneous structure of these compounds as they occur in red wines. Lastly, in Cabernet Sauvignon wines, pigmented material was isolated by preparative HPLC [43]. The spectral features of the pigmented material featured a comparatively higher 280 to 520 nm absorbance ratio compared to that of intact anthocyanins (thus indicating the presence of flavan-3-ols) and an mDP between 5 and 10 units. The tannin component of this polymeric material was also put in evidence by subjecting the polymer to protein precipitation with bovine serum albumin (BSA), further indicating the potential astringent properties of this material.

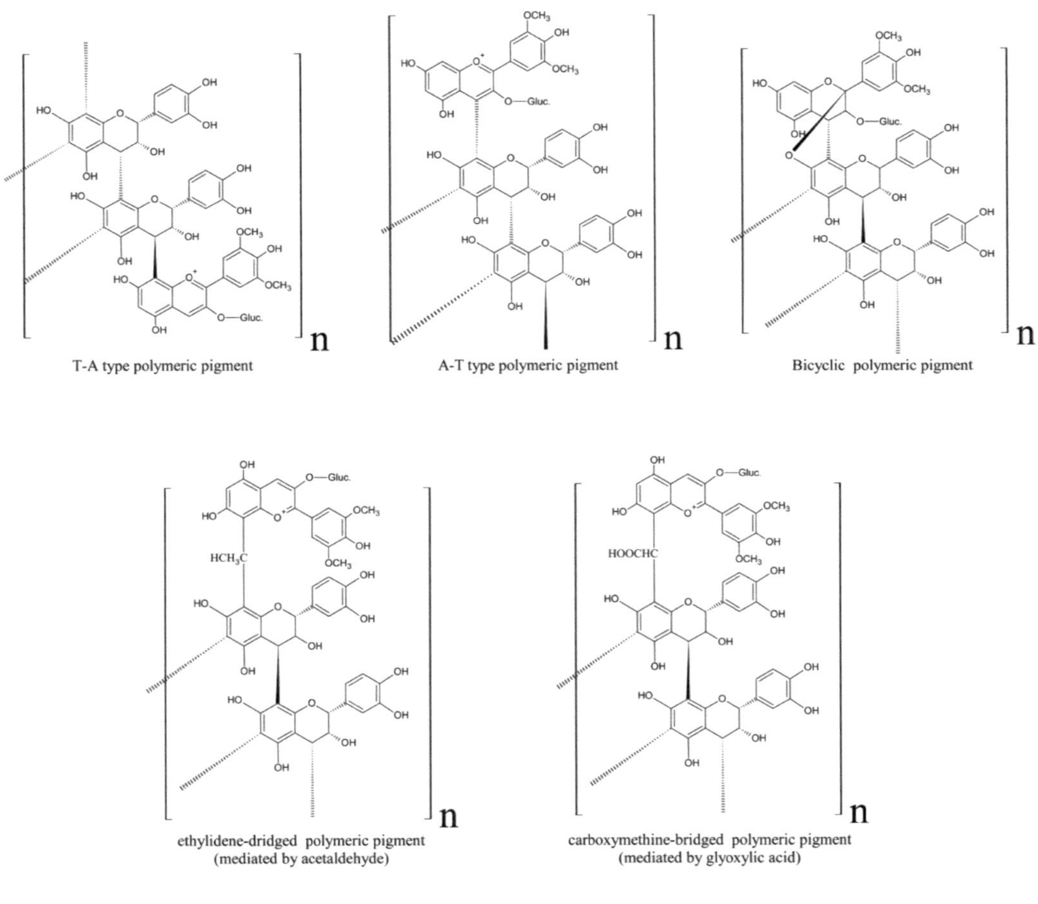

Figure 14. Potential structures of polymeric pigments resulting from the covalent reaction between anthocyanins (A) and tannins (T). Only C4 → C8 and A-type interflavanic bonds are depicted but C4 → C6 interflavanic bonds can also occur.

From a sensory standpoint, polymeric pigments affect two key aspects in red wines: color development and mouthfeel modification. Direct sensory evidence of the role of polymeric pigments in astringency changes during aging is, however, relatively recent. Indeed, the observed "lessening" of astringency during aging was thought to be the result of the reaction of anthocyanins with tannins of various sizes to give rise to polymeric pigments [222, 230]. However, the structural complexity and heterogeneity of these pigments prevented their isolation and subsequent chemical and sensory characterization. With the advent of new analytical, semipreparative, and preparative HPLC techniques, the characterization of these compounds has become possible. Work by Vidal and colleagues in 2004 found that polymeric pigments with mDP of ~3 and 9 and bearing an anthocyanin moiety were less astringent than apple tannins with the same mDP but deprived of anthocyanins [40]. Moreover, these authors showed that modifying the molecular structure by introducing an acetaldehyde bridge decreased astringency but also increased bitterness. An explanation for the comparatively lower astringency of polymeric pigments relative to that of intact tannins is that the incorporation of an anthocyanin moiety with its glycoside portion increases the polarity of the polymer [225]. As the development of astringency is partially governed by hydrophobic interactions between salivary PRPs and phenols, the higher hydrophilic character of the pigmented polymer would decrease the interaction of this polymeric material with salivary PRPs and thus reduce perceived astringency.

Recently, Weber et al., using size-exclusion chromatography on Sephadex resin, isolated 14 tannin fractions from the 2005 Cabernet Sauvignon wine [37]. Anthocyanins, mainly malvidin-3-glucoside, were found in the first 10 fractions, indicating the pigmented nature of the polymeric fraction. Fractions 1 to 3 were composed of large polymeric pigments as measured by protein precipitation, with low anthocyanin and tannin content; fractions 4 to 7 consisted of anthocyanin-rich pigmented polymers with medium tannin content; and fractions 8 to 14 consisted of small-sized, tannin-like oligomers with very low anthocyanin content but very high tannin content. Upon sensory evaluation of each fraction dissolved in model wine at iso-concentrations of 500 mg/L, fractions with a low amount of anthocyanins elicited higher astringency, suggesting that further incorporation of anthocyanins into polymers should result in a decrease in astringency. In another report, a pigmented polymer isolated from a Bordeaux red wine was fractionated into eight fractions of different molecular weights by gel permeation chromatography [229]. Upon dissolution of these fractions in 1% ethanol at iso-concentrations, astringency was found no to vary in seven of these fractions in spite of differences in mDP and degree of galloylation. However, one fraction consisting of 50% polysaccharides was found to be less astringent. Overall, current evidence suggests that the lessening of astringency along with red wine aging may not be related to a drastic change in the total amount of tannin present. Rather, the structural modification of wine tannins, primarily resulting from the incorporation of anthocyanins, and, secondarily, from the addition of other metabolites such as carbohydrates, proteins, and polysaccharides, may drive changes in perceived astringency during aging.

3.4.2. *Formation during winemaking*

Polymeric pigment formation increases progressively during maceration and aging (**Figure 15**) ultimately leading to color changes, modification of mouthfeel properties, and,

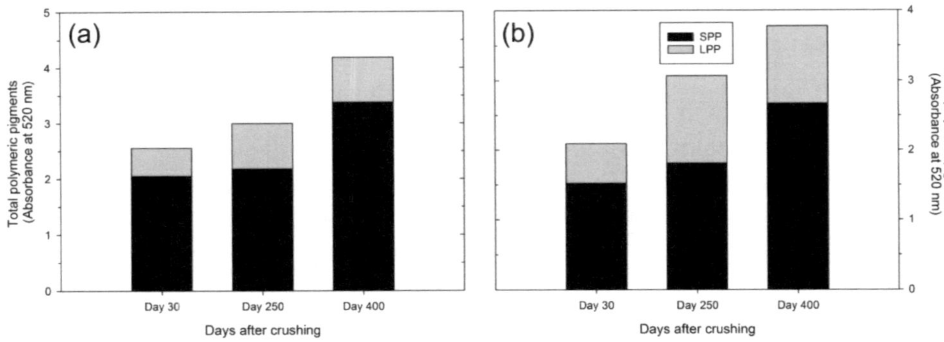

Figure 15. Overview of the formation of polymeric pigments during maceration and bottle aging of Cabernet Sauvignon wines processed with a maceration length of 10 days (control) (a) and 30 days (extended maceration) (b). SPP: small polymeric pigments; LPP: large polymeric pigments. Adapted from Ref. [43].

eventually, precipitation. Singleton and Trousdale reported that white wines produced with added tannins and anthocyanins showed a linear increase in polymeric pigment content after addition of seed tannins in the range of 0 to 1000 mg/L (gallic acid equivalents) and anthocyanins in the range of 0 to 500 mg/L [53]. Using protein precipitation, Harbertson et al. found that large polymeric pigments (LPP), which precipitate BSA, increased by 70% between pressing and 185-day postpressing in Merlot wines [58]. Small polymeric pigments (SPP), which do not precipitate BSA, and are assumed to be composed of tannin-anthocyanin dimers, either of direct condensation or mediated by acetaldehyde [231], comparatively increased 30% from pressing to 185-day postpressing. In this same experiment, wines produced with extended maceration and saignée[6] and containing a higher concentration of tannins gave rise to an enhanced formation of LPP; however, this occurred with a decline in the anthocyanin content of 43% relative to its peak concentration [58]. A similar trend was observed in Merlot wines obtained with extended maceration (30 days), in which a two-fold increase of the total polymeric pigments was observed from day 4 to day 30, along with significant losses of malvidin, delphinidin, petunidin, and peonidin anthocyanin derivatives [47]. Furthermore, this later work demonstrated that the formation of polymeric pigments alone during extended maceration was only partially responsible for the observed anthocyanin loss because an increase in the polymeric pigment content of 13 mg/L from day 4 to day 30 occurred along with a drop in wine anthocyanins of 231 mg/L in this same time frame. In summary, these results suggest a complex relationship between tannin content, anthocyanin extraction (or loss), and polymeric pigment formation during maceration. As shown in **Figure 15**, a common feature of extended maceration seems to be the formation of polymeric pigments with the ability to precipitate BSA (and by a similar mechanism to elicit astringency), but this occurs at the expense of anthocyanin loss (and, consequently, of wine color saturation) (although this anthocyanin loss is generally not fully explained by the formation

[6]The practice of saignée consists of taking a portion of the must from the bottom of the tank before the onset of alcoholic fermentation with the aim of increasing the solid to volume ratio of the remaining must and then furthering the extraction of phenolics and aroma compounds from seeds and skins.

of polymeric pigments). Altogether, these findings suggest that the presence of anthocyanins invariably leads to the formation of polymeric pigments; yet, the proportion of anthocyanins and tannins during maceration, which is expected to differ widely depending on variety, clone, and viticultural and climatic conditions, as well as winemaking technique, will condition the amount of pigmented tannins that are effectively formed during winemaking and aging.

4. Concluding remarks

The above literature review was undertaken with the aim to highlight the remarkable chemical diversity of flavonoid phenolic compounds in grapes and wines. This diversity is furthered from the very first moment the grapes are crushed, thereby allowing vacuolar and pulp components to be released into the fermenting must and wine. This chemical diversity further increases during winemaking and bottle aging, thus adding to the already present chemical diversity, a variety of new sensory dimensions, ranging from changes in wine color and aroma to modification of astringency.

For a wide number of red grape varieties, the extraction of anthocyanins peaks during the first 4 or 5 days of maceration, which is followed by a decrease in concentration along with the lengthening of maceration. This decrease in anthocyanin concentration is typically accompanied by the formation of polymeric pigments, by which formation is modulated, among others, by the molar proportion of anthocyanins and tannins. Flavan-3-ols and small oligomeric tannins from skins are extracted within the first days of maceration, whereas the extraction of seed-derived tannins requires longer maceration times. It also seems that high molecular weight tannins are not retained into wine, probably due to interactions with polysaccharides and other nonphenolic materials during winemaking. Indeed, specific matrix effects affect the rate of retention of tannins into wine, particularly at the latter stages of maceration. These include (but are not limited to) the presence of anthocyanins, polysaccharides, and other cell-wall components such as structural proteins. As phenolic and nonphenolic compounds are extracted and/or formed during maceration and aging, a dynamic set of chemical and biochemical reactions occurs, resulting in the formation of new structures not previously found in grapes. Some of these new phenolic classes, which may also contain nonphenolic material of yeast and/or grape origin, are responsible for a variety of new sensory attributes. Polymeric pigments, bearing astringent and bitter properties different from those of intact tannins of equivalent molecular weight, are candidates for the changes in the mouthfeel and textural properties of red wines during maceration and aging. Although the taste and mouthfeel attributes of polymeric pigments are starting to be clarified, their interaction with other phenolic and nonphenolic materials and the volatile fraction of the wine matrix remain to be explored.

Different maceration techniques applied during red wine production affect the rate, quantity, and sometimes the chemical composition of the phenolic compounds that end up in the wine. Control and understanding of the factors that modulate phenolic extraction and retention into wine during maceration should ultimately allow the winemaker to adjust maceration variables to meet a given wine style sought and/or to comply with commercial specifications.

Author details

L. Federico Casassa

Address all correspondence to: lcasassa@calpoly.edu

Wine and Viticulture Department, California Polytechnic State University, San Luis Obispo, CA, USA

References

[1] Frankel, E.N., et al., Inhibition of oxidation of human low-density lipoprotein by phenolic substances in red wine. The Lancet, 1993. **341**(8843): pp. 454–457.

[2] Meyer, A.S., et al., Inhibition of human low-density lipoprotein oxidation in relation to composition of phenolic antioxidants in grapes (*Vitis vinifera*). Journal of Agricultural and Food Chemistry, 1997. **45**(5): pp. 1638–1643.

[3] Frankel, E.N., A.L. Waterhouse, and P.L. Teissedre, Principal phenolic phytochemicals in selected California wines and their antioxidant activity in inhibiting oxidation of human low-density lipoproteins. Journal of Agricultural and Food Chemistry, 1995. **43**(4): pp. 890–894.

[4] Corder, R., et al., Red wine procyanidins and vascular health. Nature, 2006. **444**(7119): pp. 566.

[5] Pasteur, L., Etudes sur le vin, ses maladies, causes qui les provoquent, procédés nouveaux pour le conserver pour et le vieillir. Deuxieme ed. 1875, Paris, France: Librairie E. Savy.

[6] Willstätter, R. and E.H. Zollinger, Untersuchungen über die anthocyane. VI. Über die farbstoffe der weintraube und der heidelbeere. Justus Liebigs Annalen der Chemie, 1915. **408**(1): pp. 83–109.

[7] Pissis, M., XLVIII. Observations upon Alicant wine, and particularly the Alicant raisin wine. Philosophical Magazine Series 1, 1806. **25**(100): pp. 313–317.

[8] Manceau, E., Sur le dosage du tannins dans les vins In: Repertoire de Pharmacie, C. Crinon, Editor. Paris. 1895. **1**: p. 592.

[9] Bate-Smith, E.C., The contribution of phenolic substances to quality in plant products. Qualitas Plantarum et Materiae Vegetabiles, 1958. **3**(1): pp. 440–455.

[10] Hilgard, E.W., The extraction of color and tannin during red-wine fermentation. Bulletin of the California Agricultural Experiment Station, 1887. **Bulletin N 77**.

[11] Hilgard, E.W., Colorimetric measurement of wines. Bulletin of the California Agricultural Experiment Station, 1886. **Bulletin N 60**.

[12] Haslam, E., Practical polyphenolics: from structure to molecular recognition and physiological action. 1998, Cambridge, United Kingdom: Cambridge University Press. p. 422.

[13] Waterhouse, A.L., Wine phenolics. Annals of the New York Academy of Sciences, 2002. **957**(1): pp. 21–36.

[14] Fulcrand, H., et al., Phenolic reactions during winemaking and aging. American Journal of Enology and Viticulture, 2006. **57**(3): pp. 289–297.

[15] Waterhouse, A.L. and V.F. Laurie, Oxidation of wine phenolics: a critical evaluation and hypotheses. American Journal of Enology and Viticulture, 2006. **57**(3): pp. 306–313.

[16] Singleton, V.L., Oxygen with phenols and related reactions in musts, wines, and model systems: observations and practical implications. American Journal of Enology and Viticulture, 1987. **38**(1): pp. 69–77.

[17] Danilewicz, J.C., Review of reaction mechanisms of oxygen and proposed intermediate reduction products in wine: central role of iron and copper. American Journal of Enology and Viticulture, 2003. **54**(2): pp. 73–85.

[18] Danilewicz, J.C., Review of oxidative processes in wine and value of reduction potentials in enology. American Journal of Enology and Viticulture, 2012. **63**(1): pp. 1–10.

[19] Jeandet, P., et al., Occurrence of a resveratrol beta-D-glucoside in wine: preliminary studies. Vitis, 1994. **33**(3): pp. 183–194.

[20] Mattivi, F., F. Reniero, and S. Korhammer, Isolation, characterization, and evolution in red wine vinification of resveratrol monomers. Journal of Agricultural and Food Chemistry, 1995. **43**(7): pp. 1820–1823.

[21] Cheynier, V., et al., Must browning in relation to the behavior of phenolic compounds during oxidation. American Journal of Enology and Viticulture, 1990. **41**(4): pp. 346–349.

[22] Wildenradt, H.L. and V.L. Singleton, The production of aldehydes as a result of oxidation of polyphenolic compounds and its relation to wine aging. American Journal of Enology and Viticulture, 1974. **25**(2): pp. 119–126.

[23] Darias-Martín, J., et al., Effect of caffeic acid on the color of red wine. Journal of Agricultural and Food Chemistry, 2002. **50**(7): pp. 2062–2067.

[24] Chatonnet, P., et al., The origin of ethylphenols in wines. Journal of the Science of Food and Agriculture, 1992. **60**(2): pp. 165–178.

[25] Kennedy, J.A., C. Saucier, and Y. Glories, Grape and wine phenolics: history and perspective. American Journal of Enology and Viticulture, 2006. **57**(3): pp. 239–248.

[26] Cheynier, V., Flavonoids in wine, In: Flavonoids, chemistry, biochemistry and applications, K.R.M. Øyvind, M. Andersen, Editors. 2006, Boca Raton, FL: CRC Press, Taylor & Francis Group. pp. 263–318.

[27] He, F., et al., Mass-spectrometry evidence confirming the presence of pelargonidin-3-O-glucoside in the berry skins of Cabernet Sauvignon and Pinot Noir (*Vitis vinifera* L.). Australian Journal of Grape and Wine Research, 2010. **16**(3): pp. 464–468.

[28] Cheynier, V., et al., Structure and properties of wine pigments and tannins. American Journal of Enology and Viticulture, 2006. **57**(3): pp. 298–305.

[29] Van Buren, J.P., J.J. Bertino, and W.B. Robinson, The stability of wine anthocyanins on exposure to heat and light. American Journal of Enology and Viticulture, 1968. **19**(3): pp. 147–154.

[30] Brouillard, R., S. Chassaing, and A. Fougerousse, Why are grape/fresh wine anthocyanins so simple and why is it that red wine color lasts so long? Phytochemistry, 2003. **64** (7): pp. 1179–1186.

[31] Wrolstad, R.E., R.W. Durst, and J. Lee, Tracking color and pigment changes in anthocyanin products. Trends in Food Science & Technology, 2005. **16**(9): pp. 423–428.

[32] Brouillard, R., Chapter 1—Chemical structure of anthocyanins A2 - Markakis, Pericles, In: Anthocyanins as food colors. 1982, Academic Press. pp. 1–40. ISBN: 9780124725508/ 9780323157902

[33] Koeppen, B.H. and D.S. Basson, The anthocyanin pigments of Barlinka grapes. Phytochemistry, 1966. **5**(1): pp. 183–187.

[34] Boulton, R., The copigmentation of anthocyanins and its role in the color of red wine: a critical review. American Journal of Enology and Viticulture, 2001. **52**(2): pp. 67–87.

[35] Somers, T.C. and M.E. Evans, Grape pigment phenomena: interpretation of major colour losses during vinification. Journal of the Science of Food and Agriculture, 1979. **30**(6): pp. 623–633.

[36] Fulcrand, H., et al., Structure of new anthocyanin-derived wine pigments. Journal of the Chemical Society, Perkin Transactions 1. 1996(7): pp. 735–739.

[37] Weber, F., et al., Sensory and chemical characterization of phenolic polymers from red wine obtained by gel permeation chromatography. American Journal of Enology and Viticulture, 2013. **64**(1): pp. 15–25.

[38] Jordheim, M., et al., Molar absorptivities and reducing capacity of pyranoanthocyanins and other anthocyanins. Journal of Agricultural and Food Chemistry, 2007. **55**(26): pp. 10591–10598.

[39] Håkansson, A.E., et al., Structures and colour properties of new red wine pigments. Tetrahedron Letters, 2003. **44**(26): pp. 4887–4891.

[40] Vidal, S., et al., Taste and mouth-feel properties of different types of tannin-like polyphenolic compounds and anthocyanins in wine. Analytica Chimica Acta, 2004. **513**(1): pp. 57–65.

[41] Ortega-Regules, A., et al., A first approach towards the relationship between grape skin cell-wall composition and anthocyanin extractability. Analytica Chimica Acta, 2006. **563** (1–2): pp. 26–32.

[42] Nagel, C.W. and L.W. Wulf, Changes in the anthocyanins, flavonoids and hydroxycinnamic acid esters during fermentation and aging of Merlot and Cabernet Sauvignon. American Journal of Enology and Viticulture, 1979. **30**(2): pp. 111–116.

[43] Casassa, L.F., et al., Impact of extended maceration and Regulated Deficit Irrigation (RDI) in Cabernet Sauvignon Wines: characterization of proanthocyanidin distribution, anthocyanin extraction, and chromatic properties. Journal of Agricultural and Food Chemistry, 2013. **61**(26): pp. 6446–6457.

[44] Casassa, L.F., et al., Cofermentation of Syrah with Viognier: evolution of color and phenolics during winemaking and bottle aging. American Journal of Enology and Viticulture, 2012. **63**(4): pp. 538–543.

[45] Romero-Cascales, I., et al., The maceration process during winemaking extraction of anthocyanins from grape skins into wine. European Food Research and Technology, 2005. **221**(1–2): pp. 163–167.

[46] Bautista-Ortín, A.B., et al., The extraction kinetics of anthocyanins and proanthocyanidins from grape to wine in three different varieties. Journal international des sciences de la vigne et du vin, 2016. **50**(2).

[47] Casassa, L.F., et al., Effect of extended maceration and ethanol concentration on the extraction and evolution of phenolics, colour components and sensory attributes of Merlot wines. Australian Journal of Grape and Wine Research, 2013. **19**(1): pp. 25–39.

[48] Canals, R., et al., Influence of ethanol concentration on the extraction of color and phenolic compounds from the skin and seeds of Tempranillo grapes at different stages of ripening. Journal of Agricultural and Food Chemistry, 2005. **53**(10): pp. 4019–4025.

[49] Revilla, E., et al., Value of high-performance liquid chromatographic analysis of anthocyanins in the differentiation of red grape cultivars and red wines made from them. Journal of Chromatography A, 2001. **915**(1–2): pp. 53–60.

[50] García-Beneytez, E., E. Revilla, and F. Cabello, Anthocyanin pattern of several red grape cultivars and wines made from them. European Food Research and Technology, 2002. **215**(1): pp. 32–37.

[51] Morata, A., et al., Adsorption of anthocyanins by yeast cell walls during the fermentation of red wines. Journal of Agricultural and Food Chemistry, 2003. **51**(14): pp. 4084–4088.

[52] Kantz, K. and V.L. Singleton, Isolation and determination of polymeric polyphenols in wines using Sephadex LH-20. American Journal of Enology and Viticulture, 1991. **42**(4): pp. 309–316.

[53] Singleton, V.L. and E.K. Trousdale, Anthocyanin-tannin interactions explaining differences in polymeric phenols between white and red wines. American Journal of Enology and Viticulture, 1992. **43**(1): pp. 63–70.

[54] Kilmister, R.L., et al., A role for anthocyanin in determining wine tannin concentration in Shiraz. Food Chemistry, 2014. **152**: pp. 475–482.

[55] Casassa, L.F. and S.E. Sari, Sensory and chemical effects of two alternatives of prefermentative cold soak in Malbec wines during winemaking and bottle ageing. International Journal of Food Science & Technology, 2014. **50**(4): pp. 1044–1055.

[56] Medina, K., et al., Yeast interactions with anthocyanins during red wine fermentation. American Journal of Enology and Viticulture, 2005. **56**(2): pp. 104–109.

[57] Vasserot, Y., S. Caillet, and A. Maujean, Study of anthocyanin adsorption by yeast lees. Effect of some physicochemical parameters. American Journal of Enology and Viticulture, 1997. **48**(4): pp. 433–437.

[58] Harbertson, J.F., et al., Chemical and sensory effects of saignée, water addition, and extended maceration on high Brix must. American Journal of Enology and Viticulture, 2009. **60**(4): pp. 450–460.

[59] Rentzsch, M., M. Schwarz, and P. Winterhalter, Pyranoanthocyanins—an overview on structures, occurrence, and pathways of formation. Trends in Food Science & Technology, 2007. **18**(10): pp. 526–534.

[60] Morel-Salmi, C., et al., Effect of flash release treatment on phenolic extraction and wine composition. Journal of Agricultural and Food Chemistry, 2006. **54**(12): pp. 4270–4276.

[61] Hermosín Gutiérrez, I., Influence of ethanol content on the extent of copigmentation in a Cencibel young red wine. Journal of Agricultural and Food Chemistry, 2003. **51**(14): pp. 4079–4083.

[62] Sipiora, M.J. and M.-J.G. Granda, Effects of pre-veraison irrigation cutoff and skin contact time on the composition, color, and phenolic content of young Cabernet Sauvignon wines in Spain. American Journal of Enology and Viticulture, 1998. **49**(2): pp. 152–162.

[63] González-Neves, G., et al., Pigment profile of red wines cv. Tannat made with alternative winemaking techniques. Journal of Food Composition and Analysis, 2010. **23**(5): pp. 447–454.

[64] Su, C.T. and V.L. Singleton, Identification of three flavan-3-ols from grapes. Phytochemistry, 1969. **8**(8): pp. 1553–1558.

[65] Wu, Q., M. Wang, and J.E. Simon, Determination of proanthocyanidins in fresh grapes and grape products using liquid chromatography with mass spectrometric detection. Rapid Communications in Mass Spectrometry, 2005. **19**(14): pp. 2062–2068.

[66] Mattivi, F., et al., Differences in the amount and structure of extractable skin and seed tannins amongst red grape varieties. Australian Journal of Grape and Wine Research, 2009. **15**(1): pp. 27–35.

[67] Delcambre, A. and C. Saucier, Identification of new flavan-3-ol monoglycosides by UHPLC-ESI-Q-TOF in grapes and wine. Journal of Mass Spectrometry, 2012. **47**(6): pp. 727–736.

[68] Monagas, M., B. Bartolomé, and C. Gómez-Cordovés, Updated knowledge about the presence of phenolic compounds in wine. Critical Reviews in Food Science and Nutrition, 2005. **45**(2): pp. 85–118.

[69] Baba, S., et al., In vivo comparison of the bioavailability of (+)-catechin, (-)-epicatechin and their mixture in orally administered rats. The Journal of Nutrition, 2001. **131**(11): pp. 2885–2891.

[70] Fraga, C.G. and P.I. Oteiza, Dietary flavonoids: role of (-)-epicatechin and related procyanidins in cell signaling. Free Radical Biology and Medicine, 2011. **51**(4): pp. 813–823.

[71] Thorngate, J.H. and A.C. Noble, Sensory evaluation of bitterness and astringency of 3R(-)-epicatechin and 3S(+)-catechin. Journal of the Science of Food and Agriculture, 1995. **67**(4): pp. 531–535.

[72] Brossaud, F., V. Cheynier, and A.C. Noble, Bitterness and astringency of grape and wine polyphenols. Australian Journal of Grape and Wine Research, 2001. **7**(1): pp. 33–39.

[73] Kennedy, J.A., M.A. Matthews, and A.L. Waterhouse, Changes in grape seed polyphenols during fruit ripening. Phytochemistry, 2000. **55**(1): pp. 77–85.

[74] Prieur, C., et al., Oligomeric and polymeric procyanidins from grape seeds. Phytochemistry, 1994. **36**(3): pp. 781–784.

[75] Cadot, Y., M.T. Miñana Castelló, and M. Chevalier, Flavan-3-ol compositional changes in grape berries (*Vitis vinifera* L. cv Cabernet Franc) before veraison, using two complementary analytical approaches, HPLC reversed phase and histochemistry. Analytica Chimica Acta, 2006. **563**(1–2): pp. 65–75.

[76] Escribano-Bailón, M.T., et al., Proanthocyanidins in skins from different grape varieties. Zeitschrift für Lebensmittel-Untersuchung und Forschung, 1995. **200**(3): pp. 221–224.

[77] Souquet, J.-M., et al., Polymeric proanthocyanidins from grape skins. Phytochemistry, 1996. **43**(2): pp. 509–512.

[78] Monagas, M., et al., Monomeric, oligomeric, and polymeric flavan-3-ol composition of wines and grapes from *Vitis vinifera* L. Cv. Graciano, Tempranillo, and Cabernet Sauvignon. Journal of Agricultural and Food Chemistry, 2003. **51**(22): pp. 6475–6481.

[79] Obreque-Slier, E.a., et al., Comparative study of the phenolic composition of seeds and skins from Carménère and Cabernet Sauvignon grape varieties (*Vitis vinifera* L.) during ripening. Journal of Agricultural and Food Chemistry, 2010. **58**(6): pp. 3591–3599.

[80] Guerrero, R.F., et al., Phenolic characterisation of red grapes autochthonous to Andalusia. Food Chemistry, 2009. **112**(4): pp. 949–955.

[81] Rodríguez Montealegre, R., et al., Phenolic compounds in skins and seeds of ten grape *Vitis vinifera* varieties grown in a warm climate. Journal of Food Composition and Analysis, 2006. **19**(6–7): pp. 687–693.

[82] Kelm, M.A., et al., Mass spectral characterization of Uva Longanesi seed and skin extracts. American Journal of Enology and Viticulture, 2012. **63**(3): pp. 402–406.

[83] Kemp, B.S., R. Harrison, and G.L. Creasy, Effect of mechanical leaf removal and its timing on flavan-3-ol composition and concentrations in *Vitis vinifera* L. cv. Pinot Noir wine. Australian Journal of Grape and Wine Research, 2011. **17**(2): pp. 270–279.

[84] Boido, E., et al., Characterisation and evolution of grape polyphenol profiles of *Vitis vinifera* L. cv. Tannat during ripening and vinification. Australian Journal of Grape and Wine Research, 2011. **17**(3): pp. 383–393.

[85] Alcalde-Eon, C., et al., Pigment profiles in monovarietal wines produced in Uruguay. American Journal of Enology and Viticulture, 2006. **57**(4): pp. 449–459.

[86] Salas, E., et al., Reactions of anthocyanins and tannins in model solutions. Journal of Agricultural and Food Chemistry, 2003. **51**(27): pp. 7951–7961.

[87] Atanasova, V., et al., Structure of a new dimeric acetaldehyde malvidin 3-glucoside condensation product. Tetrahedron Letters, 2002. **43**(35): pp. 6151–6153.

[88] Saucier, C., D. Little, and Y. Glories, First evidence of acetaldehyde-flavanol condensation products in red wine. American Journal of Enology and Viticulture, 1997. **48**(3): pp. 370–373.

[89] Bark, et al., Spectroscopic studies on the oxidation of catechin in aqueous solution. Bulletin of the Korean Chemical Society (대한화학회), 2011. **32**(9): pp. 3443–3447.

[90] Carvalho, D.O., A.F. Curto, and L.F. Guido, Determination of phenolic content in different barley varieties and corresponding malts by liquid chromatography-diode array detection-electrospray ionization tandem mass spectrometry. Antioxidants, 2015. **4**(3): pp. 563–576.

[91] Guyot, S., et al., Influence of pH on the enzymic oxidation of (+)-catechin in model systems. Journal of Agricultural and Food Chemistry, 1995. **43**(9): pp. 2458–2462.

[92] Vivas de Gaulejac, N., et al., Study and quantification of monomeric flavan-3-ol and dimeric procyanidin quinonic forms by HPLC/ESI-MS. Application to red wine oxidation. Journal of the Science of Food and Agriculture, 2001. **81**(12): pp. 1172–1179.

[93] Lee, J., Degradation kinetics of grape skin and seed proanthocyanidins in a model wine system. Food Chemistry, 2010. **123**(1): pp. 51–56.

[94] Guyot, S., J. Vercauteren, and V. Cheynier, Structural determination of colourless and yellow dimers resulting from (+)-catechin coupling catalysed by grape polyphenoloxidase. Phytochemistry, 1996. **42**(5): pp. 1279–1288.

[95] Nikolantonaki, M., et al., Identification of adducts between an odoriferous volatile thiol and oxidized grape phenolic compounds: kinetic study of adduct formation under chemical and enzymatic oxidation conditions. Journal of Agricultural and Food Chemistry, 2012. **60**(10): pp. 2647–2656.

[96] Waterhouse, A.L. and M. Nikolantonaki, Quinone reactions in wine oxidation. In: Advances in wine research, S.B. Ebeler, G. Sacks, S. Vidal, P. Winterhalter, editors. American Chemical Society, 2015. **1203**: pp. 291–301.

[97] Rossi, J.A. and V.L. Singleton, Flavor effects and adsorptive properties of purified fractions of grape-seed phenols. American Journal of Enology and Viticulture, 1966. **17**(4): pp. 240–246.

[98] Robichaud, J.L. and A.C. Noble, Astringency and bitterness of selected phenolics in wine. Journal of the Science of Food and Agriculture, 1990. **53**(3): pp. 343–353.

[99] Kallithraka, S., J. Bakker, and M.N. Clifford, Evaluation of bitterness and astringency of (+)-cateching and (-)-epicatechin in red wine and in model solution. Journal of Sensory Studies, 1997. **12**(1): pp. 25–37.

[100] Gardner, R.J., Lipophilicity and the perception of bitterness. Chemical Senses, 1979. **4**(4): pp. 275–286.

[101] Charlton, A.J., et al., Tannin interactions with a full-length human salivary proline-rich protein display a stronger affinity than with single proline-rich repeats. FEBS Letters, 1996. **382**(3): pp. 289–292.

[102] Baxter, N.J., et al., Multiple interactions between polyphenols and a salivary proline-rich protein repeat result in complexation and precipitation. Biochemistry, 1997. **36**(18): pp. 5566–5577.

[103] Peleg, H., et al., Bitterness and astringency of flavan-3-ol monomers, dimers and trimers. Journal of the Science of Food and Agriculture, 1999. **79**(8): pp. 1123–1128.

[104] Poncet-Legrand, C., et al., Poly(L-proline) interactions with flavan-3-ols units: influence of the molecular structure and the polyphenol/protein ratio. Food Hydrocolloids, 2006. **20**(5): pp. 687–697.

[105] Casassa, L.F., et al., Sensory impact of extended maceration and Regulated Deficit Irrigation on Washington State Cabernet Sauvignon wines. American Journal of Enology and Viticulture, 2013. **64**(4): pp. 505–514.

[106] McManus, J.P., et al., The association of proteins with polyphenols. Journal of the Chemical Society, Chemical Communications, 1981(7): pp. 309b–311.

[107] Sun, B.S., et al., Transfer of catechins and proanthocyanidins from solid parts of the grape cluster into wine. American Journal of Enology and Viticulture, 1999. **50**(2): pp. 179–184.

[108] González-Manzano, S., J.C. Rivas-Gonzalo, and C. Santos-Buelga, Extraction of flavan-3-ols from grape seed and skin into wine using simulated maceration. Analytica Chimica Acta, 2004. **513**(1): pp. 283–289.

[109] González-Manzano, S., et al., Characterization of the mean degree of polymerization of proanthocyanidins in red wines using liquid chromatography–mass spectrometry (LC-MS). Journal of Agricultural and Food Chemistry, 2006. **54**(12): pp. 4326–4332.

[110] Gambuti, A., et al., Extraction of phenolic compounds from 'Aglianico' and 'Uva di Troia' grape skins and seeds in model solutions: influence of ethanol and maceration time. Vitis, 2009. **48**(4): pp. 193–200.

[111] Hernández-Jiménez, A., et al., Effect of ethanol on grape seed proanthocyanidin extraction. American Journal of Enology and Viticulture, 2012. **63**(1): pp. 57–61.

[112] Koyama, K., N. Goto-Yamamoto, and K. Hashizume, Influence of maceration temperature in red wine vinification on extraction of phenolics from berry skins and seeds of grape (*Vitis vinifera*). Bioscience, Biotechnology, and Biochemistry, 2007. **71**(4): pp. 958–965.

[113] Gil, M., et al., Influence of grape maturity and maceration length on color, polyphenolic composition, and polysaccharide content of Cabernet Sauvignon and Tempranillo wines. Journal of Agricultural and Food Chemistry, 2012. **60**(32): pp. 7988–8001.

[114] Peyrot des Gachons, C. and J.A. Kennedy, Direct method for determining seed and skin proanthocyanidin extraction into red wine. Journal of Agricultural and Food Chemistry, 2003. **51**(20): pp. 5877–5881.

[115] Llaudy, M.C., et al., Influence of ripening stage and maceration length on the contribution of grape skins, seeds and stems to phenolic composition and astringency in wine-simulated macerations. European Food Research and Technology, 2008. **226**(3): pp. 337–344.

[116] Casassa, L.F., et al., Influence of fruit maturity, maceration length, and ethanol amount on chemical and sensory properties of Merlot wines. American Journal of Enology and Viticulture, 2013. **64**(4): pp. 437–449.

[117] Singleton, V.L. and D.E. Draper, The transfer of polyphenolic compounds from grape seeds into wines. American Journal of Enology and Viticulture, 1964. **15**(1): pp. 34–40.

[118] Adams, D.O. and R.C. Scholz. Tannins—the problem of extraction, in 13th Australian Wine Industry Technical Conference. 2008. Adelaide, Australia: Australian Society for Viticulture and Oenology.

[119] Foo, L.Y. and L.J. Porter, The phytochemistry of proanthocyanidin polymers. Phytochemistry, 1980. **19**(8): pp. 1747–1754.

[120] Bate-Smith, E.C. and T. Swain, Flavonoid compounds, In: Comparative Biochemistry, H. S. Mason, A.M. Florkin, Editors. 1962, Academic Press: New York. pp. 755–809.

[121] Escot, S., et al., Release of polysaccharides by yeasts and the influence of released polysaccharides on colour stability and wine astringency. Australian Journal of Grape and Wine Research, 2001. **7**(3): pp. 153–159.

[122] Riou, V., et al., Aggregation of grape seed tannins in model wine—effect of wine polysaccharides. Food Hydrocolloids, 2002. **16**(1): pp. 17–23.

[123] Scollary, G.R., et al., Astringency response of red wines: potential role of molecular assembly. Trends in Food Science & Technology, 2012. **27**(1): pp. 25–36.

[124] Hanlin, R.L., et al., Detailed characterization of proanthocyanidins in skin, seeds, and wine of Shiraz and Cabernet Sauvignon wine grapes (*Vitis vinifera*). Journal of Agricultural and Food Chemistry, 2011. **59**(24): pp. 13265–13276.

[125] Kennedy, J.A. and G.P. Jones, Analysis of proanthocyanidin cleavage products following acid-catalysis in the presence of excess phloroglucinol. Journal of Agricultural and Food Chemistry, 2001. **49**(4): pp. 1740–1746.

[126] Gu, L., et al., Screening of foods containing proanthocyanidins and their structural characterization using LC-MS/MS and thiolytic degradation. Journal of Agricultural and Food Chemistry, 2003. **51**(25): pp. 7513–7521.

[127] Passos, C.P., et al., Evidence for galloylated type-A procyanidins in grape seeds. Food Chemistry, 2007. **105**(4): pp. 1457–1467.

[128] Mouls, L. and H. Fulcrand, UPLC-ESI-MS study of the oxidation markers released from tannin depolymerization: toward a better characterization of the tannin evolution over food and beverage processing. Journal of Mass Spectrometry, 2012. **47**(11): pp. 1450–1457.

[129] Esatbeyoglu, T., et al., Structure elucidation of procyanidin oligomers by low-temperature ^{1}H NMR spectroscopy. Journal of Agricultural and Food Chemistry, 2011. **59**(1): pp. 62–69.

[130] Ricardo-da-Silva, J.M., et al., Procyanidin dimers and trimers from grape seeds. Phytochemistry, 1991. **30**(4): pp. 1259–1264.

[131] Ricardo-da-Silva, J.M., et al., Procyanidin composition of Chardonnay, Mauzac and Grenache blanc grapes. Vitis, 1991. **40**(4): pp. 245–252.

[132] de Freitas, V.A.P., et al., Characterisation of oligomeric and polymeric procyanidins from grape seeds by liquid secondary ion mass spectrometry. Phytochemistry, 1998. **49**(5): pp. 1435–1441.

[133] Katalinić, V., P. Maleš, and G. Konja, Low molecular weight flavans in wine. Journal of Wine Research, 1997. **8**(1): pp. 19–27.

[134] Chira, K., et al., Grape variety effect on proanthocyanidin composition and sensory perception of skin and seed tannin extracts from Bordeaux wine grapes (Cabernet Sauvignon and Merlot) for two consecutive vintages (2006 and 2007). Journal of Agricultural and Food Chemistry, 2008. **57**(2): pp. 545–553.

[135] Spranger, I., et al., Differentiation of red winemaking technologies by phenolic and volatile composition. Analytica Chimica Acta, 2004. **513**(1): pp. 151–161.

[136] Czochanska, Z., et al., Polymeric proanthocyanidins. Stereochemistry, structural units, and molecular weight. Journal of the Chemical Society, Perkin Transactions, 1980. **1**: pp. 2278–2286.

[137] Porter, L.J., L.N. Hrstich, and B.G. Chan, The conversion of procyanidins and prodelphinidins to cyanidin and delphinidin. Phytochemistry, 1985. **25**(1): pp. 223–230.

[138] Roby, G., et al., Berry size and vine water deficits as factors in winegrape composition: anthocyanins and tannins. Australian Journal of Grape and Wine Research, 2004. **10**(2): pp. 100–107.

[139] Harbertson, J.F., J.A. Kennedy, and D.O. Adams, Tannin in skins and seeds of Cabernet Sauvignon, Syrah, and Pinot Noir berries during ripening. American Journal of Enology and Viticulture, 2002. **53**(1): pp. 54–59.

[140] Harbertson, J.F., E.A. Picciotto, and D.O. Adams, Measurement of polymeric pigments in grape berry extracts and wines using a protein precipitation assay combined with bisulfite bleaching. American Journal of Enology and Viticulture, 2003. **54**(4): pp. 301–306.

[141] Harbertson, J.F., et al., Variability of tannin concentration in red wines. American Journal of Enology and Viticulture, 2008. **59**(2): pp. 210–214.

[142] Thorngate, J.H. and V.L. Singleton, Localization of procyanidins in grape seeds. American Journal of Enology and Viticulture, 1994. **45**(2): pp. 259–262.

[143] Geny, L., et al., Composition and cellular localization of tannins in grape seeds during maturation. Journal of Agricultural and Food Chemistry, 2003. **51**(27): pp. 8051–8054.

[144] Labarbe, B., et al., Quantitative fractionation of grape proanthocyanidins according to their degree of polymerization. Journal of Agricultural and Food Chemistry, 1999. **47**(7): pp. 2719–2723.

[145] Cerpa-Calderón, F.K. and J.A. Kennedy, Berry integrity and extraction of skin and seed proanthocyanidins during red wine fermentation. Journal of Agricultural and Food Chemistry, 2008. **56**(19): pp. 9006–9014.

[146] Cortell, J.M. and J.A. Kennedy, Effect of shading on accumulation of flavonoid compounds in (*Vitis vinifera* L.) Pinot Noir fruit and extraction in a model system. Journal of Agricultural and Food Chemistry, 2006. **54**(22): pp. 8510–8520.

[147] Gagné, S., C. Saucier, and L. Gény, Composition and cellular localization of tannins in Cabernet Sauvignon skins during growth. Journal of Agricultural and Food Chemistry, 2006. **54**(25): pp. 9465–9471.

[148] Fernández, K., J.A. Kennedy, and E. Agosin, Characterization of *Vitis vinifera* L. cv. Carménère grape and wine proanthocyanidins. Journal of Agricultural and Food Chemistry, 2007. **55**(9): pp. 3675–3680.

[149] Sampaio, T.L., J.A. Kennedy, and M.C. Vasconcelos, Use of microscale fermentations in grape and wine research. American Journal of Enology and Viticulture, 2007. **58**(4): pp. 534–539.

[150] Pastor del Rio, J.L. and J.A. Kennedy, Development of proanthocyanidins in *Vitis vinifera* L. cv. Pinot noir grapes and extraction into wine. American Journal of Enology and Viticulture, 2006. **57**(2): pp. 125–132.

[151] Lee, J., et al., Effect of early seed removal during fermentation on proanthocyanidin extraction in red wine: a commercial production example. Food Chemistry, 2008. **107** (3): pp. 1270–1273.

[152] Cosme, F., J.M. Ricardo-Da-Silva, and O. Laureano, Tannin profiles of *Vitis vinifera* L. cv. red grapes growing in Lisbon and from their monovarietal wines. Food Chemistry, 2009. **112**(1): pp. 197–204.

[153] Drinkine, J., Y. Glories, and C. Saucier, (+)-Catechin–aldehyde condensations: competition between acetaldehyde and glyoxylic acid. Journal of Agricultural and Food Chemistry, 2005. **53**(19): pp. 7552–7558.

[154] Bate-Smith, E.C., Haemanalysis of tannins: the concept of relative astringency. Phytochemistry, 1973. **12**(4): pp. 907–912.

[155] Kallithraka, S., et al., Correlations between saliva protein composition and some T–I parameters of astringency. Food Quality and Preference, 2001. **12**(2): pp. 145–152.

[156] Guinard, J.-X., R.M. Pangborn, and M.J. Lewis, The time-course of astringency in wine upon repeated ingestion. American Journal of Enology and Viticulture, 1986. **37**(3): pp. 184–189.

[157] Jiang, Y., N.N. Gong, and H. Matsunami, Astringency: a more stringent definition. Chemical Senses, 2014. **39**(6): pp. 467–469.

[158] Breslin, P.A.S., et al., Psychophysical evidence that oral astringency is a tactile sensation. Chemical Senses, 1993. **18**(4): pp. 405–417.

[159] Ishikawa, T. and A.C. Noble, Temporal perception of astringency and sweetness in red wine. Food Quality and Preference, 1995. **6**(1): pp. 27–33.

[160] Mehansho, H., L.G. Butler, and D.M. Carlson, Dietary tannins and salivary proline-rich proteins: interactions, induction, and defense mechanisms. Annual Review of Nutrition, 1987. **7**(1): pp. 423–440.

[161] Kauffman, D.L. and P.J. Keller, The basic proline-rich proteins in human parotid saliva from a single subject. Archives of Oral Biology, 1979. **24**(4): pp. 249–256.

[162] Hagerman, A.E. and L.G. Butler, The specificity of proanthocyanidin-protein interactions. Journal of Biological Chemistry, 1981. **256**(9): pp. 4494–4497.

[163] Charlton, A.J., et al., Polyphenol/peptide binding and precipitation. Journal of Agricultural and Food Chemistry, 2002. **50**(6): pp. 1593–1601.

[164] Bennick, A., Structural and genetic aspects of proline-rich proteins. Journal of Dental Research, 1987. **66**(2): pp. 457–461.

[165] Murray, N.J. and M.P. Williamson, Conformational study of a salivary proline-rich protein repeat sequence. European Journal of Biochemistry, 1994. **219**(3): pp. 915–921.

[166] Pascal, C., et al., Study of the interactions between a proline-rich protein and a flavan-3-ol by NMR: Residual structures in the natively unfolded protein provides anchorage points for the ligands. Biopolymers, 2009. **91**(9): pp. 745–756.

[167] Simon, C., et al., Three-dimensional structure and dynamics of wine tannin-saliva protein complexes. A Multitechnique Approach. Biochemistry, 2003. **42**(35): pp. 10385–10395.

[168] Haslam, E., T.H. Lilley, and L.G. Butler, Natural astringency in foodstuffs—a molecular interpretation. Critical Reviews in Food Science and Nutrition, 1988. **27**(1): pp. 1–40.

[169] Lu, Y. and A. Bennick, Interaction of tannin with human salivary proline-rich proteins. Archives of Oral Biology, 1998. **43**(9): pp. 717–728.

[170] Hagerman, A.E., M.E. Rice, and N.T. Ritchard, Mechanisms of protein precipitation for two tannins, pentagalloyl glucose and epicatechin16 (4→8) catechin (procyanidin). Journal of Agricultural and Food Chemistry, 1998. **46**(7): pp. 2590–2595.

[171] Oh, H.I., et al., Hydrophobic interaction in tannin-protein complexes. Journal of Agricultural and Food Chemistry, 1980. **28**(2): pp. 394–398.

[172] Ricardo-da-Silva, J.M., et al., Interaction of grape seed procyanidins with various proteins in relation to wine fining. Journal of the Science of Food and Agriculture, 1991. **57** (1): pp. 111–125.

[173] Jöbstl, E., et al., Molecular model for astringency produced by polyphenol/protein interactions. Biomacromolecules, 2004. **5**(3): pp. 942–949.

[174] Bajec, M.R. and G.J. Pickering, Astringency: mechanisms and perception. Critical Reviews in Food Science and Nutrition, 2008. **48**(8): pp. 858–875.

[175] Gawel, R., A. Oberholster, and I.L. Francis, A 'Mouth-feel Wheel': terminology for communicating the mouth-feel characteristics of red wine. Australian Journal of Grape and Wine Research, 2000. **6**(3): pp. 203–207.

[176] Kennedy, J.A., et al., Analysis of tannins in red wine using multiple methods: correlation with perceived astringency. American Journal of Enology and Viticulture, 2006. **57**(4): pp. 481–485.

[177] Mercurio, M.D. and P.A. Smith, Tannin quantification in red grapes and wine: comparison of polysaccharide- and protein-based tannin precipitation techniques and their ability to model wine astringency. Journal of Agricultural and Food Chemistry, 2008. **56**(14): pp. 5528–5537.

[178] Mercurio, M.D., et al., Relationship between red wine grades and phenolics. 1. Tannin and total phenolics concentrations. Journal of Agricultural and Food Chemistry, 2010. **58** (23): pp. 12313–12319.

[179] Demiglio, P. and G.J. Pickering, The influence of ethanol and pH on the taste and mouthfeel sensations elicited by red wine. Journal of Food, Agriculture & Environment 2008. **6**(4): pp. 143–150.

[180] Fontoin, H., et al., Effect of pH, ethanol and acidity on astringency and bitterness of grape seed tannin oligomers in model wine solution. Food Quality and Preference, 2008. **19**(3): pp. 286–291.

[181] Smith, A.K., H. June, and A.C. Noble, Effects of viscosity on the bitterness and astringency of grape seed tannin. Food Quality and Preference, 1996. **7**(3–4): pp. 161–166.

[182] Peleg, H. and A.C. Noble, Effect of viscosity, temperature and pH on astringency in cranberry juice. Food Quality and Preference, 1999. **10**(4–5): pp. 343–347.

[183] Lyman, B.J. and B.G. Green, Oral astringency: effects of repeated exposure and interactions with sweeteners. Chemical Senses, 1990. **15**(2): pp. 151–164.

[184] Valentová, H., et al., Time–intensity studies of astringent taste. Food Chemistry, 2002. **78** (1): pp. 29–37.

[185] Carvalho, E., et al., Influence of wine pectic polysaccharides on the interactions between condensed tannins and salivary proteins. Journal of Agricultural and Food Chemistry, 2006. **54**(23): pp. 8936–8944.

[186] Sun, B., et al., Reactivity of polymeric proanthocyanidins toward salivary proteins and their contribution to young red wine astringency. Journal of Agricultural and Food Chemistry, 2013. **61**(4): pp. 939–946.

[187] Sarni-Manchado, P., et al., Analysis and characterization of wine condensed tannins precipitated by proteins used as fining agent in enology. American Journal of Enology and Viticulture, 1999. **50**(1): pp. 81–86.

[188] Vidal, S., et al., The mouth-feel properties of grape and apple proanthocyanidins in a wine-like medium. Journal of the Science of Food and Agriculture, 2003. **83**(6): pp. 564–573.

[189] Arnold, R.A., A.C. Noble, and V.L. Singleton, Bitterness and astringency of phenolic fractions in wine. Journal of Agricultural and Food Chemistry, 1980. **28**(3): pp. 675–678.

[190] Pickering, G.J., K. Simunkova, and D. DiBattista, Intensity of taste and astringency sensations elicited by red wines is associated with sensitivity to PROP (6-*n*-propylthiouracil). Food Quality and Preference, 2004. **15**(2): pp. 147–154.

[191] Dinnella, C., et al., Temporary modification of salivary protein profile and individual responses to repeated phenolic astringent stimuli. Chemical Senses, 2010. **35**(1): pp. 75–85.

[192] Hanlin, R.L., et al., Review: condensed tannin and grape cell wall interactions and their impact on tannin extractability into wine. Australian Journal of Grape and Wine Research, 2010. **16**(1): pp. 173–188.

[193] Renard, C.M.G.C., et al., Interactions between apple cell walls and native apple polyphenols: quantification and some consequences. International Journal of Biological Macromolecules, 2001. **29**(2): pp. 115–125.

[194] Bindon, K.A., P.A. Smith, and J.A. Kennedy, Interaction between grape-derived proanthocyanidins and cell wall material. 1. Effect on proanthocyanidin composition and molecular mass. Journal of Agricultural and Food Chemistry, 2010. **58**(4): pp. 2520–2528.

[195] Le Bourvellec, C., S. Guyot, and C.M.G.C. Renard, Non-covalent interaction between procyanidins and apple cell wall material: Part I. Effect of some environmental parameters. Biochimica et Biophysica Acta (BBA):General Subjects, 2004. **1672**(3): pp. 192–202.

[196] de Freitas, V., E. Carvalho, and N. Mateus, Study of carbohydrate influence on protein–tannin aggregation by nephelometry. Food Chemistry, 2003. **81**(4): pp. 503–509.

[197] Bindon, K.A., et al., Interaction between grape-derived proanthocyanidins and cell wall material. 2. Implications for vinification. Journal of Agricultural and Food Chemistry, 2010. **58**(19): pp. 10736–10746.

[198] Bindon, K.A. and J.A. Kennedy, Ripening-induced changes in grape skin proanthocyanidins modify their interaction with cell walls. Journal of Agricultural and Food Chemistry, 2011. **59**(6): pp. 2696–2707.

[199] Bindon, K.A., A. Bacic, and J.A. Kennedy, Tissue-specific and developmental modifications of grape cell walls influence the adsorption of proanthocyanidins. Journal of Agricultural and Food Chemistry, 2012. **60**(36): pp. 9249–9260.

[200] Casassa, L.F., M. Keller, and J. Harbertson, Regulated deficit irrigation alters anthocyanins, tannins and sensory properties of Cabernet Sauvignon grapes and wines. Molecules, 2015. **20**(5): pp. 7820–7844.

[201] Hazak, J.C., et al., The phenolic components of grape berries in relation to wine composition, Acta Horticulturae. 2005. (689): pp. 189–196.

[202] Alcalde-Eon, C., et al., Identification of dimeric anthocyanins and new oligomeric pigments in red wine by means of HPLC-DAD-ESI/MSn. Journal of Mass Spectrometry, 2007. **42**(6): pp. 735–748.

[203] Salas, E., et al., Demonstration of the occurrence of flavanol–anthocyanin adducts in wine and in model solutions. Analytica Chimica Acta, 2004. **513**(1): pp. 325–332.

[204] Alcalde-Eon, C., et al., Changes in the detailed pigment composition of red wine during maturity and ageing: a comprehensive study. Analytica Chimica Acta, 2006. **563**(1–2): pp. 238–254.

[205] Salas, E., et al., Characterization of pigments from different high speed countercurrent chromatography wine fractions. Journal of Agricultural and Food Chemistry, 2005. **53** (11): pp. 4536–4546.

[206] Remy, S., et al., First confirmation in red wine of products resulting from direct anthocyanin–tannin reactions. Journal of the Science of Food and Agriculture, 2000. **80**(6): pp. 745–751.

[207] Timberlake, C.F. and P. Bridle, Interactions between anthocyanins, phenolic compounds, and acetaldehyde and their significance in red wines. American Journal of Enology and Viticulture, 1976. **27**(3): pp. 97–105.

[208] Rivas-Gonzalo, J.C., S. Bravo-Haro, and C. Santos-Buelga, Detection of compounds formed through the reaction of malvidin 3-monoglucoside and catechin in the presence of acetaldehyde. Journal of Agricultural and Food Chemistry, 1995. **43**(6): pp. 1444–1449.

[209] Dallas, C., J.M. Ricardo-da-Silva, and O. Laureano, Interactions of oligomeric procyanidins in model wine solutions containing malvidin-3-glucoside and acetaldehyde. Journal of the Science of Food and Agriculture, 1996. **70**(4): pp. 493–500.

[210] Laurie, V.F. and A.L. Waterhouse, Glyceraldehyde bridging between flavanols and malvidin-3-glucoside in model solutions. Journal of Agricultural and Food Chemistry, 2006. **54**(24): pp. 9105–9111.

[211] Bakker, J. and C.F. Timberlake, Isolation, identification, and characterization of new color-stable anthocyanins occurring in some red wines. Journal of Agricultural and Food Chemistry, 1997. **45**(1): pp. 35–43.

[212] Morata, A., et al., Pyruvic acid and acetaldehyde production by different strains of *Saccharomyces cerevisiae*: relationship with vitisin A and B formation in red wines. Journal of Agricultural and Food Chemistry, 2003. **51**(25): pp. 7402–7409.

[213] Fulcrand, H., et al., A new class of wine pigments generated by reaction between pyruvic acid and grape anthocyanins. Phytochemistry, 1998. **47**(7): pp. 1401–1407.

[214] Fulcrand, H., et al., An oxidized tartaric acid residue as a new bridge potentially competing with acetaldehyde in flavan-3-ol condensation. Phytochemistry, 1997. **46**(2): pp. 223–227.

[215] Es-Safi, N.-E., V. Cheynier, and M. Moutounet, Interactions between cyanidin 3-O-glucoside and furfural derivatives and their impact on food color changes. Journal of Agricultural and Food Chemistry, 2002. **50**(20): pp. 5586–5595.

[216] Pissarra, J., et al., Reaction between malvidin 3-glucoside and (+)-catechin in model solutions containing different aldehydes. Journal of Food Science, 2003. **68**(2): pp. 476–481.

[217] Asenstorfer, R.E., et al., Formation of vitisin A during red wine fermentation and maturation. Australian Journal of Grape and Wine Research, 2003. **9**(1): pp. 40–46.

[218] Elias, R.J. and A.L. Waterhouse, Controlling the Fenton reaction in wine. Journal of Agricultural and Food Chemistry, 2010. **58**(3): pp. 1699–1707.

[219] Monagas, M. and B. Bartolomé, Anthocyanins and anthocyanin-derived compounds, In: Wine chemistry and biochemistry, M.V. Moreno-Arribas, M.C. Polo, Editors. 2009, New York, NY: Springer New York. pp. 439–462.

[220] de Freitas, V. and N. Mateus, Formation of pyranoanthocyanins in red wines: a new and diverse class of anthocyanin derivatives. Analytical and Bioanalytical Chemistry, 2011. **401**(5): pp. 1463–1473.

[221] Dueñas, M., H. Fulcrand, and V. Cheynier, Formation of anthocyanin–flavanol adducts in model solutions. Analytica Chimica Acta, 2006. **563**(1–2): pp. 15–25.

[222] Somers, T.C., The polymeric nature of wine pigments. Phytochemistry, 1971. **10**(9): pp. 2175–2186.

[223] Somers, T.C. and M.E. Evans, Spectral evaluation of young red wines: anthocyanin equilibria, total phenolics, free and molecular SO_2, "chemical age". Journal of the Science of Food and Agriculture, 1977. **28**(3): pp. 279–287.

[224] Jurd, L., Review of polyphenol condensation reactions and their possible occurrence in the aging of wines. American Journal of Enology and Viticulture, 1969. **20**(3): pp. 191–195.

[225] Singleton, V.L., Tannins and the qualities of wines, In: Plant polyphenols: synthesis, properties, significance, R.W. Hemingway, P.E. Laks, Editors. 1992, Boston, MA: Springer US. pp. 859–880.

[226] McRae, J.M., et al., Sensory properties of wine tannin fractions: implications for in-mouth sensory properties. Journal of Agricultural and Food Chemistry, 2013. **61**(3): pp. 719–727.

[227] Hayasaka, Y. and J.A. Kennedy, Mass spectrometric evidence for the formation of pigmented polymers in red wine. Australian Journal of Grape and Wine Research, 2003. **9**(3): pp. 210–220.

[228] Kennedy James, A. and Y. Hayasaka, Compositional investigation of pigmented tannin, in red wine color. 2004, American Chemical Society. pp. 247–264.

[229] Wollmann, N. and T. Hofmann, Compositional and sensory characterization of red wine polymers. Journal of Agricultural and Food Chemistry, 2013. **61**(9): pp. 2045–2061.

[230] Ira Katz George Charalambous Singleton V.L. and C. Noble, Wine flavor and phenolic substances, In: Phenolic, sulfur, and nitrogen compounds in food flavors. George Charalambous, Ira Katz, Editors. 1976, American Chemical Society. pp. 47–70.

[231] Adams, D.O., J.F. Harbertson, and E.A. Picciotto, Fractionation of red wine polymeric pigments by protein precipitation and bisulfite bleaching, In: Red wine color, revealing the mysteries. American Chemical Society, 2004. pp. 275–288.

Chapter 7

The Effect of Production Systems on Strawberry Quality

Li Fan, Claudine Dubé and Shahrokh Khanizadeh

Additional information is available at the end of the chapter

http://dx.doi.org/10.5772/67233

Abstract

Strawberry with high polyphenols is receiving much more attentions recently. Production systems were reported positive effect on improving crop quality. A complete randomized design with two replicates was employed, and the strawberry selection 'SJ8976-1' was planted under three production systems: matted row system (MRS), plastic mulch (PM) and plastic mulch with row covers (PMRC). Total yield, average fruit weight, soluble solids content (SSC), titratable acidity (TA), firmness, fruit postharvest quality, total phenolic content (TPC), total antioxidant content (TAC), oxygen radical absorbance capacity (ORAC) and phenolic composition analysed by high-performance liquid chromatography (HPLC) were evaluated at different harvest times. The results showed that PMRC advanced fruit maturity and improved average fruit weight compared to MRS. SSC, TA and firmness of the strawberry fruits from PMRC were significantly increased at all harvests. Fruit weight loss, juice leakage and presences of grey mould were lower, and fruit glossiness was higher for those harvested under PMRC than those from MRS. TPC, TAC and ORAC of strawberry were significantly changed under different production systems. The phenolic compounds assessed in strawberry were grouped into six major categories, which were anthocyanins, kaempferol, flavonols, hydroxycinnamic acids, ellagic acid and benzoic acids.

Keywords: strawberry, fruit quality, production system, plastic mulch, phenolic content

1. Introduction

Phenolic compounds including signalling molecules, pigments and flavours represent a variety of functions in plant growth, development and defence [1]. Phenolic compounds and flavonoid compounds were synthesized through a vast range of secondary metabolites in plants. These phytochemicals are structurally diverse and act as bioactive compounds or plant antioxidants in medicine, food industry and human nutrition [2]. Fruits, such as strawberry, contain abundant rich natural sources of phenolic compounds with high antioxidant capacities, which protect our bodies against free radicals and reactive oxygen species [3, 4]. The large amounts of phenolic compounds offered in strawberries have also been proved to make certain contributions in enhancing the quality of fresh fruits and extending shelf life by delaying senescence induced by oxidative degradation [5]. Therefore, due to the nutritional value and human health benefits in strawberry, the daily demand for strawberries was increasing as time went on. Research on the protective effects of plant-derived polyphenols has developed notably in recent years. In particular, their antioxidant properties have been the objective of extensive research [6–9].

Many factors including genotype, planting date, mulch type, temperature, fertilizer and production systems have been reported to affect strawberry yield and quality [10–12]. Among these factors, production systems play an important role in strawberry growth by enhancing berry size and marketable yield. Matted row system (MRS) is widely used in strawberry planting in north area in terms of lower cost and easy operation [13], but with the development of new technology, many disadvantages of MRS were reported such as poor picking efficiency, waste of fertilizer and high risk of fruits decay. Alternatively, plastic mulch (PM) and plastic mulch with row covers (PMRC) became the most recommended production systems in strawberry planting in recent years [14–16]. Fruit quality and phenolic content were improved under plastic mulch system and resulted in variety of benefits to the crop, including improved fruit cleanliness, prevention of bed erosion, extended harvest season, early ripening, increased yield, better weed control, ease of harvest and more efficient irrigation and fertilizer application [12, 17]. Furthermore, early fruit harvest under plastic culture especially with row covers provides higher income to growers in cool northern climate because of off-season fruit production. Research on bell pepper grown along the Mediterranean coast in raised soil beds (ridge) covered with plastic mulch showed that the use of impermeable plastic mulch in bell pepper cultivation affects water fluxes and may change crop water use and distribution compared to open-field conditions [18]. It also has been reported that the combination of hill plasticulture system and genotype leads a better result of higher flavonoid content and antioxidant capacity than the traditional cultivation system [19]. Plastic mulch affects the microclimate around the crop by modifying the radiation budget (absorptivity vs. reflectivity) of the mulch surface as well. Colour also affects the surface temperature of the mulch cover and consequently the underlying soil temperature [20].

Strawberries are of great nutrition amongst other fruits and have been found to contain 2- to 11-fold more antioxidant capacity than apples, peaches, pears, grapes, tomatoes, oranges or kiwifruit [21]. Therefore, to meet the everyday demands of strawberry with increasing antioxidant levels through breeding and/or cultivation systems is of great value in Canada. For the above reasons, a series of strawberry programs were set, and some cultivars were released by Agriculture and Agri-Food Canada (AAFC). 'SJ8976-1' is a new hardy strawberry selection with extra long storage life (*Fragaria* × *ananassa* Duch.) of the AAFC programme. The objective of the present study was to evaluate the effects of PM and PMRC, versus the conventional MRS, on total yield, yield per plant, average fruit weight, soluble solids content (SSC), titratable acidity (TA), firmness, fruit postharvest quality, total phenolic content (TPC), total antioxidant content (TAC), oxygen radical absorbance capacity (ORAC) and phenolic composition analysed by high-performance liquid chromatography (HPLC) in strawberry selection 'SJ8976-1', at different harvest times during the growing season.

2. Materials and methods

2.1. Strawberry material

'SJ8976-1' is a new hardy strawberry selection with large fruit size, excellent flavour and long storage life (*Fragaria* × *ananassa* Duch.). The pale fruit colour of 'SJ8976-1' in combination with firmness, resistance to fruit rot and long shelf life makes it a good candidate for fresh market and transportation. The fruit colour of 'SJ8976-1' darkens during storage [22].

2.2. Experimental design

A complete randomized design with two replicates was used, and the strawberry selection 'SJ8976-1' was grown under three production systems: (1) matted row system (MRS), (2) plastic mulch (PM) and (3) plastic mulch with row covers (PMRC). Twenty-six plants of 'SJ8976-1' were planted in a double row 30 cm × 30 cm apart in 4 m long plots, supplied with drip irrigation down the centre of each row in 2008 at Agriculture and Agri-Food Canada, L'Acadie Experimental Farm (longitude: 73'35°W; latitude: 45'32°N), in L'Acadie, Quebec, Canada. For the PM and PMRC systems, runners were removed as the runners could not root and grow into a plantlet. But for the MRS, runners were kept and placed to complete the row as recommended by Ministere de l'Agriculture, des Pecheries et de l'Alimentation du Quebec (MAPAQ) [23]. White row cover over the bed was applied under PMRC and taken away until the fruits were formed. The mean temperature from strawberry flowering to harvest was 19.32°C in 2009 and 18.95°C in 2010, respectively. The total precipitation from strawberry flowering to harvest was 241.4 mm in 2009 and 238.7 mm in 2010, respectively.

Optimum matured fruits were harvested from each plot 2–3 times per week from mid-June to mid-July of each year in 2009 and 2010 for a total of 10 harvests per season. The 10 harvests were finally pooled into three groups representing early, mid and late harvest. Total yield and average fruit weight were recorded at every harvest.

Thirty fruits at optimum maturity were harvested from each plot, rapidly put in cooler and then brought to the laboratory, where 15 of them were used to examine the shelf life and fruit quality and the remaining 15 fruits were immediately cut in halves and frozen in liquid nitrogen. Three 40-g subsamples of each treatment were stored at −80°C for chemical composition analysis [24].

2.3. Sample extraction procedures

Ten grams of fresh frozen fruits from different production systems was pestled for 2 min with 50 mL of 50% methanol at 25°C using a Polytron blender (Brinkmann Instruments, Westbury, NY, USA). The mixture was filtered firstly through Whatman no. 1 filter paper and then through a 0.45-μm Acrodisc syringe filters (Gelman Sciences, Ann Arbor, MI, USA). The filtrate was finally kept at −20°C, and the extracts resulting from this step were used in the TPC and TAC analysis [24].

2.4. Chemicals

The chemicals used for TPC, TAC, HPLC and ORAC are shown in **Table 1**.

Procedure	Chemicals	Company
TPC and TAC	Gallic acid, ferric chloride, sodium acetate, hydrochloric acid, sodium carbonate, 2,4,6-Tri(2-pyridyl)-1,3,5-triazine and the Folin-Ciocalteu reagent	Sigma Chemical Co. (St. Louis, MO, USA)
HPLC	gallic acid, p-Coumaric acid, m-Coumaric acid, o-Coumaric acid, kaempferol-3-glucoside, quercetin-3-glucoside, cyanidin-3-glucoside and pelargonidin-3-glucoside	Sigma Chemical Co. (Oakville, ON, Canada)
	Ellagic acid and pelargonidin-3-rutinoside	Apin Chemicals (Abingdon, UK)
ORAC	2,2-Azobis(2-methylpropionamidine) dihydrochloride (AAPH), fluorescein disodium and 6-hydroxy-2,5,7,8-tetramethylchroman-2-carboxylic acid (Trolox)	Sigma Chemical Co. (Oakville, ON, Canada)

Table 1. The chemicals used for TPC, TAC, HPLC and ORAC.

2.5. Soluble solids content, titratable acidity and firmness

Five strawberries were selected randomly and blended thoroughly using a Supreme Juicerator (Acme Juicer Manufacturing Co., New Hartford, CT, USA), and the SSC of the juice was measured at room temperature with a refractometer (Abbe Mark II, Reichert-Jung, Depew, NY, USA) as described previously [25]. TA was measured by diluting 1 mL of strawberry juice with 9 mL of distilled water, adjusted to pH 8.1 using 0.1 N of NaOH (Accumet AB15 Basic pH metre; Fisher Scientific) as described previously [25]. Firmness was ranked from 1 (very soft) to 5 (very firm).

2.6. Fruit postharvest quality

Five fruits per replicate randomly selected from each plot were placed on Whatman no. 1 filter paper in open Petri dishes and left at room temperature (23°C) to evaluate postharvest deterioration. Fruit weight loss and juice leakage, fruit glossiness and the postharvest incidence of grey mould (*Botrytis cinerea* Pers.) were monitored daily until the fruits were considered non marketable. Fruit weight loss was calculated according to Wang et al. [26] using the following method:

$$\text{Weight loss} = \text{weight}_{\text{day } x} - \text{weight}_{\text{day } x+1}. \quad (1)$$

where $\text{weight}_{\text{day } x}$ is the total weight of the five tested berries in each Petri dish and $\text{weight}_{\text{day } x+1}$ is the total weight of the same five berries day after.

The fruit juice leakage was ranked from 1 ('no leak or juice loss') to 4 ('significant fruit leak or juice loss'), fruit glossiness was ranked from 1 ('dull') to 3 ('shiny'), and presences of grey mould were ranked from 1 ('no disease') to 5 ('the presence of mould') [24].

2.7. Total phenolic content

The TPC was measured in accordance with the slightly modified Folin-Ciocalteu (FC) method [27]. In a 10-mL vial, a standard or sample extract (0.02 mL) and 0.1 mL of FC reagent were mixed with 1.58 mL of water for 3 min. After that, 0.3 mL of Na_2CO_3 (7.5%) was added, and the solution was allowed to stand for 30 min at 40°C. Absorption was measured at 765 nm with an Ultrospec 3100 pro UV/visible spectrophotometer (Fisher Scientific, Ottawa, ON, Canada). The result was expressed as gallic acid equivalent (GAE) in mg/g of fresh frozen weight using gallic acid as standard. The sample concentrations beyond 500 μg mL^{-1} were diluted before final analysis to suit the linear range of the standard curve.

2.8. Total antioxidant content

The TAC was determined by ferric reducing antioxidant power (FRAP) according to the method of Benzie and Strain [28]. Briefly, the FRAP reagent was freshly prepared by mixing 300 mM acetate buffer (pH 3.6), 10 mM TPTZ in 40 mM HCl and 20 mM $FeCl_3 \bullet 6H_2O$ at the ratio of 10:1:1 (v/v/v). Then add 2.4 mL FRAP reagent and 80 μl of diluted sample, and the plate was incubated at 37°C for 4 min. The absorbance readings were taken at 593 nm using Agilent 8453 spectrophotometer (Agilent Technologies, Waldbronn, Germany). The TAC value was calculated on the basis of $Fe_2SO_4 \cdot 7H_2O$ and expressed as micromoles of $Fe_2SO_4 \cdot 7H_2O$ equivalent per gram of fresh frozen weight.

2.9. Oxygen radical absorbance capacity (ORAC) assay

For the ORAC assay, 5 g of fresh frozen fruit from different production systems was completely pestled in 15 mL of 50% acetone using a Polytron blender (Brinkmann Instruments, Westbury, NY, USA). The mixture was centrifuged at 10,000 rpm for 10 min, and the supernatant was kept in a 1.5-mL vial at −20°C.

The ORAC assay measures the capacity of the antioxidant to prevent the oxidization of a free radical (peroxyl) in biological systems. The assay was performed as described previously [29] with a Synergy™ Multi-Mode Microplate Reader (Bio-Tek Instruments, Inc., Winooski, VT, USA).

2.10. HPLC analysis

Three grams of fresh frozen strawberry from different production systems was mixed with 15 mL acetone and homogenized using a Polytron® homogenizer (Brinkmann Instruments, Inc., Westbury, NY, USA) for 1 min. The extracts were then filtrated through no. 1 Whatman filter paper and concentrated under vacuum in a water bath at 35°C with a Heidolph Laborota 4000 rotary evaporator (Elk Grove village, IL, USA). The concentrated sample was dissolved in 15 mL of acidified water (3 g/100g formic acid) and then passed through a C18 Sep-Pak cartridge (Waters Ltd., Mississauga, ON, Canada), which was previously activated with methanol followed by water, and then 3 g/100 g aqueous formic acid. The anthocyanins and other phenolics were then recovered with 3 mL of acidified methanol containing 3 g/100 g formic acid. The methanol extract was passed through a 0.45-μm Acrodisc syringe filter (Sigma Chemical Co., Oakville, ON, Canada), and 10 μL of each sample was injected for the HPLC analysis. All standards except for anthocyanins were dissolved in methanol. Anthocyanins were dissolved in 1 g/100g HCl in methanol [30].

The individual phenolic compounds of berries were analysed by HPLC, using a Varian (Varian Inc., Palo Alto, CA, USA) HPLC system. A photodiode array detector (Model 335) and a quaternary pump (Model 240) were equipped in the HPLC system. Samples were injected at 40°C, and the separation was performed with a reverse phase 4.6-mm Polaris C18-A column coupled with a 4.6-mm MetaGuard Polaris C18-A guard column (Varian Inc., Palo Alto, CA, USA). The mobile phase was acidified water containing 2.5 g/100g formic acid (A) and acidified methanol containing 2.5 g/100 g formic acid (B). The flow rate was 1.0 ml/min, with a gradient profile consisting of B with the following proportions (v/v) of A: 0–15 min, 15–30% B; 15–20 min, 30% B; 20–25 min, 30–80% B; 25–42 min, 80% B; 42–42.5 min, 80–100% B; 42.5–50.25 min, 100% B; and 50.25–51.25 min, 0% B. The detector was simultaneously monitoring the different groups of phenolics at 254, 280, 320, 350 and 510 nm. Equilibration time between samples was 15 min. Data were collected and analysed by Varian Workstation Star version 6.41. Total phenolic compounds were divided into six groups and quantified as described in our previous work [30].

2.11. Statistical methods

Statistical analysis was performed using SAS software [31]. Data were analysed using GLM procedure, and the least significant difference (LSD) test was used for mean separation at 5% level when it was significant. The arcsine transformation was used for rank data before data analysis, and the results were presented as a rank for simplicity when the outcomes of transformed and non-transformed data were the same. The data from 2009 to 2010 were pooled together after analysing each year separately and testing the homogeneity of the experimental error.

3. Results and discussion

3.1. Effect of production systems on yield

Yield is an important characteristic that farmers considered a lot in strawberry production. Compared to the MRS, PMRC accelerated strawberry fruit production by 8 and 10 days in 2009 and 2010, respectively, followed by PM when fruits were harvested 4 and 5 days before MRS in 2009 and 2010, respectively, similar to those observed by Jin et al. [32]. Soria et al. [33] also reported that based on two consecutive cropping seasons data, plastic tunnels improved early marketable yield, total marketable yield and first-class fruits of strawberry. The increased yield and fruit size might be due to the increase in soil temperature for PM and PMRC. The early ripening of the fruits under PM and PMRC systems will probably generate a better income due to off-season production. Total yields of PM and PMRC were significantly lower than that of MRS, but the yield per plant of PMRC was significantly higher than PM and MRS (**Table 2**). Fruits harvested from PMRC and PM were significantly larger than those harvested from MRS (**Table 2**). This might be because of less competition between the plants since all the runners were removed under PM and PMRC but kept under MRS [24].

3.2. Effect of production systems on SSC, TA and firmness

SSC, TA and firmness dramatically affect strawberry fruit quality. The production systems had a significant effect on the SSC, TA and firmness of the strawberry fruits and the SSC increased, while TA and firmness decreased all through the harvest season (**Table 3**). PMRC significantly increased the SSC, TA and firmness of the strawberry fruits at all harvests compared to PM and MRS (**Table 3**). Meanwhile PM significantly enhanced SSC and TA of strawberry fruit but had no effect of firmness of strawberry fruit compared to MRS (**Table 3**). Strawberry planted under PMRC produce higher-quality fruits than those without row covers, possibly because the microclimate under the covers was modified and the soil temperature was elevated which maybe result in better root establishment and advanced growth stages [20].

	Total yield (g m^{-2})	Yield $plant^{-1}$ (g $plant^{-1}$)	Average fruit weight (g)
Production systems			
Matted row system	1299.14a	54.22c	9.89b
Plastic mulch	891.63b	89.16b	12.21a
Plastic mulch with row covers	982.99b	132.69a	13.23a
$LSD_{0.05}$	113.30	24.07	1.26

$LSD_{0.05}$: least significant difference at the 0.05 level.
Different lower-case letters in the same column indicate statistically significant differences between different treatments ($p < 0.05$).

Table 2. Total yield, yield per plant and average fruit weight of strawberry selection 'SJ8976-1' grown under three production systems.

	Early harvest			Mid harvest			Late harvest		
	SSC (°Brix)	TA (%)	Firmness	SSC (°Brix)	TA (%)	Firmness	SSC (°Brix)	TA (%)	Firmness
Production systems									
Matted row system	6.9b	0.75b	3.2b	7.1c	0.68b	2.4b	7.6b	0.63b	1.4b
Plastic mulch	7.8a	0.94a	3.2b	7.8b	0.88a	2.3b	8.4a	0.76a	1.5b
Plastic mulch with row covers	8.1a	0.87a	3.6a	8.7a	0.84a	2.8a	8.8a	0.78a	2.2a
$LSD_{0.05}$	0.6	0.08	0.3	0.5	0.08	0.2	0.5	0.09	0.3

$LSD_{0.05}$: least significant difference at the 0.05 level.
Different lower-case letters in the same column indicate statistically significant differences between different treatments ($p < 0.05$).

Table 3. Early-, mid- and late-harvest soluble solids content (SSC), titratable acidity (TA) and firmness of strawberry selection 'SJ8976-1' grown under three production systems.

3.3. Effect of production systems on shelf life

Photographs of disease development and fruit juice leakage of strawberry selection 'SJ8976-1' used for shelf life, juice leakage and presences of grey mould classification are presented in **Figure 1**. 'SJ8976-1' has a special longer shelf life (about 9 days) after storage at room temperature than the commercial strawberry varieties. Production systems significantly affected strawberry

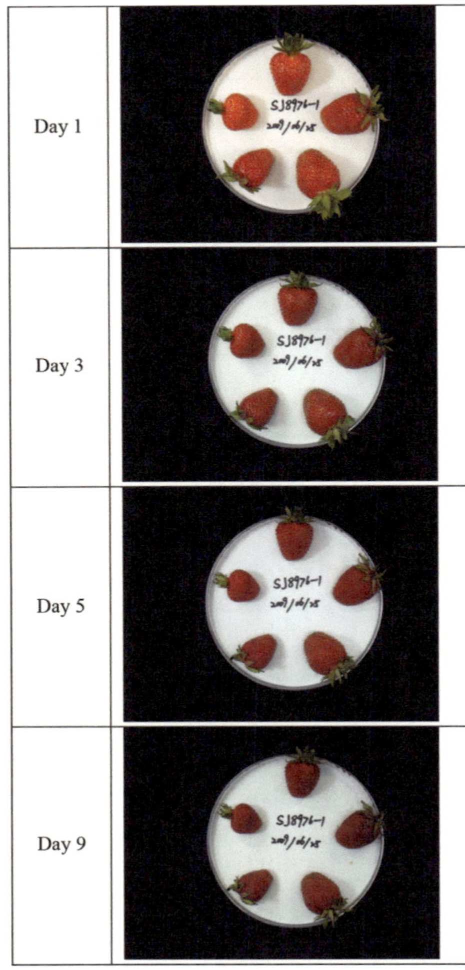

Figure 1. Photographs of disease development and fruit juice leakage of strawberry selection 'SJ8976-1' used for shelf life, juice leakage and presences of grey mould classification.

shelf life, including fruit weight loss, juice leakage, glossiness and presences of grey mould during storage (**Figure 2**). Fruit weight loss, juice leakage and presences of grey mould after storage were lower, and fruit glossiness was higher for those fruits harvested from PMRC than those from the MRS. Significant lower fruit weight loss was observed for PMRC and PM compared to the MRS, and the results were more pronounced after day 7 (**Figure 2A**). It was interesting to note that PMRC had significant lower juice leakage than the MRS, but the differences between PMRC and PM and PM and MRS were not significant until day 11 (**Figure 2B**). The fruits from PMRC kept their glossiness longer during storage at room temperature than MRS and PM. The remarkable difference among different production systems was obtained at 7 days after harvest (**Figure 2C**). Similar finding was reported in Finland that plastic production systems promoted better shelf life [11]. No significant differences were observed between the production systems on the first 5 days of harvest for presences of grey mould, but significantly less grey mould were observed on fruits harvested from PMRC than those harvested from MRS and PM from day 7 to the end of the test when all fruits were ranked as non-marketable (**Figure 2D**). Lower grey mould incidence on fruit harvested from PMRC might be due to the higher TPC and TAC content of the fruits as suggested previous [25, 34] or the protection that fruits received from the row cover against pathogen and rainfall as suggested by Freeman and Gnayem [16].

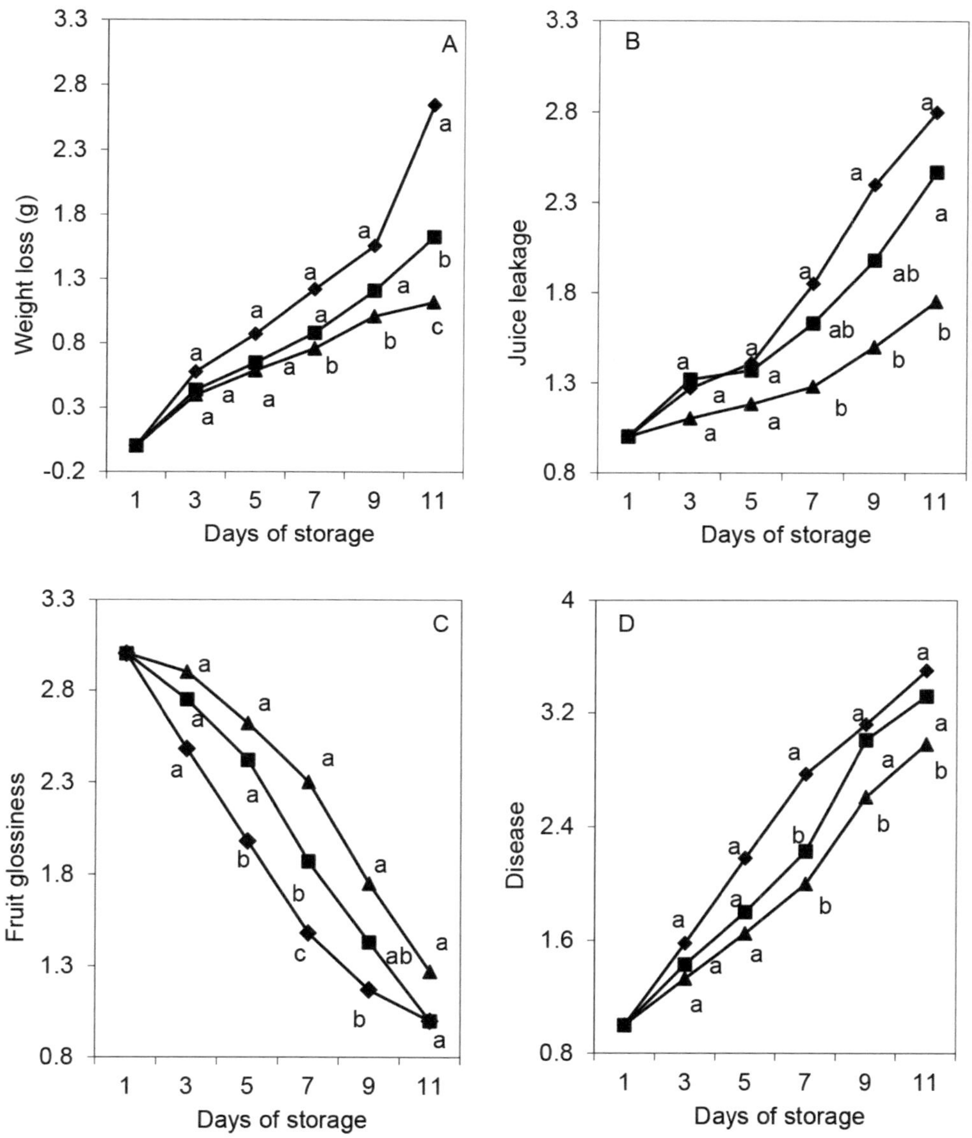

Figure 2. Early-, mid- and late-harvest fruit weight loss (A), fruit juice leakage (B), fruit glossiness (C) and incidence of grey mould (D) on strawberry selection 'SJ8976-1' during 11 days of storage at room temperature (23°C) grown under three production systems. Means with the same letter within the same day were not significantly different at 0.05 level. ◆, Matted row system; ■, plastic mulch; and ▲, plastic mulch with row covers.

3.4. Effect of production systems on antioxidant capacity

Production systems significantly changed the TPC, TAC and ORAC, but the effects varied during the different harvests (**Table 4**). PMRC promoted TPC, TAC and ORAC of strawberry fruits compared to MRS throughout the harvest season. At early harvest, TPC, TAC and ORAC were significantly higher for PMRC than MRS, but the differences of TPC and TAC between PM and MRS were not significant (**Table 4**). At mid harvest, TPC, TAC and ORAC reached the peak, and PM showed a significant change of TPC and ORAC compared to MRS. PM showed a significant change of TAC compared to MRS until the late harvest (**Table 4**). The increase of TPC and TAC might be attributed to the changes of microenvironment temperature, moisture, various enzyme activities and rhizospheric microflora of soil under the PMRC compared to the MRS which promoted the growth stages similar to the finding of Wang et al. [19] who reported that fruits from hill plasticulture had higher flavonoid contents and TAC than the MRS. The fruits harvested from PMRC not only showed the higher TPC and TAC but also the better shelf life than those from MRS in our study. The relation between antioxidant activity and fruit shelf life could be explained by our previous research that some of the antioxidants may reduce the incidence of postharvest diseases during storage and be responsible for postharvest quality [34]. Similar results were found by Wang and Millner [35] which indicated that cultural systems affected ORAC values, flavonoids and anthocyanins of 'Allstar' and 'Chandler' strawberries. ORAC is another new method of measuring antioxidant

	Early harvest			Mid harvest			Late harvest		
	TPC (μg GAE g^{-1})	TAC (μmol g^{-1})	ORAC (μmol TE g^{-1})	TPC (μg GAE g^{-1})	TAC (μmol g^{-1})	ORAC (μmol TE g^{-1})	TPC (μg GAE g^{-1})	TAC (μmol g^{-1})	ORAC (μmol TE g^{-1})
Production systems									
Matted row system	1437.9b	14.5b	20.6c	1522.8c	15.0b	22.4c	1355.8b	14.1c	21.5c
Plastic mulch	1521.1ab	14.7b	22.4b	1698.0b	16.4ab	24.1b	1603.3a	15.5b	22.4b
Plastic mulch with row covers	1610.2a	16.1a	24.3a	1867.4a	17.8a	25.2a	1631.4a	16.8a	23.6a
$LSD_{0.05}$	155.8	1.2	0.9	140.4	1.5	0.9	168.6	1.0	0.8

TPC and TAC were performed in triplicate for each of three replicates in 2009 and 2010.

TPC is expressed as micrograms of gallic acid equivalent (GAE) per gramme of fresh frozen weight; TAC is expressed as micromoles of $Fe_2SO_4 \cdot 7H_2O$ equivalent per gramme of fresh frozen weight.

LSD ($p < 0.05$): least significant difference at the 0.05 level.

Different lower-case letters in the same column indicate statistically significant differences between different treatments ($p < 0.05$).

Table 4. Early-, mid- and late-harvest total phenolic content (TPC), total antioxidant capacity (TAC) and oxygen radical absorbance capacity (ORAC) of strawberry selection 'SJ8976-1' grown under three production systems.

capacities *in vitro*. Higher ORAC value indicates higher capacity to scavenge oxygen radicals. In other words, fruit harvested from PMRC showed higher ORAC value, which means PMRC enhanced the nutritional value of strawberry fruit.

3.5. Effect of production systems on phenolic composition

Phenolic compounds, found in numerous fruits, vegetables and beverages, were not only providing supplemental functions in a plant's life cycle but also offering health benefits to humankind. They have been viewed as potential sources of dietary constituents to prevent the majority of diet-related diseases such as high cholesterol, high blood pressure and obesity. In our study, the main phenolic compounds identified by HPLC analysis in strawberry selection 'SJ8976-1' were grouped into six major categories: anthocyanins, hydroxycinnamic acids, kaempferol, flavonols, ellagic acid and benzoic acids. Among the six categories, anthocyanins accounted for 53.5% of the total phenolics and turned out to be the most predominant phenolic compounds. The physiological function of anthocyanin is responsible for the red colour of the strawberry skin and flesh. The second most abundant category was hydroxycinnamic acids, which accounted for 24.5% of the total phenolics, while kaempferol accounted for only 2.2% which was the lowest percentage of the total phenolics (**Table 5**).

In general, the strawberry fruits harvested from PMRC represented significantly higher amount of total phenolics than those from MRS and PM at all the three harvests based on the sum of the six categories of phenolic compounds (**Table 5**). In detail, significantly higher ellagic acid was assessed in the fruits from PMRC and PM than those from MRS. PMRC significantly enhanced total benzoic acids in strawberry fruit compared to MRS and PM during the early harvest, but the differences were no longer significant at mid and late harvests. Fruits harvested under PMRC showed significantly higher total hydroxycinnamic acids than MRS and PM in early harvest, and from mid to late harvest, the accumulation of hydroxycinnamic acids maximized in a peak. Those fruits from PMRC had no significantly higher amount of flavonols than those from MRS and PM at beginning, but the amount of flavonols grew quickly and achieved the peak by end of mid harvest under all three production systems; at late harvest fruit from PMRC and PM presented significantly higher total flavonols than MRS. Moreover, during all the harvests, significantly higher total anthocyanins accumulated under PMRC than the strawberry from MRS.

The modified phenolic compositions of strawberry under row covers may be primarily attributed to the elevated temperature, transmitted light spectrum and soil acidity inside the covers. Similar findings were observed in tobacco that the interaction of variety and cultural practices had significant effect on total polyphenols, rutin and chlorogenic acid in tobacco leaves [36]. Sharma et al. [37] also informed that mulching itself or combined with row covers had significant influence on plant physiology characteristics and berry yield of strawberry.

The subgroups of anthocyanins and hydroxycinnamic acids detected in strawberry selection 'SJ8976-1' were cyanidin-3-glucoside, pelargonidin-3-glucoside, pelargonidin-3-rutinoside and m-Coumaric acid, o-Coumaric acid and p-Coumaric acid, respectively. m-Coumaric acid, varied from 38.8 to 96.0 μg/g, was the major representative of the hydroxycinnamic acids.

Treatment	Ellagic acid (254 nm)	Total benzoic acids (280 nm)	Total hydroxycinnamic acids (280 nm)	Total kaempferol (280 nm)	Total flavonols (350 nm)	Total anthocyanins (510 nm)	Total phenolics—sum of 6 groups
Early harvest							
Matted row system	11.9b	18.9b	41.7b	4.2b	5.3a	97.8b	179.8b
Plastic mulch	19.3a	22.3b	52.4b	4.9b	7.2a	120.7ab	226.7ab
Plastic mulch with row cover	21.2a	28.7a	68.8a	5.8a	8.9a	144.6a	278.0a
$LSD_{0.05}$	3.8	4.5	11.9	0.8	3.9	28.0	55.3
Mid harvest							
Matted row system	15.3b	19.6b	48.8b	6.0b	11.1c	110.5b	211.3c
Plastic mulch	18.8a	23.2ab	60.8b	7.5a	22.2b	147.2a	279.7b
Plastic mulch with row cover	19.6a	24.8a	80.3a	8.2a	28.9a	168.9a	330.7a
$LSD_{0.05}$	2.7	5.1	15.8	0.8	5.9	30.1	50.3
Late harvest							
Matted row system	13.6b	20.6a	60.3b	4.4b	8.3b	144.2b	251.4b
Plastic mulch	15.4b	22.6a	81.2b	5.8a	17.4a	175.6ab	318.1ab
Plastic mulch with row cover	19.0a	24.2a	108.5a	6.2a	18.7a	206.3a	382.9a
$LSD_{0.05}$	3.2	3.7	20.6	1.0	4.7	33.5	65.2
% Total	6.3	8.3	24.5	2.2	5.2	53.5	100.0

Anthocyanins were quantified using cyanidin-3-glucoside (C3G), pelargonidin-3-glucoside (P3G) and pelargonidin-3-rutinoside (P3R); hydroxycinnamic acids were quantified using *p*-Coumaric acid (pC), *m*-Coumaric acid (mC) and *o*-Coumaric acid (oC); kaempferol was quantified using kaempferol-3-glucoside (K3G); flavonols were quantified using quercetin-3-galactoside; benzoic acids were quantified using gallic acid, and the authentic standard was used for ellagic acid.

LSD ($p < 0.05$): least significant difference at the 0.05 level.

Different lower-case letters in the same column indicate statistically significant differences between different treatments ($p < 0.05$).

Table 5. Early-, mid- and late-harvest phenolic composition (μg g^{-1} fresh frozen weight) of strawberry selection 'SJ8976-1' grown under three production systems.

In contrast, o-Coumaric acid varied from 0.5 to 4.0 μg/g from early to late harvest accounted for the lowest portion of hydroxycinnamic acids. Moreover, the content of p-Coumaric acid, m-Coumaric acid and o-Coumaric acid was significantly promoted by PMRC at all stages (**Table 6**). Pelargonidin-3-glucoside was the dominant anthocyanin measured in strawberry selection 'SJ8976-1', while cyanidin-3-glucoside and pelargonidin-3-rutinoside were found in smaller amounts. PMRC significantly promoted the contents of pelargonidin-3-glucoside and cyanidin-3-glucoside compared to MRC, while the production systems only had the significant effect on pelargonidin-3-rutinoside at the late harvest (**Table 6**).

Treatment	**Hydroxycinnamic acids (280 nm)**			**Anthocyanins (510 nm)**		
	pC	**mC**	**oC**	**C3G**	**P3G**	**P3R**
Early harvest						
Matted row system	2.4b	38.8b	0.5b	6.8b	89.3b	1.7a
Plastic mulch	3.0b	48.6b	0.8b	8.1b	110.4ab	2.2a
Plastic mulch with row cover	4.6a	62.8a	1.4a	11.8a	130.5a	2.3a
$LSD_{0.05}$	0.8	10.5	0.4	2.8	25.3	0.7
Mid harvest						
Matted row system	2.1b	46.1b	0.6b	8.9b	99.0b	2.6a
Plastic mulch	2.3b	57.6b	0.9b	14.3a	130.5a	2.4a
Plastic mulch with row cover	3.9a	74.7a	1.7a	16.8a	149.3a	2.8a
$LSD_{0.05}$	1.0	13.4	0.5	3.3	28.0	0.5
Late harvest						
Matted row system	3.3b	55.9c	1.1b	12.8b	128.6c	2.8b
Plastic mulch	3.8b	75.9b	1.5b	16.6a	155.7b	3.3ab
Plastic mulch with row cover	8.5a	96.0a	4.0a	18.1a	184.6a	3.6a
$LSD_{0.05}$	1.5	14.8	1.2	2.2	21.4	0.7

Anthocyanins were quantified using cyanidin-3-glucoside (C3G), pelargonidin-3-glucoside (P3G) and pelargonidin-3-rutinoside (P3R); hydroxycinnamic acids were quantified using *p*-Coumaric acid (pC), *m*-Coumaric acid (mC) and *o*-Coumaric acid (oC); kaempferol was quantified using kaempferol-3-glucoside (K3G); flavonols were quantified using quercetin-3-galactoside; benzoic acids were quantified using gallic acid, and the authentic standard was used for ellagic acid.

LSD ($p < 0.05$): least significant difference at the 0.05 level.

Different lower-case letters in the same column indicate statistically significant differences between different treatments ($p < 0.05$).

Table 6. Early-, mid- and late-harvest phenolic composition (μg g^{-1} fresh frozen weight) of strawberry selection 'SJ8976-1' grown under three production systems.

These results are congruent with the findings of other researchers who indicated that significantly higher amounts of cyanidin-3-glycoside and pelargonidin-3-glycoside were assessed in strawberry fruits from hill plasticulture in comparison to MRS [19]. The consecutive experiment by Wang et al. also indicated that pelargonidin was the predominant anthocyanin in strawberry fruits [26]. Furthermore, the increased antioxidant capacity and phenolic compositions in berries may offer higher nutritional values to humankind [4, 30].

4. Conclusion

In summary, PMRC advanced fruit maturity and improved average fruit weight compared to MRS. SSC, TA and firmness of the strawberry fruits from PMRC were significantly increased at all harvests. Fruit weight loss, juice leakage and presences of grey mould were lower, and fruit glossiness was higher for those harvested under PMRC than those from MRS. TPC, TAC as well as ORAC of strawberry were significantly changed under different production systems. The phenolic compounds assessed in strawberry were grouped into six major categories, which were anthocyanins, kaempferol, flavonols, hydroxycinnamic acids, ellagic acid and benzoic acids.

Our results can provide useful insight to growers in north cooler climate because PMRC not only accelerates strawberry fruit ripening but also ameliorates fruit quality by improving nutritional value. PMRC could be a recommended alternative of MRS.

Acknowledgements

We thank the China Scholarship Council and the International Scientific Cooperation Bureau of Agriculture and Agri-Food Canada for their funding and support in this project.

Author details

Li Fan[1], Claudine Dubé[2] and Shahrokh Khanizadeh[3]*

*Address all correspondence to: shahrokh.khanizadeh@agr.gc.ca

1 College of Agronomy, Inner Mongolia Agricultural University, Hohhot, PR China

2 Quebec Research and Development Centre, Gouin Boulevard Saint-Jean-sur-Richelieu, Quebec

3 Ottawa Research and Development Centre, Carling Avenue, Central Experimental Farm, Ottawa, Ontario

References

[1] Vermerris W, Nicholson R. Phenolic compound biochemistry. 1st ed. West Lafayette: Springer; 2008.

[2] Cesar G. Fraga, editor. Plant phenolics and human health: biochemistry, nutrition and pharmacology. 1st ed. New Jersey: John Wiley & Sons; 2009.

[3] Tsao R, Khanizadeh S, Dale A. Designer fruits and vegetables with enriched phytochemicals for human health. Canadian Journal of Plant Science. 2006. 86(3): 773-786.

[4] Henning S M, Seeram N P, Zhang Y, Li L, Gao K, Lee R P, Wang D C, Zerlin A, Karp H, Thames G, Kotlerman J, Li Z Heber D. Strawberry consumption is associated with increased antioxidant capacity in serum. Journal of Medicinal Food. 2010. 13(1): 116-122.

[5] Khanizadeh S, Tsao R, Rekika D, Yang R, Charles M T, Rupasinghe H P V. Advances in fruit breeding in eastern Canada—role of phytochemicals in designing specialty fruits. Acta Horticulturae. 2009. 814: 205-208.

[6] Crespo P, Giné Bordonaba J, Terry L A Carlen C. Characterisation of major taste and health-related compounds of four strawberry genotypes grown at different Swiss production sites. Food Chemistry. 2010. 122(1): 16-24.

[7] Cheplick S, Kwon Y I, Bhowmik P, Shetty K. Phenolic-linked variation in strawberry cultivars for potential dietary management of hyperglycemia and related complications of hypertension. Bioresource Technology. 2010. 101(1): 404-413.

[8] Cejudo Bastante M J, Durán Guerrero E, Castro Mejías R, Natera Marín R, Rodríguez Dodero M C, Barroso C G. Study of the polyphenolic composition and antioxidant activity of new sherry vinegar-derived products by maceration with fruits. Journal of Agricultural and Food Chemistry. 2010. 58(22): 11814-11820.

[9] Basu A, Rhone M, Lyons T J. Berries: emerging impact on cardiovascular health. Nutrition Reviews. 2010. 68(3): 168-177.

[10] Khanizadeh S. Breeding strawberries for Eastern Central Canada. Euphytica. 1994. 77(1–2): 45-49.

[11] Kivijärvi P, Parikka P, Tuovinen T. The effect of different mulches on yield, fruit quality and strawberry mite in organically grown strawberry. in Organic production of fruit and berries. 2002, The Danish Institute of Agricultural Sciences, Department of Horticulture: Årslev, Denmark.

[12] Anttonen M J, Hoppula K I, Nestby R, Verheul M J, Karjalainen R O. Influence of fertilization, mulch color, early forcing, fruit order, planting date, shading, growing environment, and genotype on the contents of selected phenolics in strawberry (*Fragaria × ananassa* Duch.) fruits. Journal of Agricultural and Food Chemistry. 2006. 54(7): 2614-2620.

[13] Hancock J F, Goulart B L, Luby J J Pritts M P. The strawberry matted row: Practical cropping system of dated anachronism? Advances in Strawberry Research. 1997. 16: 1-4.

[14] Scheel D C. The effect of clear polyethylene winter mulch on the growth and yield of strawberries. Advances in Strawberry Production. 1982. 1: 29-30.

[15] Black B L, Enns J M, Hokanson S C. Advancing the matted-row strawberry production system. in Proc. 5th North American Strawberry Conference. 2002. Alexandria, VA, USA.

[16] Freeman S, Gnayem N. Use of plasticulture for strawberry plant production. Small Fruits Review. 2005. 4(1): 21-32.

[17] Neuweiler R, Bertschinger L, Stamp P, Feil B. The impact of ground cover management on soil nitrogen levels, parameters of vegetative crop development, yield and fruit quality of strawberries. European Journal of Horticultural Science. 2003. 68(4): 183-191.

[18] Allen R G, Pereira L S, Raes D, Smith M. Crop evapotranspiration-guidelines for computing crop water requirements-FAO Irrigation and drainage paper 56. FAO, Rome. 1998. 300(9): D05109.

[19] Wang S Y, Zheng W, Galletta G J. Cultural system affects fruit quality and antioxidant capacity in strawberries. Journal of Agricultural and Food Chemistry. 2002. 50(22): 6534-6542.

[20] Filipović V, Romić D, Romić M, Borošić J, Filipović L, Mallmann F J K, Robinson D A. Plastic mulch and nitrogen fertigation in growing vegetables modify soil temperature, water and nitrate dynamics: experimental results and a modeling study. Agricultural Water Management. 2016. 176: 100-110.

[21] Scalzo J, Politi A, Pellegrini N, Mezzetti B, Battino M. Plant genotype affects total antioxidant capacity and phenolic contents in fruit. Nutrition. 2005. 21(2): 207-213.

[22] Khanizadeh S, Buszard D. SJ89264-6 and SJ8976-1 two strawberry selections with long storage life. in III International Strawberry Symposium 439. 1996.

[23] Evans W D, Dale A, Hunter C. The strawberry in Ontario. 1st ed. Ottawa: Queen's Printer for Ontario; 1988.

[24] Fan L, Yu C, Fang C, Zhang M, Ranieri M, Dubé C, Khanizadeh S. The effect of three production systems on the postharvest quality and phytochemical composition of Orléans strawberry. Canadian Journal of Plant Science. 2011. 91(2): 403-409.

[25] Khanizadeh S, Tao S, Zhang S, Tsao R, Rekika D, Yang R, Charles M T, Gauthier L Gosselin A. Profile of antioxidant activities of selected strawberry genotypes. Acta Horticulturae. 2009. 814: 551-556.

[26] Wang Q, Tury E, Rekika D, Thérèse Charles M, Tsao R, Hao Y J, Dubé C, Khanizadeh S. Agronomic characteristics and chemical composition of newly developed day-neutral strawberry lines by Agriculture and Agri-Food Canada. International Journal of Food Properties. 2010. 13(6): 1234-1243.

[27] Slinkard K, Singleton V L. Total phenol analysis: automation and comparison with manual methods. American Journal of Enology and Viticulture. 1977. 28: 49-55.

[28] Benzie I F F, Strain J J. The ferric reducing ability of plasma (FRAP) as a measure of 'antioxidant power': The FRAP assay. Analytical Biochemistry. 1996. 239(1): 70-76.

[29] Fan L, Roux V, Dubé C, Charlebois D, Tao S, Khanizadeh S. Effect of mulching systems on fruit quality and phytochemical composition of newly developed strawberry lines. Agricultural and Food Science. 2012. 21(2): 132-140.

[30] Fan L, Dubé C, Fang C, Roussel D, Charles M T, Desjardins Y, Khanizadeh S. Effect of production systems on phenolic composition and oxygen radical absorbance capacity of 'Orléans' strawberry. LWT—Food Science and Technology. 2012. 45(2): 241-245.

[31] SAS Institute. SAS User's Guide, v6. Cary, NC, USA SAS Institute; 1989.

[32] Jin P, Wang S Y, Wang C Y, Zheng Y. Effect of cultural system and storage temperature on antioxidant capacity and phenolic compounds in strawberries. Food Chemistry. 2011. 124(1): 262-270.

[33] Soria C, López-Aranda J M, Medina J J, Miranda L, Domínguez F J. Evaluation of strawberry production and fruit firmness under small and large plastic tunnels in annual crop system. Acta Horticulturae. 2009. 842: 119-124.

[34] Tao S, Zhang S, Tsao R, Charles M T, Yang R, Khanizadeh S. *In vitro* antifungal activity and mode of action of selected polyphenolic antioxidants on *Botrytis cinerea*. Archives of Phytopathology and Plant Protection. 2010. 43(16): 1564-1578.

[35] Wang S Y Millner P. Effect of different cultural systems on antioxidant capacity, phenolic content, and fruit quality of strawberries (*Fragaria* × *aranassa* Duch.). Journal of Agricultural and Food Chemistry. 2009. 57(20): 9651-9657.

[36] Yanfang D, Tiezhao Y, Yapei L, Zhixiao Y, Gang X, Xiaoquan Z, Xuhui L. Effects of ecology and variety and their interactions on polyphenols compounds in flue-cured tobacco. Chinese Tobacco Science. 2013. 34(1): 17-21.

[37] Sharma R R, Singh R, Singh D, Gupta R K. Influence of row covers and mulching interaction on leaf physiology, fruit yield and albinism incidence in 'Sweet Charlie' strawberry (*Fragaria* × *ananassa* Duch.). Fruits. 2008. 63(2): 103-110.

Chapter 8

Phenolic Compounds in Maize Grains and Its Nixtamalized Products

Yolanda Salinas-Moreno, Carolina García-Salinas,
José L. Ramírez-Díaz and Ivone Alemán-de la Torre

Additional information is available at the end of the chapter

http://dx.doi.org/10.5772/66893

Abstract

Among the cereals most consumed by humans, maize grain is in the third position, surpassed only by rice and wheat. In several countries, maize grain is the main source of carbohydrates and proteins. Maize grain is ranked as one of the cereals with the highest content of phenolic compounds. The importance for human health of the consumption of phenolic compounds is due to their proved antioxidant activity. Diets with high amount of antioxidants have been associated with a reduced probability of suffering degenerative chronic diseases. In maize grain, the phenolic acids predominate, among which the main is ferulic acid, followed by p-coumaric acid, which are highly abundant in their bound forms. However, other phenolics such as anthocyanins, flavonols, and flavanols have been identified in colored maize grains. Additionally, the processing of maize grain into different products for human consumption incorporates changes both in quantity and quality of some phenolic compounds. In the present chapter, we present the most recent information available regarding phenolic compounds in maize grain and their nixtamalized products.

Keywords: *Zea mays* L., phenolic acids, flavonoids, nixtamalization, tortilla

1. Introduction

Phenolic compounds are widely distributed in plants as products of their secondary metabolism; they are produced during plant development and some are of vital importance for their adequate functioning and interaction with the environment while others are synthesized in response to stress conditions such as infections, injuries, ultraviolet radiation, among others [1, 2]. Furthermore, phenolic compounds serve as defense mechanisms since many of them display antifeedant and antipathogenic properties [3], which contribute to their adaptation to different environments. The presence of phenolic compounds in different plant foods contributes to their distinctive characteristics and to its flavor and color. They can be found in soluble/free and bound/insoluble form. In maize grain the highest amount of phenolics (98.9%) is present in the insoluble fraction, and the remainder in the soluble fraction [4]. However, it is the soluble fraction that shows the greater chemical diversity, which depends on the color of the grain. The goal of this document is to provide a review of the various phenolic compounds present in maize grains of different colors, and the changes that occur when the grain is subject to nixtamalization processing for the elaboration of tortillas and all the diversity of nixtamalized products consumed in Mexico and in many other parts of the world.

2. The phenolic compounds

The biosynthesis of these compounds occurs via the shikimate pathway, from the amino acids phenylalanine and tyrosine and the participation of the enzyme phenylalanine ammonia lyase (PAL) that catalyzes the removal of the ammonia residue of the amino acids phenylalanine and tyrosine to produce cinnamic acid and 4-coumaric acid, respectively [5]. Both compounds further enter the phenylpropanoid pathway and it is within the various branches of this pathway where the great diversity of phenolic compounds so far identified is synthesized.

Regarding their chemical structure, phenolic compounds possess at least one aromatic ring with one or more hydroxyl groups, including their functional derivatives [6]. The polyphenols are within the group of phenolic compounds, which according to Quideau et al. [7], the term "phenolics" should be used to define compounds derived exclusively from the shikimate/phenylpropanoid pathway and/or the route of polyketides, which include more than one phenolic unit (phenol). This restriction is necessary because substances from alternative metabolic pathways may also present more than one phenolic unit. In literature, the term polyphenols and phenolic compounds are often encountered, however, if the former term is used it would not include the phenolic acids, as their structure contains only one phenol. Therefore, throughout this work we will use the term "phenolic compounds" as we will be commenting on some flavonoids and phenolic acids.

The complexity of the phenolic compounds ranges from simple molecules as phenolic acids to highly polymerized compounds as tannins. Phenolic compounds are present in plants in conjugated form with one or more sugar residues bound to the hydroxyl groups, although in some cases direct connections between a sugar molecule and an aromatic carbon may occur. The most common way to find them in nature is as glycosides, conferring them solubility in water and in organic solvents. All phenolic compounds exhibit strong absorption in the UV spectral region, and some colorful phenolic compounds absorb in the visible region as well [8].

Phenolic compounds can be classified in various ways, one proposed by Harborne and Simmonds [9] considers the number of carbons contained in their molecule. According to this

http://dx.doi.org/10.5772/66893

criterion a total of 20 different groups of phenolic compounds are considered, where simple phenols are the simplest and the phlobaphenes are the most complex group [10]. Another criterion is used by Shahidi and Naczk [6], which classifies groups of compounds based on their complexity. **Figure 1** illustrates this classification.

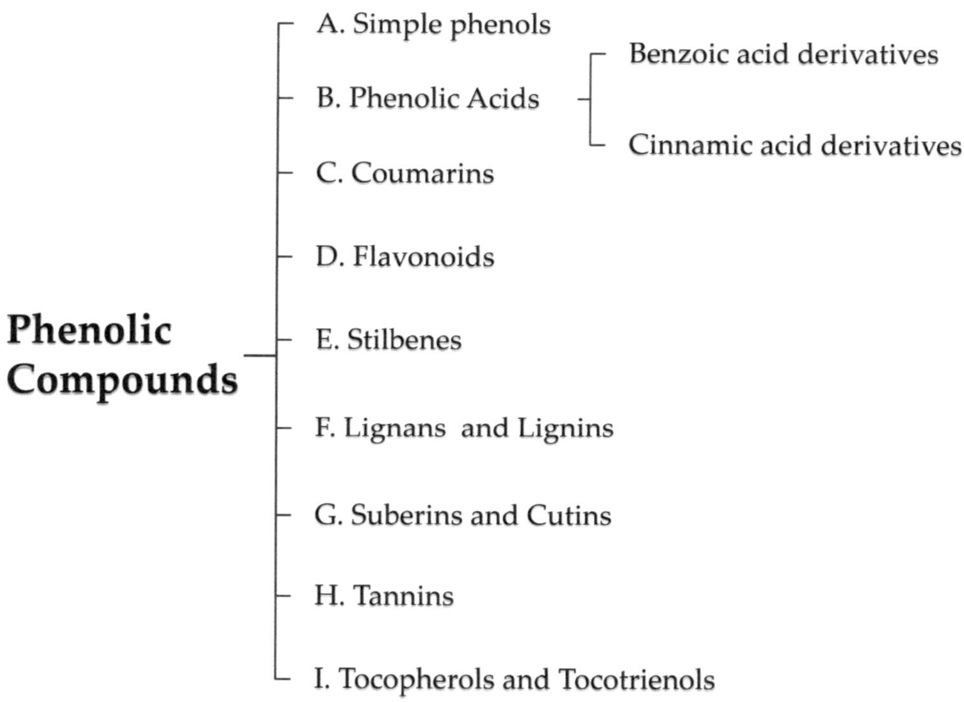

Figure 1. General classification of phenolic compounds (adapted from Shahidi and Nazck [6]).

3. Phenolic compounds in maize grain

Phenolic acids are the main phenolics in maize grain; however, other compounds like phenolic amines and some flavonoids have also been described [11]. The most abundant phenolic acids are ferulic and p-coumaric, which may be in their isomeric form cis or trans, the most common being the trans form. Both acids are present in soluble form or bound to cell wall components. Ferulic acid (3-methoxy-4-hydroxycinnamic acid) is the most abundant in the cell wall of monocots and is found in all fractions of the maize grain, but most abundantly in the pericarp and the aleurone layer. Chemically it is mostly ester-linked to plant cell wall components, hemicellulose chains, mainly in the arabinose residues, but it can also be polymerized in lignin by ether linkages [12]. When ferulic acid is oxidized, it forms dimers or trimers, which after being hydrolyzed, are capable of forming gels when linked to two pentosans or protein molecules.

Among the flavonoids present in maize grain are the flavonols, anthocyanins, and proanthocyanidins. Das and Singh [13] reported the presence of quercetin and kaempferol (flavonols) in the germ and pericarp of quality protein maize (QPM), popcorn and sweet corn. Meanwhile, Ramos et al. [14] reported the presence of kaempferol and morin in purple maize grains. The chemical structures of some phenolic compounds present in maize grain are shown in **Figure 2**.

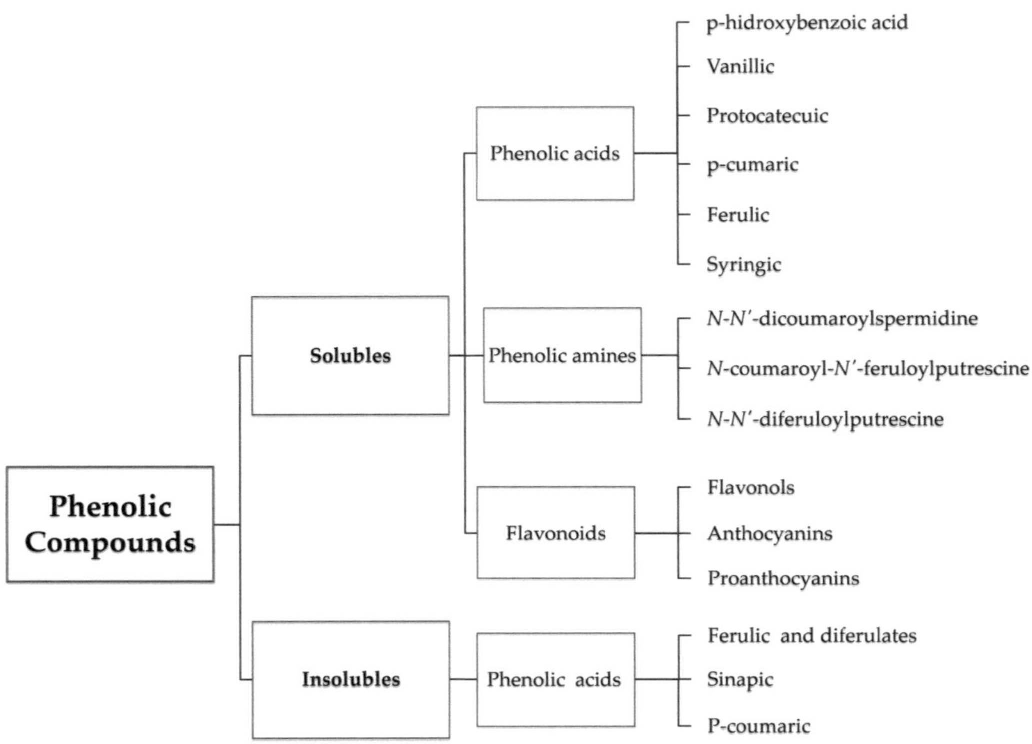

Figure 2. Phenolic compounds identified in white and pigmented maize grains.

Phenolic amines were initially identified in the pericarp of white maize grain by Sen et al. [15], who associated their presence with tolerance to storage pests; recently, Collison et al. [16] reported that the phenolic amines: N-N′-dicoumaroylspermidine, N-coumaroyl-N′-feruloylputrescine, and N-N′-diferuloylputrescine were the most abundant soluble phenolics in the methanolic extract of nixtamalized grains from red, blue, and purple maize grain. It is not known for sure what is the role of phenolic amines in maize grain.

3.1. Soluble phenolic compounds

These compounds correspond to those obtained by treating a given sample size of ground maize grain to extraction with an organic solvent, typically aqueous solutions of methanol or ethanol. Quantification is performed by the Folin-Ciocalteau method [17]. The identification of the different types of phenolic compounds is achieved by high-resolution liquid chromatography (HPLC) when standards are available or by means of mass spectrometry (MS). The

http://dx.doi.org/10.5772/66893

position in which the different moieties of the molecule are attached is elucidated by nuclear magnetic resonance (NMR) techniques.

3.1.1. Phenolic acids

Phenolic acids can be found in maize grain in soluble and insoluble form. The soluble fraction is also known as free, although in a strict sense, in this fraction phenolic acids can be present in their free soluble forms, glycosylated or esterified. This fraction is very small compared to the insoluble or bound fraction. However, the fraction of soluble phenolic acids is more diverse than the insoluble, particularly in the grains containing anthocyanin-type pigments. Phenolic acids that occur in maize grain are derived from both benzoic acid and cinnamic acid. The chemical structures of the most common are shown in **Figure 3**.

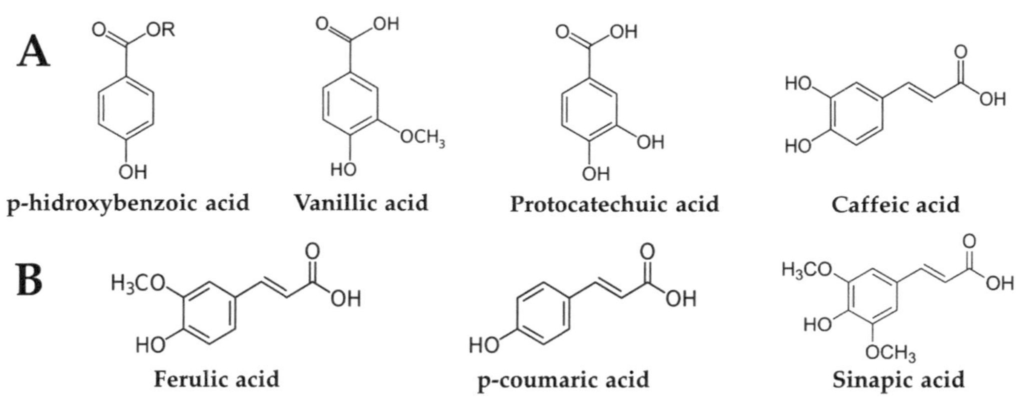

Figure 3. Phenolic acids in the maize grain. (A) Benzoic acid derivatives and (B) cinnamic acid derivatives.

Among the phenolic acids present as soluble form in the maize grain are p-hydroxybenzoic, vanillic, and protocatechuic, all derived from benzoic acid. The first two have been reported in the pericarp of popcorn and in the germ of QPM, as well as in baby corn, with greater abundance of p-hydroxybenzoic [18]. In the grain of purple maize the presence of protocatechuic, vanillic acid, and p-coumaric has been reported in amounts of 14.61 ± 0.08 and 8.46 ± 0.09 mg equivalents of ferulic acid/g sample for the former two acids, whereas the latter was found only in trace amounts [19]. Meanwhile, Sosulski et al. [20] identified in yellow maize flour, in addition to the previously mentioned acids, p-coumaric, ferulic, and syringic acids in the free phenolic fraction. This fraction was much smaller than the soluble esterified phenolic fraction, in which ferulic and syringic acids predominated. **Table 1** shows the reported concentrations of some phenolic acids in maize grain of different colors.

3.1.2. Flavonoids

They are phenolic compounds that form a very large group of which more than 5000 different members have been identified [21], many of which possess important biological activities such as antioxidant, antimutagenic, and microbicidal. Different types of flavonoids have been identified in colored maize grain with anthocyanin-type pigments. They can be found free

Concentration of phenolic acids (μg/g of dry sample)								
Maize grain color	***p*-Hydroxybenzoic**	**Vanillic**	**Caffeic**	**Syringic**	***p*-Coumaric**	**Ferulic**	**Protocatechuic**	**Reference**
White					34.4	4841.4	18.4	Del Pozo-Insfran et al. [27]
Yellow		2123.2	95.7		230.1	4543.2		Das and Singh [13]
Yellow	1.3	3.7	4.5	11.5	18.9	5.1		Sosulski et al. [20]
Red					319.90	2712.58		Zilic et al. [22]
Mexican blue					1.15	428.4		Del Pozo-Insfran et al. [27]
American blue					58.6	1237.5		Del Pozo-Insfran et al. [27]
Purple		8460			Traces		14610	Pedreschi and Cisneros [19]

Table 1. Concentrations of some phenolic acids in maize grain with different colors.

or in conjugated forms [22]. These compounds are very susceptible to hydrolysis and some losses may occur during their analysis. The presence of these phenolic has been reported in maize grains with anthocyanin-type pigments and the flavonoids reported are anthocyanins, flavonols, and proanthocyanins (flavan 3-ols).

3.1.2.1. Anthocyanins

Anthocyanins are a class of water-soluble flavonoids that are visible to the naked eyes. They are glycosides of polyhydroxy and polymethoxy derivates of 2-phenylbenzopyrylium or flavylium salts and are responsible for the red, purple, and blue colors of many fruits, vegetables, and cereal grains [23]. In maize grain, they are located in the pericarp, the aleurone layer, or in both structures [24]. Anthocyanins have sugars attached to the B-ring at the 3′ and 5′-hydroxyl position. The two most important types of glucosides are 3-monoglycoside and 3-4-diglycoside. As a rule, the 3-hydroxyl always has a sugar, except in 3-desoxypelargonidin, 3-desoxycyanidin, and 3-desoxydelphinidin [25]. The basic structure of the anthocyanidins present in maize grain is shown in **Figure 4**.

3.1.2.2. Flavonols and proanthocyanidins

In the grain of blue maize, derivatives from quercetin have been reported [27], while in the grain of purple maize, in addition to these derivatives, morin and kaempferol have also been identified and quantified [14]. These flavonols have not been reported in white maize grain, whereas in yellow maize grain, quercetin and kaempferol have been reported to be present in trace amounts [13]. Within the group of proanthocyanidins, only catechin has been reported to be present in the blue maize grain of Mexican and American origins [27].

3.2. Insoluble or bounded phenolic compounds

Insoluble phenolics from maize grain are those obtained from alkaline hydrolysis of the sample residue once soluble phenolics have been removed. The pH of the hydrolyzed sample is adjusted to 2 and liquid-liquid extractions with ethyl acetate are used to recover them. The total content is quantified by the Folin-Ciocalteau assay. They can also be analyzed by HPLC-DAD techniques or HPLC/MS/MS for a complete identification of the different phenolics present in this fraction.

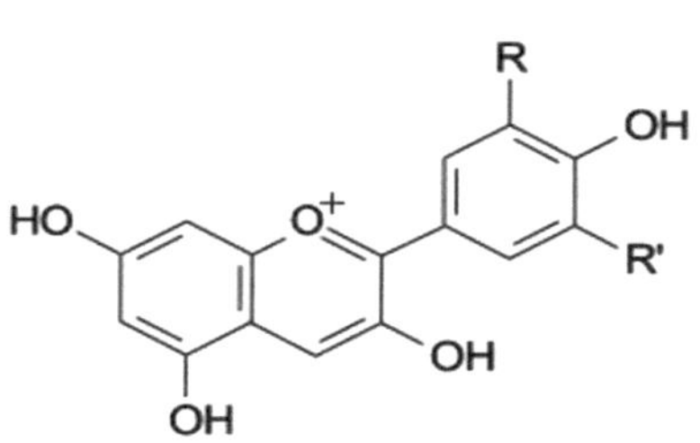

Aglicons	R1	R′
Pelargonidin	H	H
Malvidin	OCH3	OCH3
Cyanidin	OH	H
Peonidin	OCH3	H

Figure 4. Basic structure of the anthocyanidins (aglycons) present in maize grain (adapted from Rodríguez and Wrolstad [26]).

3.2.1. Ferulic acid and diferulates

Most of the phenolic acids in maize grain are found in the bound form and in this fraction, the ferulic acid (4-hydroxy-3-methoxycinnamate) predominates. It is found esterified to arabinoxylans from the hemicelluloses in the cell wall of the various structures of the maize kernel. Most of this acid in its free form is present in the germ, while its bound form is concentrated in the grain pericarp [18]. As noted above, the ferulic acid is mainly esterified to cell wall components. In the free form, it can be also found conjugated to different molecules, among which the most common are simple sugars and some amines [15, 16].

3.2.2. p-coumaric acid

After ferulic acid, the second most abundant phenolic compound in maize grain is p-coumaric acid. It is present in both forms, the soluble phenolic fraction and the insoluble, with greater presence in the latter. In the insoluble fraction it is mainly linked to lignin, and to a lesser extent to polysaccharides embedded in the cell wall [28]. In the soluble fraction of purple maize the value reported for this acid is 34.1 mg/100 g DW, and for the phenolic insoluble fraction is 573.4 mg/100 g DW. Surprisingly, ferulic acid values reported in the insoluble fraction of this maize were significantly lower (154.2 mg/100 g DW) compared to the amount of p-coumaric acid [29].

4. Phenolic compounds in maize by grain color

The content and type of phenolic compounds in the maize grain will vary depending on the color of the grain, the genetic origin, and the extraction method used. In the latter factor is determining if technologies such as ultrasound and microwave to favor extraction are used, in addition to the type of solvent used to extract them. **Table 2** contains information regarding the amount of free or bound soluble and insoluble phenolics found in maize grain. The data show very large differences in the reported values by different authors for the same maize grain color. This is undoubtedly due to the genetic variability of the biological material itself. Furthermore, in some works as that of Montilla et al. [29], the extraction of the insoluble phenolic was performed sequentially with three different hydrolysis methods and finally the phenolic compounds recovered from each were added. This method differs from what has been done, for example, by López-Martínez et al. [30], who performed alkaline hydrolysis and recovered the phenolics by liquid-liquid extraction with ethyl acetate, which is the most common method applied. Under this method, the authors reported very different values in four samples of purple grain of different genetic background.

4.1. White maize grain

In white maize grain, differences in total phenolic content (solubles + insolubles) due to grain hardness are present. These differences are supported by the role of ferulic acid to provide strength to the cell wall of different grain structures. But also, each botanical grain structure has a content and a particularly phenolic profile, which is associated with its functionality.

Maize grain color	(μg GAE/g DW) FP	SGP	SEP	IP	TP	References
Yellow	61.7 ± 0.3	163.5 ± 9.1	254.7 ± 44.2	2331.5 ± 103.3	2811.4 ±	Xu et al. [31]
Yellow	1040 ± 22.0			4470.0 ± 43	5510 ± 38	López-Martínez et al. [30]
White	334 ± 15			1360 ± 32	1700 ± 11	López-Martínez et al. [30]
White	347.0 ± 4.0			2260 ± 63	2607 ± 61	De la Parra et al. [48]
Purple	821.0			5296.0	6117.0	Montilla et al. [29]
Purple	837–6800			3810–27,200	4650–34,000	López-Martínez et al. [30]
Blue	455 ± 5			2207 ± 5	2662 ± 7	De la Parra et al. [48]

FP: free phenolics; SGP: soluble glycosylated phenolics; SEP: soluble esterified phenolics; IP: insoluble phenolics; TP: total phenolics; GAE: gallic acid equivalents; DW: dry weight.

Table 2. Contents of soluble and insoluble phenolic compounds in maize grain with different colors.

4.1.1. Differences by grain hardness

Few studies have studied the relationship between the hardness of the maize grain and its phenolic content. The hardness of the grain is a widely studied aspect, because of the importance that this feature has for the different processes to which maize grains are subject to obtain the myriad of products derived from it [32]. Among the literature on the matter, Cabrera-Soto et al. [33] analyzed the soluble and insoluble phenolic fractions in the grain of seven maize hybrids of different hardness. The authors observed a significant positive correlation between the values of soluble esterified phenolics in the three grain structures (pericarp, germ, and endosperm) and the grain hardness. In the study by Chiremba et al. [34] in maize and sorghum, a higher content of phenolic acids in grains of greater hardness was observed, however, the correlation between the content of these phenolics and hardness was higher in maize grains than in sorghum grains.

4.1.2. Differences by grain botanical structure

Phenolic compounds are distributed in all maize grain structures, however, their concentration and type varies among them. Several studies using fluorescent techniques have reported that the pericarp is the structure with the highest concentration of phenolics, followed by the germ and the endosperm [15].

Pericarp. The pericarp of maize grain is a rich source of phenolics, mainly in their bound form. However, a minor amount of phenolics can also be present in their free form. The main phenolic in the bound form is ferulic acid, followed by p-coumaric and sinapic acids. Several free phenolic acids reported in this structure are ferulic (in its conjugated form), vanillic, caffeic, p-coumaric, and p-hydroxybenzoic. However, the abundance of each one varied according to the maize type [18].

Most of the ferulic acid in the pericarp of the maize grain is linked by ester bonds to cell wall polysaccharides [35]. Nonetheless, they are also present in the form of dehydrodimers originating from the oxidative coupling of ferulate esters by means of the enzyme peroxidase. The diferulates that have been identified in maize grain are 8,5′-diferulic acid, 8,O,4′-diferulic acid, 8-8-diferulic acid, 4-O-5-diferulic acid, and 5.5′-diferulic acid [35], of which the most abundant is the first one [26]. The diferulates are linked to the arabinoxylans of the polymers that form the cell wall [36]. The resistance of the pericarp cells is attributed to the presence of these compounds and their abundance has been associated to resistance toward the development of fumonisins [37, 38] and tolerance to warehouse pests [15, 39]. The presence of ferulic acid dehydrotrimers was reported by Rouau et al. [40] in maize bran, which is a fraction that is composed of remnants of pericarp and aleurone layer. Until now, seven different ferulic acid dehydrotrimers have been identified [35]. The different ways in which ferulic acid is integrated into the cell wall components contributes to the formation of networks that support the resistance of this structure whose main function is to isolate and protect the grain from external agents.

The type of maize affects the values reported for total soluble phenolics in the grain pericarp. In dent maize grain, Cabrera-Soto et al. [33] reported a variation of 232.4–334.0 mg EAG/g DW in seven different maize varieties. In cornpop, Das and Singh [18] observed a value of 13.1 ± 0.66 μmol of FAE/g of DW, while the value for QPM maize was of 15.9 ± 0.28. Insoluble phenolic content (IPs) in this structure is between 18 and 21 times that of soluble phenolics (SPs). Das and Singh [13] reported values of SPs of 11.9 and 10.4 μmol of FAE/g of DW in the pericarp of dent and crystalline maize types, respectively: the values of IPs were 218.6 and 219.4 μmol of FAE/g of DW, for the same maize grains. In the SPs fraction, the phenolic acids presented were vanillic, caffeic, ferulic, and p-coumaric, with predominance of the latter; in IPs, ferulic acid represented between 40 and 50% of this fraction.

In maize grain with presence of anthocyanin pigments in the pericarp, the SPs content is commonly higher than the values observed in white maize grain. This is due to the presence of anthocyanins that occurred mainly in soluble form, because the amount of bounded anthocyanins reported is very low [29].

Germ. The germ concentrated the highest content of soluble phenolics of the maize grain structures. In seven varieties of white maize grain, Cabrera-Soto et al. [33] reported in the germ values of 499.1–689.2 μg GAE/g DW; in the pericarp, the values were of 232.4–334.1 μg GAE/g DW, while in the endosperm, they reported values of 124.1–194.0 μg GAE/g DW. Meanwhile, Das and Singh [13] reported 14.2 ± 0.3 for germ, 11.9 ± 1.1 for pericarp, and 0.4 ± 0.02 μmol of FAE/g of DW, in one sample of dent maize grain. In the germ, the phenolic acids: 3-hydroxybenzoic acid, caffeic, p-coumaric, ferulic in itsr cis and trans forms, and salicylic have been reported. Ferulic acid predominates in its esterified soluble form [41].

Endosperm. The concentration of phenolic compounds in this structure is very low. Cabrera-Soto et al. [33] reported values of 124.1–194.0 μg GAE/g DW. The amount of IPs is also marginal since it is less than 2% of that present in the pericarp and 3.56% of that contained in the germ. However, the endosperm represents 80–85% of the total weight of the grain, so its total contribution is close to that of the germ.

4.2. Pigmented maize grain

4.2.1. Blue/purple grain

The purple grain maize is the most studied with respect to the phenolic compounds because it is highly used for pigments extraction with application in foods [41–43]. In purple maize grain the phenolic acids have been reported: protocatechuic, vanillic, and p-coumaric acid, besides four hydroxycinnamic acid derivatives [19]. Additionally, Ramos et al. [14] found in purple maize grain the following phenolic acids: chlorogenic, caffeic, and ferulic. Of the flavonoids present in purple maize grain, the anthocyanins are the most abundant. Ramos et al. [14] reported values of 2.76 ± 0.05 g of cyanidin-3-glucoside (C3G)/kg DW, meanwhile for flavonols and flavanols the values observed were of 0.41 ± 0.02 g of rutin equivalents (RE)/kg of DW and 0.23 ± 0.01 g of catechin equivalents (CE)/kg of DW, respectively. Other flavonoids described in this maize grain color were rutin, morin, quercetin, naringenin, and kaempferol. Others phenolic compounds that have been identified in purple maize grain are those resulting from the condensation of flavanols (catechin or epicatechin) and anthocyanins. Some of these compounds are catechin-(4,8)-cyanidin-3-glucoside, catechin-(4,8)-peonidin-3-glucoside, catechin-(4,8)-pelargonidin-3-glucoside, and the corresponding derivatives in which the epicatechin is the flavonol attached [44].

4.2.2. Red grain

The red maize grain like the blue maize is especially high in phenolic compounds as compared to light colored maize genotypes. The average value of total phenolics for these grains is of 6056.9 mg/kg DW. The most abundant phenolic acids identified in red maize grain are ferulic and p-cumaric [22].

In **Table 3**, the total anthocyanins content (TAC) for different grain colors and the predominant anthocyanins in each sample are shown. It should be clarified that, within the same shade of grain color, different intensities may be present, which are evidenced by the different contents of total anthocyanins of grains of a similar shade.

Maize grain color	TAC*	Main anthocyanins	Reference
Red	85.2 ± 2.2	Cyanidin-3-glucoside Cyanidin-3,6-malonylglucoside	Zilic et al. [22] López-Martínez et al. [30]
Red	9.75 ± 0.44	Peonidin-3,6-EthylMalonylGlucoside Pelargonidin-3,6-Ethy-MalonylGlucoside	De La Parra et al. [48]
Blue	99.5 ± 1.8	Cyanidin-3-glucoside	López-Martínez et al. [30]
Blue	36.87 ± 0.71	Pelargonidin-3-glucoside Peonidin-3-glucoside	Pedreschi and Cisneros [19] De La Parra et al. [48]

* Concentration expressed in mg cyanidin-3-glucoside/100 g of sample.

Table 3. Anthocyanins content and main anthocyanins presented in the maize grain with different colors.

5. Changes of phenolic compounds during processing (nixtamalization) of maize grain

Nixtamalization results in multiple changes in the chemical components of the maize grain. Under the traditional process, the cooking water has high pH (~12) which hydrolyzes the ester bond by which ferulic acid is linked to cell wall components. The grain structure that is mostly affected by this process is the pericarp, which becomes partially or fully hydrolyzed [45, 46]. As the most abundant phenolic in the grain, the hydrolysis of the ester linkage between the ferulic acid and the cell wall components causes the soluble fraction to be higher in nixtamalized maize products relative to that found in the whole grain, white maize. However, in colored maize that contains anthocyanin pigments, the soluble fraction is reduced due to the significant loss of anthocyanins [47–49]. According to the information presented in **Table 4**, in white maize grain soluble phenolics increase by about 26% when the grains are processed into tortillas, while in red maize grain, they are reduced by 20%. The magnitude of the reduction varies according to the grain color and the origin of the genetic material. The cooking of the tortilla results in an additional loss of phenolic compounds, but much less significant that resulting from nixtamalization [47].

Phenolic compounds in maize grain	SP (mg GAE/100 g DW)	IP	TP	Ferulic (mg FAE/100 g DW)	References
WG					
Raw	34.7 ± 0.4	226.0 ± 6.3	260.7 ± 6.1	120.45	De la Parra et al. [48]
Tortilla	47.2 ± 1.8	119.0 ± 6.2	166.2 ± 6.2	85.16	
Raw			167.4	474.49	Mora-Rochin et al. [49]
Tortilla			85.4	101.66	
RG					
Raw	38.2 ± 0.4	205.6 ± 4.5	243.8 ± 4.6	130.3	De la Parra et al. [48]
Tortilla	30.5 ± 0.7	106.0 ± 3.6	136.5 ± 2.9	73.83	
Raw			149.2	532.16	Mora-Rochin et al. [49]
Tortilla			89.8	208.12	
BG					
Raw	45.5 ± 0.5	220.7 ± 0.5	266.2 ± 0.7	123.01	De la Parra et al. [47]
Tortilla	39.1 ± 1.5	122.7 ± 0.6	161.8 ± 2.1	101.36	
Raw			140.1	336.49	Mora-Rochin et al. [49]
Tortilla			86.3	187.79	

SP: soluble phenolics; IP: insoluble phenolics; TP: total phenolics; WG: white grain; RG: red grain; BG: blue grain.

Table 4. Phenolic compounds in raw grain and tortillas obtained by the traditional nixtamalization process, from maize grains with different colors.

When nixtamalization is performed by the extrusion method, total phenolic losses in maize grain are lower than those resulting from the traditional method [49]. The differences were attributed to the fact that the extrusion method does not produce losses from hydrolyzed pericarp, and there is no leaching of phenolics.

Of the phenolic acids present in maize grain, the changes in ferulic acid have been monitored, as the most abundant member of this group. In white maize grain, the losses of this acid in the process of going from grain to tortillas are between 20 and 30%; nevertheless, in red maize grain they are higher and account between 44 and 60%, while in the blue maize grain they represent 18–45%, according to the information shown in **Table 4**. It is probable that the high losses of ferulic acid observed in red or blue maize grain may be a result of the maize varieties used in the studies that have a floury grain, which is very likely to occur if the maize used were native of Mexico. Floury maize generally has a thinner pericarp as compared to that of hard grains, and thus during nixtamalization the grain is hydrated faster than hard grain. Thinner pericarps could favor the hydrolysis of the ester bond that keeps the ferulic acid linked to cell wall components, and thus the acid is leached to the cooking water, known as nejayote. There are no reports on contents of ferulic acid in the nejayote of maize of different grain hardness, processed under the same nixtamalization method. In addition, when cooking times are adjusted to achieve optimal nixtamalization for each type of maize, the values of ferulic acid found in nejayote relate more to the cooking length than with grain hardness [50] and have values of ferulic acid content greater than those found in the grain and the masa for tortillas.

6. Conclusions

Phenolic compounds are present in the maize grain in free or soluble and bound or insoluble form. The insoluble fraction is the most abundant, but the most chemically diverse is the soluble fraction. There exists greater diversity of phenolic compounds in the grain of maize with anthocyanin-type pigments than in the white maize grain. Phenolic acids are the most abundant phenolic compounds in maize grain, followed by flavonoids, particularly anthocyanins in the blue red and purple maize grain. Phenolic amines are present in the pericarp of white maize grain and grain containing anthocyanin pigments, being most abundant in the latter. Nixtamalization significantly reduces the content of phenolics present in maize grain, in white grain maize, they are lost mainly as ferulic acid, while in the maize grain that contains anthocyanin pigments, in addition to ferulic acid, anthocyanins are almost entirely lost.

Author details

Yolanda Salinas-Moreno[1]*, Carolina García-Salinas[2], José L. Ramírez-Díaz[1] and Ivone Alemán-de la Torre[1]

*Address all correspondence to: salinas.yolanda@inifap.gob.mx

1 Experimental Field "Altos de Jalisco" Center, National Institute of Forestry, Agriculture and Livestock Research (INIFAP), Tepatitlán de Morelos, Jal., Mexico

2 Monterrey Institute of Technology and Higher Education, Monterrey Campus, Monterrey, N.L., Mexico

References

[1] Nicholson RL and Hammerschmid R. Phenolic compounds and their role in diseases resistance. Annual Review of Phytopathology. 1992; 30: 369–389. DOI: 10.1146/annurev.py.30.090192.002101

[2] Boudet AM. Evolution and current status of research in phenolic compounds. Phytochemistry. 2007; 68:2722–2735. DOI: 10.1016/j.phytochem.2007.06.012

[3] Lattanzio V, Kroon PA, Quideau S, and Treutter D. Plant phenolics – secondary metabolites with diverse functions. In: Daayf F, Lattanzio V, editors. Recent advances in polyphenols research, Singapore: Wiley; Vol I; 2011. Pp. 1–35. DOI: 10.1002/9781444302400

[4] Adom KK and Liu RH. Antioxidant activity of grains. Journal of Agriculture and Food Chemistry. 2002; 50: 6182-6187. DOI: 10.1021/jf0205099

[5] Shaw NM, Bolwell GP, and Smith C. Wound-induced phenylalanine ammonia-lyase in potato (*Solanum tuberosum*) tuber discs. Biochemical Journal. 1992; 267: 163–170. DOI: 10.1042/bj2670163

[6] Shahidi F and Naczk M. Biosynthesis, classification, and nomenclature of phenolics in food and nutraceuticals. In: Phenolics in food and nutraceuticals. CRC Press, Washington, DC; 2004. pp. 1–16.

[7] Quideau S, Deffieux D, Douat-Casassus C, and Pouységu L. Plant polyphenols: Chemical properties, biological, activities, and synthesis. Angewandte Chemie International Edition. 2011; 50: 586–621. DOI: 10.1002/anie.201000044

[8] Harborne, J. B. Plant phenolics. In: Encyclopedia of plant physiology, New Series, Vol. 8. (Pirson, A. & Zimmerman, M. H. (Eds), Springer-Verlag, Berlin, 239–402.

[9] Harborne JB and Simmonds NW. Biochemistry of Phenolic Compounds. Academic Press. London; 1964. p. 101.

[10] Vermerris W and Nicholson R. Families of phenolic compounds and means of classification. Phenolic compound biochemistry. Springer, Netherlands; 2008. pp. 1–34. DOI: 10.1007/978-1-4020-5164-7

[11] Choi SW, Sung KL, Eun OK, Ji HO, Kyung SY, Parris N, Hicks KB, and Moreau RA. Antioxidant and antimelanogenic activities of polyamine conjugate from corn bran and related hydroxicinnamic acids. Journal of Agricultural and Food Chemistry. 2007; 55: 3920–3925. DOI: 10.1021/jf0635154

[12] Liyama K, Lam TBT, and Stone BA. Covalent cross-links in the cell wall. Plant Physiology. 1994; 104: 315–320. DOI: 10.1104/pp.104.2.315

[13] Das AK and Singh V. Antioxidative free and bound phenolic constituents in pericarp, germ and endosperm of Indian dent (Zea mays var. indentata) and flint (Zea mays var. indurata) maize. Journal of Functional Foods. 2015; 13: 363–374. DOI: 10.1016/j.jff.2015.01.012

[14] Ramos FE, Muñóz AM, Alvarado OC, Alvarado A, and Yañéz JA. Purple corn (Zea mays L.) phenolic compounds profile and its assessment as an agent against oxidative stress in isolated mouse organs. Journal of Medicinal Foods. 2012; 15: 206–215. DOI: 10.1089/jmf.2010.0342

[15] Sen A, Bergvinson D, Miller SS, Atkinson J, Fulcher RG, and Arnason J. Distribution and microchemical detection of phenolic acids, flavonoids, and phenolic acid amides in maize kernels. Journal of Agricultural and Food Chemistry. 1994; 42: 1879–1883. DOI: 10.1021/jf00045a009

[16] Collison A, Yang L, Dykes L, Murray S, and Awika JM. Influence of genetic background on anthocyanin and copigment composition and behavior during thermoalkaline processing of maize. Journal of Agricultural and Food Chemistry. 2015; 63: 5528-5538. DOI: 10.1021/acs.jafc.5b00798

[17] Singleton VL and Rossi JA. Colorimetry of total phenolics with phosphomolybdic-502 phosphotungstic acid reagents. American Journal of Enology and Viticulture. 1965; 16: 144–158. DOI:

[18] Das AK and Singh V. Antioxidative free and bound phenolic constituents in botanical fractions of Indian specialty maize (*Zea mays* L.). Food Chemistry. 2016; 201: 298–306. DOI: 10.1021/jf100868b

[19] Pedreschi R and Cisneros L. Phenolic profiles of Andean purple corn (*Zea mays* L.). Food Chemistry. 2007; 100: 956–963. DOI: 10.1016/j.foodchem.2005.11.004

[20] Sosulski F, Krygier K, and Hogge L. Free, esterified, and insoluble-bound phenolic acids. 3. Composition of phenolic acids in cereal and potato flours. Journal of Agriculture and Food Chemistry. 1982; 30: 337–340.DOI: 10.1021/jf00110a030

[21] Prior RL, Wu H, and Gu L. Flavonoid metabolism and challenges to understanding mechanisms of health effects. Journal of the Science of Food and Agriculture. 2006; 86: 2487-2491. DOI: 10.1002/jsfa.261

[22] Zilic S, Serpen A, Akıllıoglu G, Gokmen V, and Vanc EJ. Phenolic compounds, carotenoids, anthocyanins, and antioxidant capacity of colored maize (*Zea mays* L.) Kernels Journal of Agricultural and Food Chemistry. 2012; 60: 1224-1231.DOI: 10.1021/jf204367z

[23] Giusti MM and Wrolstad RE. Acylated anthocyanins from edible sources and their applications in food systems. Biochemical Engineering Journal. 2003; 14: 217–225. DOI: 10.1016/S1369–703X(02)00221-8

[24] Salinas-Moreno Y, Soto-Hernández M, Martínez-Bustos F, González-Hernández V, and Ortega-Paczka R. Analysis of anthocyanins in blue and red maize grains of four races. Revista Fitotecnia Mexicana. 1999; 22: 161–174.

[25] Rodríguez SL and Wrolstad RE. Extraction, isolation and purification of anthocyanins. Current Protocols in Food Analytical Chemistry. John Wiley, New York; 2001. pp. 7–17. DOI: 10.1002/0471142913.faf0101s00

[26] Yoshitama K. An acylated delphinidin-3-rutinoside-5, 3′, 5′-triglucoside from Lobelia erinus. Phytochemistry. 1977; 16: 1857–1858. DOI: 10.1016/0031-9422(71)85123-3

[27] Del Pozo-Insfran D, Brenes CH, Serna-Saldivar SO, and Talcott ST. Polyphenolic and antioxidant content of white and blue corn (*Zea mays* L.) products. Food Research International. 2006; 39: 696–703. DOI: 10.1016/j.foodres.2006.01.014

[28] Ralph J, Quideau S, Grabber JH, and Hatfield RD. Identification and synthesis of new ferulic acid dehydrodimers present in grass cell walls. Journal of the Chemical Society, Perkin Transactions. 1994; 1: 3485–3498. DOI: 10.1039/P19940003485

[29] Montilla CE, Hillebrand S, Antezana A, and Winterhalter P. Soluble and bound phenolic compounds in different Bolivian purple corn (*Zea mays* L.) cultivars. Journal of Agriculture and Food Chemistry. 2011; 59: 7068– 7074. DOI: 10.1021/jf201061x

[30] López-Martínez LX, Oliart-Ros RM, Valerio-Alfaro G, Lee CH, Parkin KL, and García HS. Antioxidant activity, phenolic compounds and anthocyanins content of eighteen strains of Mexican maize. LWT—Food Science and Technology. 2009; 42: 1187–1192. DOI: 10.1016/j.lwt.2008.10.010

[31] Xu JG, Hu QP, Wang XD, Luo JY, Liu Y, and Tian CR. Changes in the main nutrients, phytochemicals, and antioxidant activity in yellow corn grain during maturation. Journal of Agriculture and Food Chemistry. 2010; 58: 5751–5756. DOI: 10.1021/jf100364k

[32] Blandino M, Mancini MC, Peila A, Rolle L, Vanara F, and Reyneri A. Determination of maize kernel hardness: comparison of different laboratory tests to predict dry-milling performance. Journal of the Science of Food and Agriculture. 2010; 90:1870–1878. DOI: 10.1002/jsfa.4027

[33] Chiremba C, John RN, Taylor , Rooney LW, and Trust B. Phenolic acid content of sorghum and maize cultivars varying in hardness. Food Chemistry. 2012; 134: 81–88. DOI: 10.1016/j.foodchem.2012.02.067

[34] Cabrera-Soto ML, Salinas-Moreno Y, Velázquez-Cardelas GA, and Espinosa-Trujillo E. Content of soluble and insoluble phenols in the structures of corn grain and their relationship with physical properties. Agrociencia. 2009; 43: 827–839.

[35] Bunzel M. Chemistry and occurrence of hydroxycinnamate oligomers. Phytochemical Reviews. 2010; 9: 47–64. DOI: 10.1007/s11101-009-9139-3

[36] Bunzel M, Allerdings E, Ralph J, and Steinhart H. Cross-linking of arabinoxylans via 8-8-coupled diferulates as demonstrated by isolation and identification of diarabinosyl 8-8(cyclic)-dehydrodiferulate from maize bran. Journal of Cereal Science. 2008; 47: 29–40. DOI: 10.1016/j.jcs.2006.12.005

[37] Bily AC, Reid LM, Taylor JH,Johnston D, Malouin C, Burt AJ, Bakan B, Regnault-Roger C, Pauls KP, Arnason JT, and Philogène BJR. Dehydrodimers of ferulic acid in maize grain pericarp and aleurone: resistance factors to Fusarium graminearum. Phytopathology. 2003; 93: 712–719. DOI: 10.1094/PHYTO.2003.93.6.712

[38] Sampietro DA, Fauguel CM, Vattuone MA, Presello DA, and Catalán CAN. Phenylpropanoids from maize pericarp: resistance factors to kernel infection and fumonisin accumulation by Fusarium verticillioides. European Journal of Plant Pathology. 2003; 135: 105–113. DOI: 10.1007/s10658-012-0069-3

[39] Arnason JT, Gale J, Conilh de Beyssac B, Sen A, Miller SS, Philogene BJR, Lambert JDH, Fulcher RG, Serratos A, and Mihm J. Role of phenolics in resistance of maize grain to the stored grain insects, Prostephanus truncatus (Horn) and Sitophilus zeamais (Motsch.). Journal of Stored Products Research. 1992; 28: 119–126. DOI: 10.1016/0022-474X(92)90019-M

[40] Rouaua X, Cheynier V, Surget A, Gloux D, Barron C, Meudec E, Monterod JL, and Criton M. A dehydrotrimer of ferulic acid from maize bran. Phytochemistry. 2003; 63: 899–903. DOI: 10.1016/S0031-9422(03)00297–8

[41] Bakan B, Bily AC, Melcion D, Cahagnier B, Regnault- Roger C, Philogene BJR, and Richard-Molard D. Possible role of plant phenolics in the production of trichothecenes by Fusarium graminearum strains on different fractions of maize kernels. Journal of Agriculture and Food Chemistry. 2003; 51: 2826–2831. DOI: 10.1021/jf020957g

[42] Aoki H, Kuze N, and Kato Y. Anthocyanins isolated from purple corn (*Zea mays* L.). Foods & Food Ingredients Journal of Japan. 2002; 199: 41–45.

[43] Cevallos-Casals BA and Cisneros-Zevallos L. Stability of anthocyanin-based aqueous extracts of Andean purple corn and red-fleshed sweet potato compared to synthetic and natural colorants. Food Chemistry. 2004; 86: 69–77. DOI: 10.1016/j.foodchem.2003.08.011

[44] González-Manzano S, Pérez-Alonso JJ, and Salinas-Moreno Y. Flavanol-anthocyanin pigments in corn: NMR characterisation purple corn phenolic profile and assessment 213 and presence in different purple corn varieties. Journal of Food Composition and Analysis. 2008; 21: 521–526. DOI: 10.1016/j.jfca.2008.05.009

[45] Martínez-Bustos F, Martínez-Flores HE, and Sanmartín-Martínez E. Effect of the components of maize on the quality of masa and tortillas during the traditional nixtamalization process. Journal of the Science of Food and Agriculture. 2001; 81: 1455–1462. DOI: 10.1016/j.jfca.2008.05.009

[46] González R, Reguera E, Mendoza L, Figueroa JM, and Sánchez F. Physicochemical changes in the hull of corn grains during their alkaline cooking. Journal of Agriculture and Food Chemistry. 2004; 52: 3831–3837. DOI: 10.1021/jf035175h

[47] Salinas-Moreno Y, Martínez-Bustos F, Soto-Hernández M, Ortega-Paczka R, and Arellano-Vázquez JL. Efecto de la nixtamalización sobre las antocianinas del grano de maíces pigmentados. Agrociencia. 2003; 37: 617–628.

[48] De la Parra C, Serna-Saldivar SO, and Hai L. Effect of processing on the phytochemical profiles and antioxidant activity of corn for production of masa, tortillas, and tortilla chips. Journal of Agriculture and Food Chemistry. 2007; 55: 4177–4183. DOI: 10.1021/jf063487p

[49] Mora-Rochin SR, Gutiérrez-Uribe JA, Serna-Saldivar SO, Sánchez-Peña P, Reyes-Moreno C, and Milán-Carrillo J. Phenolic content and antioxidant activity of tortillas produced from pigmented maize processed by conventional nixtamalization or extrusion cooking. Journal of Cereal Science. 2010; 52: 502–508. DOI: 10.1016/j.jcs.2010.08.010

[50] Gutiérrez-Uribe JA, Rojas-García C, García-Lara S, and Serna-Saldivar SO. Phytochemicals of wastewater (nejayote) obtained after lime-cooking of different types of maize kernel processed into masa for tortillas. Journal of Cereal Science. 2010; 52: 410–416. DOI: 10.1016/j.jcs.2010.07.003

Chapter 9

Phenolic Compounds in Genus *Smilax* (Sarsaparilla)

Salas-Coronado Raúl, Hernández-Carlos Beatriz,
Llaguno-Guilberto Joseoziel and
Santos-Sánchez Norma Francenia

Additional information is available at the end of the chapter

http://dx.doi.org/10.5772/66896

Abstract

Smilax (Smilacaceae) is a genus of about 350 species, found in temperate, tropical and subtropical zones worldwide. The plants belonging to this genus are found throughout Asia, Europe, Oceania and the Americas. Species of the genus *Smilax* commonly called sarsaparilla are characterized as climbers, with long, thin thorny stem. The branches have tendrils which attach to other plants or objects and grow steadily upward. The roots of these plants have been used for centuries in Asia and the Americas as a tonic, diuretic and sudorific. The rhizome, roots, stems and leaves of sarsaparilla are used in traditional medicine. In the scientific literature, there are several reports on immunomodulatory properties, anticonvulsant, antibacterial, antifungal, anticancer, antidiabetic and antioxidant properties. However, there are no reports which explain the antioxidant activity of sarsaparilla extracts as a function of phenolic compound structures, such as flavonoids and phenylpropanoids. In this chapter, the relevance of phenolic chemical structure in antioxidant and anticancer activity of sarsaparilla extracts will be described. Special emphasis is placed on phenylpropanoid glycosides that consist of a sucrose core. These compounds are evidence of chemotaxonomy in the genus *Smilax*.

Keywords: *Smilax*, phenolic compounds, antioxidant activity, anticancer activity, phenylpropanoids, flavonoids

1. Introduction

The genus *Smilax* (Smilacaceae), commonly called sarsaparilla, consists of about 350 species. About 79 species are natives of China, 24 species are from India and 29 species are from Central America. The plants of this genus are climbers, have long, thin, thorny stems and have tendrils which attach to other plants or objects to climb steadily (**Figure 1**).

(A) (B) (C) (D) (E) (F) (G) (H) (I)

Figure 1. (A) *S. bracteata*, (B) *S. china*, (C) *S. fluminensis*, (D) *S. glyciphylla*, images from https://www.inaturalist.org/taxa. (E) *S. campestris*, image courtesy of Mauricio Bonifacino Ph.D. (Universidad de la República, Montevideo, Uruguay). (F)–(I) *S. domingensis*.

The rhizome, roots, stems and, occasionally, leaves of sarsaparilla are used as food and in traditional medicine. These plants are known to have immunomodulatory, antioxidant, antibacterial, antifungal and diuretic properties. Additionally, they are used for relief from climatery [1]. Also, the genus *Smilax* has pharmacological properties and is used to treat different types of cancer, diabetes, skin diseases, ulcers, as well as fever, gout and ophthalmic diseases [2].

In recent years, interest in the study of the genus *Smilax* has increased, mainly in Europe and Asia, due to the presence of phenolic compounds. Some species have also proven effective in the prevention and treatment of several cancers. In addition, extracts from the genus *Smilax* exhibit pro-apoptotic activity and antioxidant activity [3].

There are reports about the antioxidant property expressed as DPPH• radical scavenging activity of species of the genus *Smilax*, as *Smilax bockii* [4], *Smilax campestris* [5], *Smilax glabra* [6], *Smilax lanceifolia* [7], *Smilax perfoliata* [8], *Smilax riparia* [9], *Smilax scobinicaulis* [10] and *Smilax sebeana* [11]. This property is attributed to phenolic compounds such as stilbenes, flavones, flavanones, flavonols, smilasides, smiglasides and helionosides, among others. Phenolic compounds have a unique chemical structure for stabilizing free radicals in an aromatic system. Flavonoids and stilbenes have been identified as beneficial agents for the treatment of various diseases such as cardiovascular and neurodegenerative diseases, as well as cancers [12].

Therefore, this review will describe *Smilax* species that have been studied for their antioxidant and anticancer properties with special emphasis on reports of phenolic compounds such as smilasides, smiglasides and helionosides. These compounds are phenols with antioxidant activity and are constituted of a sucrose substituted with feruloyl and coumaril groups. These three groups of compounds are evidence of chemotaxonomy in genus *Smilax*.

2. Genus *Smilax*

The review is organized by species, and the principal uses in traditional medicine for every species discussed are described. The methods of extraction and purification of phenolic compounds are briefly mentioned. Also, the methods used to evaluate antioxidant and anticancer activities are discussed. Various reports make evident the diversity of the chemical structure of phenolic compounds and their relation to corresponding biological properties.

2.1. *Smilax aspera*

Smilax aspera has been used to treat diseases such as syphilis, rheumatism and diabetes, and as an antioxidant to reduce the discomforts of menopause [13].

Longo *et al.* isolated and identified anthocyanins from the skin of *S. aspera* berries [14]. The anthocyanins were extracted with 0.1% HCl (v/v) in methanol. Then, the extract was carried to clean process using solid phase extraction (SPE) of reverse phase C-18. This clean process allowed the removal of sugars, acids and other water-soluble compounds. Finally, the fraction, with a large quantity of phenolic compounds, was subjected to chromatographic purificationby High-Performance Liquid Chromatography-Diode Array Detector-Mass Spectrometry (HPLC-DAD-MS). The result of this study was the isolation and characterization of four anthocyanins: two pelargonidins (**1**, **3**) and two cyanidins (**2**, **4**) (**Figure 2**) [14]. The principal anthocyanin was identified as pelargonidin 3-*O*-rutinoside. The anthocyanins are responsible for the color of the *S. aspera* fruits.

2.2. *Smilax bockii*

S. bockii is a plant used in traditional Chinese medicine with anti-inflammatory and anti-rheumatic properties. Xu *et al.* prepared a 70% aqueous ethanol extract from roots of *S. bockii* [4]. Then the extract was partitioned with chloroform, ethyl acetate and butanol successively. The butanol fraction was subjected to chromatographic purification leading to the separation of four flavonols (kaempferol (**5**), kaempferol-7-*O*-β-D-glucopyranoside (**6**), quercetin (**7**) and isorhamnetin (**8**), as well as three flavanone ((+)-dihydrokaempferol (**9**), engeletin (**10**) and iso-engeletin (**11**)), and a phenylpropanoid, caffeic acid *n*-butyl ester (**12**) (**Figure 3**). Additionally, the anti-inflammatory activities of a 70% aqueous ethanol extract and chloroform, ethyl acetate and butanol fractions were evaluated and the results showed the butanol fraction had a

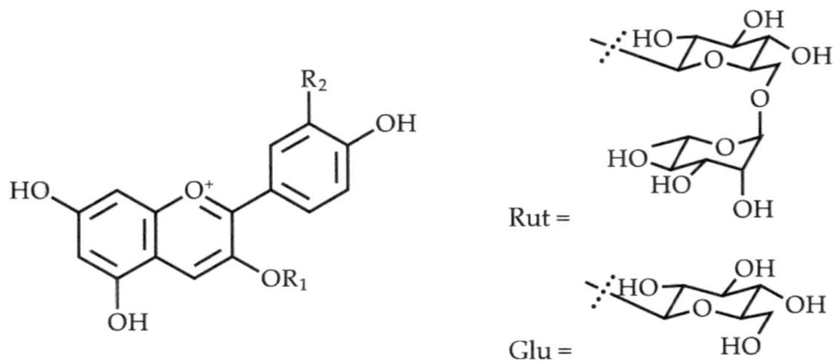

	R_1	R_2	Name
1	Rut	H	Pelargonidin 3-*O*-rutinoside
2	Rut	OH	Cyanidin 3-*O*-rutinoside
3	Glu	H	Pelargonidin 3-*O*-glucoside
4	Glu	OH	Cyanidin 3-*O*-glucoside

Figure 2. Anthocyanin glycosides isolated from *S. aspera* fruits.

relevant inhibitory activity against TNF-α-induced NF-κB activation with an IC50 value of 44.8 μg/mL. This activity can be attributed to phenolic compounds present in the butanol fraction.

2.3. *Smilax bracteata*

S. bracteata is a little-studied species. However, there are two representative chemical studies that describe the isolation and characterization of phenolic compounds. The first study was conducted by Li *et al.* who isolated and identified phenolic compounds from a methanol extract of *S. bracteata* rhizomes [15]. The air-dried and sliced rhizomes were extracted by maceration with methanol over 24 h. The extract was evaporated and re-dissolved in water and partitioned successively with dichloromethane, ethyl acetate and butanol. The butanol fraction was subjected to column chromatography and six new phenolic compounds were isolated and identified: two flavan-3-ol glucosides (**13**, **14**), one stilbene (**15**) (**Figure 4**) and three phenylpropanoid glycosides (**16–18**) (**Figure 5**). In the same study, Li *et al.* evaluated antioxidant activity of the six smilasides using the DPPH• radical scavenging activity. The smilasides J to L (**22–24**) showed an antiradical activity similar to α-tocopherol [15].

In a later study, Zhang *et al.* obtained a 95% aqueous ethanol extract from stems of *S. bracteata* [16]. The extract was concentrated and redissolved in water and successively extracted with hexane, dichloromethane and butanol. The dichloromethane fraction was purified with chromatography in several steps until smilasides G to L (**19–24**) were obtained (**Figure 5**).

2.4. *S. campestris*

S. campestris is commonly called sarsaparilla blanca [5]. Its roots and rhizomes have been used in folk medicine to treat skin diseases. An infusion from the leaves and aerial stems of *S.*

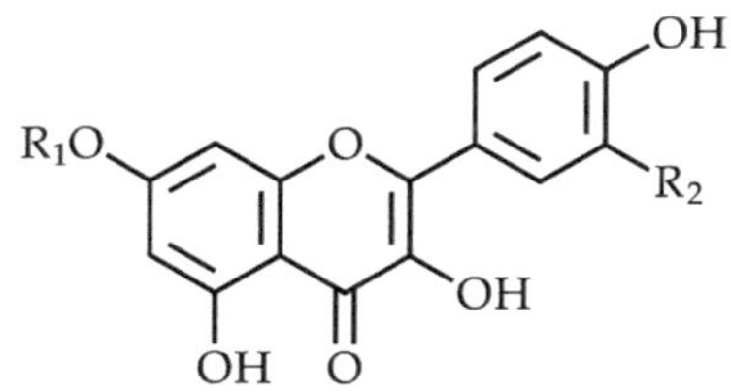

Glu =

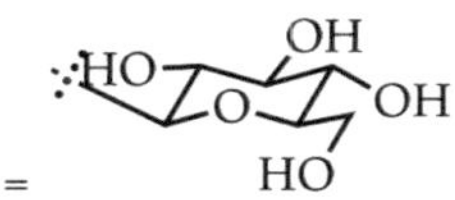

	R1	R2	Name
5	H	OH	Kaempferol
6	Glu	H	Kaempferol-7-*O*-*β*-D-glucopyranoside
7	H	OH	Quercetin
8	H	OM	Isorhamnetin

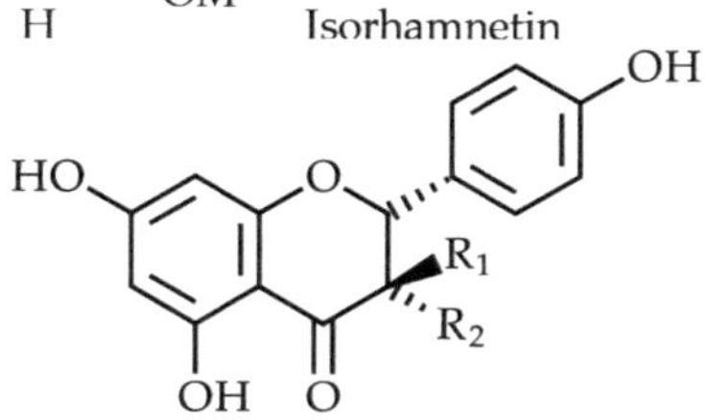

O-Rha =

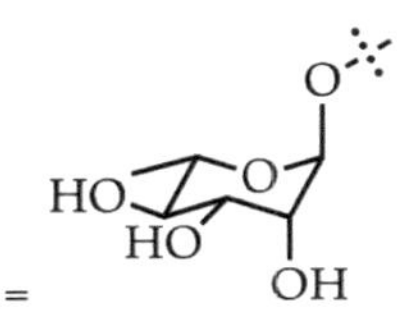

	R1	R2	Name
9	H	OH	(+)-Dihydrokaempferol
10	H	*O*-Rha	Engeletin
11	*O*-Rha	OH	Isoengeletin

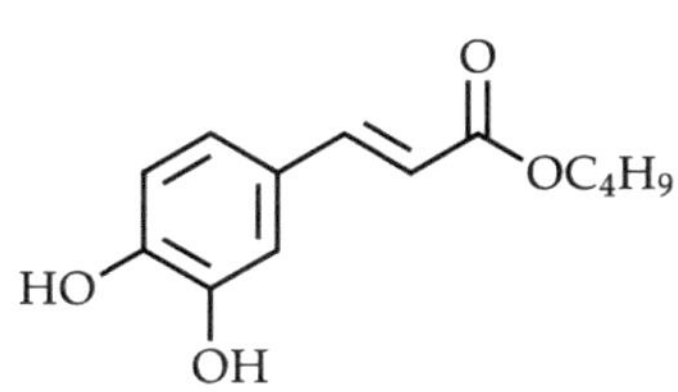

12 Caffeic acid *n*-butyl ester

Figure 3. Four flavonols (three aglycones and one glucoside), three flavanones (one aglycone and two glycosides) and one phenylpropanoid ester isolated from *S. bockii* roots.

campestris is used to relax the digestive system [5]. Rugna *et al.* reported antioxidant activity from 50% aqueous methanol extract from *S. campestris* rhizomes; the activity was expressed as total reactive potential (TRAP) [5]. Morais *et al.* obtained an ethanol extract by maceration and fractions from fresh stems of *S. campestris* [17]. The ethanol extract was concentrated, and the dried extract was re-dissolved in aqueous ethanol (7:3) and partitioned with hexane, dichloromethane, ethyl acetate and butanol. The antioxidant activity as DPPH$^{\bullet}$ radical scavenging was evaluated for all fractions. The ethanol extract and butanol fraction exhibited a strong

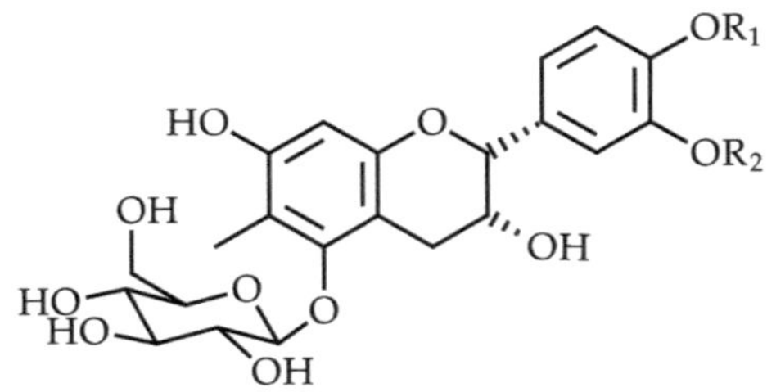

	R_1	R_2	Name
13	H	Me	(2*S*,3*S*)-5-*O*-*β*-D-glucopyranosyloxy-6-methyl-3′-methoxy-3,7,4′-trihydroxyflavan-3-ol
14	Me	H	(2*S*,3*S*)-5-*O*-*β*-D-glucopyranosyloxy-6-methyl-4′-methoxy-3,7,3′-trihydroxyflavan-3-ol

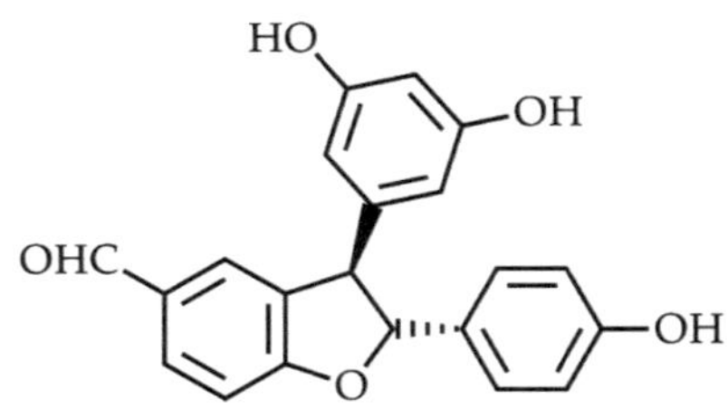

Figure 4. Two flavan-3-ol glycosides and one stilbene isolated from *S. bracteata* rhizomes.

antioxidant activity and was higher than Butylated hydroxytoluene (BHT), a commercial antioxidant. Also, Morais *et al*. reported that rutin (**25**) (**Figure 6**) and quercetin (**7**) (**Figure 3**) flavonol glycosides were the most abundant phenolic compounds present in ethanol extract and butanol fraction, respectively [17].

2.5. *Smilax china*

S. china is the most studied species of genus *Smilax*. Lee *et al*. evaluated antioxidant activity of *S. china* root using DPPH$^\bullet$ radical scavenging activity, cell viability, lipid peroxidation activity, superoxide dismutase (SOD) activity, catalase (CAT) activity and glutathione peroxidase (GPX) activity. These authors obtained a 70% aqueous methanol extract from the root of *S. china*. The extract was concentrated and partitioned with hexane, dichloromethane, ethyl acetate and butanol. The extract and its fractions were evaluated for antioxidant activity. The antiradical activity expressed as IC50 of the extract is about 8 μg/mL, while the ethyl acetate fraction exhibited the principal antiradical activity (IC50 approximately 5 μg/mL) [18]. Jeong *et al*. obtained several fractions with solvents of different polarity from *S. china* root and evaluated the antioxidant activity using DPPH$^\bullet$ radical scavenging activity, ABTS$^{\bullet+}$ radical scavenging activity, reducing power, ferric reducing/antioxidant power, ferric thiocyanate assay, malondialdehyde assay using mouse brain homogenates methods, and finally determined total phenols and phenolic composition [19]. The extraction was carried out with methanol at 70°C for 2 h. The methanol extract was evaporated to dryness. The dried extract was re-dissolved in water and the solution was consecutively partitioned with chloroform, ethyl acetate

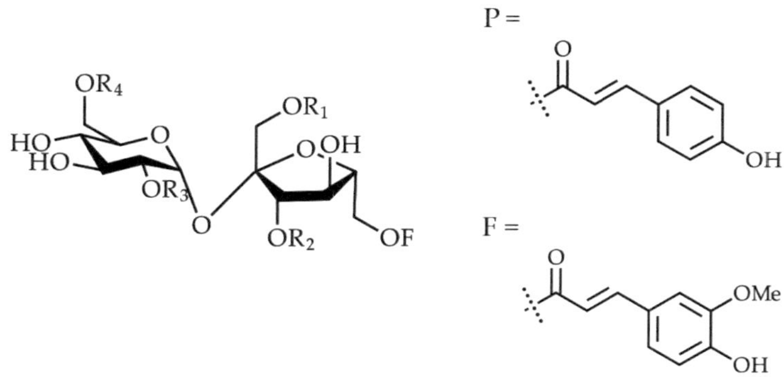

	R1	R2	R3	R4	Name
16	P	H	H	H	(1-*O*-*p*-coumaroyl-6-*O*-feruloyl)-*β*-D-fructofuranosyl-(2→1)-*α*-D-glucopyranoside
17	P	F	H	H	(1-*O*-*p*-coumaroyl-3,6-di-*O*-feruloyl)-*β*-D-fructofuranosyl-(2→1)-*α*-D-glucopyranoside, smilaside C
18	H	H	H	COMe	(6-*O*-feruloyl)-*β*-D-fructofuranosyl-(2→1)-(6-*O*-acetyl)-*α*-D-glucopyranoside
19	P	P	H	H	(1,3-*O*-di-*p*-coumaroyl-6-*O*-feruloyl)-*β*-D-fructofuranosyl-(2→1)-*α*-D-glucopyranoside, smilaside G
20	P	P	COMe	H	(1,3-*O*-di-*p*-coumaroyl-6-*O*-feruloyl)-*β*-D-fructofuranosyl-(2→1)-(2-*O*-acetyl)-*α*-D-glucopyranoside, smilaside H
21	P	P	H	COMe	(1,3-*O*-di-*p*-coumaroyl-6-*O*-feruloyl)-*β*-D-fructofuranosyl-(2→1)-(6-*O*-acety)-*α*-D-glucopyranoside, smilaside I
22	F	P	H	H	(3-*O*-*p*-coumaroyl-1,6-*O*-diferuloyl)-*β*-D-fructofuranosyl-(2→1)-*α*-D-glucopyranoside, smilaside J
23	P	F	COMe	H	(1-*O*-*p*-coumaroyl-3,6-*O*-diferuloyl)-*β*-D-fructofuranosyl-(2→1)-(2-*O*-acety)-*α*-D-glucopyranoside, smilaside K
24	F	F	H	H	(1,3,6-*O*-triferuloyl)-*β*-D-fructofuranosyl-(2→1)-*α*-D-glucopyranoside, smilaside L

Figure 5. Phenylpropanoid glycosides with a sucrose core isolated from *S. bracteata* aerial parts.

and butanol. The results obtained by Jeong *et al.* show that the ethyl acetate fraction had the highest total phenol concentration, 401.62 ± 3.13 mg GAE/g of extract and the most abundant phenols were (+)-catechin (**26**) and (−)-epicatechin (**27**) (**Figure 7**) with a concentration of 135.26 ± 10.08 and 58.10 ± 0.51 mg/100 g, respectively. Consequently, this extract showed the most important antioxidant activity.

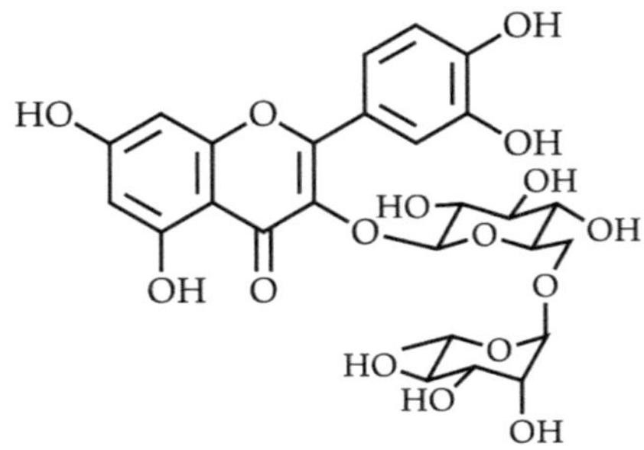

25 Rutin

Figure 6. Chemical structure of rutin, a flavonol glycoside isolated from *S. campestris* rhizomes.

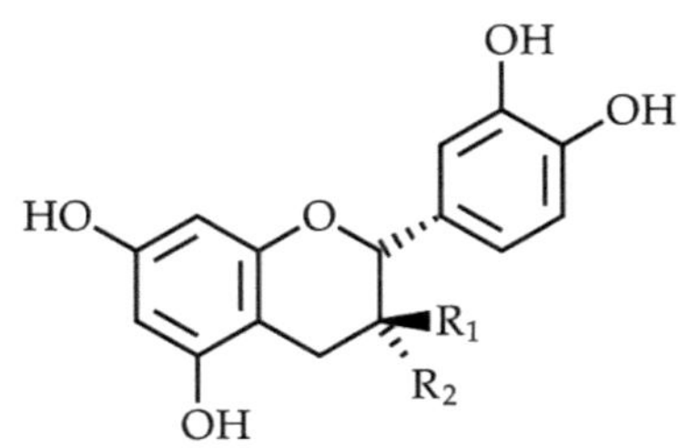

	R_1	R_2	Name
26	OH	H	(+)-Catechin
27	H	OH	(-)-Epicatechin

Figure 7. Two flavan-3-ol aglycones isolated from *S. china* roots.

Kuo *et al.* obtained a 70% aqueous ethanol extract from dried stems of *S. china*. The extracts were concentrated and suspended in water. The resulting suspension was partitioned with hexane and chloroform [20]. The chloroform fraction was purified by silica gel column chromatography. The purification conducted to isolate smilasides A to F (**Figures 5** and **8**), heloniosides A (**33**) and B (**34**) and smiglaside E (**35**) (**Figure 8**). The anticancer activity of smilasides A to F was evaluated *in vitro* and showed cytotoxic activity against cervical cancer cells (KB and HELA) and colon cancer cells (DLD-1).

Li *et al.* performed another study related to the evaluation of anticancer activity of *S. china* extracts with a high content of phenolic compounds [21]. Researchers in this study performed a bioassay-guided separation and purification of kaempferol-7-*O*-*β*-D-glucoside (**6**) from *S. china* rhizome. First, a 70% aqueous ethanol extract was obtained under reflux. Then the solvent was removed and residue was extracted with ethyl acetate, and butanol, sequentially, in a Soxhlet apparatus. Both ethyl acetate and butanol fractions were subjected to column chromatography separately. Several fractions with large amounts of flavonoid were obtained and each fraction was evaluated for *in vitro* anticancer activity. The human cells used in this study included liver cancer BEL-7402, cervical epithelial carcinoma HeLa, high metastatic lung

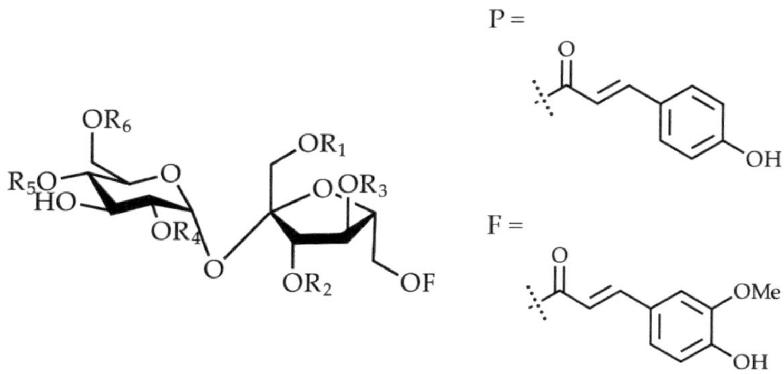

	R_1	R_2	R_3	R_4	R_5	R_6	Name
28	H	F	H	H	COMe	COMe	(3,6-*O*-diferuloyl)-*β*-D-fructofuranosyl-(2→1)-(5,6-*O*-diacetyl)-*α*-D-glucopyranoside, smilaside A
29	H	F	H	COMe	H	H	(3,6-*O*-diferuloyl)-*β*-D-fructofuranosyl-(2→1)-(2-*O*-acetyl)-*α*-D-glucopyranoside, smilaside B
30	P	F	COMe	H	H	H	(4-*O*-acety-1-*O*-*p*-coumaroyl-3,6-*O*-diferuloyl)-*β*-D-fructofuranosyl-(2→1)-*α*-D-glucopyranoside, smilaside D
31	P	F	H	H	H	COMe	(1-*O*-*p*-coumaroyl-3,6-*O*-diferuloyl)-*β*-D-fructofuranosyl-(2→1)-(6-*O*-acetyl)-*α*-D-glucopyranoside, smilaside E
32	P	P	H	COMe	H	COMe	(1,3-*O*-di-*p*-coumaroyl-6-*O*-feruloyl)-*β*-D-fructofuranosyl-(2→1)-(2,6-*O*-diacetyl)*α*-D-glucopyranoside, smilaside F
33	H	F	H	H	H	H	(3,6-*O*-diferuloyl)-*β*-D-fructofuranosyl-(2→1)-*α*-D-glucopyranoside, helonioside A
34	H	F	H	H	H	COMe	(3,6-*O*-diferuloyl)-*β*-D-fructofuranosyl-(2→1)-(6-*O*-acetyl)-*α*-D-glucopyranoside, helonioside B
35	F	P	H	COMe	H	COMe	(3-*p*-*O*-coumaroyl-1,6-*O*-diferuloyl)-*β*-D-fructofuranosyl-(2→1)-(2,6-*O*-diacetyl)-*α*-D-glucopyranoside, smiglaside E

Figure 8. Phenylpropanoid glycosides with a sucrose core (five smilasides, two helionoside and one smiglaside) isolated from *S. china* stems.

carcinoma 95-D, melanoma A375, gastric cancer MKN-45, epithelial carcinoma A431, human acute leukemia HL60, normal embryonic kidney HEK293 and normal embryonic liver L-O2. Li *et al.* found eight extracts of *S. china* tubers with anticancer activity against HeLa cells. Also, a bioassay-guided isolation of the polled extract lead to the detection of kaempferol-7-*O*-*β*-D-glucoside (**6**). This flavonoid induces apoptosis as an anti-proliferative action related to radical scavenging activity [21]. Shao *et al.* developed a specific HPLC method for determination of the six major phenolic compounds active in *S. china*: taxifolin-3-*O*-glycoside (**36**), scirpusin A (**37**), piceid (**38**), oxyresveratrol (**39**), resveratrol (**40**) (**Figure 9**) and engeletin (**10**) (**Figure 3**). These compounds were extracted from the tuber of *S. china* with 95% aqueous ethanol and the concentrated extract was partitioned with petroleum ether, ethyl acetate and butanol. The ethyl acetate fraction was subjected to repeated silica gel chromatography. Finally, the purification of phenolic compounds was performed by HPLC [22]. Wu *et al*. also reported other study related to anticancer activity of phenolic compounds from *S. china* [12]. These authors obtained a 95% aqueous ethanol extract from the tuber of *S. china*, which was concentrated and suspended in water. The suspension was partitioned with petroleum ether, ethyl acetate and butanol. The ethyl acetate was the most bioactive fraction. This fraction was subjected to chromatographic purification. Three sub-fractions and six bioactive phenolic compounds bioactives: three flavonoids (kaempferol-7-*O*-β-D-glucoside (**6**), dihydrokaempferol (**9**) and dihydrokaempferol-3-*O*-α-L-rhamnoside (**10**)) and three stilbenoids (**37, 39** and **40**), were isolated from the ethyl acetate fraction. These compounds were found to induce apoptosis in

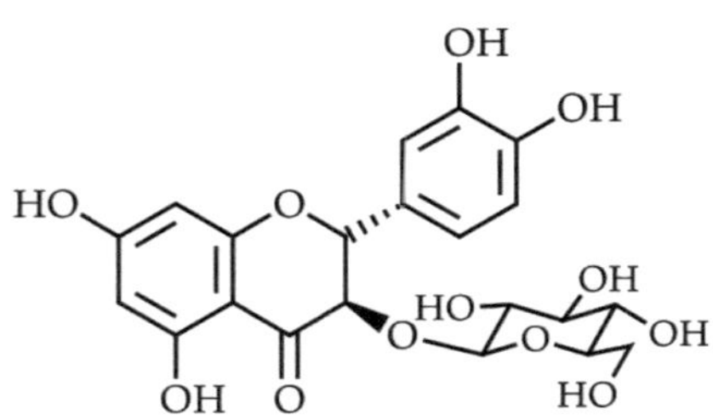

36 Taxifolin-3-*O*-glucoside

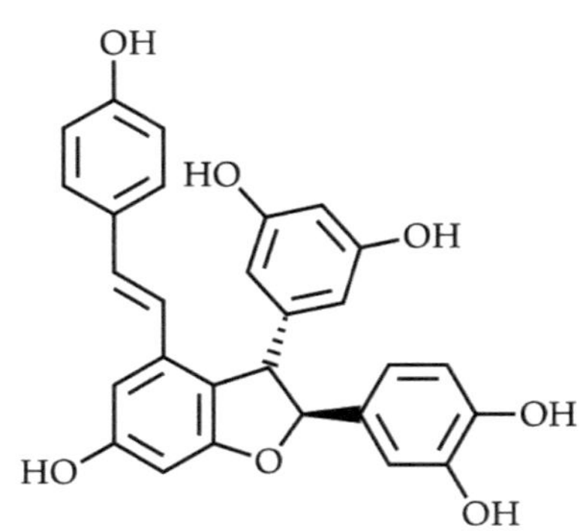

37 Scirpusin A

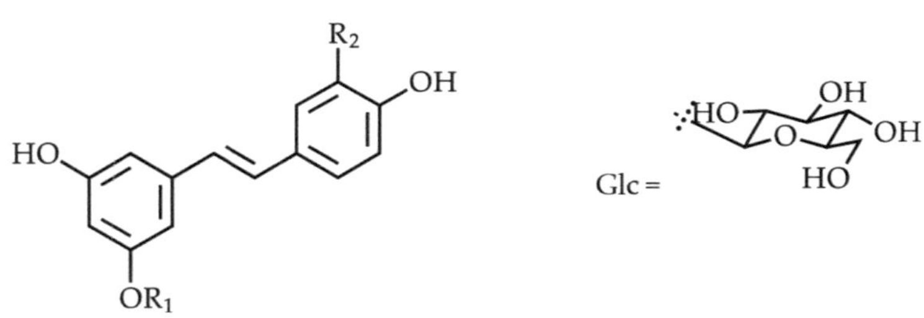

	R1	R2	Name
38	Glc	H	Piceid
39	H	OH	Oxyresveratrol (piceatannol)
40	H	H	Resveratrol

Figure 9. One flavanonol (**36**) and four stilbenes (**37–40**) isolated from 95% ethanol extracts of *S. china* tubers.

anti-breast tumor cells MCF-7 and MDA-MB-231. The results showed that resveratrol and oxyresveratrol had the highest apoptosis rates [12].

2.6. *Smilax corbularia*

S. corbularia is used in traditional Thai medicine for the ailments treatment caused by the menopause, as well as for ovarian and breast cancer. For this reason, Wungsintaweekul *et al.* isolated and characterized the phenolic compounds of methanol extract from *S. corbularia* rhizome [23]. They also evaluated the cell proliferation stimulation of the isolated compounds against human cancer cell lines MCF-7 and T47D. The major compounds present in the rhizome of *S. corbularia* were rhamnosides dihydroflavonol derivatives, which represent 15% of methanol extract by weight. The results showed that the extract did not exhibit cytotoxicity against breast cancer cell lines MCF-7 and T47D. However, the flavanonol rhamnosides (engeletin (**10**) and isoengeletin (**11**)) (**Figure 3**); as well as, astilbin (**41**), isoastilbin (**42**), neoastilbin (**43**) and neoisoastilbin (**44**) (**Figure 10**), showed a suppressive effect on estradiol at concentration of 1 μM as evidenced by human breast cancer cell proliferation.

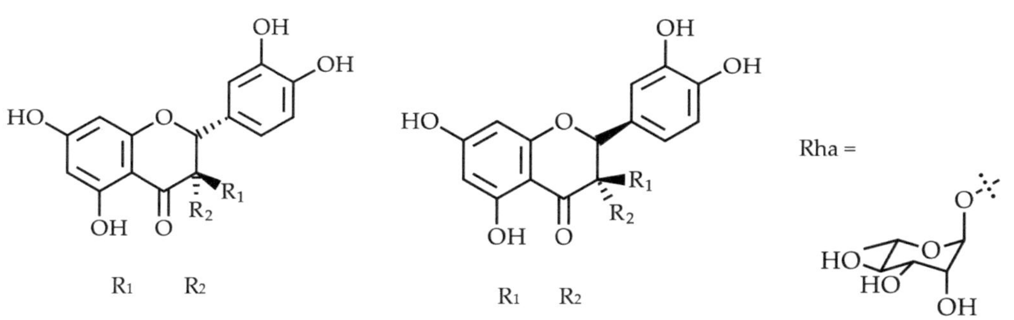

	R1	R2	
41	Rha	H	Astilbin
42	H	Rha	Isoastilbin

	R1	R2	
43	H	Rha	Neoastilbin
44	Rha	H	Neoisoastilbin

Figure 10. Flavononol rhamnosides (**41–44**) with activity against human breast cancer isolated from methanol extract of *S. corbularia* rhizomes.

2.7. *Smilax domingensis*

S. domingensis is used in Central America by the pharmaceutical and cosmetics industries. The most representative studies of *S. domingensis* are related to cytotoxicity to cancer cells [24], inhibitory activity of estrogen [25] and antioxidant activity [26]. The chemical studies of *S. domingensis* only cover qualitative identification of flavonoids and anthocyanins using thin-layer chromatography (TLC) [27].

2.8. *Smilax excelsa*

S. excelsa is used in Turkey's traditional medicine to treat breast cancer, stomach pain and bloating [28]. Ozsoy *et al.* evaluated antioxidant activity of infusion, decoction, ethanol and ethyl acetate extracts from *S. excelsa* leaves using the inhibition of lipid peroxidation, metal ion chelating, reducing power, DPPH• radical scavenging, superoxide, hydroxyl radicals and hydrogen peroxide [29]. Also, total phenols, total flavonoid and anthocyanin were quantified in the extracts. The content of total phenols and total flavonoid was found in the intervals of 8.8–35.7 GAE mg/g of dry matter and 0.61–28.7 catechin equivalents mg/g of dry matter, respectively. The extract with the highest content of total phenols was the infusion. The decoction and infusion showed major DPPH• radical scavenging. These results agree with the high content of phenols and flavonoids present in the infusion and decoction. On other hand, Khaligh *et al.* isolated and elucidated three phenol compounds (*trans*-resveratrol (**40**) (**Figure 9**), 5-*O*-caffeoylshikimic acid (**45**) and 6-*O*-caffeoyl-*β*-D-fructofuranosyl-(2→1)-*α*-D-glucopyranoside (**46**)) from ethyl acetate extract of *S. excelsa* (**Figure 11**) [30]. The extraction was a maceration performed at room temperature. After the solvent was removed, the extract was separated using silica gel column chromatography. In these study, also was evaluated the cytotoxicity of isolated compounds against human breast adenocarcinoma MCF-7cell lines. The 6-*O*-caffeoyl-*β*-D-fructofuranosyl-(2→1)-*α*-D-glucopyranoside showed a promising activity against MCF-7 cell lines [30].

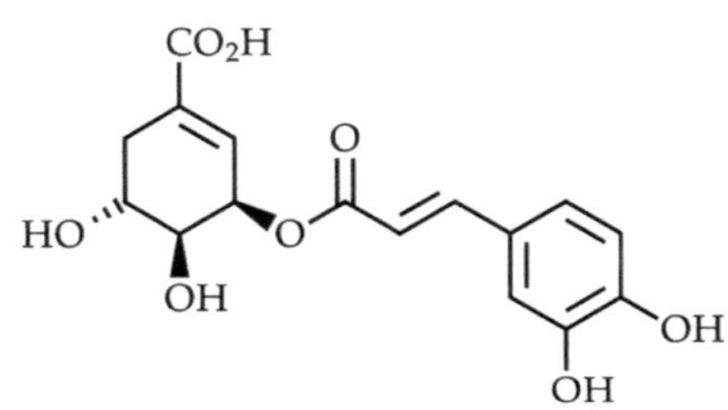

45 5-*O*-Caffeoylshikimic acid

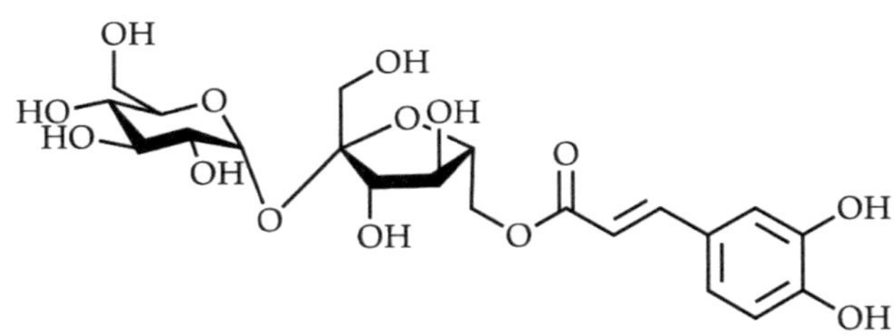

46 6-*O*-Caffeoyl-*β*-D-fructofuranosyl-(2→1)-*α*-D-glucopyranoside

Figure 11. Two phenylpropanoids (**45** and **46**) derivates of caffeic acid with activity against human breast cancer.

2.9. *Smilax fluminensis*

S. fluminensis has a wide geographical distribution in Brazil, and propagation studies have shown it to grow relatively easily. Hence, it is a promising species for growing demand from the pharmaceutical industry [31, 32] have published the only chemical study of *S. fluminensis* thus far. They obtained extracts from leaves and isolated phenolic compounds. Two flavonol glycosides were isolated and characterized, rutin (**25**) (**Figure 6**) and quercetin-3-*O*-*β*-D-galactopyranoside (**47**) (**Figure 12**).

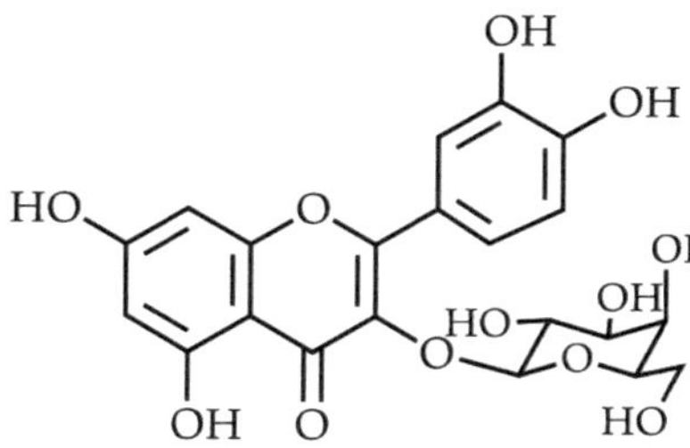

47 Quercetin-3-*O*-*β*-D-galactopyranoside

Figure 12. Flavonol glycoside (**47**) isolated from branches of *S. fluminensis*.

2.10. *Smilax glabra*

The second most-studied species, after *S. china* is *S. glabra*. This species has been used in Chinese folk medicine for the treatment of acute bacterial dysentery, syphilis, acute and chronic nephritis [33], hyperinsulinemia [34] and cancer [35]. She *et al.* evaluated the effect of aqueous extract of *S. glabra* on cancer cell adhesion, migration and invasion of HepG2, MDA-MB-231 and T24 cells *in vitro* and the metastasis suppression of MDA-MB-231 cells *in vivo* [36]. Gao *et al.* showed 95% aqueous ethanol extracts of *S. glabra* rhizomes to be effective against cancer via mitochondrial apoptosis in human breast cancer MCF7, colon carcinoma HT-29 and gastric cancer cell line BGC-823 [3]. The results obtained by these authors point out that the aqueous extract of *S. glabra* possibly promotes cell adhesion by increasing the size and strength of focal adhesions and inhibits the invasion of HepG2, MDA-MB-231 and T24 cells.

Xia *et al.* performed an evaluation of the protective effect of 60% aqueous ethanol extract of *S. glabra* rhizome against lead-induced oxidative stress in rats and quantified total phenols and total flavonoids [37]. The results of this study proved that the extract of *S. glabra* could minimize damages caused by the lead. The protective effect can be attributed to high concentrations of total phenols and total flavonoids in the extract. Total phenols reported were 262 ± 12.7 mg gallic acid equivalents (GAE)/g dry weight of the extract and total flavonoids were 203.4 ± 9.1 mg rutin equivalents/g dry weight of the extract [37].

Trinh *et al.* evaluated the antioxidant activity of 95% aqueous ethanol extract of *S. glabra* roots. They obtained two fractions from a partitioning with hexane and ethyl acetate, and astilbin (**41**) (**Figure 10**) isolated from the ethyl acetate fraction [38]. The fractions were obtained from 95% aqueous ethanol previously dried. The extract was suspended in 50% aqueous ethanol and partitioned with hexane and ethyl acetate. The extract, hexane and ethyl acetate fractions, and **41** were subjected to evaluation of DPPH• radical scavenging activity, thiobarbituric acid-reactive species (TBARS) and 3-(4,5-dimethylthiazol-2-yl)-2,5-diphenyltetrazolium bromide (MTT) cell viability assay for hepatoprotective effect via H_2O_2-injured mouse hepatocytes. The astilbin was isolated from the ethyl acetate fraction. The results of this study showed the ethyl acetate fraction has the principal antioxidant activity and MTT, attributed to the presence of astilbin, the main phenolic compound present in *S. glabra* rhizome [38]. Astilbin (**41**), a rhamnosyl flavanonol, besides exhibiting antioxidant activity, has coenzyme

A reductase-inhibiting [39], aldose reductase-inhibiting [40], hepatoprotective [41], anti-oedemaogenic [42] and anti-arthritis [43] activities.

Zhang *et al*. obtained two aqueous and methanolic extracts, from *S. glabra* rhizomes and evaluated their antioxidant activity using DPPH• radical scavenging, ABTS radical cation scavenging, reducing power, superoxide anion radical scavenging activity and antioxidant activity in a linoleic acid emulsion system. They carried out quantification of total phenols [44]. The results showed that the methanol extract had the highest content of total phenols (152.28 ± 10.57 mg GAE/g of extract) and astilbin (245.65 ± 8.21 mg/g of extract). In general, the methanol extract has more antioxidant activity than aqueous extract and this behavior is attributed to the presence of **41** in the methanol extract [44]. Lu *et al*. evaluated antioxidant and anti-inflammatory activities of 70% aqueous ethanol extract from *S. glabra* rhizomes [45]. The methods used to evaluate antioxidant activity were DPPH• radical scavenging, ABTS radical cation scavenging and reducing power. Anti-inflammatory activity was evaluated with MTT cell viability, measuring of nitric oxide/nitrite and enzyme-linked immunosorbent assay for IL-6 and TNF-α cytokines detection. The study also involved quantification of total phenols and total flavonoids. This was done by separation and identification of major phenols using ultrahigh pressure liquid chromatography coupled to electrospray mass spectrometry (U-HPLC-ESI-MS). The results obtained showed the 70% aqueous ethanol extract of *S. glabra* rhizome has a radical scavenging on DPPH• statistically equal to ascorbic acid ($P > 0.05$). The results regarding anti-inflammatory activity showed that accumulation of NO, IL-6, and TNF-α in lipopolysaccharides (LPS)-stimulated groups was higher than the group used as the positive control. Dexamethasone was employed as a positive control. Finally, 17 phenolic compounds were isolated and identified from the 70% aqueous ethanol extract of *S. glabra* rhizome, including engeletin (**10**) and isoengeletin (**11**) (**Figure 3**); astilbin (**41**), isoastilbin (**42**), neoastilbin (**43**), neoisoastilbin (**44**) (**Figure 10**); and 5-*O*-caffeoylshikimic acid (**45**) (**Figure 11**).

Moreover, there have been several chemical studies to isolate and characterize phenolic compounds from different parts of *S. glabra*. Chen *et al*. obtained a methanol extract of *S. glabra* rhizome. After solvent removal, the dry extract was suspended in water and partitioned with ether petroleum and ethyl acetate. The ethyl acetate extract was purified by chromatographic column to separate the phenolic compounds. The compounds isolated were five flavonoids (engeletin (**10**), astilbin (**41**), smitilbin (**48**), taxifolin or dihydroquercetin (**49**) and eucryphin (**50**), **Figure 13**), and two phenylpropanoids, resveratrol (**40**) and 5-*O*-caffeoylshikimic acid (**45**) [33]. Cheng *et al*. isolated new five phenylpropanoid glycosides, containing a sucrose core, smiglasides A–E (**35**, **51–54**, (**Figures 8** and **13**) from *S. glabra* rhizomes, [46]. The extraction and isolation procedures were followed exactly as was described by Chen *et al*. [33].

Xu *et al*. conducted a comprehensive chemical study of *S. glabra* rhizomes [47]. The air-dried and powdered rhizomes of *S. glabra* were extracted with 95% aqueous ethanol and 50% aqueous ethanol under reflux, consecutively. The extracts were combined and evaporated until to dryness and the residue was suspended in water and partitioned with petroleum ether, ethyl acetate and butanol. The ethyl acetate and butanol fractions were subjected to chromatographic separation. The purification allowed to the isolation 13 flavanones (dihydrokaempferol (**9**), engeletin (**10**), astilbin (**41**), isoastilbin (**42**), neoastilbin (**43**), neoisoastilbin

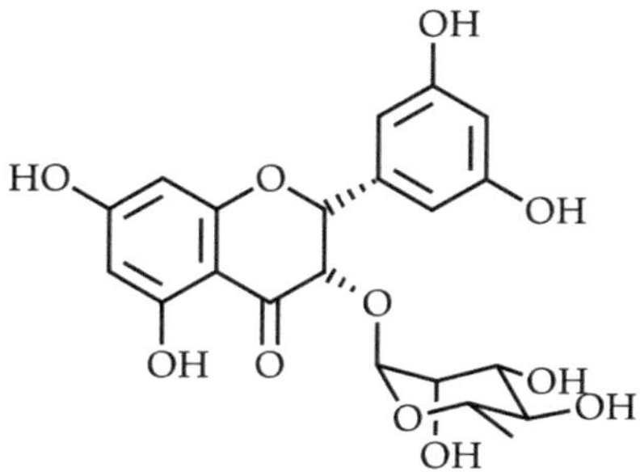

48 Smitilbin

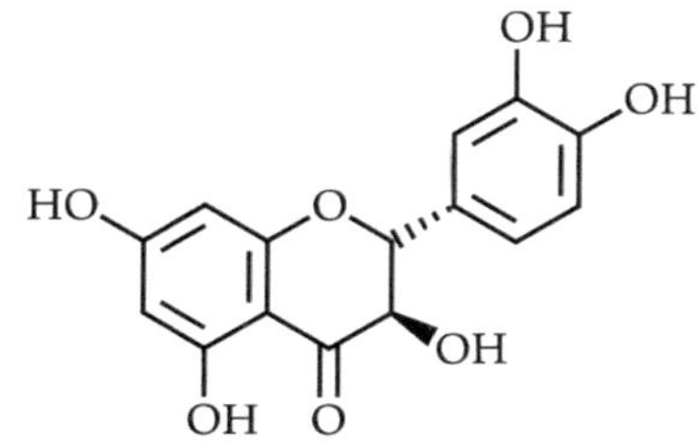

49 Taxifolin

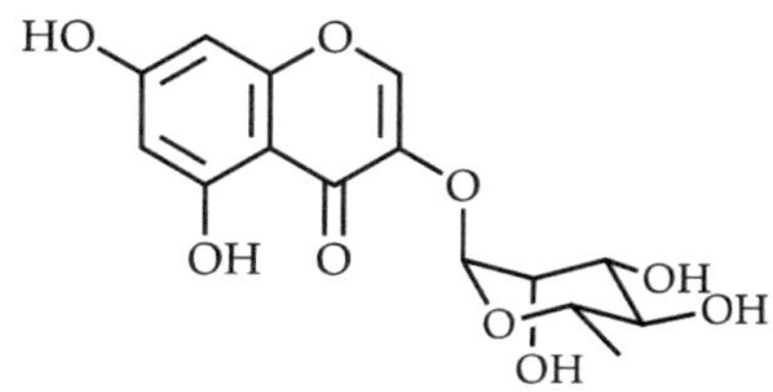

50 Eucryphin

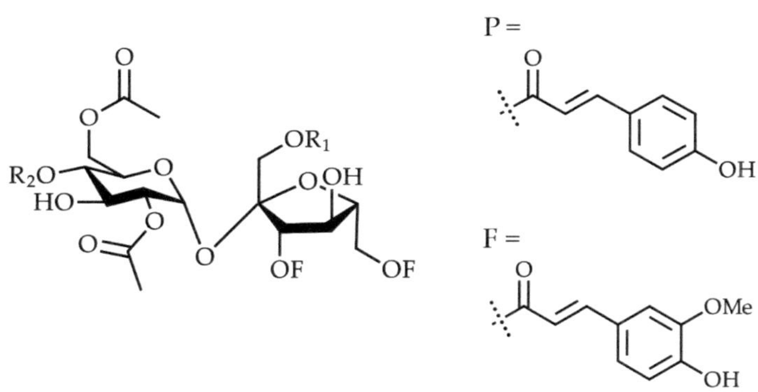

	R1	R2	Name
51	F	COMe	(1,3,6-*O*-triferuloyl)-*β*-D-fructofuranosyl-(2→1)-(2,4,6-*O*-triacetyl)-*α*-D-glucopyranoside, smiglaside A
52	F	H	(1,3,6-*O*-triferuloyl)-*β*-D-fructofuranosyl-(2→1)-(2,6-*O*-diacetyl)-*α*-D-glucopyranoside, smiglaside B
53	H	COMe	(3,6-*O*-diferuloyl)-*β*-D-fructofuranosyl-(2→1)-(2,4,6-*O*-triacetyl)-*α*-D-glucopyranoside, smiglaside C
54	P	COMe	(1-*O*-*p*-coumaroyl-3,6-*O*-diferuloyl)-*β*-D-fructofuranosyl-(2→1)-(2,4,6-*O*-triacetyl)-*α*-D-glucopyranoside, smiglaside D

Figure 13. Flavonoids and smiglasides isolated from *S. glabra* rhizome.

(**44**), taxifolin (**49**), naringenin (**55**), sakuranetin (**56**), arthromerin B (**57**), sinesin (**58**), (2*R*,3*R*)-taxifolin-3′-*O*-*β*-D-glucopyranoside (**59**) and (2*S*,3*S*)-glucodistylin (**60**); 3 flavanes: (+)-catechin

(**26**), (−)-epicatechin (**27**) and cinchonain 1b (**60**)); 2 flavanones (luteolin (**62**) and apigenin (**63**)); two flavonols (quercetin (**7**) and myricetin (**64**)); 1 chalcone, kukulkanin B (**65**); 3 stilbenes (piceid (**38**), piceatannol (**39**), and resveratrol (**40**)); 6 phenylpropanoids (5-*O*-caffeoylshikimic acid (**45**), caffeic acid (**66**), 3-*O*-*p*-coumaroylshikimic acid (**67**), smiglycerol (**68**), juncusyl ester B (**69**) and 1-*O*-p-coumarylglycerol (**70**), **Figure 14**). It is noteworthy that in this study no smiglasides were detected, despite an intensive separation of phenolic compounds having conducted. One explanation for these results is that possibly the high temperatures used for extraction caused smiglasides degradation.

2.11. *Smilax glycyphylla*

S. glycyphylla is a plant endemic to Australia and its leaves and fruits have a sweet taste like honey grass. The sweet principle of *S. glycyphylla* is called glycyphyllin A (**71**) and identified as a phenol compound with a structure of dihydrochalcone, phloretin-2′-α-L-rhamnose [48]. Cox *et al.* evaluated aqueous extracts of leaves and stems of *S. glyciphylla* [49]. The methods used to evaluate antioxidant activity were lipid peroxidation using thiobarbituric acid reactive substances (TBARS), superoxide quenching by coupling superoxide generation to the reduction of nitroblue tetrazolium (NBT), inhibition of deoxyribose-driven fenton degradation and total radical-antioxidant potential (TRAP) using free radicals derived from ABTS (2,2′-azinobis(3-ethylbenzothiazoline 6-sulphonate). The results showed that *S. glycyphylla* extract inhibited deoxyribose degradation. The total radical-antioxidant potential was seven times that the trolox a water-soluble analog of vitamin E with a high antioxidant activity.

Huang *et al.* performed a study of phenolic profile and antioxidant activity of *S. glycyphylla* leaves [50]. The leaves of *S. glycyphylla* previously dried and blended were extracted with 80% aqueous ethanol. Then, the extract was concentrated and partitioned with hexane and butanol, consecutively. The butanol fraction was subjected to chromatographic separation. The separation produced eight phenolic compounds such as the glycyphyllin A previously mentioned, two new dihydrochalcones (glycyphyllin B (**72**) and C (**73**)) and five flavonoids (catechin (**26**), (2*R*,3*R*)-dihydrokaempferol-3-*O*-β-D-glucopyranoside (**57**), kaempferol-3-*O*-β-D-glucopyranoside (**74**), quercetin-3-*O*-β-D-glucopyranoside (**75**) and kaempferol-3-*O*-β-neohesperidoside (**76**), **Figures 14** and **15**). The antioxidant activity of pure compounds was evaluated using ferric reducing ability of plasma (FRAP) and DPPH$^{\bullet}$ radical scavenging. The flavonoids showed a good antioxidant activity; contrary to this, the dihydrochalcones showed weak antioxidant activity [50].

2.12. *Smilax lanceaefolia*

The root of *S. lanceifolia* prepared as a decoction is used in traditional Indian medicine to soothe stomach pain and rheumatism. The boiled extract is used to expel gallbladder and kidney stones. It was also found that the aqueous extract of *S. lanceifolia* contains compounds with a high affinity for binding proteins, specifically by active sites joining reverse transcriptase. Therefore, it can be used to inhibit proliferation of retroviruses-agents in viral diseases such as AIDS and T-cell leukemia [51]. Laintojam and Kongbrailatpam performed a study on the chemical constituents and antioxidant activiy of *S. lanceifolia* roots extracts [7]. The extracts were obtained from petroleum ether, chloroform and methanol, successively. Antioxidant

http://dx.doi.org/10.5772/66896

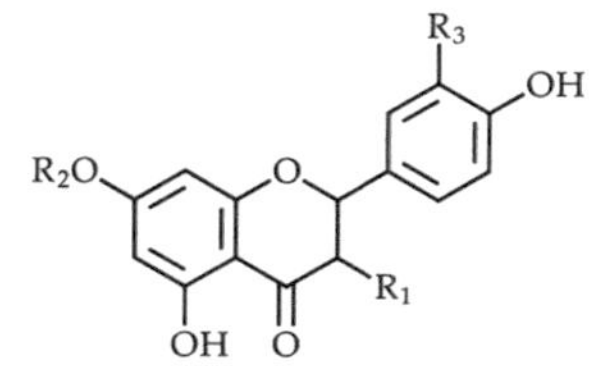

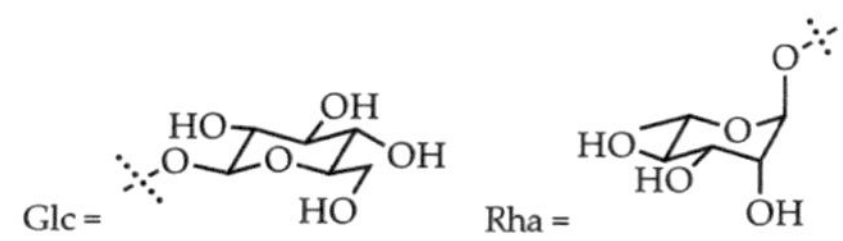

	R1	R2	R3	Name
55	H	H	H	Naringenin
56	H	Me	H	(2*S*)-Sukaranetin
57	Glc	H	H	(2*R*,3*R*)-Arthromerin B
58	OH	Glc	H	(2*R*,3*R*)-Sinensin
59	OH	H	Glc	(2*R*,3*R*)-Taxifolin-3′-*O*-*β*-D-glucopyranoside
60	Glc	H	OH	(2*S*,3*S*)-Glucodistylin

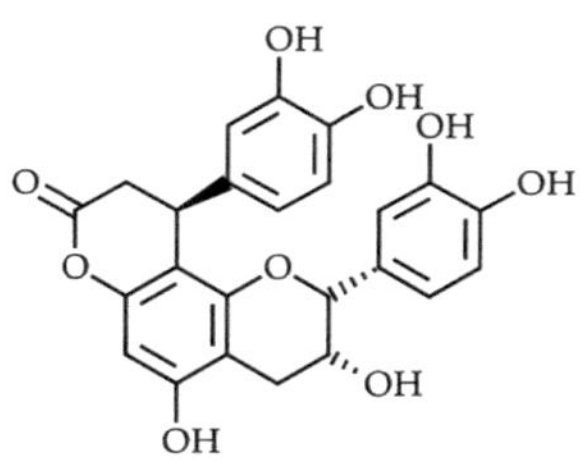

61 Cinchonain Ib

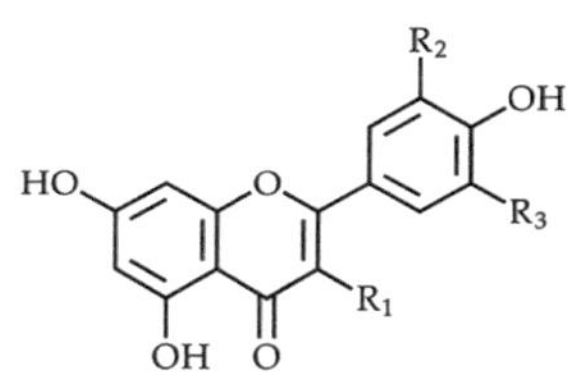

	R1	R2	R3	Name
62	H	OH	H	Luteolin
63	H	H	H	Apigenin
64	OH	OH	OH	Myricetin

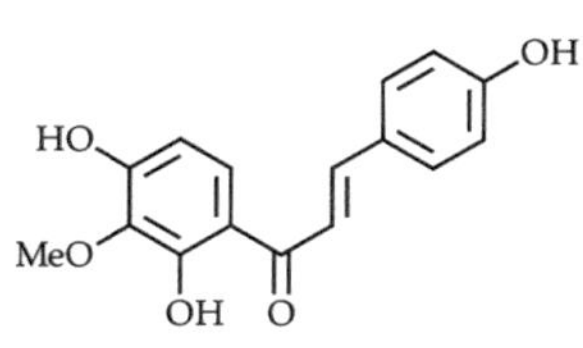

65 Kukulkanin B

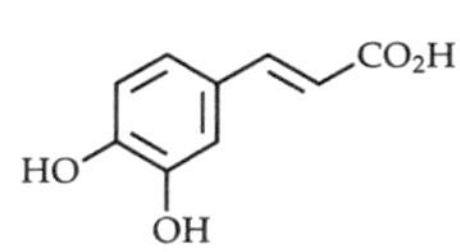

66 Caffeic acid

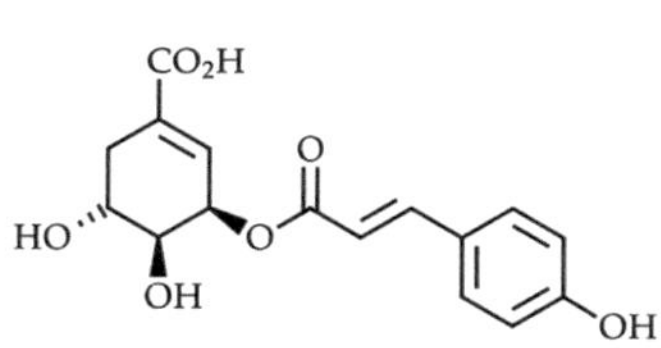

67 *O*-*p*-coumaroylshikimic acid

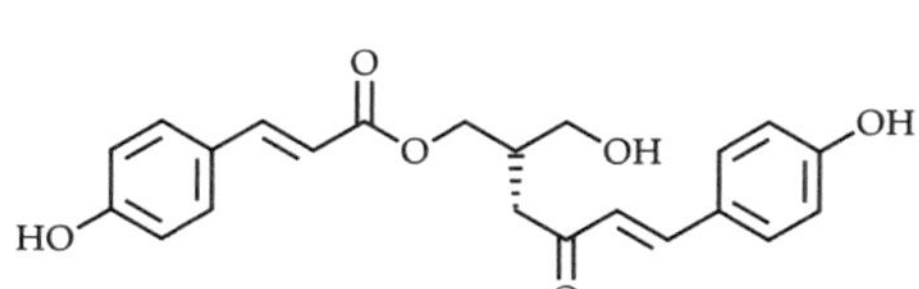

68 Smiglycerol

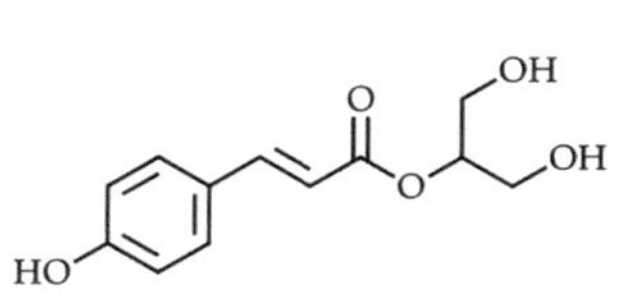

69 Juncusyl ester B

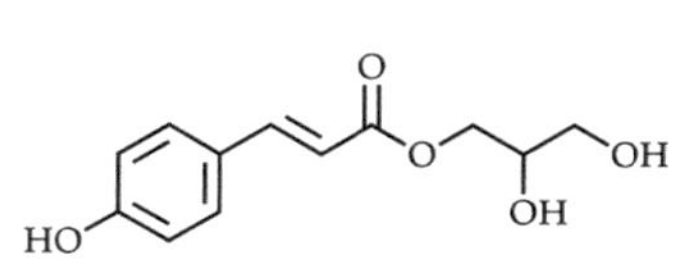

70 1-*O*-*p*-coumarylglycerol

Figure 14. Flavonoids and phenylpropanoids isolated from *S. glabra* rhizome.

	R_1	R_2	Name
71	H	H	Glycyphyllin A
72	H	OH	Glycyphyllin B
73	Glu	H	Glycyphyllin C

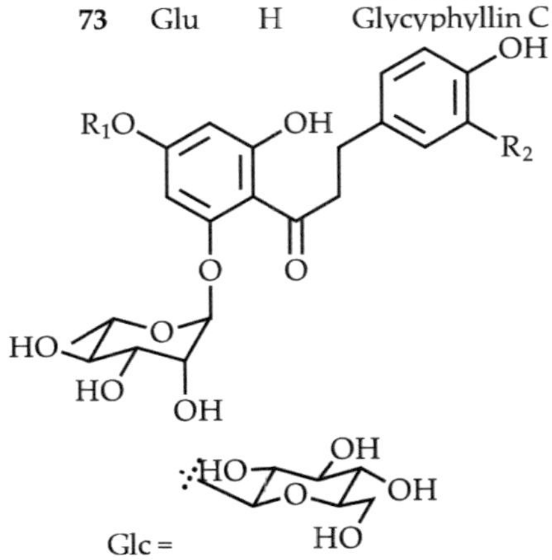

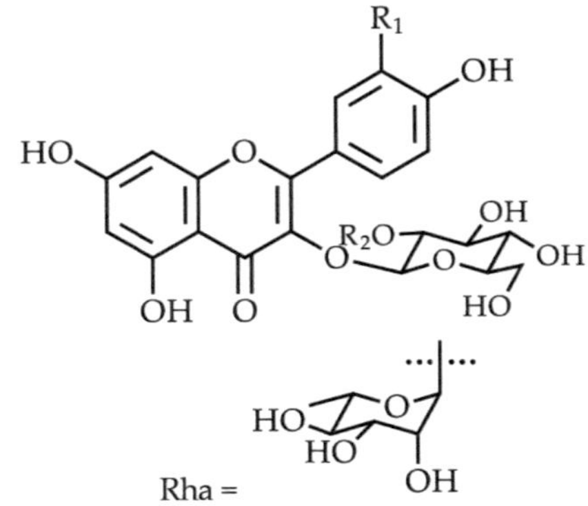

	R_1	R_2	Name
74	H	H	Kaempferol-3-*O*-*β*-D-glucopyranoside
75	OH	H	Quercetin-3-*O*-*β*-D-glucopyranoside
76	H	Rha	Kaempferol-3-*O*-*β*-D-neohesperidoside

Figure 15. Phenolic compounds isolated from 80% ethanol extract of *S. glycyphylla* leaves.

activity of the extracts was evaluated by DPPH• radical scavenging and it was found that the methanol extract had the principal antiradical activity attributed to phenolic compounds. The only compound isolated from methanol extract was flavanonol glycoside, quercitrin (**77**, **Figure 16**) [7].

2.13. *Smilax riparia*

The roots and rhizomes of *S. riparia*, commonly called "Niu-Wei-Cai" in China, are used in traditional Chinese medicine as diuretics, treatments for inflammation and cancer [52], and in some cases as food [53]. Sun *et al.* isolated three phenylpropanoid glycosides (the smilasides M (**78**) and N (**79**), and 2′,6′-diacetyl-3,6-diferuloylsucrose (**80**), **Figure 16**) from a 95% aqueous ethanol extract of *S. riparia* roots and rhizomes. The concentrated extract was suspended

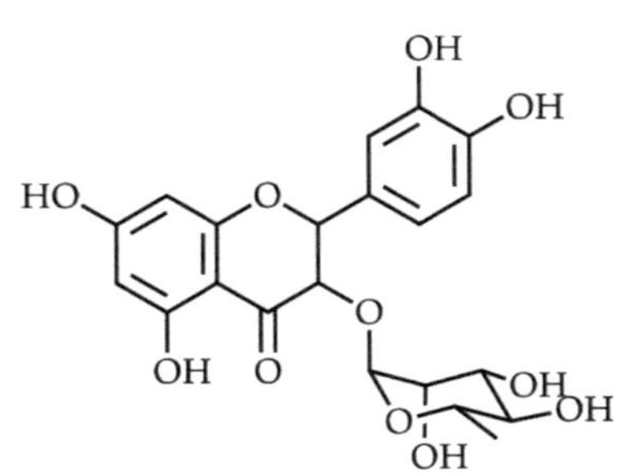

77 Quercitrin

Figure 16. Flavanonol glycoside from methanol extract of *S. lanceifolia* roots.

in water and subjected to D101 macroporous resin column chromatography and eluted with water and 30% and 70% ethanol, successively. The fraction eluted with 70% ethanol was suspended in water and partitioned with chloroform, ethyl acetate and butanol. The ethyl acetate fraction was chromatographed over a silica gel column, followed by thin-layer chromatography (TLC) and finally, the fractions were subjected to C18 reversed-phase silica gel column to obtain the afore-mentioned compounds, [54].

Wang *et al.* isolated five phenylpropanoids with a sucrose core (helonioside B (**18**), smiglaside A (**51**), smiglaside B (**52**), and smilaside P (**81**), **Figures 5**, **13** and **17**) [55]. The phenylpropanoid compounds were obtained from a 95% aqueous ethanol extract of *S. riparia* roots and rhizomes. The extract was subjected to macroporous resin HPD-600 and eluted with 95% aqueous ethanol and ethyl acetate. Subsequently, the ethyl acetate fraction was subjected to silica gel column chromatography, and from this fraction were obtained the phenylpropanoids compounds. The five compounds were subjected to cytotoxicity test against human promyelocytic leukemia (HL-60), human hepatocellular carcinoma (SMMC-7721), human lung cancer (A-549), human breast cancer (MCF-7) and human colon cancer (SW480) using 3-(4,5-dimethylthiazol-2-yl)-5-(3-carboxymethoxy-phenyl)-2-(4-sulfophenyl)-2H-tetrazolium (MTS). The five compounds were evaluated for anticancer activity. Antioxidant activity was also evaluated using DPPH$^\bullet$ radical scavenging activity. The results showed the phenylpropanoid compounds with three feruloyl and acetyl groups exhibited the primary antitumoral and antioxidant activities. The proposed explanation of these results was that the feruloyl and

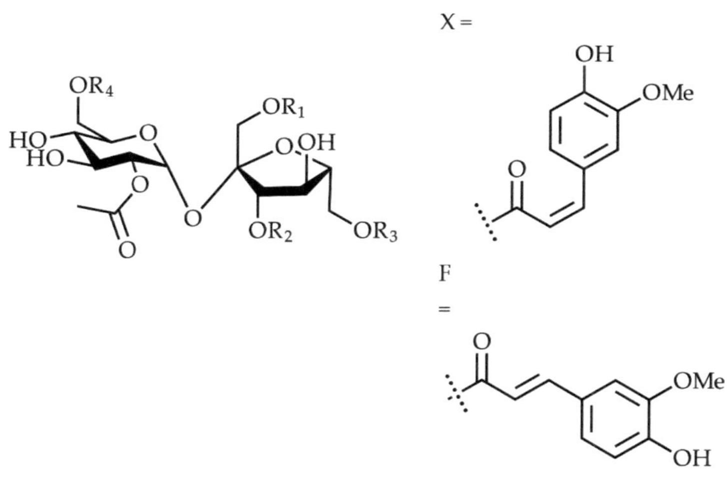

	R1	R2	R3	R4	Name
78	H	X	F	COMe	2′,6′-diacetyl-3-Z-feruloyl-6-feruloylsucrose, smilaside M
79	H	F	X	COMe	2′,6′-diacetyl-3-feruloyl-6-Z-feruloylsucrose, smilaside N
80	H	F	F	COMe	2′,6′-diacetyl-3,6-diferuloylsucrose
81	F	F	F	H	Smilaside P

Figure 17. Phenylpropanoids from 95% ethanol extract of *S. riparia* roots and rhizomes.

acetyl groups confer a minor polarity to the compounds with the most activity and are key to inducing antitumoral activity [55].

2.14. *Smilax scobinicaulis*

The roots of *S. scobinicaulis*, also called "Hei Ci Ba Quia" in Chinese, are used in traditional Chinese medicine for the treatment of arthritis, gout and inflammatory diseases [10]. Zhang *et al.* studied the chemical composition of *S. scobinicaulis*. These researchers first obtained a 95% aqueous ethanol extract from *S. scobinicaulis* roots and rhizomes. The extract was concentrated, suspended in water and partitioned with petroleum ether, ethyl acetate and butanol, successively. The ethyl acetate fraction was subjected to column chromatography. The purification was done to isolate and characterize two new flavones (7,3′,5′-trihydroxy-5,6,4′-trimethoxyflavone (**82**) and 7-hydroxy-5,6,3′,5′-pentamethoxyflavone (**83**), **Figure 18**). The new flavones were evaluated for cell proliferation and viability assay against human breast adenocarcinoma (MCF-7) and human lung carcinoma (H520). The results showed a weak activity for **82**, but 7-hydroxy-5,6,3′,5′-pentamethoxyflavone, **83** was inactive [10].

2.15. *Smilax sebeana*

S. sebeana is used in traditional Japanese and Chinese medicine to treat syphilis, arthritis and gout [11]. Ao *et al.* evaluated antioxidant activity in isolated phenolic compounds from methanol extracts of *S. sebeana* rhizomes and roots [11]. The methanol extract was obtained from maceration of fresh rhizomes and roots with methanol at room temperature. The methanol extract was concentrated, suspended in water and partitioned with hexane, ethyl acetate and butanol, successively. The ethyl acetate fraction was purified in two steps. First, it was subjected to column chromatography packed with sephadex LH-20. Some of the fractions collected were selected to evaluate by HPLC to evaluate their components. Finally, the fractions with similar composition were pooled and purified by preparative HPLC. Also, the total phenol content and antioxidant activity expressed by DPPH radical scavenging were evaluated for methanol, ethyl acetate and butanol fractions. The major content of total phenols was

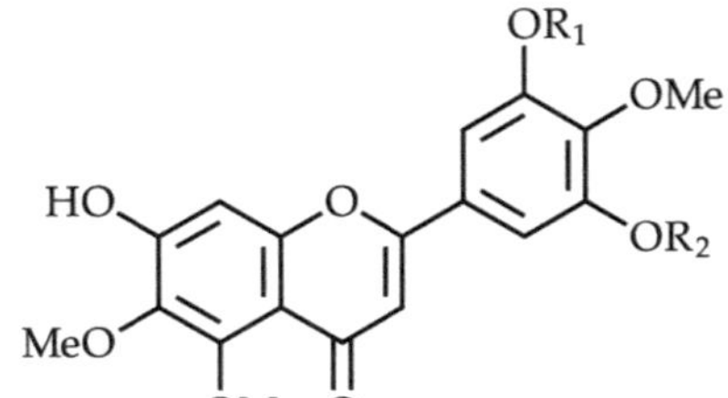

	R_1	R_2	Name
82	H	H	7,3′,5′-trihydroxy-5,6,4′-trimethoxyflavone
83	Me	Me	7-hydroxy-5,6,3′,5′-pentamethoxyflavone

Figure 18. Flavones from 95% ethanol extract of *S. scobinicaulis* roots.

found in the ethyl acetate fraction, 238.5 mg catechin/g extract. This fraction also showed the principal DPPH• radical scavenging activity, IC_{50} of approximately 10.4 μg/mL. The compounds isolated from the ethyl acetate fraction were one phenylpropanoid (chlorogenic acid (**84**), **Figure 19**) and three cinchonain (1b (**61**), 1a (**85**) and 11a (**86**), **Figures 14** and **19**) [11].

2.16. *Smilax trinervula*

S. trinervula is a plant used in traditional Chinese medicine. The rhizomes and roots are sources of the Chinese drug "Ba-Quia" used as a diuretic and to treat pelvic inflammation [56]. Shu *et al.* carried out the first chemical designed to isolate phenolic compounds from rhizomes of *S. trinervula*. These researchers obtained a 70% aqueous ethanol extract and removed the solvent [56]. The extract was partitioned with ethyl acetate and butanol. The butanol fraction was subjected to macroporous resin column chromatography. The fractions obtained from this separation were subjected to repeated silica gel and sephadex LH-20 column chromatography. Finally, the fractions were subjected to semipreparative HPLC. From this separation process were isolated eight phenolic compounds, three phenylpropanoids and five neolignans. These

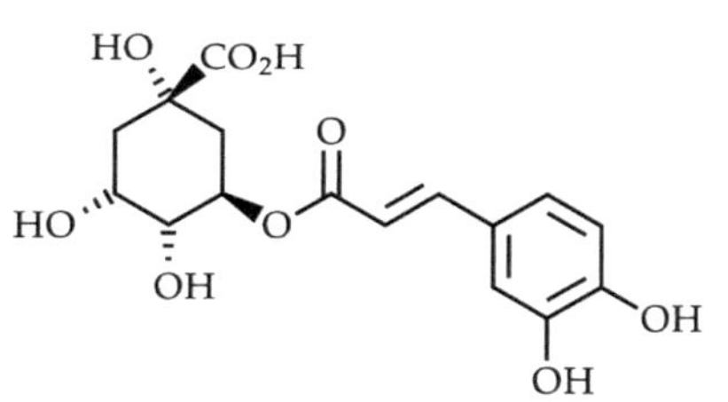

84 Chlorogenic acid

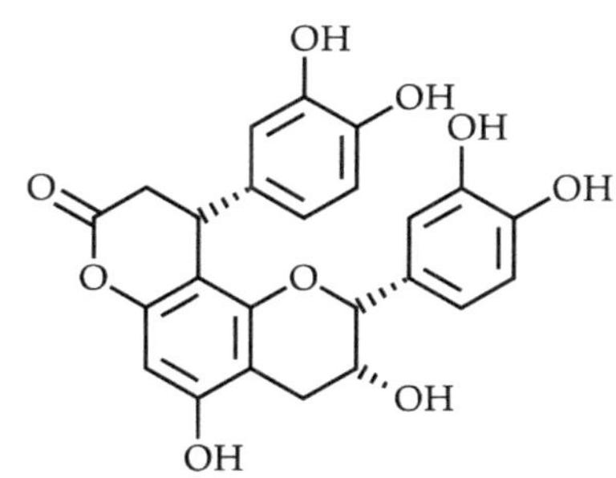

85 Cinchonain Ia

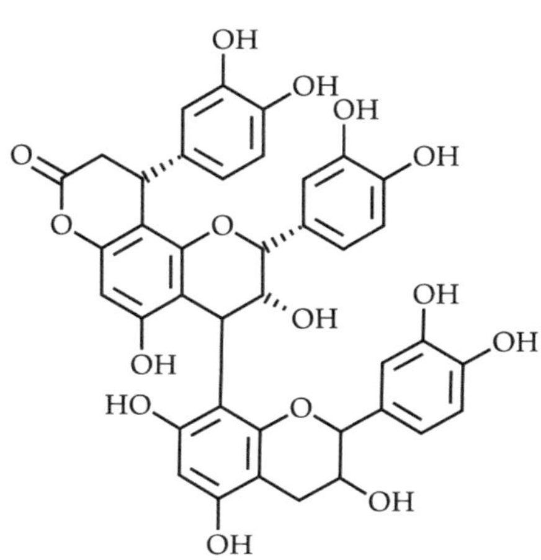

86 Cinchonain 11a

Figure 19. Phenolic compounds from methanol extract of *S. sebeana* rhizomes and roots.

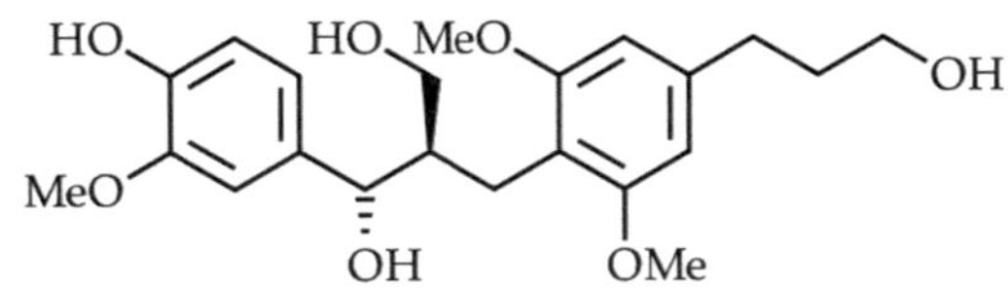

87 (7*S*,8*R*)-4,7,9,9′-tetrahydroxy-3,3′-dimethoxy-8-*O*-4′-neolignan

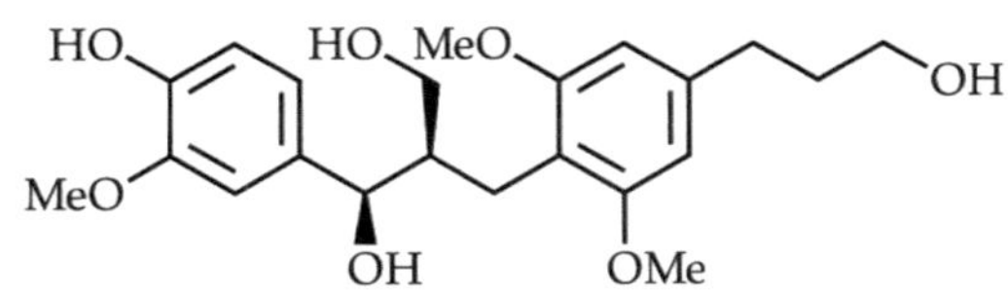

88 (7*R*,8*R*)-4,7,9,9′-tetrahydroxy-3,3′-dimethoxy-8-*O*-4′-neolignan

Figure 20. Two neolignans (**87** and **88**) with colon anticancer activity isolated of 70% ethanol extract from *S. trinervula* rhizomes.

compounds were evaluated against five human cell lines (SH-SY5Y, SGC-7901, HCT-116, Lovo and Vero) using the MTT method. The anticancer evaluation showed (7*S*,8*R*)-4,7,9,9′-tetrahydroxy-3,3′-dimethoxy-8-*O*-4′-neolignan (**87**) and (7*R*,8*R*)-4,7,9,9′-tetrahydroxy-3,3′-dimethoxy-8-*O*-4′-neolignan (**88**) (**Figure 20**) had cytotoxic activity against Lovo [56].

3. Conclusions

The polar extracts of *Smilax* species have high concentrations of phenolic compounds with high antioxidant activity. The main investigations of *Smilax* species are oriented to evaluate cytotoxicity against human cancer of cervical, lung, breast adenocarcinoma, liver and colon. The results of this review chapter showed flavonol, astilbin and phenylpropanoids binding to the sucrose nucleus (three moieties of ferulic acids and acetyl group to maximize activity) have a high potential as anticancer compounds. The anticancer activity with high concentrations of phenol compounds is attributed to antioxidant activity that induces cell apoptosis. The flavonols and phenylpropanoids can be isolated from ethyl acetate fractions obtained from a 95% aqueous ethanol extract of rhizomes and roots. Also, areal parts of *Smilax* plants contain phenolic compounds, for example, leaves and fruits. **Table 1** shows a resume of total phenols content and DPPH$^{\bullet}$ radical scavenging activity of extracts and fractions obtained from five *Smilax* species. The leaves and fruits of *Smilax* plants contain phenolic compounds. The fruits are also a source of anthocyanins. The chemical studies based on isolation and evaluation of phenolic compounds are few and do not cover more than 10% of the total *Smilax* species. Most studies have been carried out on species that grow in Asia. Hence, it is necessary continue studying the phenolic compound content of *Smilax* species, as well as evaluating their antioxidant, antidiabetic and anticancer properties.

http://dx.doi.org/10.5772/66896

Specie (part of plant) Extract/fraction	Total phenols (mg GAE/g dm)	DPPH• Radical scavenging activity IC_{50} (μg/mL)
S. campestris (aerial)		Ref. [17]
Ethanol extract	–	13.6
Hexane fraction	–	405.5
Dichloromethane fraction	–	298.9
Ethyl acetate fraction	–	108.9
Butanol fraction	–	2.1
S. china (*roots*)	Ref. [19]	Ref. [19]
Methanol extract	–	–
Chloroform fraction	142.6	302.2
Ethyl acetate fraction	401.6	85.5
Butanol fraction	206.8	210.9
Water fraction	97.3	224.9
S. excelsa (leaves)	Ref. [29]	Ref. [29]
Water extract	30.6	1190
Infusion	35.7	1240
Ethanol extract	30.1	1490
Ethyl acetate extract	8.8	2660
S. glabra (rhizomes)	Ref. [44]	Ref. [44]
Water extract	29.41	236
Ethanol fraction[a]	109.8	58
Methanol extract	152.3	43
S. riparia (rhizomes and roots)		Ref. [55]
95% ethanol extract	–	2520
95% ethanol fraction	–	1460
Ethyl acetate fraction	–	1330
Methanol fraction	–	1000

GAE mg: milligrams of gallic acid equivalent, dm: dry matter, IC_{50}: extract concentration necessary to decrease 50% the initial concentration of DPPH•. The fractions were obtained from a partition of corresponding extract.This fraction was obtained from water extract.

Table 1. Total phenols, antioxidant activity expressed as DPPH radical scavenging activity of *Smilax* genus species.

Acknowledgements

The authors thank Carol Ann Hayenga for her English assistance in the preparation of this manuscript. Support was provided by the Universidad Tecnológica de la Mixteca.

Conflict of interest

The authors have no conflict of interest to declare and are responsible for the content and writing of the manuscript.

Ethical approval

This chapter does not contain any studies with human participants or animals performed by any of the authors.

Author details

Salas-Coronado Raúl, Hernández-Carlos Beatriz, Llaguno-Guilberto Joseoziel and Santos-Sánchez Norma Francenia*

*Address all correspondence to: nsantos@mixteco.utm.mx

Institute of Agroindustry, Technological University of the Mixteca, Huajuapan de León, Oaxaca, México

References

[1] Jiang J, Xu Q. Immunomodulatory activity of the aqueous extract from rhizome of *Smilax glabra* in the later phase of adjuvant-induced arthritis in rats. J Ethnopharmacol. 2003; 85:53–59. DOI: 10.1016/S0378-8741(02)00340-9

[2] Zubair M, Rizwan K, Rashid U, Saeed R, Saeed AA, Rasool N, Riaz M. GC/MS profiling, *in vitro* antioxidant, antimicrobial and haemolytic activities of *Smilax macrophylla* leaves. Arabian J Chem. 2013. DOI: 10.1016/j.arabjc.2013.04.024

[3] Gao Y, Su Y, Qu L, Xu S, Meng L, Cai SQ, Shou C. Mitochondrial apoptosis contributes to the anti-cancer effect of *Smilax glabra* Roxb. Toxicol Lett. 2011; 207:112–120.

[4] Xu J, Li X, Zhang P, Li ZL, Wang Y. Antiinflammatory constituents from the roots of *Smilax bockii* warb. Arch Pharm Res. 2005; 4:395–399. DOI: 10.1007/BF02977667

[5] Rugna A, Polo J, Evelson P, Gurni AA, Llesuy S, Wagner ML. Antioxidant activity in rhizomes from *Smilax campestris* Griseb. *Smilacaceae*. Mol Med Chem. 2003; 1:21–25.

[6] Chen T, Li JX, Xu Q, Komatsu K. A new flavanone isolated from rizoma *Smilacis glabrae*, and the structural requirements of its derivatives for preventing immnulogical hepatocyte damage. Plant Med. 1999; 65:56–59. DOI: 10.1055/s-1999-13963

[7] Laitonjam WS, Kongbrailatpam BD. Studies on the chemical constituents and antioxidant activities of extracts from the roots of *Smilax lanceaefolia* Roxb. Nat Prod Res. 2010; 24:1168–1176. DOI: 10.1080/14786410903040402

[8] Cheng YB, Zhang DM, Yu SS. Chemical constituents of *Smilax perfoliata*. Acta Bot Sin. 2004; 46:618–620.

[9] Sun TT, Zhang DW, Han Y, Dong FY, Wang W. Smilasides M and N, two new phenypropanoid glycosides from *Smilax riparia*. J Asian Nat Prod. 2012; 2:165–170. DOI: 10.1080/10286020.2011.641536

[10] Zhang CL, Feng SX, Wang Q, Wang P, Xu J, Chen T. Flavonoids and phenolic compounds from *Smilax scobinicaulis*. Chem Nat Comp. 2014; 50:254–257. DOI: 10.1007/s10600-014-0925-9

[11] Ao C, Higa T, Khanh TD, Upadhyay A, Tawata S. Antioxidant phenolic compounds from *Smilax sebeana* Miq. LWT-Food Sci Technol. 2011; 44:1681–1686. DOI: 10.1016/j.lwt.2011.02.001

[12] Wu LS, Wang XJ, Wang H, Yang HW, Jia H, Jia AQ, Ding Q. Cytotoxic polyphenols against breast tumor cell in *Smilax china* L. J Ethnopharmacol. 2010; 130:460–464. DOI: 10.1016/j.jep.2010.05.032

[13] Ivanova A, Mikhova B, Batsalova T, Dzhambazov B, Kostoka I. New furostanol saponins from *Smilax aspera* L. and their *in vitro* cytotoxicity. Fitoterapia. 2011; 82:282–287. DOI: 10.1016/j.fitote.2010.10.012

[14] Longo L, Vasapollo G. Extraction and identification of anthocyanins from *Smilax aspera* L. berries. Food Chem. 2006; 96:226–231. DOI: 10.1016/j.foodchem.2004.11.008

[15] Li SY, Fuchino H, Kawahara N, Sekita S, Satake M. New phenolic constituents from *Smilax bracteata*. J Nat Prod. 2002; 65:262–266. DOI: 10.1021/np010338m

[16] Zhang L, Liao C-C, Huang H-H, Shen Y-C, Yang L-M, Kuo Y-H. Antioxidant phenylpropanoid glycosides from *Smilax bracteata*. Phytochemistry. 2008; 69:1398–1404. DOI: 10.1016/j.phytochem.2008.01.002

[17] Morais MI, Amaral Pinto ME, Guimarães Araújo S, Fonsêca Castro AH, Duarte-Almeida JM, Rosa LH, Rosa CA, Johann S, Rodrigues dos Santos Lima LA. Antioxidant and antifungal activities of *Smilax campestris* Griseb. (Smilaceae). Nat Prod Res. 2014; 28:1275–1279. DOI: 10.1080/14786419.2014.895728

[18] Lee SE, Ju EM, Kim JH. Free radical scavenging and antioxidant enzyme fortifying activities of extracts from *Smilax china* root. Exp Mol Med. 2001; 33:263–268. DOI: 10.1038/emm.2001.43

[19] Jeong C-H, Jeong HR, Kwak JH, Choi GN, Kim D-O, Lee U, Heo HJ. Phenolic Composition and in vitro antioxidant activities of smilax china root. J Food Biochem. 2013; 37:98–107. DOI:10.1111/j.1745-4514.2011.00610.x

[20] Kuo Y-H, Hsu Y-W, Liaw C-C., Lee JK, Huang H-C, Kuo L-MY. Cytotoxic phenylpropanoid glycosides from the stems of *Smilax china*. J Nat Prod. 2005; 68:1475–1478. DOI: 10.1021/np050109q

[21] Li Y-L, Gan G-P, Zhang H-Z, Wu H-Z, Li C-L, Huang Y-P, Liu Y-W, Liu J-W. A flavonoid glycoside isolated from *Smilax china* L. rhizome *in vitro* anticancer effects on human cancer cell lines. J Ethnopharmacol. 2007; 113:115–124. DOI: 10.1016/j.jep.2007.05.016

[22] Shao B, Guo H-Z, Cui Y-J, Liu A-H, Yu H-L, Guo H, Xu M, Guo D-A. Simultaneous determination of six major stilbenes and flavonoids in *Smilax china* by high performance liquid chromatography. J Pharm Biomed Anal. 2007; 44:737–742. DOI: 10.1016/j.jpba.2007.03.008

[23] Wungsintaweekul B, Umehara K, Miyase T, Noguchi H. Estrogenic and anti-estrogenic compounds from the Thai medicinal plant, *Smilax corbularia* (Smilacaceae). Phytochemistry. 2011; 72:495–502. DOI: 10.1016/j.phytochem.2010.12.018

[24] Calderón AI, Vázquez Y, Solís PN, Caballero-George C, Zacchino S, Gimenez A, Pinzón R, Cáceres A, Tamayo G, Correa M, Gupta MP. Screening of Latin American plants for cytotoxic activity. Pharm Biol. 2006; 44:130–140. DOI: 10.1080/13880200600592285

[25] Michel J, Duarte RE, Yao P, Bolton JL, Huang Y, Cáceres A, Veliz M, Soejarto DD, Mahady GB. Medical potential of plants used by the Q'eqchi Maya of Livingston, Guatemala for the treatment of women's health complaints. J Ethnopharmacol. 2007; 114:92–101. DOI: 10.1016/j.jep.2007.07.033

[26] Cáceres A, Lange K, Cruz SM, Velásquez R, Lima S, Menéndez MC, Dardón R, Córdova D, González J. Assessment of antioxidant activity of 24 native plants used in Guatemala for their potential application in natural product industry. Acta Hortic. 2012; 964:85–92. DOI: 10.17660/ActaHortic.2012.964.10

[27] Cáceres A, Cruz SM, Martínez V, Gaitán I, Santizo A, Gattuso S, Gattuso M. Ethnobotanical, pharmacognostical, pharmacological and phytochemical studies on *Smilax domingensis* in Guatemala. Rev Bras Farmacogn. 2012; 22:239–248. DOI: 10.1590/S0102-695X2011005000211

[28] Yeşilada E, Sezik E, Honda G, Takaishi Y, Takeda Y, Tanaka T. Traditional medicine in Turkey IX: folk medicine in north-west Anatolia. J Ethnopharmacol. 1999; 64:195–210. DOI: 10.1016/S0378-8741(98)00133-0

[29] Ozsoy N, Can A, Yanardag R, Akev N. Antioxidant activity of *Smilax excelsa* L. leaf extracts. Food Chem. 2008; 110:571–583. DOI 10.1016/j.foodchem.2008.02.037

[30] Khaligh P, Salehi P, Farimani MM, Ali-Asgari S, Esmaeili MA, Ebrahimi SM. Bioactive compounds from *Smilax excelsa* L. J Iran Chem Soc. 2016; 13:1055–1059. DOI: 10.1007/s13738-016-0819-9

[31] Nascimento Soares A, Coelho Novembre ADL, Redondo Martins A, de Stefano Piadade SM, Appezzato-da-Glória B. Propagation studies in *Smilax fluminensis* Stued. (Smilaceae). Cienc Rural. 2011; 41:1762–1768. DOI: 10.1590/S0103-84782011001000014

[32] Petrica EEA, Sinhorin AP, Sinhorin VDG, Júnior GMV. First phytochemical studies of japecanga (*Smilax fluminensis*) leaves: flavonoids analysis. Rev Bras Farmacogn. 2014; 24:443-445. DOI: 10.1016/j.bjp.2014.07.020

[33] Chen T, Li J, Cao J, Xu Q, Komatsu K, Namba T. A new flavanone isolated from rhizoma *Smilacis glabrae* and the structural requirements of its derivatives for preventing immunological hepatocyte damage. Planta Med. 1998; 65:56–59. DOI: 10.1055/s-1999-13963

[34] Gunn J, Che CT, Farnsworth N. Diabetes and natural products. In: Watson RR, Preedy RV, editor. Bioactive food as dietary interventions for diabetes. 2013. Elsevier, Inc.; 381–394. DOI: 10.1016/B978-0-12-397153-1.00042-1

[35] Sa F, Gao JL, Fung KP, Zheng Y, Lee SM. Anti-proliferative and pro-apoptotic effect of *Smilax glabra* Roxb. extract on hepatoma cell lines. Chem Biol Interact. 2008; 171:1–14. DOI: 10.1016/j.cbi.2007.08.012

[36] She T, Zhao C, Feng J, Wang L, Qu L, Fang K, Cai S, Shou C. Sarsaparilla (*Smilax glabra* Rhizome) extract inhibits migration and invasion of cancer cells by suppressing TGF-β1 pathway. PLoS One. 2015; 10:1–16. DOI: 10.1371/journal.pone.0118287

[37] Xia D, Yu X, Liao S, Shao Q, Mou H, Ma W. Protective effect of *Smilax glabra* extract against lead-induced oxidative stress in rats. J Ethnopharmacol. 2010; 130:414–420. DOI: 10.1016/j.jep.2010.05.025

[38] Trinh TTV, Vu VC, Pham TH, Nguyen QV. Antioxidant activity of extracts and astilbin from the root of *Smilax glabra* of Vietnam. Malaysian J Chem. 2015; 17:12–19.

[39] Chen TH, Liu JC, Chang JJ, Tsai MF, Hsieh MH, Chan P. The *in vitro* inhibitory effect of flavonoid astilbin on 3-hydroxy-3-methylglutaryl coenzyme A reductase on Vero cells. Zhonghua Yi Xue Za Zhi (Taipei). 2001; 64:382–387.

[40] Haraguchi H, Ohmi I, Fukuda A, Tamura Y, Mizutani K, Tanaka O, Chou WH. Inhibition of aldose reductase and sorbitol accumulation by astilbin and taxifolin dihydroflavonols in *Engelhardtia chrysolepis*. Biosci Biotechnol Biochem. 1997; 61:651–654. DOI: 10.1271/bbb.61.651

[41] Closa D, Torres M, Hotter G, Bioque G, Leon OS, Gelpi E, Roselo-Catafau J. Prostanoids and free radicals in Cl4C-induced hepatotoxicity in rats: effect of astilbin. Prostaglandins Leukot Essent Fatty Acids, 1997; 56:331–334.

[42] Cechinel-Filho V, Vaz ZR, Zunino L, Calixto JB, Yunes RA. Antinociceptive and anti-oedematogenic properties of astilbin, taxifolin and some related compounds. Arzneimittelforschung. 2000; 50:281–285. DOI: 10.1055/s-0031-1300200

[43] Cai Y, Chen T, Xu Q. Astilbin suppresses collagen-induced arthritis via the dysfunction of lymphocytes. Inflamm Res. 2003; 52:334–340. DOI: 10.1007/s00011-003-1179-3

[44] Zhang Q-F, Zhang Z-R, Cheung H-Y. Antioxidant activity of rhyzoma *Smilacis glabrae* extracts and its key constituent—astilbin. Food Chem. 2009; 115:297–303. DOI:10.1016/j.foodchem.2008.11.053

[45] Lu C-L, Zhu W, Wang M, Xu X-J, Lu C-J. Antioxidant and anti-inflammatory activities of phenolic-enriched extracts of *Smilax glabra*. Evid Based Complement Alternat Med. 2014; 910438, 8 pages. DOI: 10.1155/2014/910438

[46] Cheng T, Li J-L, Xu Q. Phenylpropanoid glycosides from *Smilax glabra*. Phytochemistry. 2000; 53:1051–1055.

[47] Xu S, Shang M-Y, Liu G-X, Xu F, Wang X, Shou C-C, Cai S-Q. Chemical constituents from the rhizomes of *Smilax glabra* and their antimicrobial activity. Molecules. 2013; 18:5265–5287. DOI: 10.3390/molecules18055265

[48] Williams AH. Phenolic compounds of *Smilax glyciphylla*. Phytochemistry. 1967; 6:1583–1584.

[49] Cox SD, Jayasinghe C, Markham JL. Antioxidant activity in Australian sarsaparilla (*Smilax glyciphylla*). J Ethnopharmacol. 2005; 101:162–168. DOI: 10.1016/j.jep.2005.04.005

[50] Huang A-C, Wilde A, Ebmeyer J, Skouroumounis GK, Taylor DK. Examination of the phenolic profile and antioxidant activity of the leaves of the Australian native plant *Smilax glycipylla*. J Nat Prod. 2013; 76:1930–1936. DOI: 10.1021/np4005163

[51] Kusumoto JT, Shimada I, Kakiuchi N, Hattori M, Namba T. Inhibitory effects of Indonesi a plant extracts on reverse transcriptase of an RNA tumour virus (1). Phytother Res. 1992; 6:241–244. DOI: 10.1002/ptr.2650060504

[52] Zhang, SL, Han, ZH. Textural research of categories and functions on Nian-Yu-Xu recorded in Cai-Yao-Lu from Ben-Cao-Gang-Mu-Shi-Yi. J Zhejiang Univ Trad Chin Med. 2012;36:484–486.

[53] Wang, WH, Shao, MN, Han, YG. The nutrition composition analysis of *Smilax riparia* A. DC. Spec Wild Econ Anim Plant Res. 2000; 3:46–47.

[54] Sun T-T, Zhang D-W, Dong F-Y, Wang W. Smilasides M and N, two new phenylpropanoid glycosides from *Smilax riparia*. J Asian Nat Prod Res. 2012; 14:165–170. DOI: 10.1080/10286020.2011.641536

[55] Wang W-X, Li T-X, Ma H, Zhang J-F, Jia A-Q. Tumoral cytotoxic and antioxidative phenylpropanoid glycosides in *Smilax riparia* A. DC. J Ethnopharmacol. 2013; 140:527–532. DOI: 10.1016/j.jep.2013.07.011

[56] Shu J, Liang F, Liang J, Liang Y, Li F, Shao F, Liu R, Huang H. Phenylpropanoids and neolignans from *Smilax trinervula*. Fitoterapia. 2015; 105:64–68. DOI: 10.1016/j.fitote.2015.05.010

Chapter 10

Genotype, Environment and Management Practices on Red/Dark-Colored Fruits Phenolic Composition and Its Impact on Sensory Attributes and Potential Health Benefits

Fernanda Cosme, Berta Gonçalves,
Eunice A. Bacelar, António Inês,
António M. Jordão and Alice Vilela

Additional information is available at the end of the chapter

http://dx.doi.org/10.5772/66881

Abstract

Phenolic compounds are secondary metabolites abundant in our diet. These compounds may affect positively or negatively the sensory characteristics of food with important impacts on color, flavor, and astringency. An adequate consumption of phenolic compounds may also offer health benefits. After the consumption of fruits, the colon is the main site of microbial fermentation, where high molecular weight phenolic compounds are transformed into low molecular weight phenolic compounds such as phenolic acids or lactone structures by intestinal microbiota, which produce metabolites with biological and antioxidant activity, with evidence on health benefits for humans. A large amount of different phenolic compounds are responsible for physicochemical and sensory characteristics of table grapes and wines. Also, sweet cherry (*Prunus avium* L.) is one of the most popular temperate table fruits; they contain flavonoids, flavan-3-ols, and flavonols in addition to non-flavonoid compounds. Anthocyanins are the major polyphenols in blueberries, and this group of phytochemicals is thought to be responsible for many of the health benefits of berry consumption. Therefore, considering the importance of red/dark-colored fruits phenolic composition, the purpose of this chapter is to make a review of the most recent publications about these fruits' phenolic composition and their impact on sensorial properties as well as the effect of microorganisms on fruit phenolic composition.

Keywords: phenolic compounds, grapes, sweet cherries, blueberries, sensorial characteristics

1. Introduction

Phenolic compounds (phenolic acids, flavonoids, and stilbenes) are today among the most important classes of phytochemicals, since they are responsible for disease protection conferred from diets rich in these compounds [1]. Some fruits with high content of phenolic compounds, including flavonols, flavones, anthocyanins, and phenolic acids are grapes, sweet cherries, and blueberries. Polyphenolic compounds form complexes with salivary proteins, playing a role in the sensation of astringency, due to delubrication of oral surfaces. For astringency, the tannin molecular weight seems to be important for its perception and to the interactions with salivary proteins. Flavor and color are also important factors for the selection of fruit by consumers. Sweetness and bitterness are mutually suppressed in mixtures, but astringency and bitterness tend to be perceived as negative attributes. Polyphenols' sensory properties are related to molecules specific structures, including pigments correlated to fruit color [2]. This richness in phenolic compounds is also directly related with the positive effects on human health. However, the phenolic composition of the red/dark-colored fruits depends on cultivar, maturity, growing environment, cultural practices, postharvest conditions, and processing techniques [3].

2. Phenolic composition of red/dark-colored fruits

2.1. Phenolic composition of wine grapes and table grapes

Grapevine (*Vitis vinifera* L.) is the most important Mediterranean fruit crop, used to produce wine, table grapes, and raisins. The phenolic compounds in grapes include two classes of phenolic compounds: non-flavonoids and flavonoids. The major C_6-C_3-C_6 flavonoids in grapes include conjugates of flavonols, quercetin, and myricetin; flavan-3-ols (+)-catechin and (-)-epicatechin; and malvidin-3-*O*-glucoside and other anthocyanins. Non-flavonoids include C_6-C_1 hydroxybenzoic acids, and gallic acid, C_6-C_3 hydroxycinnamates caffeic, caftaric, and *p*-coumaric acids; and C_6-C_3-C_6 stilbenes *trans*-resveratrol, *cis*-resveratrol, and *trans*-resveratrol glucoside. Polyphenols are a diverse group of secondary metabolites, which exist in different grape bunch fraction, such as stems, skins, pulp, and seeds [4–8]. According to Pastrana-Bonilla et al. [6], the average concentration of total phenolic compounds in wine grapes is around 2178.8 mg/g gallic acid equivalent, in seeds, 374.6 mg/g gallic acid equivalent, in skins, and 23.8 mg/g gallic acid equivalent, in pulps. In addition, for table grapes, several authors also reported high levels of global and individual phenolic compounds [9]. Also for grape raisins, several works reported high levels of phenolic compounds [10, 11]. Thus, Sério et al. [12] reported levels of total phenolic compounds from several commercial red raisins (namely from Cardinal and Moscatel of Alexandria grape varieties) that ranged from 110.8 to 406.9 mg/100 g raisin. Phenolic compounds play an important role in wine quality and also in sensorial characteristics of table grapes, such as color, astringency, bitterness, and aroma. However, it is important to note that the phenolic composition of grape berries depends on grape variety, environmental factors, and viticultural practices [8, 13–15]. Consequently, all these isolated or combined factors will be critical for the composition of grape phenolic compounds, grape variety being one of the most important [16, 17]. Thus, genotypic differences among different varieties have a great influence in grape phenolic synthesis and accumulation during grape fruit maturation and development [18]. However, the interaction between the genotype, environment, and management practices heavily influences the overall phenolic composition. Recently, Costa et al. [8] analyzed the phenolic composition of several grape varieties cultivated at the same time in two Portuguese regions with distinct climatic conditions and reported that in general significantly higher global phenolic composition was obtained in the grapes collected in one of the regions. In addition, other work recently published [15] analyzed the adaptability of several red grape varieties from French origin to the other specific "terroirs" and compare their characteristics with native grape varieties. These authors reported that French grape varieties studied showed a higher degree of adaptation of the climate and soil conditions from the Portuguese vineyards, especially for phenolic composition. Thus, grape phenolic characteristics are strongly influenced by environmental conditions specific from each place and consequently each grape variety produced in a specific terroir reflects the locality in its chemical composition, including in phenolic composition. According to several works, the geological and soil conditions [19], vineyard altitude [20], sunlight exposition [21], climate [21, 22], and solar radiation [23] of a region are important environmental factors that determine grape phenolic composition. Finally, there are also other factors that directly or indirectly may determine the grape phenolic composition, namely cultivation practices [22], exposure to diseases [24], and the degree of grape ripeness [4, 17].

2.2. Phenolic composition of sweet cherry

Cherries are an excellent source of antioxidants, particularly phenolics, such as flavonoids, flavan-3-ols, and flavonols in addition to non-flavonoid compounds such as hydroxycinnamic and hydroxybenzoic acids, which are concentrated in the epicarp and mesocarp of the fruit [25, 26]. The most abundant phenolic compounds are anthocyanins such as cyanidin-3-*O*-rutinoside, cyanidin-3-*O*-glucoside, peonidin-3-*O*-rutinoside and glucoside, as well as pelargonidin-3-*O*-rutinoside are the most important anthocyanins in cherries [27]. The total anthocyanin content ranged from 6.21 to 94.20 mg cyanidin-3-*O*-glucoside equivalents/100 g fresh weight in 24 sweet cherry cultivars grown on the mountain sides of the Etna volcano (Sicily, Italy) [28]. Other phenolics in cherries include neochlorogenic acid, *p*-coumaroylquinic acid, and chlorogenic acid as the main hydroxycinnamic acids [26, 29, 30], the flavonol rutin and the flavan-3-ols (+)-catechin and (-)-epicatechin (**Figure 1**) [26, 31]. The total phenol content ranged from 84.96 to 162.21 mg gallic acid equivalents/100 g fresh weight in 24 sweet cherry cultivars grown in Italy [28]. Moreover, several studies reported higher phenolic content [26, 32] and antioxidant activity [32] in ripe cherries than in partially ripe. However, other pre- and postharvest factors, such as rootstock, cultivar, climate, soil type, storage conditions, and processing can significantly alter the amounts of bioactive compounds. In fact, levels of chlorogenic acid, neochlorogenic acid, *p*-coumaric acid, and quercetin-3-rutinoside were higher in fruits grown on Weiroot 13 and PiKu 1 rootstocks compared to MaxMa 14, Weiroot 158, F12/1 and Gisela 5 rootstocks [31]. According to Gonçalves et al. [26], the cherry cultivars have the same phenolic pattern, however, with large variation on content as presented in **Table 1**. The climatic conditions have great influence on phenolic levels. Indeed, Gonçalves et al. [26] stated that higher temperature and solar irradiation favored the biosynthesis of phenolic acids and decreased the content of anthocyanins. However, the phenolic content tends to reach highest levels in the late stage of final maturity as refereed by Stöhr et al. [33]. In recent research, the preharvest application of several products to improve cherry quality, such as the oxalic acid (2 mM), has been studied, which increased anthocyanins, flavonols, neochlorogenic, and chlorogenic acids [34]. All the phenolic compounds and the antioxidant activity increased in several sweet cherry cultivars during cold storage [26, 27, 32, 35]. Also, the level of phenolics in "Canada Giant" and "Ferrovia" cherries increased during 8 days of shelf life [36]. Nevertheless, Esti et al. [37] detected a total anthocyanin content decrease of 41–52% in two sweet cherry cultivars after 15 days at 1°C and 95% RH. The use of edible coatings has been used to extend the postharvest storage of cherries. Petriccione et al. [38] specified that chitosan-coated sweet cherries presented higher total phenolic, flavonoid, and anthocyanin levels. Moreover, increasing health-promoting properties of cherry fruit can be achieved with the addition of methyl salicylate treatment to cherry trees. This compound also delays the fruit postharvest senescence process by increasing the activity of the enzymes involved in ROS scavenging [39].

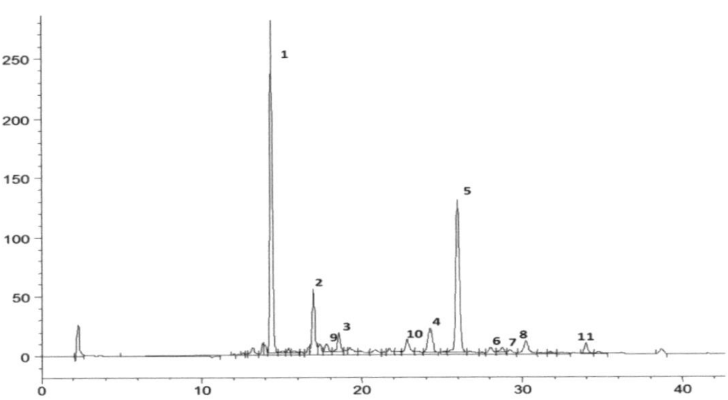

Peaks legend:
(1) neochlorogenic acid;
(2) *p*-coumaroylquinic acid;
(3) chlorogenic acid;
(4) cyanidin 3-*O*-glucoside;
(5) cyanidin 3-*O*-rutinoside;
(6) peonidin 3-*O*-glucoside;
(7) pelargonidin 3-*O*-rutinoside;
(8) peonidin 3-*O*-rutinoside;
(9) (+) - catechin;
(10) (-) - epicatechin;
(11) rutin

Figure 1. HPLC chromatogram of the Van sweet cherry cultivar extracts recorded at 280 nm. Adapted from Gonçalves et al. [26].

Cultivar	Hydroxycinnamic acids			Flavan-3-ols		Flavonols	Anthocyannis				
	NcAc	*p*CqAC	CAc	Cat	Epi	Rut	cy-3-glu	cy-3-rut	pn-3-glu	plg-3-rut	pn-3-rut
Burlat	23.8	24.7	3.8	7.2	6.7	4.8	23.2	44.6	<1.0	<1.0	2.1
Saco	153.5	12.2	9.8	10.5	10.3	11.8	5.1	38.6	n.d.	<1.0	<1.0
Summit	34.4	27.5	7.2	5.8	8.2	3.1	2.4	26.0	<1.0	<1.0	<1.0
Van	65.6	5.6	4.8	3.5	4.5	4.0	3.4	28.2	<1.0	<1.0	1.5

NcAc, neochlorogenic acid; *p*CqAC, *p*-coumaroylquinic acid; CAc, chlorogenic acid; Cat, catechin; Epi, epicatechin; Rut, Rutin; cy-3-glu, cyanidin-3-*O*-glucoside; cy-3-rut, cyanidin-3-*O*-rutinoside; pn-3-glu, peonidin-3-*O*-glucoside; plg-3-rut, pelargonidin-3-*O*-rutinoside; pn-3-rut, peonidin-3-*O*-rutinoside; n.d., not detected.
Adapted from Gonçalves et al. [26].

Table 1. Content of several phenolic compounds in four sweet cherry cultivars (mg /100g fresh weight).

Almost all phenolic compounds in sweet cherry show strong antioxidant activity [35, 40, 41]. Adequate consumption of phenolic compounds may offer health benefits that include inhibition of tumor cells growth [41], inhibition of inflammation [42], and protection against neurodegenerative diseases [43]. According to Matias et al. [44], a phenolic-rich extract derived from sweet cherries could be an attractive candidate to formulate an agent for the prevention of oxidative stress-induced disorders such as intestinal inflammation disorders. In spite of the large variations in the phenolic compounds content observed among several cherry cultivars, the levels of health-promoting compounds are relevant to human health. Sweet cherries might therefore be considered as a functional food [41]. In fact, cyanidin-3-*O*-rutinoside can slow down the absorption of carbohydrates by the inhibition of α-glucosidase which may therefore be useful as inhibitor to prevent or treat diabetes mellitus [45]. Cyanidin-3-*O*-glucoside showed cardioprotective effects by reducing blood lipid levels in rats [46]. The oxygen radical absorbance capacity (ORAC) assay indicated that the fruit of all genotypes possessed considerable antioxidant activity [28]. Moreover, several cherry cultivars were effective in inhibiting human cancer cells derived from colon (HT29) and stomach (MKN45) [41]. Finally, cherry phenolic, mainly anthocyanins, also protects neuronal PC 12 cells from cell-damaging oxidative stress (antineurodegenerative activity). However, this protection is dose-dependent [43].

2.3. Phenolic composition of blueberries

Blueberries are flowering plants of the genus *Vaccinium* with dark-purple berries, whose anthocyanins are considered to be nature's most potent antioxidants [47]. The genus *Vaccinium* belongs to the *Ericaceae* family [48] and includes many popular berries consumed around the world including blueberries, huckleberries, cranberries, lingonberries, and bilberries [49]. Of the more than 400 species in the genus *Vaccinium*, highbush, lowbush, and rabbiteye blueberries (*V. corymposum* L., *V. augustifolium* Ait., and *V. ashei* Reade, respectively) are of high economic importance [50]. In fact, in recent years the production of these fruits has increased rapidly in Europe and across the globe, as a result of the recognition of their high nutritive value, characteristic taste, and flavor but also due to recent press regarding the health benefits of fresh berries consumption [51, 52]. Over 89,820 acres of land are growing cultivated blueberries with an estimated annual production of 280,000 tons [53]. Blueberries are both a food product and a dietary supplement, consumed not only as fresh fruits but also as frozen fruits, or in dried or preserved form in bakery products. Blueberry anthocyanins are used as a natural food colorant [54] and blueberry extract can be used as a prebiotic [49]. The fruit quality traits and the phytochemical content of blueberries are of increasing importance to researchers in the field of food and health [55]. Blueberries are a source of vitamins, minerals, dietary fiber, phenolics, and flavonoids and they are very low in fat and sodium [56]. Anthocyanins, which provide blueberry with their characteristic colors, are the major polyphenols in blueberries and this group of phytochemicals is thought to be responsible for many of the health benefits of berry consumption [57]. The anthocyanins detected in blueberries are 3-glycosidic derivatives of cyanidin, delphinidin, malvidin, petunidin, and peonidin [49]. Nevertheless, anthocyanins vary in their quantity and composition among genotypes and also depend on the environmental growth conditions, postharvest storage conditions, and the method of analysis. Anyway, malvidin-3-glucoside and malvidin-3-galactoside have been found to be the two most predominant anthocyanins in many cases [58]. Blueberries also contain varying amounts of other polyphenols, and chlorogenic acid is particularly high as compared with other food sources [59]. It is accompanied by small amounts of quercetin glycosides [60].

3. Impact of fruit phenolic compounds on sensorial characteristics

Regarding fruit's oral sensory characteristics, there are six oral sensory attributes of fruit: sourness, sweetness, bitterness, spiciness, aroma, and astringency. For many people, the oral sensory properties of fruit have a great impact on their choice, acceptability, and consumption. Phenolic compounds, apart from possessing valuable biological properties, impart a high sensory activity to foods [61]. They are closely associated with the sensory and nutritional quality of fresh and processed plant foods and may affect positively or negatively the sensory characteristics of food with impacts on color, flavor, and astringency. This impact becomes important for consumer's acceptance, so that health-promoting products can be palatable and largely consumed [2]. Fruit preservation also influences the quantity and quality of fruits' phenolic content. For instances, during thawing of fruits, oxidation of phenolic compounds takes place and is negatively correlated with the acceptance level of fruits [62]. However, in a study comparing different pretreating processes of strawberries, samples with the highest phenolic content were also the most pleasant ones [63]. Specific structures are described to be related to polyphenols' sensory properties, namely color perception. Color, in fruits, is derived from natural pigments that change through plant ripening. Chlorophylls (green), carotenoids (yellow, orange, and red), anthocyanins (red and blue), flavonoids (yellow), and betalains (red) are the primary pigments responsible for fruit color [64]. Also, water-soluble brown-, gray-, and black-colored pigments may occur due to enzymatic and non-enzymatic browning reactions [65]. Many polyphenol pigments in plants are reactive anthocyanins, yellow flavanols, and flavones [66]. Anthocyanins can be used in food industry to color food. The six anthocyanins that can be found in the following red/dark-colored fruits are cyanidin (cherries, blackcurrants, raspberries, and elderberries), delphinidin (blackcurrants and blueberries), malvidin (grapes), pelargonidin (strawberries and radishes), peonidin (cranberries), and petunidin (blueberries)—**Figure 2**. Due to their water solubility, anthocyanins are applicable for dyeing low pH systems. Increasing pH leads to a lesser color intensity and a bluer tone appears at pH higher than 4.5, giving its bluish color to blackcurrant. Proanthocyanidins react with anthocyanins to form new red pigments [68]. Loss or stabilization of color and increases in the range of available hues are resulted by the conversion of anthocyanins to other compounds during food processing [2]. The color of fruits is a sensory attribute that can really change consumers' fruit acceptance. It is considered the most important product-intrinsic sensory cue leading the sensory expectations that the consumer holds concerning the foods that they may consume [69] and, according to Piqueras-Fizman et al. [70], humans' experience of taste/flavor is determined by the expectations that they often generate prior to tasting. Consumers inspect fruits, visually, before deciding on whether or not to buy them. People associate certain colors with certain flavors. For instances, red/dark fruit coloring also appears to be a particularly good inducer of sweetness [71].

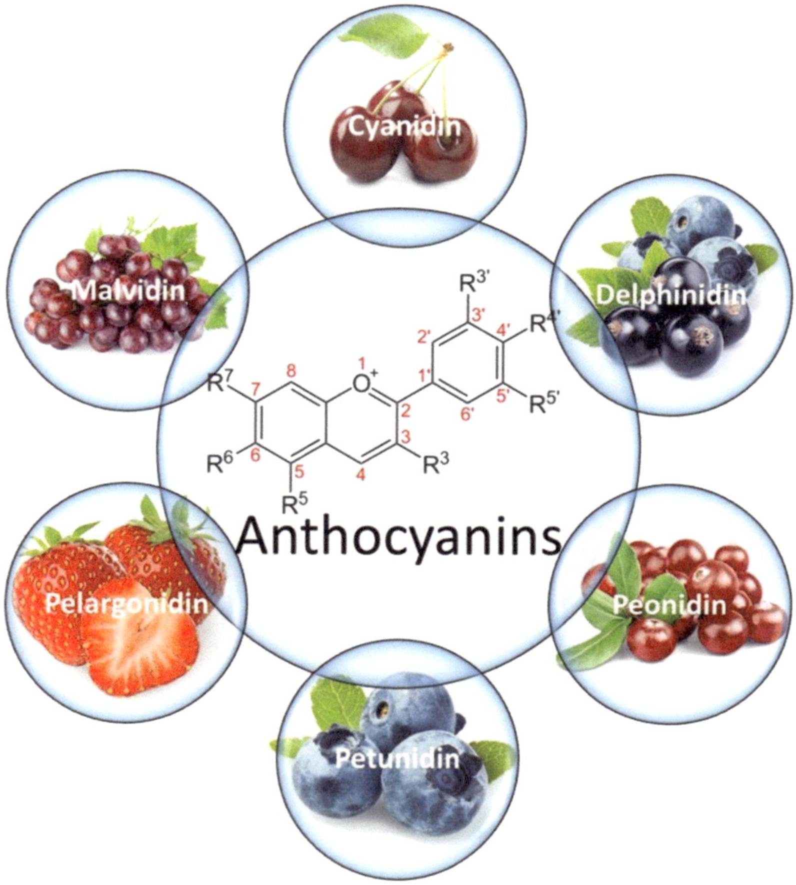

Figure 2. Anthocyanins in red/dark-colored fruits. Adapted from Just the Berries [67].

Gavrilova et al. [72] studied the phenolic profile of four blueberry varieties (*V. corymbosum* L., cv. Toro, Legacy, Duke, and Bluecrop) and two varieties (Rosenthal and Rovada) of red currants (*Ribes rubrum* L.) and black currants (*R. nigrum* L.) cultivated in Macedonia. They found that anthocyanins comprised the highest content of total phenolic compounds in currants (>85%), namely in the dark (black) currents, and lower and variety dependent in blueberries (35–74%). Hydroxycinnamic acid derivatives comprised 23–56% of total phenolics in blueberries and 1–6% in currants (**Table 2**). Besides bitterness, astringency, and color, some volatile polyphenols are strong odorants [66]. However, in dark-colored fruits, phenolic compounds present an almost insignificant role in fruit flavor profile. In raspberry fruit (*Rubus idaeus* L.), phenolic compounds only represent 1% of the total flavor compounds (**Figure 3**), whose concentration varies between "trace amount" and 0.3 mg/kg [73]. Nevertheless, in wild berries, several volatile phenolic compounds were identified by Honkanen et al. [73], such as 2-methoxy-4-vinylphenol, 2-methoxy-5-vinylphenol, 3,4-dimethoxybenzaldehyde, and 4-vinylsyringol, none of which have been reported in cultivated varieties [74]. An important fact stated by Honkanen et al. [73] is that with the exception of ionones, the amounts of individual volatile compounds in wild raspberries were generally three to four times higher than in the cultivated varieties. Moreover, the higher amounts of volatile compounds, in wild raspberry, may have contributed to their characteristic aroma. Also, the increased berry size, hybridization, and/or fertilization lead to worsening in the aroma profile of cultivated raspberries.

	Red currants		**Black currants**		**Blueberries**			
Compounds (total)	**Rosenthal**	**Rovada**	**Rosenthal**	**Rovada**	**Toro**	**Legacy**	**Duke**	**Bluecrop**
Phenolic compounds.	18.05 ± 0.58	17.97 ± 0.31	207.77 ± 1.14	187.69 ± 1.84	94.60 ± 0.93	137.74 ± 1.05	113.02 ± 1.28	120.14 ± 1.02
Anthocyanins	15.93 ± 0.95	14.73 ± 0.29	180.44 ± 3.59	162.83 ± 2.46	56.35 ± 1.04	68.55 ± 2.35	83.64 ± 3.16	41.99 ± 0.25
Flavonols	1.89 ± 0.08	0.48 ± 0.005	7.36 ± 0.57	6.95 ± 0.92	2.28 ± 0.80	5.17 ± 0.03	3.41 ± 0.16	6.08 ± 0.45
Flavan-3-ols	n.d.	1.60 ± 0.002	13.35 ± 0.90	11.02 ± 1.23	2.85 ± 0.54	1.75 ± 0.07	n.d.	4.52 ± 0.43
Hydroxycinnamic acid derivatives	0.23 ± 0.002	1.16 ± 0.10	6.62 ± 0.18	6.89 ± 0.24	33.12 ± 1.78	62.27 ± 1.97	25.97 ± 3.21	67.54 ± 3.03

n.d., not detected.

Table 2. Contents of phenolic compounds in red currants (*Ribes rubrum* L.), black currants (*Ribes nigrum* L.), and blueberries (*Vaccinium corymbosum* L.) determined by HPLC-DAD and expressed in mg per 100 g fresh weight±SD (n = 3). Adapted from Gavrilova et al. [72].

Plant-based phenol compounds, flavonoids, isoflavones, terpenes, and glucosinolates are almost bitter and astringent [75]. These substances provide defense against predators by making the plants unpalatable [75]. But also humans reject foods that are perceived to be excessively bitter [76]. Flavonoid phenols have been indicated as the main responsible for the taste of bitterness and the mouth-fell sensation of astringency in several types of fruits and in beverages [2, 77]. Several works suggested that some polyphenols can be responsible for the bitterness of fruits even if they are present in very low concentrations [78]. The bitterness and astringency of red wines and red/dark-colored fruits are mainly given by the flavanols. The mechanisms through which bitter taste perception occurs are not well understood; however, it is known that these mechanisms involve the activation of distinct human bitter taste receptors [77, 78]. While lower-molecular-weight phenolic compounds tend to be likely bitter, higher-molecular-weight polymers are perceived as astringent. Astringency or drying/puckering mouth-feel detectable throughout the oral cavity is due to a complex reaction between polyphenols and proteins of the mouth and saliva [79]. Interaction between tannins and saliva proteins plays an important role in astringency perception in wine [80]; however, the physiological and physicochemical mechanisms for this phenomenon are not fully understood and more studies focusing on this subject must be done in wines and fruits.

Figure 3. Volatile compounds reported in raspberry fruit (*Rubus idaeus* L.) according to chemical class. Adapted from Aprea et al. [74].

Total concentration, mean degree of polymerization [81], subunit composition, and distribution [82] are some of the variables related to tannins, highly correlated with the perception of astringency in fruits. Tannins vary in size, from dimers up to oligomers, with more than 30 subunits [83]. Polymer size affects astringency correlating positively with the perception of astringency [84]. Increased galloylation can be responsible for increased "abrasiveness" while trihydroxylation of the B-ring can decrease it [85]. As referred by He et al. [86], the synthesis of astringent substances controlled by a variety of structural and regulatory genes must be studied. Moreover, these authors state that "(…) cloning and functional identification of genes, in the astringency metabolic pathway, and their spatio-temporal expression patterns as well as tannin biosynthesis-related transcription factor genes must be considered in future work to finally make it possible to control fruit astringent substances quantitatively (…)"[86].

4. Effect of microorganisms on fruit phenolic compounds

After the consumption of fruits, the colon is the main site of microbial fermentation, where high molecular weight phenolic compounds are transformed into low molecular weight phenolic compounds such as phenolic acids or lactone structures by intestinal microbiota. The human healthy adult gut microbiota already identified can be classified into three dominant phyla: *Bacteroidetes, Firmicutes* and *Actinobacteria*. This highly complex and diverse bacterial ecosystem is mainly composed by a dominant group(> 109 Colony Forming Units (CFU)/g) of anaerobic bacteria, including genera *Bacteroides, Eubacterium, Bifidobacterium, Peptostreptococcus,*

Precursors			Major metabolites	Bacteria	Ref.
Flavonoids		Myricetin	2-(3,5-Dihydroxyphenyl) acetic acid 2-(3-Hydroxyphenyl) acetic acid	*Clostridium orbiscidens, Eubacterium oxidoreducens*	[90–92]
		Quercetin	3-(3,4-Dihydroxyphenyl) propionic acid 3-(3-Hydroxyphenyl)propionic acid		[91–93]
		Kaempferol	2-(4-Hydroxyphenyl)propionic acid 2-(3,4-Dihydroxyphenyl)acetic acid 2-(3-Hydroxyphenyl)acetic acid		[90]
	Flavan-3-ols	Catechin	3-(3-Hydroxyphenyl)propionic acid 5-(3',4'-Dihydroxyphenyl)-γ-valerolactone	*Clostridium coccoides, Bifidobacterium* spp.	[94–97]
		Epicatechin	5-(3,4-Dihydroxyphenyl) valeric acid 3-(3,4-Dihydroxyphenyl)propionic acid		
		Epigallocatechin	5-(3',4'-Dihydroxypheny[l)-γ-valerolactone 5-(3',5'-Dihydroxyphenyl)-γ-valerolactone		
	Anthocyanins	Malvidin	3,4-Dimethoxybenzoic acid	*Lactobacillus (plantarum, casei, acidophilus LA-5) Bifidobacterium lactis BB-12*	[98, 99]
		Cyanidin	3,4-Dihydroxybenzoic acid		
		Peonidin	3-Methoxy4-hydroxybenzoic acid		
		Pelargonidin	4-Hydroxybenzoic acid		
Non-flavonoids	Hydroxycinnamates	Caffeic, ferulic, and *p*-coumaric acids linked to a quinic acid to form, respectively, caffoylquinic feruloylquinic, and *p*-coumaroylquinic acids	3-Hydroxyphenyl propionic acid Benzoic acid 3-(4-Hydroxyphenyl) propionic acid Vanillin	*Escherichia coli, Bifidobacterium lactis, Lactobacillus gasseri*	[100–102]

Table 3. Major metabolites resulting by phenolic compounds (flavonoids and non-flavonoids) biodegradation and bacteria implicated in their transformation (adapted from Marín et al. [88]).

Ruminococcus, Clostridium and Propionibacterium, and sub-dominant groups (< 109CFU/g), of bacteria of the Enterobacteriaceae family, especially E. *coli*, and the genera *Streptococcus, Enterococcus, Lactobacillus, Fusobacterium, Desulfovibrio* and *Methanobrevibacter* [89]. Thus, the microbial metabolism (**Table 3**) of most of the phenolic classes such as flavonoids, isoflavonoids, lignans, phenolic acids, and tannins may produce metabolites with biological activity, presenting increased antioxidant activity, with evidence on health benefits for consumers. As most dietary polyphenolic compounds occur in glycosylated form in plants [87], for acquiring bioactivity in human body after being absorbed at enterocytes, these compounds must suffer various intestinal transformations, including the activities of digestive and microbial enzymes [88]. After cleavage of sugar responsible for glycosylation, the final absorbed compounds enter the vein circulation toward liver (**Figure 4**). Other enzymatic transformations occur from the liver to other organs, including digestive tract or via blood being excreted by urine [88].

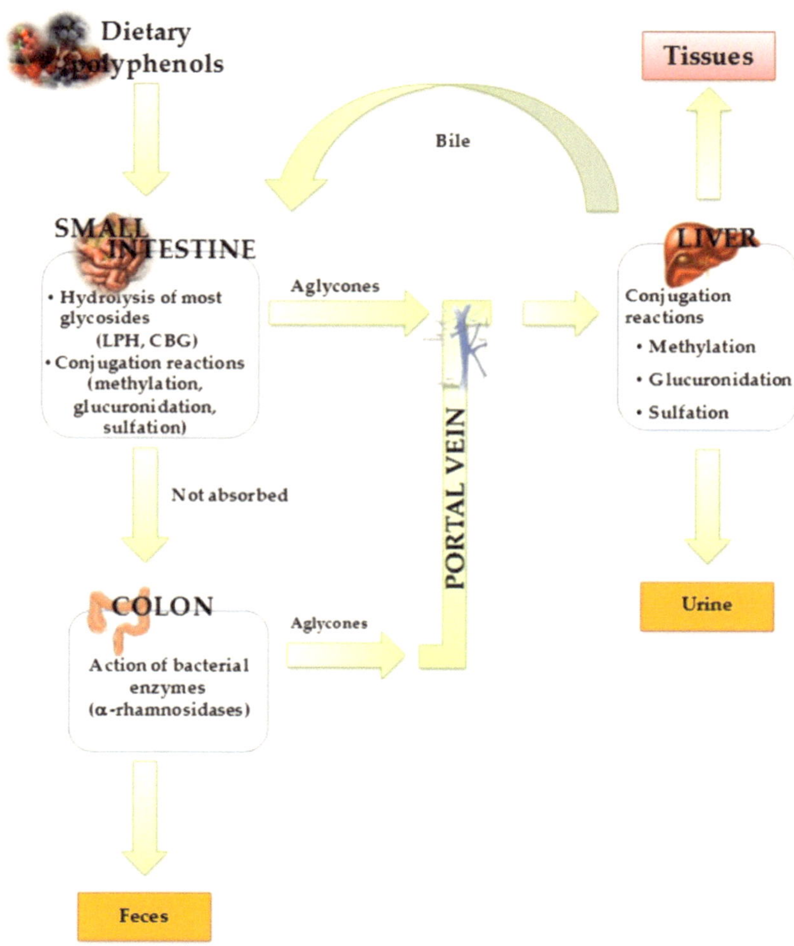

Figure 4. Absorption and metabolism routes for dietary polyphenols and their derivatives in humans. Adapted from Marín et al. [88].

5. Final remark

Red/dark-colored fruits are considered healthy and nutritious, the major potential health benefits being a reduced risk for cardiovascular and neurodegenerative diseases. Phytochemicals from red/dark-colored fruits are also shown to prevent body weight gain, lower blood cholesterol, and reduce cancer risk. Nevertheless, further rigorous, prospective studies are needed in order to better understand the benefits included in red/dark-colored fruits in our diet. There is also an emergent interest in the study of red/dark-colored fruits astringency because of the healthy properties of astringent substances found in red/dark-colored fruits including antibacterial, antiviral, anti-inflammatory, antioxidant, anticarcinogenic, antiallergenic, hepatoprotective, and vasodilating. The role of phenolic compounds and their metabolites as prebiotics, contributing to beneficial gastrointestinal health effects by modulating gut microbial balance with the simultaneous inhibition of pathogens and stimulation of beneficial bacteria, should also be highlighted.

Author details

Fernanda Cosme[1]*, Berta Gonçalves[2], Eunice A. Bacelar[2], António Inês[1], António M. Jordão[3] and Alice Vilela[1]

*Address all correspondence to: fcosme@utad.pt

1 Department of Biology and Environment, School of Life Sciences and Environment, CQ-VR, Chemistry Research Centre, University of Trás-os-Montes and Alto Douro, Vila Real, Portugal

2 Department of Biology and Environment, School of Life Sciences and Environment, CITAB, Centre for the Research and Technology of Agro-Environmental and Biological Sciences, University of Trás-os-Montes and Alto Douro, Vila Real, Portugal

3 Polytechnic Institute of Viseu (CI&DETS), Agrarian Higher School, Estrada de Nelas, Quinta da Alagoa, Ranhados, Viseu, Portugal

References

[1] Arts IC, Hollman PC: Polyphenols and disease risk in epidemiologic studies. American Journal of Clinical Nutrition, 2005, **81** (1 Suppl.): 317S–325S.

[2] Oliveira LL, Carvalho MV, Melo L: Health promoting and sensory properties of phenolic compounds in food. Revista Ceres, 2014, **61** (Suppl.): 764–779. DOI: 10.1590/0034-737x201461000002

[3] Kårlund A, Moor U, Sandell M, Karjalainen RO: The impact of harvesting, storage and processing factors on health-promoting phytochemicals in berries and fruits. Processes, 2014, **2**: 596–624. DOI: 10.3390/pr2030596

[4] Jordão AM, Ricardo-da-Silva JM, Laureano O: Evolution of proanthocyanidins in bunch stems during berry development (*Vitis vinifera* L.). Vitis, 2001, **40**: 17–22.

[5] Jordão AM, Ricardo-da-Silva JM, Laureano O: Evolution of catechins and oligomeric procyanidins during grape maturation of Castelão Francês and Touriga Francesa. American Journal of Enology and Viticulture, 2001, **53**: 231–234.

[6] Pastrana-Bonilla E, Akoh CC, Sellappan S, Krewer G: Phenolic content and antioxidant capacity of muscadine grapes. Journal of Agricultural and Food Chemistry, 2003, **51**: 5497–5503. DOI: 10.1021/jf030113c

[7] Cosme F, Ricardo-Da-Silva JM, Laureano O: Tannin profiles of *Vitis vinifera* L. cv. red grapes growing in Lisbon and from their monovarietal wines. Food Chemistry, 2009, **112**: 197–204. DOI: 10.1016/j.foodchem.2008.05.058

[8] Costa E, Cosme F, Rivero-Pérez MD, Jordão AM, González-SanJosé ML: Influence of wine region provenance on phenolic composition, antioxidant capacity and radical scavenger activity of traditional Portuguese red grape varieties. European Food Research and Technology, 2015, **241**: 61–73. DOI: 10.1007/s00217-015-2434-x

[9] Eyduran SP, Akin M, Ercisli S, Eyduran E, Maghradze D: Sugars, organic acids, and phenolic compounds of ancient grape cultivars (*Vitis vinifera* L.) from Igdir province of Eastern Turkey. Biological Research, 2015, **48**: 1–8. DOI: 10.1186/0717-6287-48-2

[10] Meng J, Fang Y, Zhang A, Chen S, Xu T, Ren Z, Han G, Liu J, Li H, Zhang Z, Wang H: Phenolic content and antioxidant capacity of Chinese raisins produced in Xinjiang Province. Food Research International, 2011, **44**: 2830–2836. DOI: 10.1016/j.foodres.2011.06.032

[11] Ghrairi F, Lahouar L, Amira A, Brahmi F, Ferchichi A, Achour L, Said S: Physicochemical composition of different varieties of raisins (*Vitis vinifera* L.) from Tunisia. Industrial Crops and Products, 2013, **43**: 73–77. DOI: 10.1016/j.indcrop.2012.07.008

[12] Sério S, Rivero-Pérez MD, Correia AC, Jordão AM, González-San José ML: Analysis of commercial grape raisins: phenolic content, antioxidant capacity and radical scavenger activity. Ciência e Técnica Vitivinícola, 2014, **29**: 1–8. DOI: 10.1051/ctv/20142901001

[13] Jordão AM, Ricardo-da-Silva JM, Laureano O: Influence of irrigation on the phenolic composition of red grapes from the Touriga Francesa variety (*Vitis vinifera* L.). CyTA - Journal of Food, 1998, 2: 60–73. DOI: 10.1080/11358129809487584.

[14] Pérez-Magariño S, González-Sanjosé ML: Polyphenols and colour variability of red wines made from grapes harvested at different ripeness grade. Food Chemistry, 2006, **96**: 197–208. DOI: 10.1016/j.foodchem.2005.02.021

[15] Costa E, Da Silva JF, Cosme F, Jordão AM: Adaptability of some French red grape varieties cultivated at two different Portuguese terroirs: comparative analysis with two Portuguese red grape varieties using physicochemical and phenolic parameters. Food Research International, 2015, **78**: 302–312. DOI: 10.1016/j.foodres.2015.09.029

[16] Monagas M, Gómez-Cordovés C, Bartolomé B, Laureano O, Ricardo-da-Silva JM: Monomeric, oligomeric, and polymeric flavan-3-ol composition of wines and grapes from *Vitis vinifera* L. Cv. Graciano, Tempranillo, and Cabernet Sauvignon. Journal of Agricultural and Food Chemistry, 2003, **51**: 6475–6481. DOI: 10.1021/jf030325+

[17] Obreque-Slier E, López-Solís R, Castro-Ulloa L, Romero-Díaz C, Peña-Neira A: Phenolic composition and physicochemical parameters of Carménère, Cabernet Sauvignon, Merlot and Cabernet Franc grape seeds (*Vitis vinifera* L.) during ripening. LWT - Food Science and Technology, 2012, **48**: 134–141. DOI: 10.1016/j.lwt.2012.02.007

[18] He JJ, Liu YX, Pan QH, Cui XY, Duan CQ: Different anthocyanin profiles of the skin and the pulp of Yan73 (Muscat Hamburg× Alicante Bouschet) grape berries. Molecules, 2010, **15**: 1141–1153. DOI: 10.3390/molecules15031141

[19] Ubalde JM, Sort X, Zayas AA, Poch RM: Effects of soil and climatic conditions on grape ripening and wine quality of Cabernet Sauvignon. Journal of Wine Research, 2010, **21**: 1–17. DOI: 10.1080/09571264.2010.495851

[20] Mateus N, Machado JM, De Freitas V: Development changes of anthocyanins in *Vitis vinifera* grapes grown in the Douro Valley and concentration in respective wines. Journal of the Science of Food and Agriculture, 2002, **82**: 1689–1695. DOI: 10.1002/jsfa.1237

[21] Bergqvist J, Dokoozlian N, Ebisuda N: Sunlight exposure and temperature effects on berry growth and composition of Cabernet Sauvignon and Grenache in the Central San Joaquin Valley of California. American Journal of Enology and Viticulture, 2001, **52**: 1–7.

[22] Downey MO, Dokoozlian NK, Krstic MP: Cultural practice and environmental impacts on the flavonoid composition of grapes and wine: a review of recent research. American Journal of Enology and Viticulture, 2006, **57**: 257–268.

[23] Madeira FC: Effect of ultraviolet radiation b on vine-growing behavior of the variety Cabernet Sauvignon (*Vitis Vinífera* L.) grown on São Joaquim, Santa Catarina. Master dissertation (Master in Plant Genetic Resources), University Federal of Santa Catarina, Florianópolis, Brazil, 2011.

[24] Bruno G, Sparapano L: Effects of three esca-associated fungi on *Vitis vinifera* L: V. Changes in the chemical and biological profile of xylem sap from diseased cv. Sangiovese vines. Physiological and Molecular Plant Pathology, 2007, **71**: 210–229. DOI: 10.1016/j.pmpp.2008.02.005

[25] Gao L, Mazza G: Characterization, quantification and distribution of anthocyanins and colorless phenolics in sweet cherry. Journal of Agricultural and Food Chemistry, 1995, **43**: 343–346. DOI: 10.1021/jf00050a015

[26] Gonçalves B, Landbo AK, Knudsen D, Silva AP, Moutinho-Pereira J, Rosa E: Effect of ripeness and postharvest storage on the phenolic profiles of cherries (*Prunus avium* L.). Journal of Agricultural and Food Chemistry, 2004, **52**: 523–530. DOI: 10.1021/jf030595s

[27] Gonçalves B, Silva AP, Moutinho-Pereira J, Bacelar E, Rosa E, Meyer AS: Effect of ripeness and postharvest storage on the evolution of colour and anthocyanins in cherries (*Prunus avium* L.). Food Chemistry, 2007, **103**: 976–984. DOI: 10.1016/j.foodchem.2006.08.039

[28] Ballistreri G, Continella A, Gentile A, Amenta M, Fabroni S, Rapisarda P: Fruit quality and bioactive compounds relevant to human health of sweet cherry (*Prunus avium* L.) cultivars grown in Italy. Food Chemistry, 2013, **140**: 630–638. DOI: 10.1016/j.foodchem.2012.11.024

[29] Usenik V, Fajt N, Mikulic-Petkovsek M, Slatnar A, Stampar F, Veberic R: Sweet cherry pomological and biochemical characteristics influenced by rootstock. Journal of Agricultural and Food Chemistry, 2010, **58**: 4928–4933. DOI: 10.1021/jf903755b

[30] Liu Y, Liu X, Zhong F, Tian R, Zhang K, Zhang X, Li T: Comparative study of phenolic compounds and antioxidant activity in different species of cherries. Journal of Food Science, 2011, **76**: 633–638. DOI: 10.1111/j.1750-3841.2011.02150.x

[31] Jakobek L, Marijan S, Vóca S, Šindrak Z, Dobričević N: Flavonol and phenolic acid composition of sweet cherries (cv. Lapins) produced on six different vegetative rootstocks. Scientia Horticulturae, 2009, **123**: 23–28. DOI: 10.1016/j.scienta.2009.07.012

[32] Serrano M, Díaz-Mula HM, Zapata PJ, Castillo S, Guillén F, Martínez-Romero D: Maturity stage at harvest determines the fruit quality and antioxidant potential after storage of sweet cherry cultivars. Journal of Agricultural and Food Chemistry, 2009, **57**: 3240–3246. DOI: 10.1021/jf803949k

[33] Stöhr H, Mosel H, Herrmann KZ: VII. The phenolics of cherries and plums and the changes in catechins and hydroxycinnamic acid derivatives during the development of fruits. Lebensmittel-Untersuchung und Forschung, 1975, **159**: 85–91.

[34] Martínez-Esplá A, Zapata PJ, Valero D, García-Viguera C, Castillo S, Serrano M: Preharvest application of oxalic acid increased fruit size, bioactive compounds, and antioxidant capacity in sweet cherry cultivars (*Prunus avium* L.). Journal of Agricultural and Food Chemistry, 2014, **62**: 3432–3437. DOI: 10.1021/jf500224g

[35] Gonçalves B, Landbo AK, Let M, Silva AP, Rosa E, Meyer AS: Storage affects the phenolic profiles and antioxidant activities of cherries (*Prunus avium* L.) on human low-density lipoproteins. Journal of the Science of Food and Agriculture, 2004, **84**: 1013–1020. DOI: 10.1002/jsfa.1752

[36] Goulas V, Minas IS, Kourdoulas PM, Lazaridou A, Molassiotis AN, Gerothanassis IP, Manganaris GA: 1H NMR metabolic fingerprinting to probe temporal postharvest changes on qualitative attributes and phytochemical profile of sweet cherry fruit. Front. Plant Science, 2015, **6**: 1–11. DOI: 10.3389/fpls.2015.00959

[37] Esti M, Cinquanta L, Sinesio F, Moneta E, Di Matteo M: Physiochemical and sensory fruit characteristics of two sweet cherry cultivars after cool storage. Food Chemistry, 2002, **76**: 399–405. DOI: 10.1016/S0308-8146(01)00231-X

[38] Petriccione M, De Sanctis F, Pasquariello MS, Mastrobuoni F, Rega P, Scortichini M, Mencarelli F: The effect of chitosan coating on the quality and nutraceutical traits of sweet cherry during postharvest life. Food and Bioprocess Technology, 2014, 8(2): 1–15. DOI: 10.1007/s11947-014-1411-x

[39] Valverde JM, Giméneza MJ, Valero D, Martínez-Romero D, Serrano M: Methyl salicylate treatments of sweet cherry trees increase antioxidant systems in fruit at harvest and during storage. Postharvest Biology and Technology, 2015, **109**: 106–113. DOI: 10.1016/j.postharvbio.2015.06.011

[40] Kelebek H, Selli S: Evaluation of chemical constituents and antioxidant activity of sweet cherry (*Prunus avium* L.) cultivars. Journal of Food Science and Technology, 2011, **46**: 2530–2537. DOI: 10.1111/j.1365-2621.2011.02777.x

[41] Serra AT, Duarte RO, Bonze MR, Duarte CMM: Identification of bioactive response in traditional cherries from Portugal. Food Chemistry, 2011, **125**: 318–325. DOI: 10.1016/j.foodchem.2010.07.088

[42] Seeram NP, Momin RA, Nair MG, Bourquin LD: Cyclooxygenase inhibitory and antioxidant cyanidin glycosides in cherries and berries. Phytomedicine, 2001, **8**: 362–369. DOI: 10.1078/0944-7113-00053

[43] Kim DO, Heo HJ, Kim YJ, Yang HS, Lee CY: Sweet and sour cherry phenolics and their protective effects on neuronal cells. Journal of Agricultural and Food Chemistry, 2005, **53**: 9921–9927. DOI: 10.1021/jf0518599

[44] Matias AA, Rosado-Ramos R, Nunes LS, Figueira I, Serra AT, Bronze MR, Santos CN, Duarte C: Protective effect of a (poly)phenol-rich extract derived from sweet cherries culls against oxidative cell damage. Molecules, 2016, **21**: 1–16. DOI: 10.3390/molecules21040406

[45] Adisakwattana S, Yibchok-Anun S, Charoenlertkul P, Wongsasiripat N: Cyanidin-3-rutinoside alleviates postprandial hyperglycemia and its synergism with acarbose by inhibition of intestinal α-glucosidase. Journal of Clinical Biochemistry and Nutrition, 2011, **49**: 36–41. DOI: 10.3164/jcbn.10-116

[46] Xia M, Hou M, Zhu H, Ma J, Tang Z, Wang Q, Li Y, Chi D, Yu X, Zhao T, Han P, Xia X, Ling W: Anthocyanins induce cholesterol efflux from mouse peritoneal macrophages. The role of peroxisome proliferators-activated receptor Ó-Liver X Receptor α-ABCA1 Pathway. Journal of Biological Chemistry, 2005, **280**: 36792–36801. DOI: 10.1074/jbc.M505047200

[47] Srivastava A, Akoh CC, Fischer J, Krewer G: Effect of anthocyanin fractions from selected cultivars of Georgia-grown blueberries on apoptosis and phase II enzymes. Journal of Biological Chemistry, 2007, **55**: 3180–3185. DOI: 10.1021/jf062915o

[48] Vander Kloet SP, Dickinson TA: The taxonomy of *Vaccinium* section *Myrtillus* (*Ericaceae*). Brittonia, 1999, **51**: 231–254. DOI: 10.2307/2666632

[49] Routray W, Orsat V: Blueberries and their anthocyanins: factors affecting biosynthesis and properties. Comprehensive Reviews in Food Science and Food Safety, 2011, **10**: 303–320. DOI: 10.1111/j.1541-4337.2011.00164.x

[50] Martin RR, Polashock JJ, Tzanetakis E: New and emerging viruses of blueberry and cranberry. Viruses, 2012, **4**: 2831–2852. DOI: 10.3390/v4112831

[51] Liu RH: Health-promoting components of fruits and vegetables in the diet. Advances in Nutrition, 2013, **4**: 384S–392S. DOI: 10.3945/an.112.003517

[52] Villata M: Trends in world blueberry production. Available online: http://www.growingproduce.com/article/26272/trends-in-world-blueberry-production/2016 (accessed on 31 August 2016).

[53] US Department of Agriculture (USDA): 2016. Noncitrus Fruits and Nuts—2015 Summary. Available online: http://usda.mannlib.cornell.edu/usda/current/NoncFruiNu/NoncFruiNu-07-06-2016.pdf (accessed on 31 August 2016).

[54] Espin JC, Soler-Rivas C, Wichers HJ, Garcia-Viguera C: Anthocyanin-based natural colorants: a new source of antiradical activity for foodstuff. Journal of Agricultural and Food Chemistry, 2000, **48**: 1588–1592. DOI: 10.1021/jf9911390

[55] Stevenson D, Scalzo J: Anthocyanin composition and content of blueberries from around the world. Journal of Berry Research, 2012, **2**: 179–189. DOI: 10.3233/JBR-2012-038

[56] Nile SH, Park SW: Edible berries: bioactive components and their effect on human health. Nutrition, 2014, **30**: 134–144. DOI: 10.1016/j.nut.2013.04.007

[57] Yousef GG, Brown AF, Funakoshi Y, Mbeunkui F, Grace MH, Ballington JR, Loraine A, Lila MA: Efficient quantification of the health-relevant anthocyanin and phenolic acid profiles in commercial cultivars and breeding selections of blueberries (*Vaccinium* spp.). Journal of Agricultural and Food Chemistry, 2013, **61**: 4806–4815. DOI: 10.1021/jf400823s

[58] Huang WY, Liu YM, Wang J, Wang XN, Li CY: Anti-inflammatory effect of the blueberry anthocyanins malvidin-3-glucoside and malvidin-3-galactoside in endothelial cells. Molecules, 2014, **19**: 12827–12841. DOI: 10.3390/molecules190812827

[59] Borges G, Degeneve A, Mullen W, Crozier A: Identification of flavonoid and phenolic antioxidants in black currants, blueberries, raspberries, red currants, and cranberries. Journal of Agricultural and Food Chemistry, 2010, **58**: 3901–3909. DOI: 10.1021/jf902263n

[60] Cho MJ, Howard LR, Prior RL, Clark JR: Flavonoid glycosides and antioxidant capacity of various blackberry, blueberry and red grape genotypes determined by high-performance liquid chromatography/mass spectrometry. Journal of the Science of Food and Agriculture, 2004, **84**: 1771–1782. DOI: 10.1002/jsfa.1885

[61] Troszyñska A, Estrella I, Lamparski G, Hernández T, Amarowicz R, Pegg RB: Relationship between the sensory quality of lentil (*Lens culinaris*) sprouts and their phenolic constituents. Food Research International, 2011, **44**: 3195–3201. DOI: 10.1016/j.foodres.2011.08.007

[62] Blanda G, Cerretani L, Bendini A, Cardinali A, Scarpellini A, Lercker G: Effect of vacuum impregnation on the phenolic content of Granny Smith and Stark Delicious frozen apple cv. European Food Research and Technology, 2008, **226**: 1229–1237. DOI: 10.1007/s00217-007-0624-x

[63] Blanda G, Cerretani L, Cardinali A, Barbieri S, Bendini A, Lercker G: Osmotic dehydrofreezing of strawberries: polyphenolic content, volatile profile and consumer acceptance. LWT – Food Science and Technology, 2009, **42**: 30–36. DOI: 10.1016/j.lwt.2008.07.002

[64] Barrett DM, Beaulieu J, Shewfelt R: Color, flavor, texture, and nutritional quality of fresh-cut fruits and vegetables: desirable levels, instrumental and sensory measurement, and the effects of processing. Critical Reviews in Food Science and Nutrition, 2010, **50**: 369–389. DOI: 10.1080/10408391003626322.

[65] Kasım MU, Kasım R: Color quality of fresh-cut fruits and vegetables. Conference paper in the Sixth International Scientific Agricultural Symposium "Agrosym 2015", Jahorina, Bosnia and Herzegovina, October 15–18. Book of Proceedings, 2015, **47**: 1007–1014.

[66] El Gharras H: Polyphenols: food sources, properties and applications – a review. International Journal of Food Science and Technology, 2009, **44**: 2512–2518. DOI: 10.1111/j.1365-2621.2009.02077.x

[67] Just the Berries. What Are Blackcurrants? A Brief History of the World's Best Super-Fruit. Available online: http://jtbpd.com/information.php?information_id=1. 2013 (accessed on 22 August 2016).

[68] Cadot Y, Caillé S, Samson A, Barbeau G, Cheynier V: Sensory representation of typicality of Cabernet franc wines related to phenolic composition: impact of ripening stage and maceration time. Analytica Chimica Acta, 2012, **732**: 91–99. DOI: 10.1016/j.aca.2012.02.013

[69] Spence C: On the psychological impact of food colour. Flavour, 2015, **4**: 16. DOI: 10.1186/s13411-015-0031-3

[70] Piqueras-Fizman B, Spence C: Sensory expectations based on product-extrinsic food cues: an interdisciplinary review of the empirical evidence and theoretical accounts. Food Quality and Preference, 2015, **40**: 165–179. DOI: 10.1016/j.foodqual.2014.09.013

[71] Spence C: Multisensory flavour perception. Current Biology (special issue), 2013, **23**: R365–R369. DOI: 10.1016/j.cub.2013.01.028

[72] Gavrilova V, Kajdzanoska M, Gjamovski V, Stefova M: Separation, characterization and quantification of phenolic compounds in blueberries and red and black currants by HPLC-DAD-ESI-MSn. Journal of Agricultural and Food Chemistry, 2011, **59**: 4009–4018. DOI: 10.1021/jf104565y

[73] Honkanen E, Pyysalo T, Hirvi T: The aroma of Finnish wild raspberries, *Rubus idaeus*. Zeitschrift für Lebensmittel-Untersuchung und Forschung, 1980, **171**: 180–182. DOI: 10.1007/BF01042646

[74] Aprea E, Biasioli F, Gasperi F: Volatile compounds of raspberry fruit: from analytical methods to biological role and sensory impact (Review). Molecules, 2015, **20**: 2445–2474. DOI: 10.3390/molecules20022445

[75] Bravo L: Polyphenols: chemistry, dietary sources, metabolism, and nutritional significance. Nutrition Reviews, 1998, **56**: 317–333.

[76] Rouseff RL: Bitterness in food products: an overview. In: Bitterness in foods and beverages. Developments in Food Science, Rouoseff R.L., ed. 1990, 25. Amsterdam: Elsevier, pp. 1–14.

[77] Vilela A, Cosme F, Jordão AM: Wine phenolics: looking for a smooth mouthfeel (Review Article). SDRP Journal of Food Science &' Technology, 2015, **1**: 8.

[78] Soares S, Kohl S, Thalmann S, Mateus N, Meyerhof W, De Freitas V: Different phenolic compounds activate distinct human bitter taste receptors. Journal of Agricultural and Food Chemistry, 2013, **61**: 1525–1533. DOI: 10.1021/jf304198k

[79] Noble CA: Bitterness in wine. Physiology and Behavior, 1994, **56**: 1251–1255.

[80] Ma W, Guo A, Zhang Y, Wang H, Liu Y, Li H: A review on astringency and bitterness perception of tannins in wine. Trends in Food Science and Technology, 2014, **40**: 6–19. DOI: 10.1016/j.tifs.2014.08.001

[81] Preys S, Mazerolles G, Courcoux P, Samson A, Fischer U, Hanafi M, Bertrand D, Cheynier V: Relationship between polyphenolic composition and some sensory properties in red wines using multiway analyses. Analytica Chimica Acta, 2006, **563**: 126–136. DOI: 10.1016/j.aca.2005.10.082

[82] Quijada-Morin N, Regueiro J, Simal-Gandara J, Tomas E, Rivas-Gonzalo JC, Escribano-Bailon MT: Relationship between the sensory-determined astringency and the flavanolic composition of red wines. Journal of Agricultural and Food Chemistry, 2012, **60**: 12355–12361. DOI: 10.1021/jf3044346

[83] Adams DO: Phenolics and ripening in grape berries. American Journal of Enology and Viticulture, 2006, **57**: 249–256.

[84] Vidal S, Francis L, Guyot S, Marnet N, Kwiatkowski M, Gawel R, Cheynier V, Waters EL: The mouth-feel properties of grape and apple proanthocyanidins in a wine-like medium. Journal of the Science of Food and Agriculture, 2003, **83**: 564–573. DOI: 10.1002/jsfa.1394

[85] Peleg H, Gacon K, Schlich P, Noble AC: Bitterness and astringency of flavan-3-ol monomers, dimers and trimers. Journal of the Science of Food and Agriculture, 1999, **79**: 1123–1128. DOI: 10.1002/(SICI)1097-0010(199906)79:8<1123::AID-JSFA336>3.0.CO;2-D

[86] He M, Tian H, Luo X, Qi X, Chen X: Molecular progress in research on fruit astringency (Review). Molecules, 2015, **20**: 1434–1451. DOI:10.3390/molecules20011434

[87] Cardona FC, Andrés-Lacueva S, Tulipani F, Tinahones J, Queipo-Ortuño MI: Benefits of polyphenols on gut microbiota and implications in human health. Journal of Nutritional Biochemistry, 2013, **24**: 1415–1422. DOI: 10.1016/j.jnutbio.2013.05.001

[88] Marín L, Miguélez EM, Villar CJ, Lombó F: Bioavailability of dietary polyphenols and gut microbiota metabolism: antimicrobial Properties. BioMed Research International, 2015, **905215**: 18. DOI: 10.1155/2015/905215

[89] Mariat D, Firmesse O, Levenez F: The Firmicutes/Bacteroidetes ratio of the human microbiota changes with age. BMC Microbiology, 2009, **9**: 123. DOI: 10.1186/1471-2180-9-123

[90] Winter J, Moore LH, Dowell Jr VR, Bokkenheuser VD: C-ring cleavage of flavonoids by human intestinal bacteria. Applied and Environmental Microbiology, 1989, **55**: 1203–1208. DOI: 0099-2240/89/051203-06$02.00/0

[91] Krumholz LR, Bryant MP: *Eubacterium oxidoreducens* sp. nov. requiring H_2 or formate to degrade gallate, pyrogallol, phloroglucinol and quercetin. Archives of Microbiology, 1986, **144**: 8–14. DOI: 10.1007/BF00454948

[92] Schneider H, Simmering R, Hartmann L, Pforte H, Blaut M: Degradation of quercetin-3-glucoside in gnotobiotic rats, associated with human intestinal bacteria. Journal of Applied Microbiology, 2000, **89**: 1027–1037. DOI: 10.1046/j.1365-2672.2000.01209.x

[93] Rechner AR, Smith MA, Kuhnle G, Gibson GR, Debnam ES, Srai SKS, Moore KP, Rice-Evans CA: Colonic metabolism of dietary polyphenols: influence of structure on microbial fermentation products. Free Radical Biology and Medicine, 2004, **36**: 212–225. DOI: 10.1016/j.freeradbiomed.2003.09.022

[94] Monagas M, Urpi-Sarda M, Sanchez-Patán F, Gómez-Cordovés C, Andres-Lacuevab C, Bartolomé B: Insights into the metabolism and microbial biotransformation of dietary flavan-3-ols and the bioactivity of their metabolites. Food and Function, 2010, **1**: 233–253. DOI: 10.1039/C0FO00132E

[95] Tzounis X, Vulevic J, Kuhnle GGC, George T, Leonczak J, Gibson GR, Kwik-Uribe C, Spencer JPE: Flavanol monomer-induced changes to the human faecal microflora. British Journal of Nutrition, 2008, **99**: 782–792. DOI: 10.1017/S0007114507853384

[96] Schneider H, Blaut M: Anaerobic degradation of flavonoids by *Eubacterium ramulus*. Archives of Microbiology, 2000, **173**: 71–75.

[97] Yamakoshi J, Tokutake S, Kikuchi M, Kubota Y, Konishi H, Mitsuoka T: Effect of proanthocyanidin-rich extract from grape seeds on human fecal flora and fecal odor. Microbial Ecology in Health and Disease, 2001, **13**: 25–31. DOI: 10.3402/mehd.v13i1.7996

[98] Miladinović B, Kostić M, Šavikin K, Đordević B, Mihajilov-Krstev T, Živanović S, Kitić D: Chemical profile and antioxidative and antimicrobial activity of juices and extracts of 4 black currants varieties (*Ribes nigrum* L.). Journal of Food Science, 2014, **79**: C301–C309. DOI: 10.1111/1750-3841.12364

[99] Burdulis D, Sarkinas A, Jasutiené I, Stackevicené E, Nikolajevas L, Janulis V: Comparative study of anthocyanin composition, antimicrobial and antioxidant activity in bilberry (*Vaccinium myrtillus* L.) and blueberry (*Vaccinium corymbosum* L.) fruits. Acta Poloniae Pharmaceutica—Drug Research, 2009, **66**: 399–408.

[100] Gonthier MP, Remesy C, Scalbert A, Cheynier V, Souquet JM, Poutanen K, Aura AM: Microbial metabolism of caffeic acid and its esters chlorogenic and caftaric acids by human faecal microbiota in vitro. Biomedicine and Pharmacotherapy, 2006, **60**: 536–540. DOI: 10.1016/j.biopha.2006.07.084

[101] Andreasen MF, Kroon PA, Williamson G, Garcia-Conesa MT: Esterase activity able to hydrolyze dietary antioxidant hydroxycinnamates is distributed along the intestine of mammals. Journal of Agricultural and Food Chemistry, 2001, **49**: 5679–5684. DOI: 10.1021/jf010668c

[102] Couteau D, McCartney AL, Gibson GR, Williamson G, Faulds CB: Isolation and characterization of human colonic bacteria able to hydrolyse chlorogenic acid. Journal of Applied Microbiology, 2001, **90**: 873–881. DOI: 10.1046/j.1365-2672.2001.01316.x

Chapter 11

Perspective on Co-feeding of Phenolic Compounds into Existing Refinery Units

Thuan Minh Huynh, Udo Armbruster and
Andreas Martin

Additional information is available at the end of the chapter

http://dx.doi.org/10.5772/66883

Abstract

Replacement of fossil materials by renewable feedstocks is forced by depletion and environmental concerns but requires new technologies for energy generation or production of chemicals. Co-processing of petroleum with renewable feedstocks in current refinery infrastructure is an attractive option in the mid-term to increase renewable fuel capacity, as the capital investment and operational costs would be marginal. In this chapter, various strategies for admixing of phenolic compounds as renewable feeds into standard refineries are described. Starting from the role of renewable resources (e.g. biomass, lignin and bio-oil) in the current and future energy and chemical community, an overview on the present energy supply situation and the role of phenolic compounds are discussed. Later, a summary on co-feeding of phenolic model compounds with conventional feeds in refineries are illustrated. The co-processing of upgraded bio-oil in refinery units [e.g. fluid catalytic cracking (FCC), hydrotreating] is summarized, showing the potential utilisation of bio-feeds via such processes. Finally, some concluding remarks address the perspectives for further research and development to overcome future challenges.

Keywords: bio-feeds, bio-oil, co-feeding, lignin, refinery

1. Introduction

Fossil fuel (e.g. crude oil, coal and natural gas) reserves are limited, but they still share a significant proportion in the worldwide energy consumed (i.e. more than 85% in 2014). Particularly, 86% and 81% of primary energy in the US and Germany are from those sources in 2014, respectively [1]. A minor portion stems from other resources (e.g. nuclear and hydroelectric power, wind, solar, geothermal and biomass) [2]. The current share of renewable feedstock supplied to chemical industry looks similar, e.g. only 8–10% of the raw materials of the

European chemical industry are bio-based. It is projected that energy demand increases in the coming decades in spite of improved energy efficiency. Power plants based on photovoltaics and wind energy will continuously emerge for primary energy supply. At the same time, demand for transportation fuels will grow, but the production of renewable fuels is an even more challenging task. No single renewable source can provide sufficient energy to close the gap between the supply and demand of energy.

Another driving force for replacing petroleum-derived liquid fuels is the concerns about environmental pollution, as the production and combustion of fossil fuel add more CO_2, SO_x and NO_x to the atmosphere. Hence, there is a strong motivation for research on alternatives for fossil fuels. Many researchers have recently turned attention to the massive biomass resources due to several reasons. First, some types of biomass like vegetable oils already fit quite well into the present carbon-based fuel infrastructure. Second, biomass production is based on short-time carbon cycles and overall CO_2 neutral. Additionally, biomass is a cheap, abundant and sustainable raw material. Moving the world market dependence away from fossil-based resources to renewable ones will definitely contribute to the climate protection and sustainable economy [3–5].

Current production of first-generation biofuels (e.g. bioethanol and biodiesel) and blending in conventional fuels up to 10 vol% are steps in the right direction. However, the use of edible oils and seeds for the biofuels might compete with the food value chain, affecting material availability and prices. Furthermore, only part of biomass is converted into fuels. Consequently, the next step aims at the utilisation of complete biomass, leading to second-generation biofuels. The access to biofuels from biomass resources offered by forestry, agriculture and industry have great potential for the production of fuels and chemicals [6]. As a result, the governments of many countries have set ambitious goals and set the mandatory legislation for partly replacing fossil fuels to promote the implementation of renewable energy, e.g. the U.S Department of Energy sets a target to expect use 20% of transportation fuel from biomass.

The three most important plant biomass constituents are as follows: (i) cellulose, a polymer of glucose; (ii) hemicellulose, also a polymer of different sugars; and (iii) lignin, a highly aromatic polymer consisting of an irregular array of variously hydroxyl- and methoxy-substituted phenylpropane units. Such biomass has low volumetric and energy densities, resulting in high costs for collecting and transportation. As a result, converting biomass either chemically or thermally into liquid crudes is necessary as a first step. Fast pyrolysis (FP) or liquefaction (LF) seems to be potential technologies for liquefying biomass. Usually, such crudes possess oxygen contents varying in a range of 35–45 wt%, which has to be lowered prior to any use as a transportation fuel. Otherwise undesired properties like low specific energy content or limited shelf life will be serious drawbacks for application as fuels compared to conventional fuels.

Fortunately, the processes for upgrading such crudes already exist. Petroleum industry is mature all over the world and the use of the existing infrastructure (e.g. storage, refining units, blending and distribution systems) for production of biofuels requires little capital investment cost. As a result, research and development of the co-processing of biomass-derived feeds into refinery have been proposed. Three insertion points have been proposed: (i) feeding into crude oil before the crude distillation units; (ii) blending in near finished fuel and (iii) feeding into facilities within the refinery. The first option might be ruled out as the separation in distillation units does not chemically alter the materials and the oxygen-containing contaminants would be spread throughout the refinery. The second option requires converting the biomass into blending components which must meet all standards for transportation fuels. This is really challenging and needs higher costs. The last option receives more and more attention from academia and industrial partners, as various material streams are usually processed in a refinery and different bio-crudes with similar properties can be fed to the most suited unit operation.

This book chapter summarizes the main aspects involved in the co-feeding of liquefied lignocellulosic biomass feedstock based on phenolic compounds together with conventional hydrocarbon feeds into standard refinery units.

2. Overview of bio-feeds and conventional feed for standard refinery

Bio-feeds can be generally categorized based on the following sources: (i) food crops such as corn, wheat, barley, sugar crops, vegetable oils and hydrocarbon plants; (ii) waste materials such as agricultural residues, wood, urban wastes and crop residues; and (iii) aquatic biomass such as algae and seaweed. The use of biomass-derived feedstocks for a petroleum refinery can be classified into three categories according to the sources: lignocellulosic biomass, starch- and sugar-derived biomass (or edible biomass) and triglyceride-based biomass. There are several issues to identify what kind of bio-feeds is suited for refinery, among which price, availability and conversion costs play important roles. Generally, the cost of biomass increases in the order: lignocellulosic biomass < starch (and sugar)-based biomass < triglyceride-based biomass. However, the investment cost of conversion technology raises in the reverse order [7]. Naturally, the cost is also linked to supply and demand and thus finding new uses for biomass-derived products will result in higher prices.

For comparison between renewable and fossil feeds, hydrogen-to-carbon (H/C) and oxygen-to-carbon (O/C) atomic ratios are generally evaluated. Particularly, H/C ratios of crude oil are typically between 1.6 and 2.1 and the O/C ratios range between 0 and 0.03. In contrast, wood-based biomass typically has O/C and H/C ratios higher than 0.61 and 1.4, respectively. Of the biomass components, lignin is markedly different in structure and composition from hemicellulose and cellulose, being highly aromatic and containing less oxygen and is thus the one most similar to petroleum. Lignin has lower O/C and H/C ratios compared to wood-based biomass and thus making it to be a potential source for fuels production [8]. Naturally, lignin is a cross-linked macromolecule and consists of three basic monomers such as p-coumaryl alcohol, coniferyl alcohol and synapyl alcohol. Lignin from softwoods is mostly made-up of coniferyl alcohol-derived components, but lignin from hardwoods consists of mixtures of coniferyl- and syringyl-derived structures. Nowadays, the utilisation of lignin is continuously growing. Large amounts of lignin and lignin containing residues originate from the pulp and paper industry. The expected growth of the production capacity of second generation biofuels (e.g. bioethanol) from lignocellulosic biomass will lead to another source of lignin and lignin containing residues.

It should be highlighted that the complex nature of lignin polymer and its stability make it difficult to convert it into valuable monomeric chemicals. As mentioned above, FP or LF is widely used to convert biomass or lignin into liquid bio-crude or bio-oil. Under these conditions, biomass is converted into more than 200 oxygenated compounds, having various types of functional groups (e.g. acids, alcohols, phenols, sugars, aldehydes, ketones and esters) with specific chemistry. Lignin is preferably converted into phenolic compounds such as phenol, anisole, guaiacol, cresol and syringol. These compounds are highly recalcitrant to further treatment and require severe reaction conditions. As a result, such phenolic compounds have attracted attention as model compounds to develop effective treatment processes. **Figure 1** illustrates the structure of the three main biomass components and a variety of commonly detected monomeric oxygenates in bio-oil; in addition, phenolic dimers are also represented largely in lignin-derived bio-oil [9].

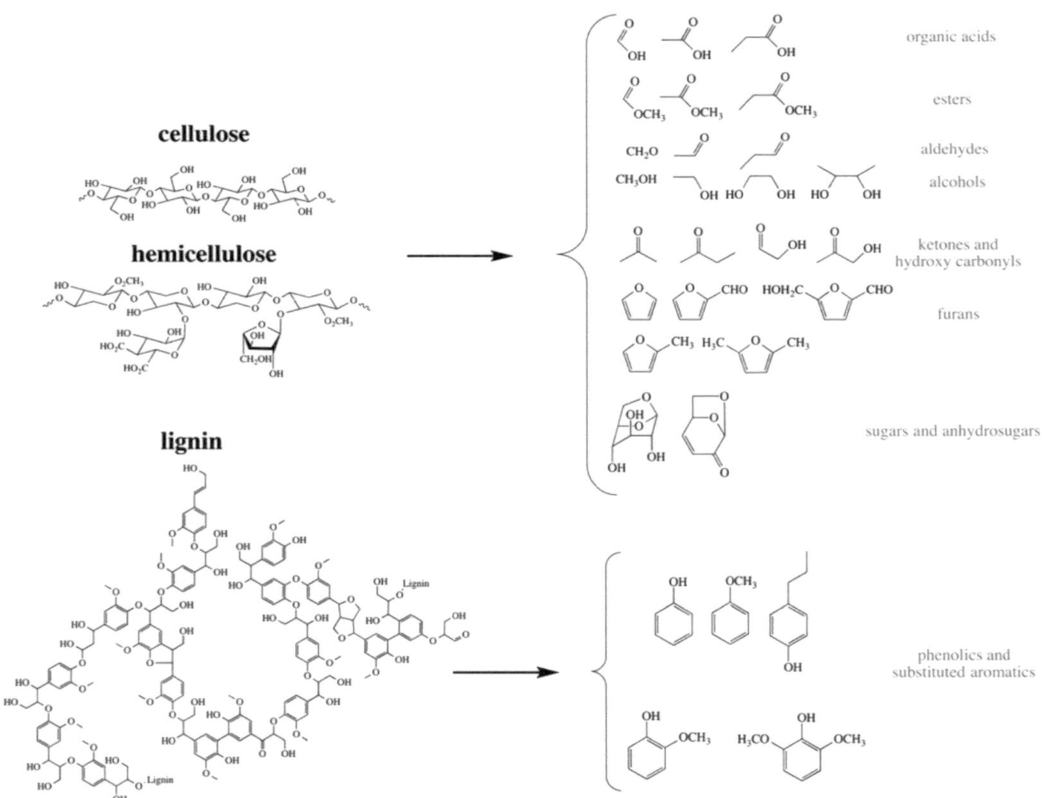

Figure 1. Typical products formed from FP of lignocellulosic biomass. Adapted from Ref. [9].

Details on the nature of conventional petroleum feeds and a block scheme of a typical refinery are presented elsewhere [10]. It should be noted that there are five major types of hydrocarbons in petroleum feedstocks such as paraffins, iso-paraffins, aromatics, naphthenens and olefins (PIANO). The main objective of refineries are (i) to transform crude oil into a set of refined products in accordance with precise specification and in quantities corresponding as closely as possible to the market requirement. For specific purpose, crude oil is first fractionated (distilled) into fractions with a specified range of carbon number. Following that, such large fractions (referred to gas oil and residue) are further processed in order to reduce molecular weight and to increase the H/C ratios.

It is suggested that refineries are well-suited to handle FP oil or phenolic compounds, in particular. However, the significant difference in the quality of biomass-derived liquids and petroleum feeds are obvious. For example, FP oil reveals a general sum formula of $CH_{1.4}O_{0.6}$ in contrast to hydrocarbon fuels, showing a sum formula close to CH_2. In addition, the higher heating values of FP or LF oils amount to approximately 16-34 MJ/kg, in contrast to heavy fuel oil that offers 40 MJ/kg (**Table 1**).

The different properties definitely cause some problems [9]: (i) the high oxygen content is not accommodated by refineries, usually dealing with oxygen contents in the crude oil far below 1 wt%; (ii) oxygenated compounds typically have higher boiling points than hydrocarbon with the same carbon number; (iii) water is considered a contaminant in conventional refineries; (iv) the acidity of FP oil is much higher than that of crude oil; and (v) the presence of various reactive oxygen-related functionalities allows thermal polymerization and might subsequently cause a high coking rate.

Therefore, downstream removal of the remaining oxygen from the bio-crude is needed; this can be done by using the existing refinery infrastructure or standalone units [12, 13]. Among

Properties	**FP oil**	**LF oil**	**Heavy fuel oil**
Water content (wt%)	15–30	5.1	0.1
pH	2.5	–	–
Specific gravity[a]	1.2	1.1	0.94
Elemental composition (wt%)			
Carbon	54–58	73	85
Hydrogen	5.5–7.0	8	11
Oxygen	35–40	16	1.0
Nitrogen	0–0.2	–	0.3
Ash	0–0.2	–	0.1
HHV (MJ/kg)	16–19	34	40

[a] Ratio of the density of the substance to the density of water.

Table 1. Typical properties of wood-based bio-oil (via FP, LF) compared to heavy fuel oil. Adapted from Ref. [11].

the available upgrading strategies, fluid catalytic cracking (FCC), hydrotreating, and hydrocracking supported by catalysts are considered as most effective technologies provided by the refinery [14–16]. However, these unit operations are tuned to upgrade fossil fuels. On the other side, recently developed standalone processes are definitely tailored to lower the oxygen content in biocrudes most effectively. They are often discussed as deoxygenation or hydrodeoxygenation (HDO) processes. A detailed review on the deoxygenation of liquefied biomass and related model compounds in standalone units have been reported in Ref. [17]. The focus of the present review is now set on the co-feeding of phenolic model compounds with hydrocarbons and later on blending of (pre-treated) bio-crudes with conventional refinery feeds. This latter strategy might represent a kind of third way, tailoring the bio-crudes to make them suited co-feeds and to benefit from existing technology.

3. Co-feeding of model compounds into existing refinery units

Several options are available for converting oxygen-containing biomass-derived feeds into biofuels in a petroleum refinery: (i) thermal conversion (e.g. visbreaker and coker); (ii) catalytic conversion (e.g. FCC, hydrotreating and hydrocracking) [18].

Nevertheless, the obtained organic liquid product from thermal units would contain a high fraction of oxygenates and thus those units seem to be unsuitable choices. In contrast, in presence of catalyst (FCC unit), catalytic cracking is much faster and more selective than thermal cracking and it allows working under milder reaction conditions.

The main objective of hydrotreating in conventional refineries is to remove impurities (e.g. sulphur, nitrogen and oxygen) being present in petroleum feedstock via the addition of hydrogen (hydrodesulfurisation = HDS, hydrodenitrogenation = HDN). Therefore, hydrotreating is also expected to remove the high content of oxygenates in bio-feeds. Hydrocracking, on the other hand, combines hydrotreating and catalytic cracking, thereby transforming hydrocarbon feedstocks in the presence of hydrogen into lighter products. Hydrocracking typically is carried out using other catalysts than for hydrotreating, and is run at more severe operating conditions (higher temperatures and pressures).

3.1. HDO of co-feed of phenolic model compounds with hydrocarbon

The individual HDO of bio-oil and related oxygenated model compounds has been studied extensively. In the past, this process was considered to provide hydrocarbon fractions that might be blended directly with conventional fuels. However, this needs huge efforts to achieve the necessary hydrogenation depth and oxygen removal efficiency. Recently, it is often discussed as a pre-treatment (or upgrading) step to make bio-crudes suited for co-processing. Details are summarized in Ref. [17] and related reviews [2, 19]. We also studied the HDO of phenol and intermediates on monometallic and bimetallic Ni-based catalysts (Ni, Ni-Co, Ni-Cu) supported on different acidic materials (H-ZSM-5, H-Beta, H-Y and ZrO_2) at comparatively mild conditions (250°C, 50 bar initial H_2 pressure) [20, 21]. Hydrocarbons (e.g. cyclohexane and benzene) can be mostly produced from deoxygenation of phenol. Similarly, guaiacol and its derivatives, which possess hydroxyl and methoxyl groups attached to the aromatic ring, have been investigated extensively as model compounds, e.g. Refs. [22, 23]. Various pathways have been reported for guaiacol conversion towards a variety of products such as phenol, catechol, benzene, cyclohexane, and methyl-substituted phenols. Such a reaction network for the guaiacol catalytic cracking has been proposed in Ref. [32]. Besides, phenolic dimers have been involved in HDO studies due to their large amount in lignin derived bio-oils. During the aqueous phase, HDO of phenolic dimers on bifunctional catalysts (Pd/C, H-ZSM-5, or Ni/H-ZSM-5) hydrocarbon yield were observed up to 95–100% at 64–100% conversion [24, 25].

Co-processing of guaiacol to straight-run gas oil (SRGO) was studied in a conventional hydrotreating process [26]. In the presence of SRGO and under severe HDS conditions, no inhibiting effect on HDS activity was observed; however, at mild reaction temperature (below 320°C) and low space velocity, inhibition of HDS became relevant, likely due to competitive adsorption of intermediate phenols on the catalyst active sites. By increasing the temperature, these adsorbates are rapidly deoxygenated into hydrocarbons which did not affect HDS reactions. Otherwise, hydrogen sulphide from HDS suppresses hydrogenolysis and hydrogenation (HDO) of phenols, especially with NiMo and CoMo catalysts, via competitive adsorption of phenol and H_2S [27]. Similarly, ammonia stemming from HDN not only depresses the activity of NiMo and CoMo catalysts in HDS process, but also the conversion of carboxylic and methoxy groups, while ketones were not affected [28]. The presence of other compounds, such as water, has little influence on HDO reaction but does affect the lifespan of HDS catalyst.

3.2. Co-feeding of phenolic model compounds with conventional feeds at FCC conditions

Co-processing of oxygenated model compounds with conventional feeds at FCC conditions has been studied in lab-scale FCC units [29–32]. A maximum amount of 10 wt% of oxygenated compounds (related to gasoil) could be fed to a FCC without major problems. Additionally, the authors indicated that catalytic cracking of oxygenate compounds consists of a complex net of reaction pathways.

Either phenol or guaiacol was co-fed with hydrocarbon (e.g. n-heptane or methylcyclohexane) for cracking reactions over HZSM-5 and HY zeolites [31]. The severe slow-down of the cracking reaction of methylcyclohexane and n-heptane was observed in the presence of the named oxygenates. The authors proposed the observation due to strong adsorption of phenolic species on the catalysts and thus it could be competitive with the absorption of hydrocarbon. The increased coke formation in the presence of phenolic compounds also led to a slightly changed product distribution compared with hydrocarbons cracking.

Co-feeding of oxygenates (including guaiacol, acetic acid, phenol and hydroxyacetone) with gas oil over an equilibrated FCC catalyst [32] lead to an increase in yields of fuels gas, liquefied petroleum gas (LPG) and gasoline, however, this was possible mostly because boiling point range of those oxygenates and their products match these fractions. Additionally, some aromatic products were obtained from dehydration and alkylation of both phenol and guaiacol in the gasoline fraction products.

Recently, it was shown that small amounts of m-cresol at low reactant concentrations caused fast deactivation of an FCC catalyst [33]. Nevertheless, increasing the paraffin concentration hindered the deactivating effect of m-cresol. The authors postulated a hydride transfer between the phenolic compound and the paraffins. The interaction of the phenolic pool and the conventional feed (paraffin) via hydride transfer is summarized in **Scheme 1**.

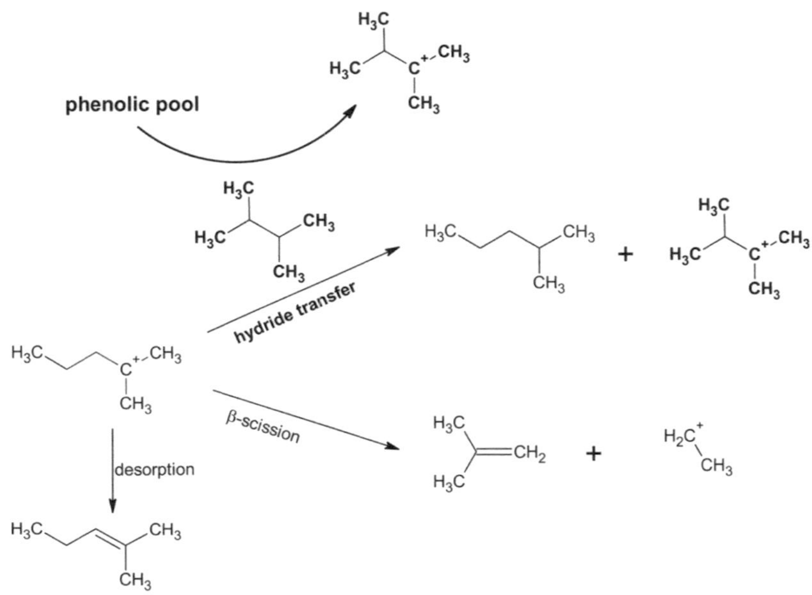

Scheme 1. Interaction of the m-cresol and paraffin transformation via hydride transfer. Adapted from Ref. [33].

In sum, the cited studies on co-feeding of phenolic model compounds with hydrocarbons give some insight (e.g. competitive adsorption and hydride transfer) that should be taken into account for the development of effective catalyst and revise the processes later on.

4. Co-processing of upgraded bio-oil as a phenolic feed into refineries

As mentioned above, bio-oils obtained from FP or LF of solid biomass have some peculiar properties (high oxygenate (35–50 wt%) and water content (15–30 wt%), high acidity and immiscibility with petroleum fuels) being different from those of conventional refinery streams [34]. Conversion of pure FP oil over conventional FCC catalysts has been studied already in the nineties [35, 36]. However, major challenges were identified (e.g. nozzle plugging and irreversible catalyst deactivation) owing to significant formation of coke, tar and char [37]. This leads to a more severe catalyst deactivation compared to regular FCC process. Thus, the direct use of an untreated bio-oil in standard refinery units needs large efforts in catalyst and process design that might make this route less attractive. Instead, blending of FP oil with conventional feed (e.g. vacuum gas oil) before introduction into FCC unit is the logical alternative due to the interest of petroleum oil companies.

http://dx.doi.org/10.5772/66883

The standard lab-scale techniques for evaluation of FCC catalysts [e.g. micro-activity test (MAT) or advanced cracking evaluation (ACE)] may also simulate the co-processing of FP oil with conventional FCC feeds. Such tests are known to elucidate the actual behavior of commercial FCC units quite well and various parameters [e.g. catalyst-to-oil (CTO) ratios, temperature, conversion and product distribution] can be systematically investigated. For example, UOP reported the first results for such blending tests in an ACE test unit [38]. **Table 2** provides typical results for pure vacuum gas oil (VGO) cracking in comparison with conversion of a blend of 20 wt% of FP oil and 80 wt% of VGO.

The results indicate that significant amounts of carbon are transferred to the gasoline, gas, LPG and coke, but less to LCO and slurry oil fraction. As a result, replacement of 20% of conventional feed by FP oil reduces the total amount of carbon fed to the FCC unit by 13% (due to the oxygen in the FP oil), but the gasoline yield dropped only by less than 5%. This might point to a synergetic effect between VGO and FP oil and the VGO seems to act as a hydrogen donor to the FP oil. Otherwise, the FP oil appears to increase the crackability of the VGO and shifts the product range towards desired light ends. In general, the co-feeding of FP oil to FCC units is not beneficial, with only an estimated 10% of the carbon from the liquids ending up in useable products (LPG and liquids). Much of the recent advances to obtain a better understanding of the co-processing of untreated FP oil in oil refineries have been conducted in BIOCOUP project within the 6th European Framework Program [39]. Particularly, various upgrading routes have been studied: (i) HDO to remove oxygen as water under high hydrogen pressure with a catalyst; (ii) high pressure thermal treatment (HPTT), in which FP oil is thermally treated to obtain an oil with a higher energy density [40]; and (iii) treatment without hydrogen, leading to decarboxylated oil (DCO). Comprehensive data on the use of FP oil either pure or as co-feed with VGO along all these routes are not published, but it is mentioned that despite lower oxygen content, a FP oil upgraded without oxygen (DCO route) could not be effectively co-processed without catalysts or hydrogen (HPTT route). An important criterion for successful co-feeding of such oils is a low-coking tendency (measured as micro carbon residue testing – MCRT), high H:C ratio, and a low average molecular weight [41].

Product yields (wt%)	VGO FCC	(20 wt% FP oil + 80 wt% VGO) FCC
Ethylene	2.0	3.3
Propylene	5.9	5.9
Propane	1.2	2.1
Butane	11.1	13.5
Gasoline	42.7	40.6
Light cycle oil (LCO)	14.8	9.1
Slurry oil	18.5	4.8
Coke	3.9	7.1
Water and CO_2	0.0	13.4

Table 2. Product yields from co-feeding of VGO and FP oil at FCC conditions. Data from Ref. [38].

Many efforts have been made in the recent years on HDO for upgrading of FP oil to deoxygenate the organic compounds effectively into so-called HDO oils or upgraded bio-oil (UBOs). HDO of bio-oil with various catalysts (e.g. Ru/C, Ru/Al_2O_3, Ru/TiO_2, Pd/C, Pt/C, NiMo/Al_2O_3, CoMo/Al_2O_3 and Ni-based catalysts) in the past decades has been comprehensively described in reviews [42, 43]. Besides, modified strategies for HDO of bio-oil have been proposed, e.g. a mild HDO process, non-isothermal hydrotreatment, low-severity HDO [44, 45], two-stage HDO [46] and aqueous phase HDO [47].

The co-feeding of such upgraded HDO oils (20 wt%) and 80 wt% standard feedstock (Long residue) is successful in laboratory-scale even if oxygen-rich HDO oils (17–28 wt% on dry basis) are used. Product yields, e.g. for gasoline (44–46 wt%) and light cycle oil (LCO) (23–25 wt%) were retained compared to the base feed [48, 49]. The authors also carried out the co-processing of 80 wt% of SRGO + 10 wt% HDO oil + 10 wt% isopropanol (to reduce viscosity) in a lab-scale HDS reactor, but the competition between HDS and HDO was observed and the efficiency of HDS was reduced [50]. Tests on co-feeding of hydrotreated bio-oil with an aromatic hydrocarbon feedstock (15/85 wt/wt) with two commercial FCC catalysts (ReUSY1, ReUSY2) showed that the conversion was slightly lower than that of the ordinary VGO [51]. The limited crackability of the aromatic feedstock seems to be the primary reason. On the other hand, the conversion obtained from co-processing of hydrotreated bio-oil with VGO was reported to be higher than that obtained from pure VGO feed experiment [52].

Own studies on the HDO of FP oil over bimetallic catalysts (10%Ni-10%Co/HZSM-5; 300 °C and 60 bar initial H_2 pressure) resulted in an UBO, which was co-fed with conventional FCC feed (atmospheric distillation residue of Dung Quat refinery-Vietnam) in a lab-scale MAT unit [53, 54]. Several runs with the same equilibrated FCC catalyst and various fractions of UBO (10, 20, 30 wt%) in the feed and different CTO ratios were performed at FCC conditions (520 °C, 1 bar, CTO = 2.5 or 3 g/g). **Figure 2** shows that the conversion is similar for both the co-processed feeds and the 100% conventional feed, whereas a reduction of HCO yield and slight increase of gasoline, gas and LCO fraction is evident for the co-processed feeds at the CTO ratio = 3 g/g. However, at a CTO ratio of 2.5 (g/g), which correlates to somewhat milder reaction conditions in terms of residence time and respective catalyst load, the conversion decreased gradually with the increase of the UBO fraction from 80% to 65% (with the 20UBO sample). This indicates that oxygenates in the UBO are more recalcitrant to cracking due to the many O-containing functional groups and the lower H-content (e.g. phenols, guaiacols, syringols and dimers). This observation is in line with literature [44], showing that a slightly higher CTO ratio is required for co-processing of UBO with conventional feed (Long residue) in order to obtain an equivalent conversion.

The gasoline fraction is the primary objective of a FCC unit and thus its composition obtained with the 4 samples tested at a CTO ratio of 3 (g/g) was analysed and showed in **Figure 3**. Obviously, co-processed feeds give larger amounts of aromatic compounds in the gasoline as compared to 100% conventional feed. In addition, the iso-paraffin and olefin fractions were reduced compared to 100% conventional feed, while the n-paraffin and naphthene fractions were more or less of the same size.

http://dx.doi.org/10.5772/66883

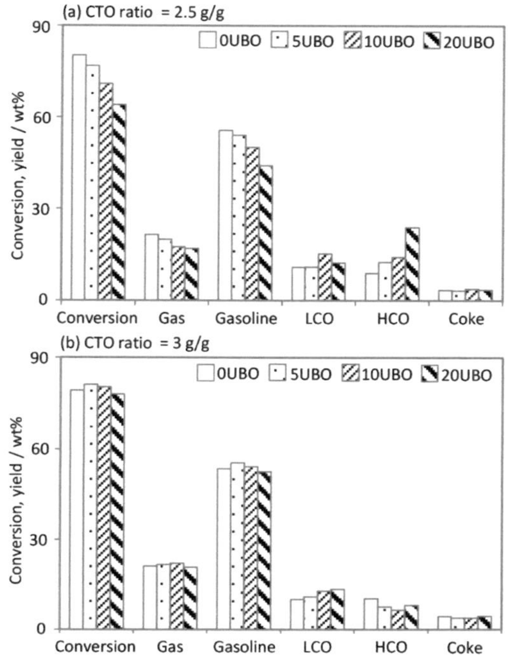

Figure 2. Performance of co-feeding tests at different feed compositions and CTO ratios in MAT unit. Adapted from Ref. [50].

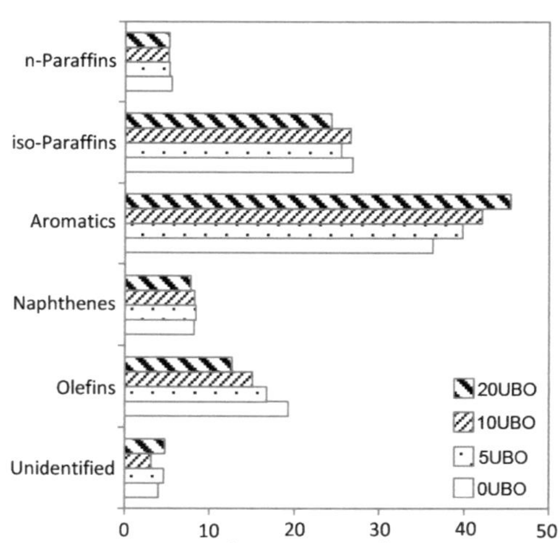

Figure 3. Gasoline composition in the products from co-feed tests at 520 °C and CTO = 3 (g/g). Adapted from Ref. [54].

On the other side, Petrobras implemented a near commercial FCC unit to co-feed pure FP oil with VGO [55]. The FP and VGO were fed into the riser reactor at two different heights. The feed rate was 150 kg/h and the results are shown in **Table 3**.

Product yields (wt%)	Feedstock		
	VGO	90% gas oil and 10% FP oil	80% gas oil and 20% FP oil
Fuel gas	3.9	2.8	2.5
LPG	15.2	12.9	9.9
Naphtha	40.4	40.7	37.7
LCO	18.1	17.4	16.5
Decanted oil	14.8	14.0	13.7
Coke	7.4	7.5	8.5
CO	0.1	1.9	3.1
CO_2	0.1	0.5	0.8
Water	0	2.3	7.3

Table 3. Product yields from co-feeding of VGO and FP oil by Petrobras at 540 °C. Data from Ref. [55].

The results indicate that the liquid yields from the blend VGO-FP oil did not significantly drop compared to FCC of VGO, whereas the yield of fuel/LPG was dramatically decreased. The introduction of 10 wt% of FP oil did not change the gasoline yields; however, the fraction was reduced significantly when co-feeding 20 wt% of FP oil.

It can be concluded that there are substantial differences in the conversion and product patterns obtained at laboratory-scale, pilot plant and semi-commercial scale. This is understandable as different FP oil, conventional feeds and reaction conditions were used [56].

5. Summary and perspective

Co-feeding of biomass derived liquids with conventional feeds into refinery units has potential for partial replacement of fossil crudes by renewable and sustainable resources in the short-term. In addition, it might be economically advantageous for biofuels production as the capital costs could be reduced due to the use of available existing infrastructure of petroleum refineries. Various tests with both FP oil and upgraded bio-oil (UBO) not only at lab-scale, but also at the semi-demonstration FCC scale showed promising results.

Studies with phenolic model compounds provide insight into the effect of oxygenates during co-feeding on elementary steps such as hydride transfer or competitive adsorption of phenolic compounds and hydrocarbon. It seems as if hydrocarbons might act as hydrogen donor for oxygen removal from the bio-feeds. The tests with FPO or UBO indicate some crucial aspects: (i) co-feeding possibly reduces the acidity and oxygenate content in the co-feed; (ii) upgrading helps to reduce oxygen content and to increase yields of usable products (e.g. naphtha, LCO and LPG); (iii) separate injection of conventional and bio-feeds could be a suitable choice in order to take advantage of the different reactivity of those feeds ruled by the aforementioned elementary steps.

However, the challenges of processing such bio-feeds in oil refineries are still significant and need to be further studied. As it is not expected due to economics that FCC catalyst and process design will be modified, the co-processing should be more deeply investigated using more standard conventional feeds and commercial FCC catalysts. On the other side, there might be some potential for optimisation of the upgrading step to make the UBOs more suited. As for the upstream FP process, the greater the improvement of FP, the higher the quality of bio-oil during storage and transportation and the easier the upgrading steps.

From a refiner's perspective, the important properties are the boiling-range distribution and the acidity. The high oxygen content of FP oil and UBO might cause corrosion and augmented coking of catalyst surfaces as well as downstream contamination risks. Thus, the upgrading of bio-feed to what extent should be adapted to the requirement of the refinery. It is likely that the degree of deoxygenation correlates with the oil yield and the heating value of UBO. Besides, another issue is to identify the best inlets for bio-feeds into the refinery and the requirements for venting of oxygenated gases (e.g. CO and CO_2) should be considered as it is not usual in conventional refinery.

Finally, one question might be open for the reader: who will responsible for the control and the management of bio-feeds and their co-processing into refinery? A realistic scenario will be that both industries cooperate, one producing the biofuel precursors and the other processing and converting them into valuable fuels.

Author details

Thuan Minh Huynh[1]*, Udo Armbruster[2] and Andreas Martin[2]

*Address all correspondence to: thuanhm.pvpro@vpi.pvn.vn

1 Vietnam Petroleum Institute – R&D Center for Petroleum Processing, Ho Chi Minh City, Vietnam

2 Leibniz Institute for Catalysis at the University of Rostock, Rostock, Germany

References

[1] BP Statistical Review of World Energy. 2015; Available from: www.bp.com/en/global/corporate/about-bp/energy-economics/statistical-review-of-world-energy.html [Accessed: 2015-06-20].

[2] Wang H, Male J, Wang Y. Recent advances in hydrotreating of pyrolysis bio-oil and its oxygen-containing model compounds. ACS Catalysis. 2013; **3**: 1047–1070.

[3] Kuparinen K, Heinimö J, Vakkilainen E. World's largest biofuel and pellet plants- geographic distribution, capacity share, and feedstock supply. Biofuels Bioproducts and Biorefining. 2014; **8**: 747–754.

[4] Carpenter D, Westover TL, Czernik S, Jablonski W. Biomass feedstocks for renewable fuel production: a review of the impacts of feedstock and pretreatment on the yield and product distribution of fast pyrolysis bio-oils and vapors. Green Chemistry. 2014; **16**: 384–406.

[5] Saxena RC, Adhikari DK, Goyal HB. Biomass-based energy fuel through biochemical routes: a review. Renewable Sustainable Energy Review. 2009; **13**: 167–178.

[6] Regalbuto JR. Cellulosic biofuels – got gasoline?. Science, 2009; **325**: 822–824.

[7] Huber GW, Corma A. Synergies between bio- and oil refineries for the production of fuels from biomass. Angewandte Chemistry International Edition. 2007; **46**: 7184–7201.

[8] Saidi M, Samimi F, Karimipourfard D, Nimmanwudipong T, Gates BC, Rahimpour MR. Upgrading of lignin-derived bio-oils by catalytic hydrodeoxygenation. Energy and Environmental Science. 2014; **7**: 103–129.

[9] Talmadge MS, Baldwin RM, Biddy MJ, McCormick RL, Beckham GT, Ferguson GA, Czernik S, Magrini-Bair KA, Foust TD, Metelski PD, Hetrick C, and Nimlos MR, A perspective on oxygenated species in the refinery integration of pyrolysis oil. Green Chemistry. 2014; **16**: 407–453.

[10] Wauquier JP. Petroleum Refining. Vol. 1 Crude Oil. Petroleum Products. Process Flowsheets, IFP, Paris, 1995, ISBN: 9782710806851.

[11] Huber GW, Iborra S, Corma A. Synthesis of transportation fuels from biomass: chemistry, catalysts, and engineering. Chemical Reviews. 2006; **106**: 4044–4098.

[12] Elliott DC. Historical developments in hydroprocessing bio-oils. Energy Fuels. 2007; **21**: 1792–1815.

[13] Zacher AH, Olarte MV, Santosa DM, Elliott DC, Jones SB. A review and perspective of recent bio-oil hydrotreating research. Green Chemistry. 2014; **16**: 491–515.

[14] Mortensen PM, Grunwaldt JD, Jensen PA, Knudsen KG, Jensen AD. A review of catalytic upgrading of bio-oil to engine fuels. Applied Catalysis A: General. 2011; **407**: 1–19.

[15] He Z, Wang X. Hydrodeoxygenation of model compounds and catalytic systems for pyrolysis bio-oils upgrading. Catalysis for Sustainable Energy. 2013; **1**: 28–52 (DOI: 10.2478/cse-2012-0004).

[16] Ruddy DA, Schaidle JA, Ferrell JR, Wang J, Moens L, Hensley JE. Recent advances in heterogeneous catalysts for bio-oil upgrading via "ex situ catalytic fast pyrolysis": catalyst development through the study of model compounds. Green Chemistry. 2014; **16**: 454–490.

[17] Huynh TM, Armbruster U, Martin A. Deoxygenation of Liquefied Biomass. In: Cavani F, Albonetti S, Basile F, Gandini A, editors. Chemicals and Fuels from Bio-Based Building Blocks. Wiley-VCH Verlag GmbH & Co. KGaA, Weinheim, 2016, pp. 403–430.

[18] Melero JA, Iglesias J, Garcia A. Biomass as renewable feedstock in standard refinery units. Feasibility, opportunities and challenges. Energy and Environmental Science. 2012; **5**: 7393–7420.

[19] Bu Q, Lei H, Zacher AH, Wang L, Ren S, Liang J, Wei Y, Liu Y, Tang J, Zhang Q, Ruan R. A review of catalytic hydrodeoxygenation of lignin-derived phenols from biomass pyrolysis. Bioresource Technology. 2012; **124**: 470–477.

[20] Huynh TM, Armbruster U, Pohl MM, Schneider M, Radnik J, Hoang DL, Phan BMQ, Nguyen DA, Martin A. Hydrodeoxygenation of phenol as a model compound for bio-oil on non-noble bimetallic nickel-based catalysts. ChemCatChem. 2014; **6**: 1940–1951.

[21] Huynh TM, Armbruster U, Phan BMQ, Nguyen DA, Martin A. The influence of cobalt in bimetallic Ni-Co catalyst supported on H-Beta for phenol hydrodeoxygenation. Chimica Oggi-Chemistry Today. 2014; **32**: 40–44.

[22] Bykova MV, Ermakov DY, Kaichev VV, Bulavchenko OA, Saraev AA, Lebedev MY, Yakovlev VA. Ni-based sol–gel catalysts as promising systems for crude bio-oil upgrading: guaiacol hydrodeoxygenation study. Applied Catalysis B: Environmental. 2012; **113–114**: 296–307.

[23] Sun J, Karim AM, Zhang H, Kovarik L, Li XS, Hensley AJ, McEwen JS, Wang Y. Carbon-supported bimetallic Pd–Fe catalysts for vapor-phase hydrodeoxygenation of guaiacol. Journal of Catalysis. 2013; **306**: 47–57.

[24] Zhao C, Lercher JA, Upgrading pyrolysis oil over Ni/HZSM-5 by cascade reactions. Angewandte Chemistry International Edition. 2012; **51**: 1–7.

[25] Zhao C, Lercher JA. Selective hydrodeoxygenation of lignin-derived phenolic monomers and dimers to cycloalkanes on Pd/C and HZSM-5 catalysts. ChemCatChem. 2012; **4**: 64–68.

[26] Bui VN, Toussaint G, Laurenti D, Mirodatos C, Geantet C. Co-processing of pyrolysis bio-oils and gas oil for new generation of bio-fuels: hydrodeoxygenation of guaiacol and SRGO mixed feed. Catalysis Today. 2009; **143**: 172–178.

[27] Senol O, Ryymin EM, Viljava TR, Krause AOI. Effect of hydrogen sulphide on the hydrodeoxygenation of aromatic and aliphatic oxygenates on sulphided catalysts. Journal of Molecular Catalysis A: Chemical. 2007; **277**: 107–112.

[28] Laurent E, Delmon B. Influence of water in the deactivation of a sulfide NiMo/Al_2O_3 catalyst during hydrodeoxygenation. Journal of Catalysis. Illinois, 1994; **146**: 281–285.

[29] Corma A, Huber GW, Sauvanaud L, O'Connor P. Processing biomass-derived oxygenates in the oil refinery: catalytic cracking (FCC) reaction pathways and role of catalyst. Journal of Catalysis. 2007; **247**: 307–327.

[30] Domine ME, van Veen AC, Schuurman Y, Mirodatos C. Coprocessing of oxygenated biomass compounds and hydrocarbons for the production of sustainable fuel. ChemSusChem. 2008; **1**: 179–181.

[31] Graça I, Ribeiro FR, Cerqueira HS, Lam YL, de Almeida MBB. Catalytic cracking of mixtures of model bio-oil compounds and gasoil. Applied Catalysis B: Environment. 2009; **90**: 556–563.

[32] Graça I, Lopes JM, Ribeiro MF, Ribeiro FR, Cerqueira HS, de Almeida MBB. Catalytic cracking in the presence of guaiacol. Applied Catalysis B: Environment. 2011; **101**: 613–621.

[33] To AT, Resasco DE. Hydride transfer between a phenolic surface pool and reactant paraffins in the catalytic cracking of m-cresol/hexanes mixtures over an HY zeolite. Journal of Catalysis. 2015; **329**: 57–68.

[34] Vispute TP, Huber GW. Production of hydrogen, alkanes and polyols by aqueous phase processing of wood-derived pyrolysis oils. Green Chemistry. 2009; **11**: 1433–1445.

[35] Adjaye JD. Bakhshi NN. Production of hydrocarbons by catalytic upgrading of a fast pyrolysis bio-oil. Fuel Processing Technology. 1995; **45**: 185–202.

[36] Vitolo S, Bresci B, Seggiani M, Gallo MG. Catalytic upgrading of pyrolytic oils over HZSM-5 zeolite: behaviour of the catalyst when used in repeated upgrading–regenerating cycles. Fuel. 2001; **80**: 17–26.

[37] Al-Sabawa M, Chen J, Ng S. Fluid catalytic cracking of biomass-derived oils and their blends with petroleum feedstocks: a review. Energy Fuels. 2012; **26**: 5355–5372.

[38] Marinangeli R, Marker T, Petri J, Kalnes T, McCall M, Mackowiac D, Jerosky B, Regan B, Nemeth L, Krawczyk M, Czernik S, Elliott D, Shonnard D. 2006. Opportunities for Biorenewables in Oil Refineries, UOP, Report No. DE-FG36-o5GO15085.

[39] http://www.biocoup.com/ [Accessed: 2016-09-09].

[40] de Miguel Mercader F, Groeneveld MJ, Kersten SRA., Venderbosch RH, Hogendoorn JA. Pyrolysis oil upgrading by high pressure thermal treatment. Fuel. 2010; **89**: 2829–2837.

[41] Karatzos S, McMillan JD, Saddler JN. 2014. The potential and Challenges of Drop-in fuels, IEA Bioenergy Task 39, Available from: http://task39.org/2014/01/the-potential-and-challenges-of-drop-in-fuels-members-only/ [Accessed: 2016-09-20].

[42] Elliott DC, Biller P, Ross AB, Schmidt AJ, Jones SB. Hydrothermal liquefaction of biomass: developments from batch to continuous process. Bioresource Technology. 2015; **178**: 147–156.

[43] Choudhary TV, Phillips CB, Renewable fuels via catalytic hydrodeoxygenation. Applied Catalysis A: General. 2011; **397**: 1–12.

[44] Elliott DC, Baker EG, Piskorz J, Scott DS, Solantausta Y. Production of liquid hydrocarbon fuels from peat. Energy Fuels. 1988; **2**: 234–235.

[45] Baldauf W, Balfanz U, Rupp M. Upgrading of flash pyrolysis oil and utilization in refineries. Biomass Bioenergy. 1994; **7**: 237–244.

[46] French RJ, Stunkel J, Baldwin RM. Mild hydrotreating of bio-oil: effect of reaction severity and fate of oxygenated species. Energy Fuels. 2011; **25**: 3266–3274.

[47] Vispute TP, Zhang H, Sanna A, Xiao R, Huber GW. Renewable chemical commodity feedstocks from integrated catalytic processing of pyrolysis oils. Science. 2010; **330**: 1222–1227.

[48] de Miguel Mercader F, Groeneveld MJ, Kersten SRA, Way NWJ, Schaverien CJ, Hogendoorn JA. Production of advanced biofuels: co-processing of upgraded pyrolysis oil in standard refinery units. Applied Catalysis B: Environmental. 2010; **96**: 57–66.

[49] de Miguel Mercader F, Groeneveld MJ, Kersten SRA, Geantet C, Toussaint G, Way NWJ, Schaverien CJ, Hogendoorn KJA, Hydrodeoxygenation of pyrolysis oil fractions: process understanding and quality assessment through co-processing in refinery units. Energy and Environmental Science. 2011; **4**: 985–997.

[50] de Miguel Mercader F, Koehorst PJJ. Competition between hydrotreating and polymerization reactions during pyrolysis oil hydrodeoxygenation. AIChE Journal. 2011; **57**: 3160–3170.

[51] Samolada MC, Baldauf W, Vasalos IA. Production of a bio-gasoline by upgrading biomass flash pyrolysis liquids via hydrogen processing and catalytic cracking. Fuel. 1998; **77**: 1667–1675.

[52] Fogassy G, Thegarid N, Toussaint G, van Veen AC, Schuurman Y, Mirodatos C. Biomass derived feedstock co-processing with vacuum gas oil for second-generation fuel production in FCC units. Applied Catalysis B: Environmental. 2010; **96**: 476–485.

[53] Huynh TM, Armbruster U, Nguyen LH, Nguyen DA, Martin A. Hydrodeoxygenation of bio-oil on bimetallic catalysts: from model compound to real feed. Journal of Sustainable Bioenergy Systems. 2015; **5**: 151–160.

[54] Huynh TM, Armbruster U, Atia H, Bentrup U, Phan BMQ, Eckelt R, Nguyen LH, Nguyen DA, Martin A. Upgrading of bio-oil and subsequent co-processing under FCC conditions for fuel production. Reaction Chemistry & Engineering. 2016; **1**: 239–251.

[55] Pinho AR, de Almeida MBB, Mendes FL, Ximenes VL, Casavechia LC. Co-processing raw bio-oil and gasoil in an FCC Unit. Fuel Processing Technology. 2015; **131**: 159–166.

[56] Al-Sabawi M, Chen J, Ng S. Fluid catalytic cracking of biomass-derived oils and their blends with petroleum feedstocks: a review. Energy Fuels. 2012; **26**: 5355–5372.

Section 2

Inhibitors and Treatment Methods

Chapter 12

Modified Byproduct of Coke Phenols as Effective and Prospective Inhibitors for Petrochemical Industry

Alexey Fedorovich Gogotov

Additional information is available at the end of the chapter

http://dx.doi.org/10.5772/66886

Abstract

Experimental author results on the use of technogenic phenols as inhibitors are presented to undesirable polymerization in petrochemical manufactures. The basic ways of increase inhibiting properties of coke-chemical phenols are shown, such as rectification for separation of two-nuclear derivatives or some examples of their chemical modification, allowing to increase the inhibiting efficiency of phenolic mixes, for example, by novolak fenol-formaldehyde condensation or oxidative coupling reactions. The received derivatives of phenols are recommended both for processing pyrocondensates and for rectification of styrene.

Keywords: pyrocondensates, styrene, phenolic inhibitors, sterically hindered phenols, technological phenols, phenols from wood, coke-chemical phenols, increase inhibiting properties, rectification, chemical modification, phenol formaldehyde condensation, oxidizing coupling

1. Introduction

Petrochemical industry can be represented in general as two main branches, and they are (a) pyrolysis industry and (b) manufacture of individual monomers, for example, butadiene, styrene, and so on. Both of them have one strong problem that is undesirable polymerization of olefinic monomers during distillation and purification. Thermal impacts during fractioning process are usually accompanied with the formation of active alkyl radicals to be prone to polymerize, which results in decreasing of yields of main products, contamination of equipment with polymer residues, and impairment in the control of distillation process. Such ever-present phenomenon during distillation process could be removed by several methods, and the most reasonable one is the use of special additives, the so-called inhibitors. Inhibitors are chemicals, which stabilize reactive monomers, they could react fast with free radicals, and so the latter are withdrawn and could not activate polymerization, notably, inhibitors stop spontaneous polymerization. Inhibitors cut off every chain reaction until they are spent.

2. The basic classes of inhibitors

The following substances of different chemical classes can be used as inhibitors, for example:

(1) quinones

(2) nitro- and nitroso compounds;

(3) amines and hydroxyl amines and their derivatives;

(4) stable nitroxyl radicals;

(5) different phenolic compounds;

(6) salts of aliovalent metals;

(7) sulfur- and phosphor-containing compounds, as well as

(8) high-molecular-weight inhibitors on the basis of the above classes.

Every compound class has its specific action on monomer stabilizations, and so its own intrinsic advantages and disadvantages. The most common among them are the representatives of four and five groups.

It should be noted that phenol inhibitors demonstrate some technological advantages, than others (**Figure 1**).

(a) (b)

Figure 1. Surface appearances of heat exchangers T-58 EP-300 utility after application of phenol inhibitors (a) and nitroxyl radicals (b).

3. Phenolic inhibitors

Among the phenol inhibitors, two main series are differentiated:

(1) Sterically hindered phenols (SHPs)

(2) Technogenical phenols.

The latter are multicomponent phenol mixtures produced during thermal treatment of organic raw materials—wood and coal. These mixtures are different in their compositions and properties.

Unlike the other inhibitor classes, the phenolic compounds advantageously combine their antioxidant properties together with the capacities to trap radicals [1].

Practical applications reveal the monoatomic SHP—Ar-OH and their derivatives to be the most effective.

Diatomic phenols $Ar(OH)_2$ (e.g., hydroquinone, pyrocatechol, pyrogallol, trimethylhydroquinone, 1,6-naphthalene diol, etc.) are more effective as radical process inhibitors in anoxic environment when compared with others. Three and more atomic phenols show their best antioxidant properties.

Phenol inhibitors can be divided into small number of classes depending on their properties and structures, according to the following characteristics:

(1) The number of aromatic fragment Ar: they are true phenols or naphthols—the oxy derivatives of condensed aromatic hydrocarbons;

(2) The number of hydroxyl groups in one aromatic circle (nuclei): they are mono- and polyphenols;

(3) The number of $Ar(OH)_n$ fragments in one molecule of antioxidant: they are mononuclear and polynuclear phenols (*bis*-phenols, *tris*-phenols, etc.); and

(4) The so-called shielded SHP phenols, where both *o*-substituents of OH-group in the aromatic circle are *tert*-alkyl groups, usually they are *tert*-butyl ones.

Sterically hindered phenols dramatically differ from other phenols. They are often used as industrial antioxidants, stabilizers, and anti-polymerizers, and many of them possess very high inhibiting activities.

At present, the most common in use are SHP having *tert*-butyl substituents, and the most well-known among them are as follows:

Ionol (2,6-di-*tert*-butyl-*p*-cresol, butyloxytoluene, BOT, alkophen, BP, topanol-O, DBC, topanol-OC, topanol-OF, 4M26B, CAO, Agidol-1, etc.), ***tert*-butyl pyrocatechine** (TBPC), and di-*tert*-butylhydroquinone.

4. The mechanism of phenol action as inhibitors

Involving phenols in radical reactions results in detachment of hydrogen atom from hydroxyl group on the first stage of the process with the formation of phenoxyl radical:

$$PhOH + R^{*} \rightarrow RH + PhO^{*} \qquad (1)$$

The formation of phenoxyl radicals was confirmed by electron paramagnetic resonance (EPR) method, which was supported with kinetic isotopic effect, when hydrogen atom is changed for deuterium.

Phenoxylic radicals are formed in the reaction of phenols with peroxide radicals: alkoxyl, alkyl, carboxyl radicals, as well as with molecular oxygen. The subsequent behaviors of phenols in radical reactions depend on properties of the phenoxyl radicals formed.

The reaction of peroxide and phenolic radicals leads to the formation of quinolide peroxides. Another active radicals add to PhO^{*} similarly. The quinolide peroxides so formed could be decomposed at O-O bond to give active radicals even at moderate temperatures:

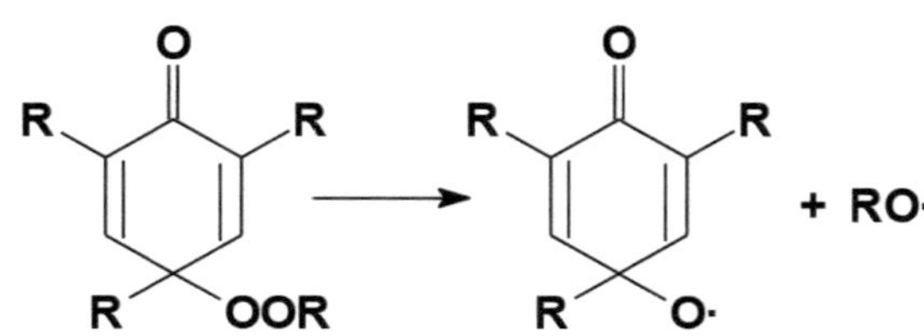

The alkoxyl radical could be subsequently dissociated to give quinone:

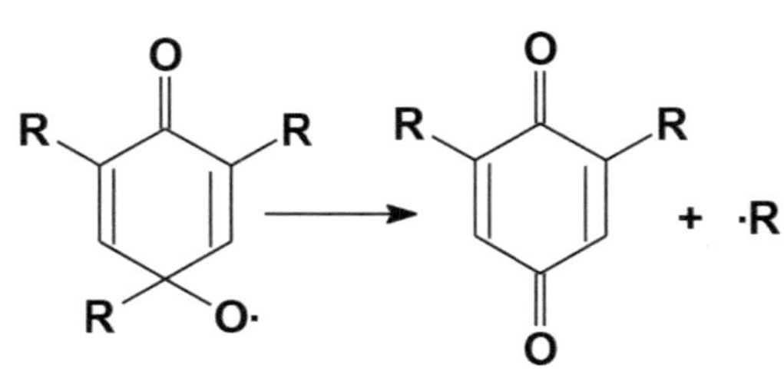

Hydroquinone, when reacting step by step with two peroxide radicals, transforms into quinone:

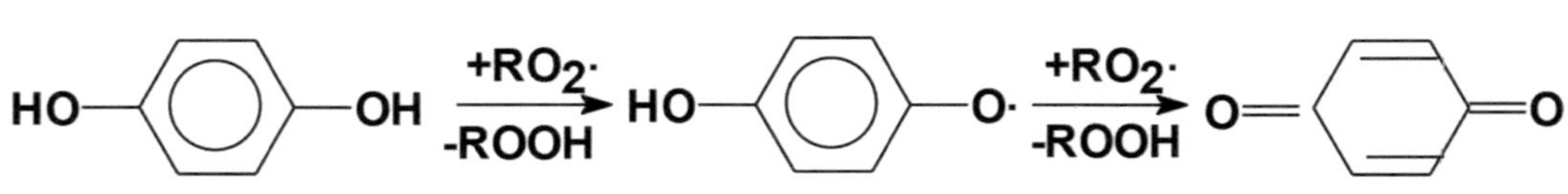

Together with the interaction of phenoxyl and peroxide radicals, the competitive reaction bimolecular loss of phenoxyl radicals takes place:

$$PhO^* + PhO^* \rightarrow \text{Products} \quad (2)$$

The reaction direction depends on a structure of phenoxyl radical. Some radicals having at least one non-substituted "active" position are typically recombined into dimeric phenol compounds, for example:

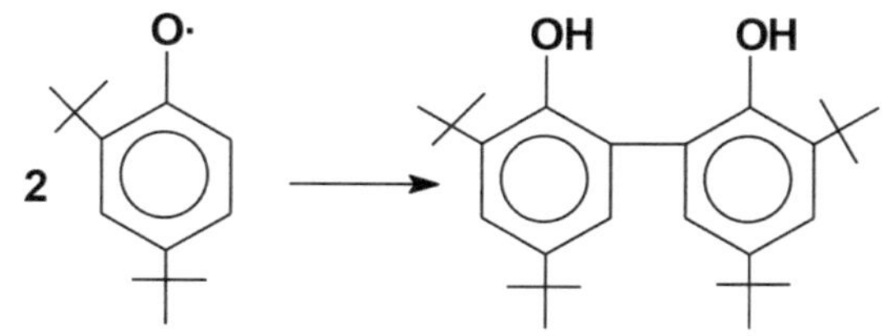

2,4,6-Trisubstituted phenoxyls do not form stable dimers as a rule. If the radical has *n*-, or *sec*-alkyl substituents, the reaction of disproportionation is possible to obtain methylene quinone and to recover the initial phenol:

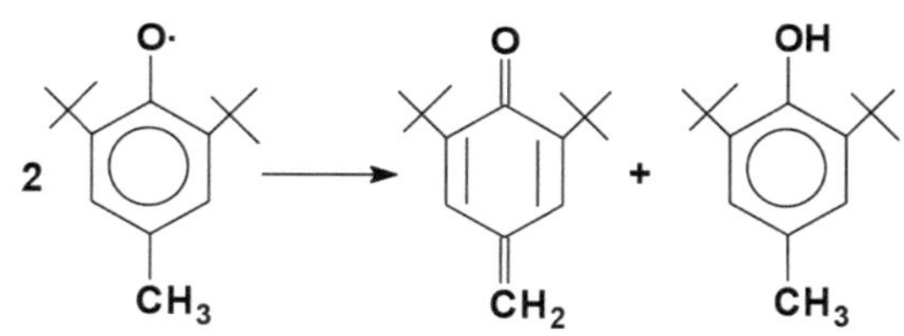

Methylene quinones are capable of dimerization with the formation of polynuclear phenols:

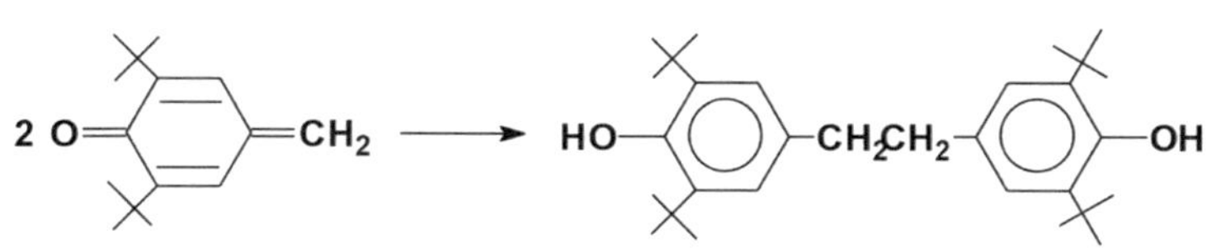

Methylene quinones could react with radicals to be added to unsaturated double methylene bond, generating phenoxyl radicals:

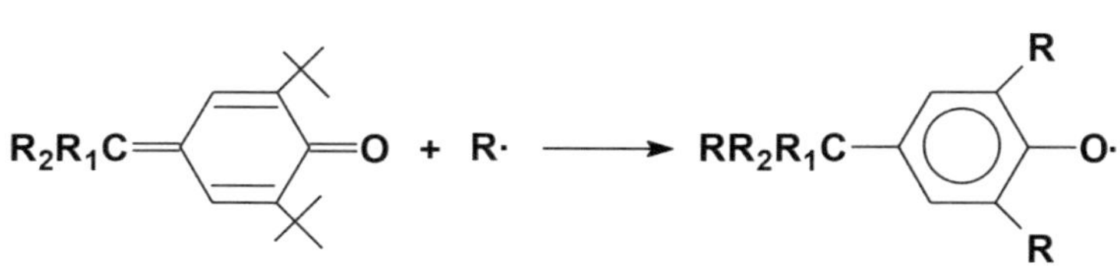

The activities of methylene quinones are almost equal to the same for the most reactive alkenes (diene, vinyl aromatic compounds), as well as sterically hindered quinones, but unlike the latter, they are very active to peroxides.

The disproportionations involving H-atoms of OH-groups are rather typical for phenol radicals, which have two hydroxyl groups in the circle, the so-called diatomic phenols—pyrocatechol and hydroquinone. The stabilization of phenoxyl radicals formed is possible owing to *ortho-* and *para-*quinones' generation. The disproportionation products in this case are quinoid compound and initial phenol:

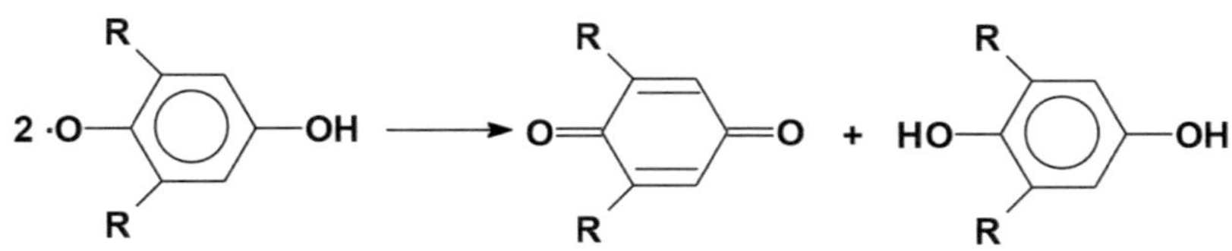

The quinones obtained when compared with the starting ones are effective acceptors of alkyl radicals, which add the radicals according to the scheme. So, the products of phenol inhibitor oxidation turn to be the generator of secondary inhibitors.

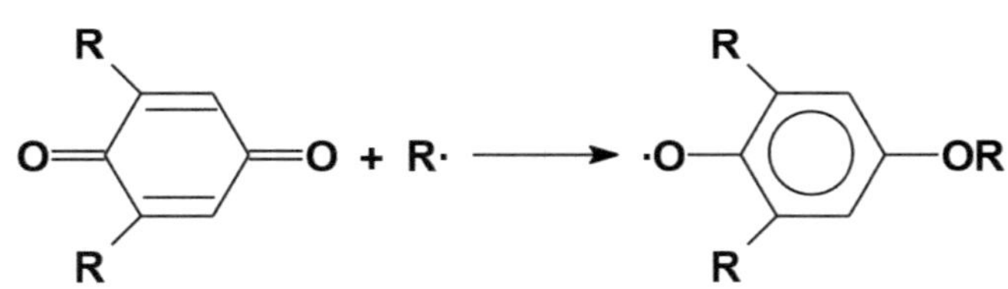

Phenoxyl radicals could react with some molecules. The most important reactions are as follows:

$$PhO^{*} + ROOH \rightarrow PhOH + RO_{2}^{*} \quad (3)$$

$$PhO^{*} + RH \rightarrow PhOH + R^{*} \quad (4)$$

$$PhO^{*} + PhOH \rightarrow PhOH + PhO^{*} \quad (5)$$

The inhibitor becomes ineffective, if the equilibrium of the phenoxyl radical and peroxide reaction is shifted to the formation of RO_2^{*}.

The reaction, where phenoxyl radical attacks C-H bond of hydrocarbon, is the most unwelcome for inhibition process. This reaction results in the formation of active alkyl radical and regeneration of a chain.

The processes of addition of phenoxyl radicals to unsaturated compounds lead to the formations of dimeric cyclohexadienones:

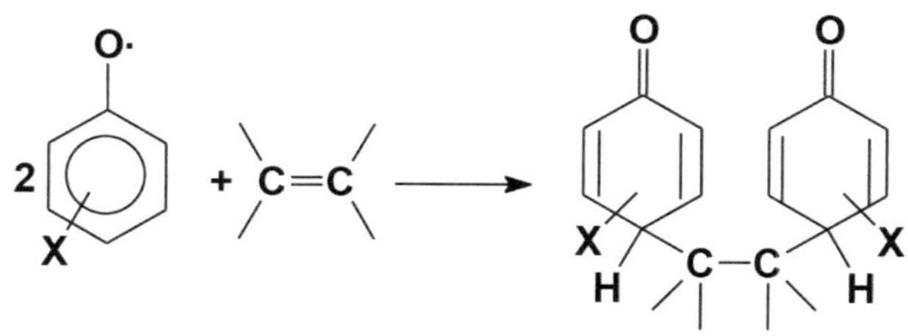

As the result of the reaction of phenoxyl radicals with molecular oxygen, the symmetric quinolide peroxides are formed:

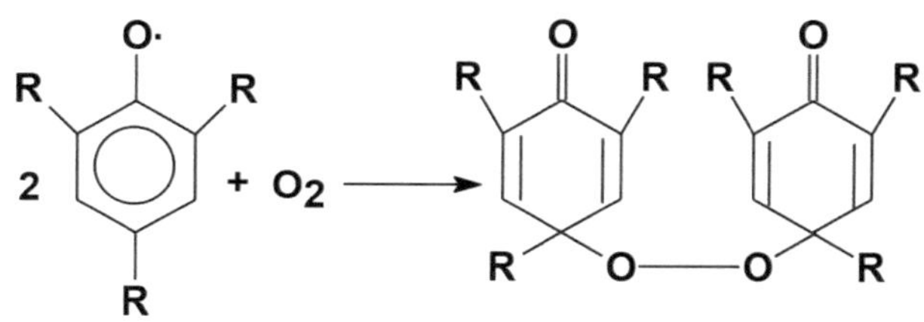

Some reactions of phenoxyl radical dissociation with emissions of active radicals are realizable.

As it is demonstrated on the schemes shown earlier, the application of phenol compounds allows to increase the inhibiting effect due to partial regeneration in the system "phenol-quinone" [1].

The stability of forming phenoxyl radicals depends fundamentally on substituents in *ortho*-position to hydroxyl group, and it increases with the increasing branches and bulks of the substituents [2]. The properties of SHP become apparent, if the substituent is *tert*-alkyl at least. For the phenols having methyl substituents or having no substituents in *ortho*-position of hydroxyl group, the inhibiting effects are considerably low, and often it occurs to be missing or even having a negative value. That is why monoatomic phenols unsubstituted in *ortho*-position to hydroxyl group are practically unknown as inhibitors.

Diatomic phenols of resorcin series reveal their inhibiting activity as compounds with CH-activity, that is, according to another mechanism [3]. The distinctive features of the resorcin series compounds are their extreme manifestations of inhibiting activity, but in a very narrow concentration interval and in considerable less (1–2 order of magnitude) concentrations, than for other diatomic phenols.

During pyrolysis of plant raw materials besides monoatomic phenols, cresols and xylenols (32.8%) and insufficient amounts of pyrocatechine and methylpyrocatechine (7.1%), the derivatives of guaiacol (20.1%) and syringol (31.3%) are dominated (**Table 1**).

Component	Component composition, %
Phenol	3.5
Crezols	13.9
Xylenols	15.4
Guaiacol	7.1
4-Methylguaiacol	5.3
4-Ethylguaiacol	4.2
4-Propylguaiacol	2.1
Eugenol	1.4
Pyrocatechine	5.7
Methylpyrocatechine	1.6
1,3-Dimethyl ether of 5-methylpyrogallol	12.7
1,3-Dimethyl ether of 5-ethylpyrogallol	10.3
1,3-Dimethyl ether of 5-propylpyrogallol	6.5

Table 1. Composition of phenols for pyrolysis of wood.

It is found that guaiacol and syringol derivatives in non-polar media do not reveal their inhibiting activities, so the hydrogen atom of phenol hydroxyl group forms the hydrogen binding with oxygen atom of neighboring (in *ortho*-position) methoxyl group, which prevents the phenoxyl radical formation:

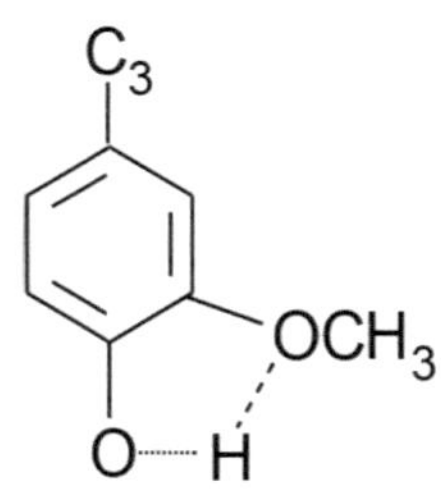

5. Coke-chemical phenols

During thermal processing of coals, the main components of coke-chemical phenol (CCP) mixtures are different mono- and diatomic phenols, where compositions are dependent on coalfield (**Table 2**).

Table 2 demonstrates the essential differentiations in mono- and diatomic phenol proportions in different storage reservoirs. These proportions play very important role to produce effective inhibitors from CCP. It is experimentally established that the optimal proportion is <2:1.

http://dx.doi.org/10.5772/66886

As an example, the composition of phenols from storage reservoir No. 2 is given in **Table 3**.

According to **Table 3**, diatomic phenols are predominantly presented with pyrocatechine and his gemologists, as well as with insignificant amount of resorcin and its derivatives and with minor amount of methyl and dimethyl hydroquinone.

That is why the first alternative to increase the inhibiting activity of CCP is the fractionating of the phenol mixture for the purpose of isolating diatomic phenols from monoatomic ones. During fractionating, the distillate of diatomic phenols, the so-called "pyrocatechine fraction"

Reservoir No	Monoatomic phenols, %mass	Diatomic phenols, % mass	Nonphenolic admixtures, % mass	Proportions of mono- and diatomic phenols
1	72.3	25.9	1.8	2.79
2	51.7	39.9	8.4	1.30
3	68.3	25.3	6.3	2.70
4—Overall (Summary) mixture (res.1: res.2: res.3 = 1.2:1.7:1.5)	62.4	32.5	5.1	1.92

Table 2. Comparative composition of CCP from different storage reservoirs.

Monoatomic phenols			**51.67**
They include:			
Phenol	3.30	2-Ethyl-4-methyl phenol	0.16
2-Methyl phenol	10.14	4-Ethyl-3-methyl phenol	0.69
4-Methyl phenol	13.32	3-Ethyl-5-methyl phenol	0.20
2,6-Dimethyl phenol	4.34	2,3,5-Trimethyl phenol	0.27
2-Ethyl phenol	1.03	4-Propyl phenol	0.15
2,4-Dimethyl phenol	11.57	2-Methyl-5-(1-methylethyl phenol	0.32
4-Ethyl phenol	3.74	2,3,6-Trimethyl phenol	0.10
3,5-Dimethyl phenol	0.75	4-(1-methylpropyl)phenol	0.10
2,4,6-Trimethyl phenol	0.42	2-Butyl phenol	0.13
2-Propyl phenol	0.15	4-Ethyl-2-methoxyphenol	0.09
2-Ethyl-6-methyl phenol	0.66	2-Naphthol	0.04
Diatomic phenols			**39.90**
They include:			
Pyrocatechine	17.16	4-Ethyl resorcin	2.21
3-Methyl pyrocatechine	9.67	4,5-Dimethyl resorcin	0.29
4-Methyl pyrocatechine	4.96	Methyl hydroquinone	0.17
Resorcin	1.69	2,3-Dimethyl hydroquinone	2.17
2-Methyl resorcin	1.51	3,4-Methylene dioxyanisole	0.07

Table 3. Component composition of phenol mixture in storage reservoir No. 2.

(PCF), has been isolated. The fraction is analyzed and tested as polymerization inhibitor of industrial condensates for EP-300 utility of Angarsk Polymer Plant (APP) of "ROSNEFT" Joint-Stock Company, Irkutsk region, Russia [4]. Inhibiting activities of PCF and other inhibitors have been tested in accordance with the standard protocol of 8489-85 RU State Standard. The industrial samples of liquid pyrolysis products (pyrocondensates) of straight-run distillation of gasoline and hydrocarbon gases have been used as investigation objects.

The estimation of their inhibiting activities has been modeled under very closed conditions, or even under more hard conditions, than that for real-system "column-boiler" at the rectification process of pyrolysis semi-products, where the example is chosen from pyrocondensate of typical utility of "Ethylene/propylene-300" of APP, where production is benzene produced by hydrodealkylation.

Laboratory experiment on modeling polymer formation has included the following:

(1) Initiation of polymer formation in pyrocondensate by temperature impact with the addition of investigated inhibitor.

For this purpose, pyrocondensate of 100 cm^3 was loaded into a metal autoclave, and the inhibitor was added (or not added in blank experiment); the autoclave was sealed hermetically and placed into a preliminary heated thermostat at 130°C for 1 h.

(2) The amounts of the polymer formed products were detected by RU State Standard 8489-85 by the equipment "POS-77M."

Effectiveness of inhibiting was estimated in accordance with the formula:

$$\text{Effectiveness} = \frac{C_{\text{blank}} - C_{\text{inhib}}}{C_{\text{blank}}} 100\% \tag{6}$$

where C_{inhib} and C_{blank} are the amounts of experimental resins in pyrocondensates in inhibit and blank experiments, correspondingly.

6. Increase inhibiting properties of phenols by rectification

During the rectification process of total coke-chemical phenols (TCCPs), the isomers of diatomic phenols were concentrated in the distillation residue after monoatomic phenol separation (DRAMPS). Some samples of the DRAMPS product were taken from the operating "FCh-16" phenol inhibitor production utility of JSC "Angarsk Petrochemical Company" (APCC), and then pyrocatechine fractions were distilled (up to 300°C) from them. Their inhibiting properties were tested. The laboratory results served as a basis for the design of technical requirements for the prototype production of the inhibitor, and then technical regulations for the production were developed. The pilot butches of PCF inhibitor were produced on the plant utility. The composition of the laboratory and pilot plant samples was proved by chromato-mass spectrometry method, and they are presented in **Table 4**, and their quality factors in accordance with the State Standard demands are presented in **Table 5**.

http://dx.doi.org/10.5772/66886

Components	DRAMPS	PCF samples	
		Laboratory	Plant
Phenol	4.7	2.5	1.4
Methylphenol	3.9	1.9	1.2
Dimethyl and ethylphenols	6.8	3.3	2.3
Trimethyl and propylphenols	5.1	3.7	3.0
Tert-butyl phenol	1.5	1.6	1.2
Tert-butylmetylphenol	1.0	–	0.9
Sum of monatomic phenols	23.0	13.0	10.0
Pyrocatechine*	14.3	15.2	18.2
Methyl pyrocatechine	26.8	37.8	37.9
Dimethyl- and ethyl pyrocatechines	11.4	20.8	14.2
Trimethyl- and propyl pyrocatechines	1.1	1.4	1.6
Tert-butyl pyrocatechine (TBPC)	2.0	0.9	1.6
Tert-butyl- and methyl pyrocatechines	2.7	0	1.6
Tert-butyl-dimethyl pyrocatechine	0.7	0	1.5
Sum of diatomic phenols	59.0	76.1	76.6
Nonindentified (resins)	18.0	10.9	13.4

*minor amounts of resorcin and its derivatives were identified by GLC-method together with pyrocatechine

Table 4. Compositions of DRAMPS and PCF, % mass.

Factors	Standard with State regulations. 38.40154-90	Experimental results
Appearance	Liquid of light yellow up to brown color	
Density at 20°C, g/sm^{3}sup>	Not more than 1.2	1.1
Fraction composition, % vol.:		
Boiled-off before 240°C	Not more than 10	7.5
Boiled-off before 285°Cs	Not more than 96	96.2
Water content, % mass	Not more than 0.5	0.5

Table 5. Quality factors of the "PCF" pilot batch.

The results of the laboratory investigations of different phenol samples are presented in **Table 6**.

Thus, it is shown that fine fractionating phenol mixtures to remove inactive monoatomic phenols allow to obtain special preparation of diatomic phenols "PCF," having its inhibiting activity close to the well-known individual and expensive preparation such as *tert*-butyl pyrocatechine (TBPC), belonged to sterically hindered phenols (SHPs).

No.	Inhibitors	Practical yields of resins, mg/100 sm^3	Effect of inhibiting, %
1	Blank experiment	233	–
2	Initial overall CCP	177	24
3	PCF fraction	77.6	66.7
4	PFR on the basis of CCP	76.4	67.2
5	CCP, modified with oxidized coupling	64.8	72.2
6	5-Methyl resorcin (from resorcin fraction)—0.009%	78.8	66.2
7	Ionol	139.8	40.0
8	TBPC	75.0	67.8

Table 6. Effectiveness of inhibiting polymer formations for EP-300 pyrocondensates at the inhibitor consumption of 0.025 %.

The obtained PCF preparation was introduced into EP-300 production unit of Angarsk Polymer Plant of Petrochemical Company "ROSNEFT" and was employed for 2 years up to its run-out in the company stocks.

The problem of utilization of CCP without the stage of power cost-based phenol vacuum fractionating has been arisen. But the problem is not unequivocal: its main disadvantages, for example, good water solubility and unpleasant organoleptic properties are not removed by this way of utilization, and in addition to that, the major part of overall (summarized) phenols (more than 2/3 of total amount) is not found in any application.

7. Increase inhibiting properties of phenols by chemical modification

That is why new methods of utilization of summarized CCP (SCCP) have been elaborated to avoid the stage of preliminary power cost-based fractionating.

Analysis of literary sources reveals some ways to decrease water solubility of phenols, they are as follows:

(1) Phenol formaldehyde condensation (PFC) under acid catalysis (by novolak method)

(2) Phenol alkylation

(3) Oxidative cross-linking of phenols

8. Phenols modification by phenol formaldehyde condensation

During the process of novolak polycondensation, the phenols become insoluble in water, but phenol formaldehyde resins (PFRs) have a good solubility in organic solvents, for example, in alcohols. The first step of phenol transformation is the chemical modification of SCCP without their preliminary fractionating to synthesize FFR with the view of obtaining inhibitor having the highest activity.

http://dx.doi.org/10.5772/66886

Indeed, the standard novolak PFR from monoatomic phenols has demonstrated inhibiting effect of 30%, when it is tested as phenol inhibitor at 0.03% mass consumption. Novolak polycondensate, obtained from summarized coke-chemical phenols (SCCP, No. 4., **Table 2**), demonstrates ~67% effect at the same consumption [5]. Unmodified SCCPs show 24% effect at the consumption of 0.03% mass (**Figure 2**).

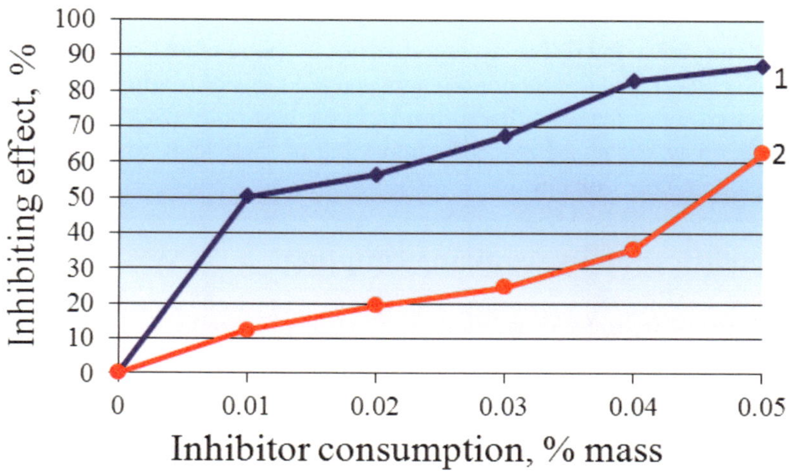

Figure 2. Dependency of inhibit effect on the consumption of PFR (1) and SCCP(2).

It must be accentuated that PFR, synthesized from separate fractions of monoatomic phenols obtained from summarized phenols, or diatomic phenols are worse than for PFR, produced from non-fractionated phenols, where inhibiting effects are 30–35%, at the same as above consumption of 0.03% mass. This fact is the witness of synergism of the mixtures of monoatomic and diatomic phenols after their phenol formaldehyde polycondensation. The effect is found by us early [6] during inhibiting the thermo-polymerization process of styrene with ionol and TBPC inhibitors. The inhibiting factors of the resins obtained from phenols of different stock reservoirs are quite different (**Table 7**).

Reservoir	No. 1	No. 2	No. 3	Summarized
Effectiveness, %	33.2	74.7	38.5	67.2

Table 7. Comparative effect of PFR produced from phenols of different stock reservoirs at the consumption of 0.03% mass.

The results of **Table 7** support the considerable inhibiting effect of synthesized novolak resins that depends on the ratios of mono- and diatomic phenols in the initial mixtures. Indeed, the increase of monoatomic phenol amounts leads to the decrease in inhibiting effect, and just the reverse, it leads to increasing this effect at the increasing diatomic phenol amounts (reservoir No. 2).

However, it should be noted that inhibiting activity of synthesized PFR from SCCP is fundamentally dependent on molecular mass (viscosity) of the resin taken.

Dependency of the inhibiting effectiveness from viscosity of PFR at the consumption of 0.03 % mass is as follows:

Viscosity of resins, sec.	70	135	240	540	720
effectiveness, %	56.3	74.7	43.9	8.1	0

According to these dates, the resins having viscosity in the rage of 135 s (Viscosimeter VZ-4) are the most effective. There is dependence between amounts of residual phenols in synthetic tar-water and viscosity of the resin. For example, in the best point (74.7%) the amount of non-reacted phenols on water phase reached almost 5% of their load, and all the phenols are presented by monoatomic isomers.

9. Phenol modification by oxidizing coupling

That is why it is necessary to find conditions for the perfect binding of phenols. It appears to be the method of oxidative cross-linking of phenols by their treatment with persulfates of alkali metals. The mechanism of the modification includes dimerization of phenols to form C-C bond involving unsubstituted (free) *ortho-* and *para-*positions of hydroxyl group, the mechanism is accompanied by a series of different reactions. The method of oxidative cross-linking is presented by a scheme:

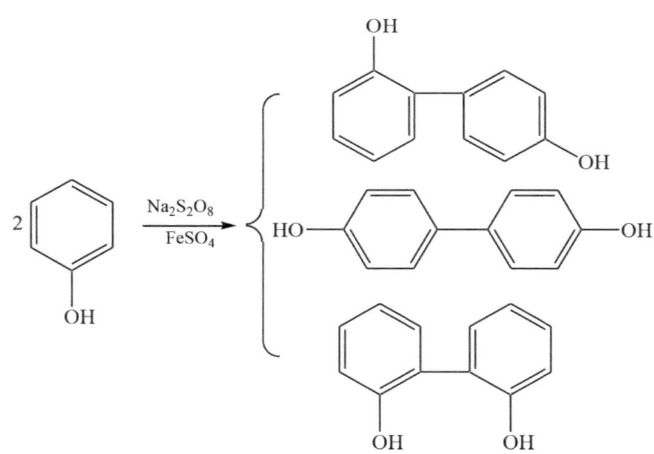

And it was employed as the method of modification of SCCP to obtain the inhibitor of thermo-polymerization for processing semi-products of pyrolysis [5]. As far as dimeric phenols practically are non-water soluble, it is supposed that oxidative dimerization of phenols allows to remove one of the main disadvantages of coke-chemical phenols—their high water solubility.

During the first stage of the reaction of the oxidizer and phenol, the generation of phenoxyl radical occurs (for monoatomic phenols) and then its mesomerization with the transfer of radical center from oxygen atoms to carbon atoms of aromatic circle and subsequent recombination of the formed radicals takes place [7]. During oxidation of di- and polyatomic phenols, the mechanism is more complicated, and the possibility of the generation of quinoid compounds is not excluded, as well as some other secondary inhibitors. During the oxidation reaction, quinone methides, polymers, and subsequent oxidative products of primary formed binuclear compounds could be formed. It is evident, the yields of individual product at so diversity of the products obtained in the oxidation are not high. The secondary reactions between oxidized substrates and the starting phenols are not precluded. That is why the summary effect of oxidized phenol mixture essentially depends on the structures of initial phenols and amounts of the oxidizer used (**Figure 3**). As it was stated above during the one-step cross-linking modification of phenols under action of persulfates of alkali metals, the dimers of phenols are formed, which are not soluble in water, but well soluble in alcohols and alkyl acetates.

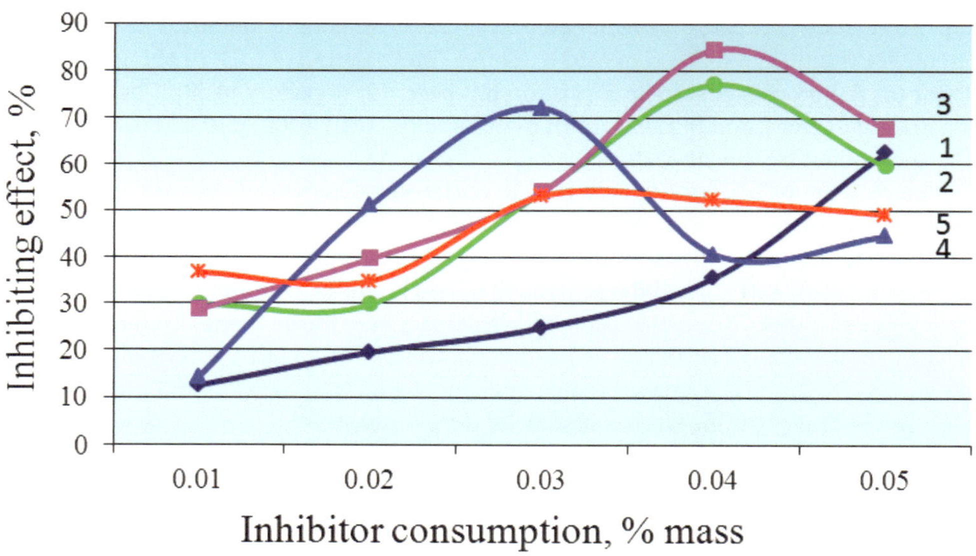

Figure 3. Inhibiting activity of SCCP after treatment with sodium persulfate; oxidizer consumption: 1—0%, 2—10%, 3—25%, 4—50%, 5—75%mass.

According to the data of chromato-mass-spectrometry and nuclear magnetic resonance (NMR)-spectroscopy ^{1}H and ^{13}C for the products obtained by the oxidation of monoatomic phenol with sodium persulfate, the expected dimeric products of "*ortho-ortho,*" "*para-para,*" and "*ortho-para*" cross-linking are obtained in a combination of 9:50:19%, respectively.

The analysis of inhibiting activity of oxidative cross-linking SCCP mixtures demonstrates that the maximal effectiveness of 72–84% is manifested for oxidized mixtures with the consumption of oxidizer of 25% mol relatively to the overall phenol mass and with the inhibitor consumption of 0.03–0.04% mass.

Thus, we have developed single stage and rather simple methods of chemical modifications of CCP, which make it possible to obtain high-performance phenol inhibitors for the processing of pyrolysis liquid products of petrochemical industry.

10. Application of CCP in inhibitors synthesis for styrene produce

The approaches to inhibiting process for pyrocondensates and individual monomers are rather different. For example, as a rule for the inhibiting of thermo-polymerization of liquid products of pyrolysis, it is necessary to apply individual compounds as inhibitors, for example, TBPC or ionol, whereas for the prevention of undesirable polymerization of individual monomers, such as styrene, for example, it is advisable to use more effective so-called "inhibiting compositions."

It is known [8] that the phenol oligomer, obtained by phenol formaldehyde condensation, manifests rather low effectiveness as inhibitor in the production process of styrene, so it means that all production processes of phenol modifications for obtaining inhibitors are considerably different. For example, a method of acid-oxidative condensation of CCP to obtain effective co-inhibitor for the process of styrene distilation has been developed by us [9]. Chemical modification of CCP to give "modificated phenol co-inhibitor" (MPC) is the synthesis one-pot reaction by the subsequent treatment of the substrate with catalytic amounts of concentrated sulfuric acid at heating up to 95°C for 1 h, then the hot acidic phenol solution is treated with hydrogen peroxide in the presence of catalyst that is iron (II) sulfate for 1 h, and thereafter the oxidate is cooled and neutralized with sodium nitrite at room temperature. The phenol oxidate is extracted with alkyl acetate, and the solution obtained is employed for the inhibiting of styrene thermo-polymerization. The estimation of the efficiency is made by thermal processing of pure styrene in the standard conditions at heating to 120 ± 1.5°C, with the blowing of nitrogen under stirring for 2.5 h. The amount of the polymer formed is detected by sedimentation of the styrene probe (2 sm^3) from autoclave and its dilution with ethyl alcohol (8 sm^3) and then centrifugation and separation of the sediment, the latter's drying and weighting according to the State Standard RU 10003-90.

The results of the experiments with different inhibiting compositions of styrene thermo-polymerizations are presented in **Table 8**.

It is clear from **Table 8** that neither itself MPC nor itself MB (Mannich Base - 2,6-di-*tert*-butyl-4-dimethyl-aminometyl phenol, N,N'-dimethyl(3,5-di-*tert*-butyl-4-oxybenzyl)amine) (entry 2) possess satisfactory inhibiting properties, but their compositions "MCP + MB" in different combinations are rather competitive in comparison with the best known inhibitors, and they could be recommended as inhibitors for the styrene production.

Accordingly, the task-oriented chemical modification of coke phenols could solve the very complicated ecological proble the utilization of toxic technogeneous phenols transforming them into competitive reagents for petrochemical industry, namely high effective phenolic inhibitors of polymerization.

Chemical modification of phenols in this line could be realized by other methods, for example, by the reaction of SHP with bulky bi- and tricyclic substituents, for example, diisobornylphenols [2] and diadamantylphenols [9], or by the generation of the so-called inhibiting systems employing phenols, solvent, and turpentine [10]. That is another subject for investigations.

The system applied for the styrene mass, %	Time, min			
	60	90	120	150
	Formation of the polymer, mass %			
1. MB (0.03) + MPC (0.045)	0.08	0.12	0.23	0.35
2. MB (0.03)	0.88	1.51	3.68	8.81
3. MPC (0.045)	0.21	0.68	1.93	4.61
4. MB (0.025) + MPC (0.05)	0.10	0.19	0.32	0.44
5. AHM-700 (0.017)	0	0.10	0.23	0.55
6. "Nalco EC 3264A" 0.0200%	0.14	0.17	0.25	0.36
7. Kawakami (98BH+99BH), with 0.025% each	0	0.39	0.42	0.45
8. IPON* 0.025%	0.10	0.18	0.33	0.48
9. DOQ**(0.006)+0,2[MB (0.025) + PNP (0.0125)] + 0.8(MB +PNP)-1 h	0.33	0.66	1.35	1.91

*IPON - Russian analog Nalco inhibitor on base nitroxilyc radicals **DOQ - Quinone-dioxime; PNP - p-nitro-phenol; AHM - isobutyl-2,4-dinitro-phenol

Table 8. Comparative effect of PFR produced from phenols of different stock reservoirs at the consumption of 0.03% mass.

11. Conclusions

(1) The possibility of the task-oriented utilization of anthropogenic phenols without their preliminary fractionating to produce the inhibitors of undesirable polymerization, which are demanded products for different branches of petrochemical industry: high-effective and competitive on qualitative and cost-oriented properties has been demonstrated and the effectiveness has been proved experimentally.

(2) It is shown that multicomponent mixtures of diverse mono- and diatomic coke-chemical phenols could be purposefully transformed by known single-staged methods into high-effective phenol inhibitors of thermo-polymerization applicable to the processing liquid pyrolysis products of petrochemical hydrocarbons, and the inhibitors designed are able to become the competitors of the known expensive inhibitors that are pure-isolated sterically hindered phenols.

(3) The new one-pot tree-stage method of coke-chemical phenols' processing into "modified phenol co-inhibitor" that is the component of highly effective compositions of styrene distillation has been offered.

Author details

Alexey Fedorovich Gogotov

Address all correspondence to: alfgoga@mail.ru

A.E. Favorsky Irkutsk Institute of Chemistry, Siberian Branch of Russian Academy of Sciences, Irkutsk, Russian Federation

References

[1] Ershov V., Nikiforov G., Volod'kin F. Prostranstvenno zatrudnennye fenoly (Sterically hindered phenols). Moscow: Khimia, 1972. 351 p.

[2] Method of Inhibiting Thermopolymerisation When Processing Liquid Pyrolysis Products. RU Patent 2375342. [Internet]. 2009. Available from: https://worldwide.espacenet.com/publicationDetails/originalDocument?FT=D'date=20091210'DB='locale=en_EP'CC=RU'NR=2375342C2'KC=C2'ND=4 [Accessed: 2016-10-12].

[3] 3,5-Dihydroxytoluene (Orcinol) For Inhibiting Polymerization In Liquid Pyrolysis Product Processing. RU Patent 2215728. [Internet]. 2003. Available from: https://worldwide.espacenet.com/publicationDetails/originalDocument?CC=RU'NR=2215728C2'KC=C2'FT=D'ND=3'date=20031110'DB='locale=en_EP[Accessed: 2016-10-12].

[4] Method of Reducing Polymer Formation in Processing of Pyrolysis Intermediate Products on Plants of The Ethylene And Propylene Production Type. RU Patent 2127750. [Internet]. 1999. Available from: https://worldwide.espacenet.com/publicationDetails/originalDocument?CC=RU'NR=2127750C1'KC=C1'FT=D'ND=3'date=19990320'DB='locale=en_EP [Accessed: 2016-10-12].

[5] Gogotov A., Levchuk A., Do Chiem Tai, Kanitskaya L., Stankevich V. Mint: Coal tar phenols as a promising intermediate for producing high-performance polymerization inhibitors in petrochemical industry. Petroleum Chemistry. 2015; 55: 530–536. DOI: 10.7868/S0028242115050044.

[6] Method of Inhibiting Styrene Thermopolymerization. RU Patent 2129112. [Internet]. 1999. Available from: https://worldwide.espacenet.com/publicationDetails/originalDocument?CC=RU'NR=2129112C1'KC=C1'FT=D'ND=3'date=19990420'DB='locale=en_EP [Accessed: 2016-10-12].

[7] Hains A.H. Metody okisleniya organicheskikh soedinenii (Methods for Oxidation of Organic Compounds. Alkanes, Alkenes, Alkynes, and Arenes. Academic Press), Moscow: Mir, 1988. p. 193–194.

http://dx.doi.org/10.5772/66886

[8] Batura I., Gogotov A., Cherepanov V., et al. Mint: Utilizing the phenol byproducts of coke production. 3. Phenols as co-inhibitors of thermopolymerization during styrene production. Coke and Chemistry. 2009; **52**: 23–27. © Allerton Press, Inc., 2009.

[9] Method of Inhibiting Thermopolymerisation When Processing Liquid Pyrolysis Products. RU Patent 2460714. [Internet]. 2012. Available from: https://worldwide.espacenet.com/publicationDetails/originalDocument?CC=RU'NR=2460714C1'KC=C1'FT=D'ND=3'-date=20120910'DB='locale=en_EP [Accessed: 2016-10-12].

[10] Gogotov A., Do Chiem Tai, Kanitskaya L., Stankevich V. Mint: recycling of total phenols —by-products of coke-chemical plant. 5. Increase inhibiting activity of coke-chemical phenols by thermal processing in system "phenol–turpentine–solvent". Coke and Chemistry. 2014; **57**: 208–212. © Allerton Press, Inc., 2014.

Chapter 13

Phenolic Wastewaters: Definition, Sources and Treatment Processes

Yana Batista Brandão,
Julierme Gomes Correia de Oliveira and
Mohand Benachour

Additional information is available at the end of the chapter

http://dx.doi.org/10.5772/66366

Abstract

This chapter aims the state of the art concerning the development of advanced oxidation processes (AOPs) for treatment of organic-aqueous effluent for the reuse of liquid water. It presents the major oxidative processes applied for industrial and domestic treatment, where the effluents are often contaminated by phenolic compounds. A special emphasis is given to a relatively new technique called direct contact thermal treatment (DiCTT) that has the advantages of conventional AOP without its inconveniences. The DiCTT process is characterized by the generation of hydroxyl radicals (•OH) by combustion of natural gas, its compact installation and easy operation, being able to be used in offshore oil-exploration platforms, where natural gas is available and the space is reduced. Also, in this chapter, original results on the treatment of the DiCTT technique are presented, which are considered unconventional, by evaluating the oxidation and the conversion of the total organic carbon (TOC) of phenolic compounds at low temperature and atmospheric pressure, with identification and quantification of the intermediate compounds, using high-performance liquid chromatography (HPLC), which may be more toxic than the original pollutants.

Keywords: phenolic compounds, natural gas, AOP, DiCTT, TOC

1. Introduction, definitions and concepts

Water is an important renewable natural resource; however, it is also reusable. For humans, its employment becomes a problem due to demographic growth and its application in agricultural and industrial enterprises. Thus, the limited availability of water for an eventual reuse appears to be the unique solution. In the past few years, wastewater treatment is the adopted solution in the majority of countries. Traditionally, industrial effluents are frequently contaminated by hazardous organic substances, such as phenolic compounds and aromatic intermediates, as well as other halogenated or volatile organic substances, metals (mainly, Sb, Cu, Pb, Zn, Cd, Cr, Ni, Hg) and other chemical species, such as cyanide (CN^-), benzene and chloroform [1]. The presence of these compounds in wastewater and drinking water, due to their toxicity, become a serious problem affecting the ecosystem and causing, for example, health problems [2].

The phenolic compounds are harmful to human health, causing necrosis, digestive problems, and liver and kidney damages. The presence of phenols in drinking water may cause serious public health problems, and the death of fishes, even at concentrations in the range of 1 mg L^{-1}. At concentrations of less than 1 mg L^{-1} (ppm), they are also toxic to other biological species and destroy the aquatic environment [3].

Phenolic substances are widely employed as industrial chemicals for wood preservation; in petroleum refineries and petrochemical plants; coke gasifiers; in manufacturing of pulp and paper; pharmaceutical plants; food industry; minerals; plastics; metals and organic chemical plants, as well as in agricultural activities as pesticides. However, several research studies indicated that some of these phenolic organic substances are recalcitrant and persistent in treated water as they are refractory to conventional treatment [4, 5]. The most detected organic pollutants in those wastewater are presented in **Table 1** [1].

Phenols are compounds derived from aromatic hydrocarbons by replacing hydrogen atoms with hydroxyls. These compounds are generally solids and are obtained from coal tar distillation and heating of chloro-benzene with water [6]. Phenols can be classified according to the number of hydroxyls into monophenols, diphenols and triphenols. Phenol is less volatile than water and sparingly soluble in it, since the phenol-water system forms an azeotrope to 9.2% by mass of phenol [7].

The toxicity of these organic substances contaminants in different water bodies, including wastewater, surface water, groundwater and drinking water, at environmental levels of mg L^{-1}

Groups	Phenolic compounds
Phenol (C_6H_5OH)	2-Chloro-phenol; 2,4-di-chloro-phenol; 2,6-di-chloro-phenol; 2,4,5-tri-chloro-phenol; 2,4,6-tri-chloro-phenol; 2,3,4,5-tetra-chloro-phenol; 2,3,4,6-Tetra-chloro-phenol; 2,3,5,6-tetra-chloro-phenol; penta-chloro-phenol
Cresols (C_7H_8O)	2-Methyl-phenol; 3-methyl-phenol; 4-methyl-phenol; 2,4-di-methyl-phenol
Nitrophenols ($C_6H_5NO_3$)	2-Nitro-phenol; 4-nitro-phenol; 2,4-di-nitro-phenol
Aminophenols (C_6H_7NO)	4-Amino-phenol
Others	2-Chloro-3-methyl-phenol; 4-chloro-3-methyl-phenol; 2-methyl-4,6-di-nitro-phenol; 2-cycle-hexyl-4,6-di-nitro-phenol; 2-sec-butyl-4,6-di-nitro-phenol (Dinoseb)

Table 1. The main phenolic pollutants more detected in effluents.

significantly affect the organoleptic properties of water [8]. Resolution 430 of the Conselho Nacional do Meio Ambiente—CONAMA (National Council on the Environmental), Brazil set a maximum of total phenols concentration at 0.5 mg L^{-1} for all effluents originating from any polluting source that can be disposed of in water bodies as of 13 May 2011 [9].

According to literature review, phenolic and petroleum wastewater are recalcitrant compounds. Wastewater is generally characterized by the biological oxygen demand (BOD), chemical oxygen demand (COD), pH and total organic carbon (TOC) [10]. Petroleum wastewater contains a COD range from 850 to 1020 mg L^{-1}, a BOD range from 570 mg L^{-1}, a TOC range from 300 to 440 mg L^{-1} and a pH range from 8 to 8.2, showing that it contains large amounts of non-biodegradable organic matter [11]. Petro-chemical wastewater contains an initial COD range of 300–600 mg L^{-1}, a BOD range of 150–360 mg L^{-1} and a pH range of 7–8 [12]. Conventional treatment of these substances is difficult because biologically resistant organic compounds do not induce oxygen depletion in receiving water [13].

Several biological and chemical methods have shown a low efficiency to degrade the contaminants completely. As biological and chemical methods degrade only up to 60% of the recalcitrant components and in addition they require larger operation area and more chemical processes to reduce the sludge [14].

There are a number of studies of industrial effluents treatments via conventional methods that are combined with chemical, biological and physical methods and also advanced oxidation processes (AOPs), shared with reactors for complete degradation of highly recalcitrant industrial wastewater. Section 2 discusses the AOP theory and the possibility of enhancement in reactor performance when implemented in the processes.

2. Theory of advanced oxidation processes (AOPs)

The need for an efficient treatment of phenolic compounds in water is very important. When conventional treatment methods such as biological processes fail due to the recalcitrant nature of the contaminants, physical and chemical methods are a good solution. Therefore, oxidation processes are preferred to degrade such organics present. For example, the chemical processes are commonly used to degrade recalcitrant substances. High degradation efficiencies are possible with direct oxidation methods. However, pollution load, process limitation and operation condition are the more important factors considered during the selection for most oxidation processes [15]. The main treatment methods of industrial effluents with biological processes are aerobic, anaerobic and enzymatic. The physical processes are decantation, filtration and adsorption. The chemical methods are incineration, electrochemical and advanced oxidation processes(AOP), for example, photocatalysis, ozonation, Fenton/photo-Fenton and direct contact thermal treatment (DiCTT) [4].

The AOPs have been a viable alternative method for the wastewater treatment containing toxic and refractory organic pollutants, being studied in various combinations and are mainly based on intermediate reactions of hydroxyl radicals (•OH), an unstable and very reactive species, resulting in the degradation of toxic organic contaminants due to its high oxidizing potential of 2.8 V under acidic conditions and these processes have the major advantage of being a destructive treatment. Depending on the species to be degraded the hydroxyl radicals of reactive species, which attack the main part of organic molecules with a rate constant frequently in the order of 10^6–10^9 L mol^{-1}s^{-1}, and reacts 10^6–10^{12} times faster than ozone [16]. According to Tisa et al. [10] apud Bach et al. [17] and Garrido-Ramírez et al. [18], the principles of •OH generation is based on various combinations of strong oxidants, such as oxygen, ozone, hydrogen peroxide (H_2O_2), ultra-violet (UV) and electron beam.

The AOPs are classified according to the reactive phase (homogeneous and heterogeneous). The homogeneous including: ozone (O_3), ozone/ultraviolet (O_3/UV), ozone/hydrogen peroxide (O_3/H_2O_2), (O_3/H_2O_2/UV), and (H_2O_2/UV) processes, as well as Fenton (Fe^{2+}/H_2O_2), Fenton-like (Fe^{3+}/m^{n+}/H_2O_2), photo-Fenton (UV/Fe^{2+}/H_2O_2), sono-Fenton (US/Fe^{2+}/H_2O_2), electro-Fenton, sono-electro-Fenton, photo-electro-Fenton, sono-photo-Fenton. The heterogeneous includes: (TiO_2/ZnO/CdS+UV), (TiO_2/H_2O_2), (H_2O_2+Fe^{2+}/Fe^{3+}/m^{n+}-solid), (H_2O_2+Fe^0/Fe-nano-zero valente iron) and (H_2O_2+immobilized nano-zero valente iron). The non-conventional AOPs include ionizing radiation, microwaves and pulsed plasma techniques. Depending on the matrix and on the pollutant, degradation kinetics of AOPs can be zero order, first order and second order. First-order kinetics constant is achieved for pollutant degradation due to concentration of hydroxyl radicals within 1–10^{-4} s^{-1} [15, 4]. The mechanisms of different AOPs are presented in **Table 2**.

In the AOPs, oxygen and their reactive species (O_x, HO_x, $x = 1, 2, 3, 4$) act as main precursors during the oxidation that occurs in step of degradation of the organic component [23].

Devlin and Harris [24] proposed in their experimental trials that levels of O_2 decrease quickly in accordance with the rates of degradation of aromatic compounds, due to the high concentrations of phenol, in the temperature range from 420 to 498 K, near or above the stoichiometric conditions. These results led to demonstrate that the concentration of radicals (•O) is the dominant mechanism for this temperature range, in which the intermediate rings resulting from the oxidation of phenol are degraded.

Section 3 indicates that the direct contact thermal treatment technology is recent, with a limited investigation and was initially developed in Canada by Benali et al. [25, 26], being the unique experimental results available in the literature, until this moment. The DiCTT process provides a promising novel means to induce degradation and mineralization of organic pollutants in water being an AOP treatment method with respective advantages and limitations. Since combining AOP treatment process with reactors can be more promising in industrial applications, this research area needs to be explored further. Section 3 describes the DiCTT technique with applications in degradation and mineralization of the organic pollutant, being elaborated to highlight the important factors and their effects, as also a detailed investigation of the effective design and operating parameters are summarized.

The actual work evaluated the liquid phase flow rate (Q_L) of 100 and 170 L h^{-1} and the effect of initial phenol concentration (C_{Ph0}) of 500, 2000 and 3000 mg L^{-1}. The experiments studies were performed using a molar stoichiometric ratio of phenol/hydrogen peroxide ($R_{P/H}$) of 50%, an air excess (E) of 40%, a recycle rate of gaseous thermal wastes (Q_{RG}) of 50%, and, a natural gas

http://dx.doi.org/10.5772/66366

Name of the AOP	Types	Mechanism reaction	Highlights	References
Fenton oxidation	Homogeneous	$H_2O_2 + Fe^{2+} \rightarrow Fe^{3+} + \bullet OH + OH^-$	-Degradation of pollutant happens in acidic aqueous mixture	Lucas and Peres [19]
			- High efficiency	
	Heterogeneous	$Fe^{2+} + H_2O_2 \rightarrow Fe^{3+} + \bullet OH + OH^-$	- Reaction is possible in pH 5–7	Bach et al. [17]
		$Fe^{3+} + H_2O_2 \rightarrow FeHOO^{2+} + H^+$	- Catalyst reusability is possible	Garrido-Ramírez et al. [18]
		$FeHOO^{2+} + H^+ \rightarrow Fe^{2+} + \bullet HO_2$ $\bullet OH$ + organics $\rightarrow$ products	- Lag phase is observed that happens in activating the catalyst	
		$H_2O_2 + \bullet OH \rightarrow H_2O + \bullet HO_2$		
		$Fe^{2+} + \bullet OH \rightarrow Fe^{3+} + \bullet OH$		
		$Fe^{2+} + \bullet HO_2 \rightarrow Fe^{3+} + OH_2^-$		
		$Fe^{3+} + \bullet HO_2 \rightarrow Fe^{2+} + O_2 + H^+$		
Ozone	O_3/H_2O_2	$H_2O_2 + 2O_3 \rightarrow 2\bullet OH + 3O_2$	- This combined AOP works better in higher pH values	Esplugas et al. [20]
			- Degrades the pollutants into smaller cycle particles	
			- Inactive in reducing COD	
			- Oxidation rate is higher	
			- Follows direct pathway for pollutant degradation	
	O_3/UV	$O_3 + H_2O \rightarrow 2\bullet OH + O_2$	- UV accelerates ozone molecules and produces oxidizing radical	Esplugas et al. [19]
UV	UV/H_2O_2	$H_2O_2 \xrightarrow{\lambda > 300nm} 2 \bullet OH$	- UV irradiation supplies energy to the chemical compounds as radiation	Esplugas et al. [19]

Name of the AOP	Types	Mechanism reaction	Highlights	References
			- Reaction molecules reaches their excited state absorbing UV and promotes further reaction	
	UV/Fe^{2+}	$Fe(OH)^{2+} + UV \rightarrow Fe^{2+} + \bullet OH$	- Effectively can be applied in variety of pollutant degradation [21]	
		$Fe^{2+} + H_2O_2 \rightarrow Fe(OH)^{2+} + \bullet OH$	- High synergy effect between the ozone and the UV radiation has been noted	
Photo catalysts	Photo catalysts	$TiO_2 \xrightarrow{\lambda<400nm} e^-_{(cb)} + h^-_{(vb)}$	-Degradation of pollutant takes place by redox Oxidation [22]	Chen and Chou (1994)
		$e^-_{(cb)} + h^-_{(vb)} \rightarrow$ heat	- First-order kinetics are observed	
		$H_2O_2 + h^-_{(vb)} \rightarrow \bullet OH + H^+$		
		$\bullet OH$ + dyes $\rightarrow$ colourless		

Table 2. Mechanisms of different advanced oxidation processes.

flow (Q_{GN}) of 4 m^3 h^{-1}, on the oxidation of phenolic effluents by DiCTT process. The phenol concentration and mineralization content were obtained by high-performance liquid chromatography (HPLC) and mineralization Total Organic Carbon (TOC), respectively. A new data bank was compiled in this work by optimizing the operation conditions for the degradation/mineralization of phenol by the DiCTT process. (Phenol was used as a model compound for liquid organic wastes.)

3. Concepts of DiCTT-AOPs in degradation and mineralization of organic pollutants

3.1. Pilot plant and experimental procedure

Currently, a non-conventional AOP called Direct Contact Thermal Treatment (DiCTT) process has been investigated, whose main attraction are the use of natural gas as the energy source, the demonstrated ability to oxidize phenolic compounds at low temperatures and atmospheric pressure, and the generation of free radicals (•OH, •H, •CH_3 and •CHO) resulting from combustion of natural gas (methane). The installation of the DiCTT technology presents a compact reactor configuration that involves maintaining the reactor in a vertical position, favouring the immediate application of this technology in off-shore oil platforms, where natural gas is available and space is limited [4, 27].

The experimental unit used is mainly composed of a vertical stainless steel reactor, being 1.359 mm high and 203 mm internal diameter, a gas-liquid separator, a first tank (Tank 1) for preparation of phenol synthetic solutions, 400 L volume, a second tank (Tank 2) for feeding in polluted waters, also of 400 L volume, and a natural gas burner with 50 kW maximum power. The combustion air is provided by an axial fan with 0.3 HP power. The output pressure of the natural gas supply is 2 × 10^5 Pa, followed by a reduction to 1 × 10^5 Pa for combustion in the burner.

In this process, the effluent is tangentially injected into the reactor to produce a helical flow in its inner walls. The helical flow allows an intimate contact between the effluent and free radicals produced in the combustion flame of the natural gas, resulting in a thermochemical oxidation in the liquid phase, while avoiding their incineration. The high flame temperature contributes to the oxidation performance of the effluent in the presence of free radicals and works for the oxidation process to be carried out in completely liquid phase, by transferring some of the excess oxygen present in the flame.

The DiCTT process is a thermochemical oxidation method in aqueous medium, and generating free radicals resulting from the combustion of natural gas (methane) according to the reaction mechanism described by the whole of Eq. (1) following [27]:

$$\begin{aligned} &CH_4 \rightarrow \bullet CH_3 + \bullet H \\ &\bullet CH_3 + O_2 \rightarrow CH_2O + \bullet OH \\ &\bullet OH + CH_4 \rightarrow \bullet CH_3 \\ &\bullet OH + CH_4 \rightarrow \bullet CH + 2 \bullet OH \end{aligned} \tag{1}$$

This technique presents operational and capital costs 2.5 times lower than those of wet air oxidation (WAO) and 4.1 times lower than those of electric plasma oxidation (EPO) [27].

Figure 1 show a schematic representation of the pilot plant used in the experiments that was composed of a vertical, stainless-steel reactor and a gas-liquid separator.

The phenol solution was prepared in Tank 1. The operation of the system was stabilized by heating the water to almost 70°C for an hour and a half; phenol was subsequently added to Tank 1, and the synthetic effluent was transferred from Tank 1 to Tank 2. The reactor had an internal helical groove with a rectangular shape, in the axial direction, through which the liquid effluent flowed. Wastewater polluted with phenol was injected into the reactor

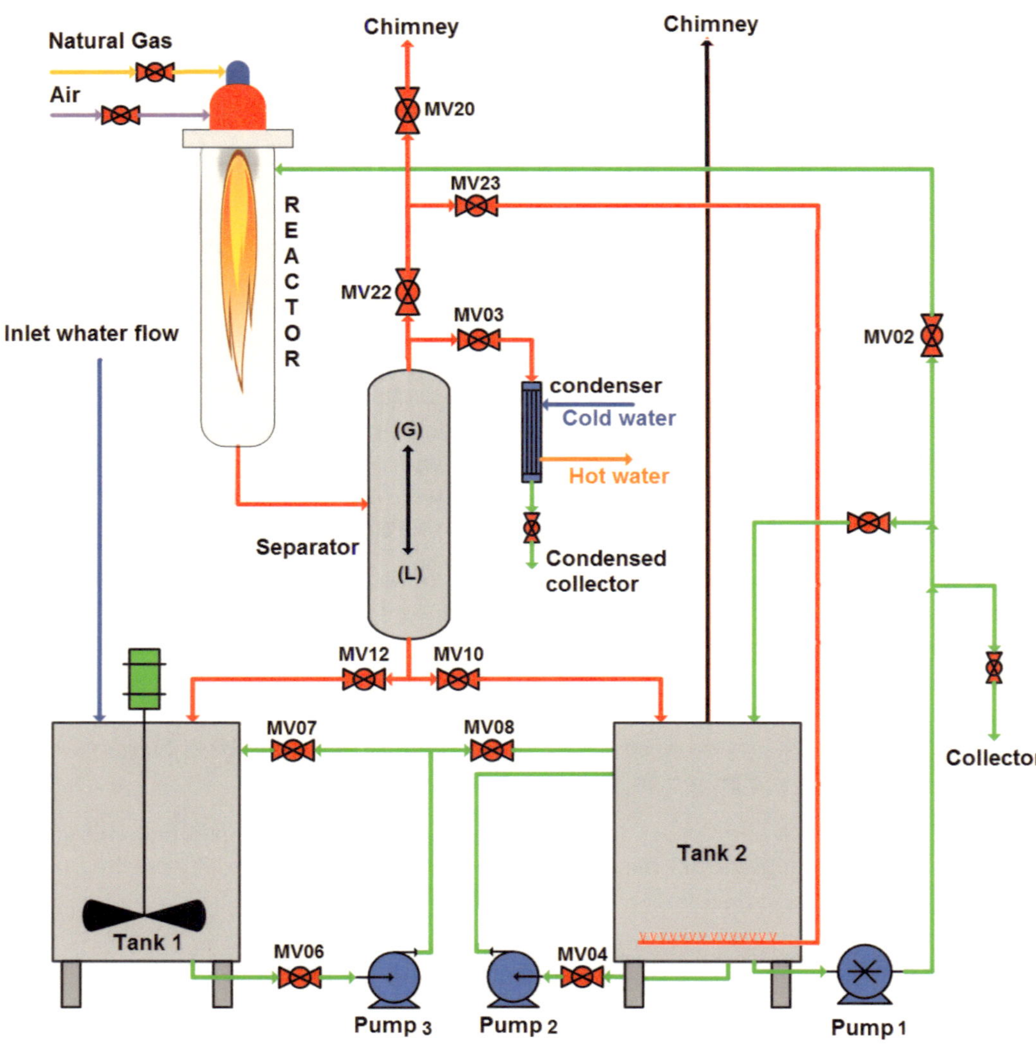

Figure 1. Pilot plant using the DiCTT process.

tangentially to produce a liquid helical stream on its inner walls. The combustion gases were vented to the atmosphere through a chimney; a fraction of recycled combustion gases (of the total flow rate Q_{RG}) was immediately injected into Tank 2 by adjusting an open valve to heat the solution in the recirculation tank (Tank 2) more rapidly and to dissolve a fraction of the residual oxygen from combustion into the reaction liquid, thereby inducing the thermochemical oxidation of the phenolic compounds. For the experiments, 250L of effluent was prepared. For each experiment 250mL samples of effluent in black plastic bottles were collected at previously chosen points and put to cool in a refrigerator. For the analyses, 250mL of treated drinking water was employed as a reference. To initiate the oxidation reaction, some millilitres of phenol/hydrogen peroxide with a mole ratio, R_{PH} was introduced into Tank 2. Liquid samples were withdrawn for analysis through a collector located at the entrance of the tubes connecting the feed tank (Tank 2) to the reactor inlet (**Figure 1**).

3.2. Analytical methods

For the experiments, a phenol solution of analytical grade and oxygenated water 35% PA were employed. For the chromatographic analysis, methanol UV/HPLC grade and for TOC analysis phosphoric acid 25% PA were used.

The concentrations of phenol, catechol and hydroquinone were monitored using an HPLC instrument (Shimadzu, model LC-20AT, with integrated data acquisition using a UV detector and a CLC-ODS column (M)/(C-18) that was 250 mm in length and 4.6 mm in diameter, also from Shimadzu). An isocratic elution mode was used under the following conditions: oven temperature of 35°C; flow rate of 0.75 mL min^{-1} for the mobile phase; injection volume of 20 μL; mobile phase consisting of 10% methanol and 90% phosphoric acid/deionised water with pH adjusted to 2.2; and operation of the UV detector at a wavelength of 270 nm to detect phenol, catechol and hydroquinone [4].

The TOC content was measured using a TOC analyser (TOC-Vcsh model, Shimadzu) to analyse phenolic mineralisation quantitatively [28].

3.3. Parameters and calculated data

Phenol oxidation is a common reaction stoichiometry described in Eq. (2):

$$C_6H_5OH + 14\,H_2O_2 \rightarrow 6\,CO_2 + 17\,H_2O \tag{2}$$

From Eq. (2) others relations were calculated.

Molar ratios other than 100% were calculated proportionally using the reaction stoichiometry in Eq. (2).

In this work, natural gas (89.24% of methane) supplied by COPERGAS (Pernambuco, Brazil) was employed [29]. Stoichiometrically 9.881 mol of air reacts with one mol of methane. So, the excess air (E) in the combustion and the equivalent ratio (Φ) may be evaluated employing Eqs. (3) and (4) [30, 31].

$$E = \frac{1}{9.881}\left(\frac{Q_{AR}}{Q_{GN}}\right) - 1, \tag{3}$$

$$\varphi = 9.881\left(\frac{Q_{GN}}{Q_{AR}}\right), \tag{4}$$

where Q_{AR} denotes the volumetric flow rate of air, and Q_{GN} denotes the volumetric flow rate of natural gas.

The power dissipated by the burner (P) was calculated using Eq. (5):

$$P = Q_{GN} \cdot PCM, \tag{5}$$

where PCM denotes the average heat of combustion of natural gas, which has a value of 34,740 kJ m^{-3} [29].

The percent degradation of phenol (X_F) was calculated using Eq. (6):

$$X_F = \left(\frac{Q_L \cdot C_{Ph_0} - Q_L \cdot C_{Ph} - F_G \cdot C_{Phv}}{Q_L \cdot C_{Ph_0}}\right) \times 100\,, \tag{6}$$

where Q_L represents the volumetric flow rate, C_{ph0}, the initial phenol concentration, C_{ph}, the phenol concentration at time t, F_G the dry air mass flow rate, C_{phv}, the phenol concentration in the condensate at time t. TOC conversion was evaluated via Eq. (7).

The TOC conversion (X_T) was evaluated via Eq. (7):

$$X_T = \left(\frac{Q_L \,.\, TOC_0 - Q_L \,.\, TOC - F_G \,.\, TOCv}{Q_L \cdot (TOC_0 - TOC_B)}\right).100\,, \tag{7}$$

where TOC_0 denotes the initial total organic carbon concentration, TOC and TOC_V denote the total organic carbon and the total organic carbon in the condensate, respectively, at a time point t of the process and TOC_B denotes the total organic carbon in the blank.

3.4. Effect of the liquid phase flow rate

This work was to evaluate the influence of the liquid phase flow rate on the level of phenol oxidation, settling other process variables. These operational parameters are listed in **Table 3**.

Tests	Q_L(L h^{-1})	Q_{GN}(m^3 h^{-1})	E(%)	C_{ph0}(mgL^{-1})	Q_{RG}(%)	$R_{P/H}$(%)
E1	170	4	40	500	50	50
E2	100	4	40	500	50	50

Table 3. Operational parameters for the study of the influence of the liquid phase flow rate.

Figure 2a and **b** show, respectively, the evolution of the temperature and pH of the liquid effluent present in the feed tank (Tank 2) during the process, varying the volumetric flow of the same.

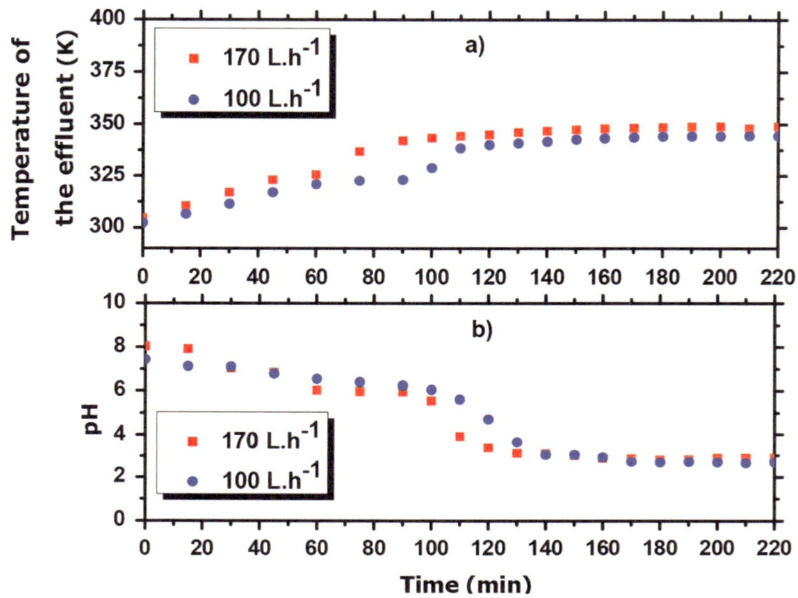

Figure 2. (a) Evolution of temperature of the liquid effluent as a function of the operating time. (b) Evolution of pH as a function of the operating time. E = 40%, Q_{GN} = 4 m^3 h^{-1}, C_{Ph0} = 500 mg L^{-1}, $R_{P/H}$ = 50% and Q_{RG} = 50%.

Figure 2a indicates that the elevation of effluent flow affects the heating curve of liquid effluent, getting it faster to obtain steady-state temperature in Tank 2. **Figure 2a** also show that for the larger effluent flow (170 L h^{-1}), the same reaches a maximum temperature of 348 K, quite greater than the temperature of the effluent flow of the 100 L h^{-1} (344 K), featuring a small effect, since the difference in these temperatures reaches an order of magnitude lower than the measurement uncertainty by thermocouples.

The oxidation process indicates that the heating profile is characterized by two distinct steps, a first, approximately 110 min, characterized by a rapid temperature rise and a second, after 110 min of operation, showing a temporal increase rate of low temperature. **Figure 2b** show that the evolution of pH also identifies these two steps, the first of about 110 min, characterized by a decreased less than hydrogen potential as a function of time, and the second-fastest where the temporal decay profile of pH is more significant. It is shown a low influence in liquid flow in the dynamics of acid formation [24].

Figure 3a and **b** show, respectively, the profile of phenol degradation and residual fraction of the TOC as a function of time for the two flows of liquid effluent studied, 100 and 170 L h^{-1}.

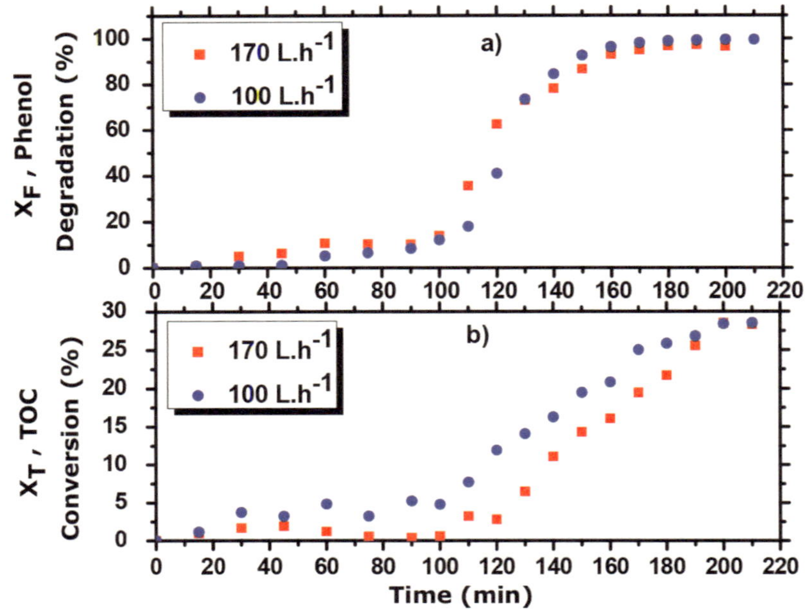

Figure 3. (a) Evolution of phenol degradation as a function of the operating time. (b) Evolution of TOC conversion as a function of the operating time. $E = 40\%$, $Q_{GN} = 4\ m^3\ h^{-1}$, $C_{Ph0} = 500\ mg\ L^{-1}$, $R_{P/H} = 50\%$ and $Q_{RG} = 50\%$.

Figure 3a indicates that, in the range studied, the variation of the effluent flow does not affect the profile of phenol degradation and total degradation of the same is almost completely achieved in 180 min of operation, reaching values of 99.5 and 97.4%, respectively, at effluent flow of 100 and 170 L h^{-1}. **Figure 3b** indicates that the increase of the effluent flow of 100 to 170 L h^{-1} allows a higher speed of phenol mineralization, due the acceleration of the lowest value of the pH, but not interfering in the maximum value TOC conversion, around 28% with an operating time of 210 min.

Figure 4a and **b** indicates, respectively, the time profiles of the hydroquinone and catechol formed by thermochemical phenol oxidation, for the two flows of liquid effluent, 100 and 170 L h^{-1}. Analysing the same figures, it can be seen that the rate of formation of these species becomes appreciable after the induction period, approximately 110 min, previously observed by the curves of the evolution of pH, phenol degradation and TOC conversion.

The evolution of the hydroquinone and catechol concentrations happened quickly because of the thermochemical oxidation reaction of phenol with high speed, regardless of the flows of liquid effluent studied. It has been observed that hydroquinone and catechol concentrations are reached when phenol consumption rate is maximum, which is identified in the process time between 140 and 150 min. After reaching the maximum hydroquinone and catechol formation, an immediate reduction in the concentration of these two species is observed, indicating that to achieve the maximum consumption of phenol, the oxidation rate of these two organic compounds becomes greater than its rate of formation, enabling to be degraded, thus favouring the formation of other organic compounds that are not acids, because the pH remained almost constant at 2.5–2.8, after

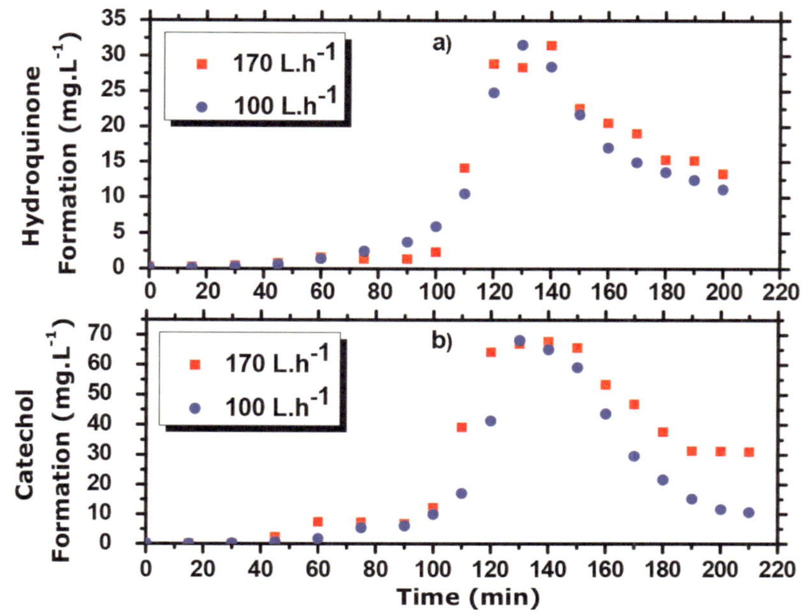

Figure 4. (a) Evolution of hydroquinone formation as a function of the operating time. (b) Evolution of catechol formation as a function of the operating time. $E = 40\%$, $Q_{GN} = 4\ m^3\ h^{-1}$, $C_{Ph0} = 500\ mg\ L^{-1}$, $R_{P/H} = 50\%$ and $Q_{RG} = 50\%$.

140–150 min of operation. The products resulting from the oxidation of hydroquinone and catechol are probably aldehydes (Glyoxal, for example, in the case of hydroquinone and catechol) and alkenes (1,4-dioxo-2-butene, for example, in the case of hydroquinone) [24].

Figure 4a and presents the results obtained in the quantification of concentrations of hydroquinone and catechol, respectively. It also show a speed of formation and consumption of these species not significantly affected by variation of the flows of liquid effluent. It is also observed a rate of production and disappearance slightly larger (especially in the case of catechol) with use of effluent flow rate 170 L h^{-1}, similarly what was evidenced in the evolution of pH, being less for the flow rate of 170 L h^{-1}, allowing more oxidation of phenol to hydroquinone and catechol. However, regardless of the flow of the liquid studied, catechol concentrations were approximately two times higher compared to those obtained in relation to the hydroquinone.

3.5. Effect of initial phenol concentration

In order to evaluate the effect of initial concentration of the organic pollutant on the efficiency of the process DiCTT on thermochemical phenol oxidation, three initial concentrations of the aromatic contaminant (C_{Ph0}): 500, 2000 and 3000 mg L^{-1}, were employed keeping all other variables constant. The operating conditions used in this study are presented in **Table 4**.

Figure 5a and **b** presents, respectively, the evolution of the temperature and pH of the liquid effluent in the perfect mixing tank (Tank 2) during the process, varying only the initial concentration of phenol. From **Figure 5a** it can be seen that the concentration of phenol does

Tests	C_{PH0}(mgL^{-1})	Q_{GN}(m^3 h^{-1})	E(%)	Q_L(Lh^{-1})	Q_{RG}(%)	$R_{P/H}$(%)
E3	500	4	40	170	50	50
E4	2000	4	40	170	50	50
E5	3000	4	40	170	50	50

Table 4. Operational parameters for the study of the influence of the initial concentration of phenol.

not influence the heating curve of the liquid phase, reaching a temperature of approximately 350 K (77°C). An expected behaviour since the variation of concentration, in different experiments, it is not enough to significantly change the chemical and thermophysical properties of the effluent, since the natural gas flow in the process is the same for all cases, as well as the excess air and effluent flow which remained constant during these essays.

Figure 5b show the curves of the evolution of pH for concentrations of phenol 2000 and 3000 mg L^{-1}. It can be seen that the initial pH value already show low values, 4 and 3, respectively, while for a C_{Ph0} equal 500 mg L^{-1}, pH presents an initial value of 8. This can be explained due to the amount of hydrogen peroxide added. In the procedure adopted for the preparation of the synthetic effluent, the peroxide is mixed with the phenol solution in the preparation tank, causing uncontrolled reactions. As the molar stoichiometric ratio of the mix is kept constant, to higher concentrations of phenol oxidant availability in the reaction medium is greater, increasing the effect and decreasing the initial pH due to a possible premature oxidation of phenol to form organic acids.

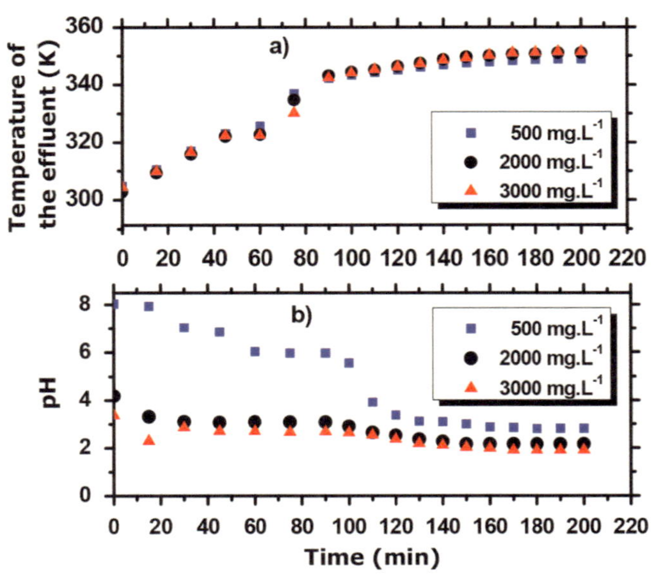

Figure 5. (a) Evolution of temperature of the liquid effluent as a function of the operating time. (b) Evolution of pH as a function of the operating time. $E = 40\%$, $Q_{GN} = 4$ m^3 h^{-1}, $Q_L = 170$ L h^{-1}, $R_{P/H} = 50\%$ and $Q_{RG} = 50\%$.

http://dx.doi.org/10.5772/66366

Figure 6a and **b** presents, respectively, the profile of phenol degradation and TOC conversion as a function of time, for different initial concentrations of phenol studied.

Figure 6a show that the increase of the initial concentration of phenol from 500 to 3000 mg L^{-1} does not affect the duration of the first step of the reaction, called induction period, and does not have a significant effect on the phenol degradation after a time of approximately 130 min. After the induction period, around 110 min, the speed of the reaction becomes more pronounced, as was expected, reaching values of X_F practically the same after an operating time of around 130 min, regardless of the initial concentration of phenol. Phenol degradation around 99% is obtained after an operating time of 180 min.

Figure 6b show the evolution of TOC conversion, identifying a time of induction period also approximately 110 min, and show a slight increase in the TOC conversion with increasing of the initial phenol concentration, reaching X_T values of 27.5; 31.5 and 33.5% to initial concentration of phenol of 500, 2000 and 3000 mg L^{-1}, respectively, after an operating time of 210 min. Regardless of the value of the initial phenol concentration, the process presents maximum rates of phenol degradation almost 100% after 170 min of operation, in addition to providing a TOC conversion, between a range of 27.5–33.5%, after 210 min, within the range of the initial concentration of phenol, being the air excess used of 40% and a combustion gases recycling rate of 50%.

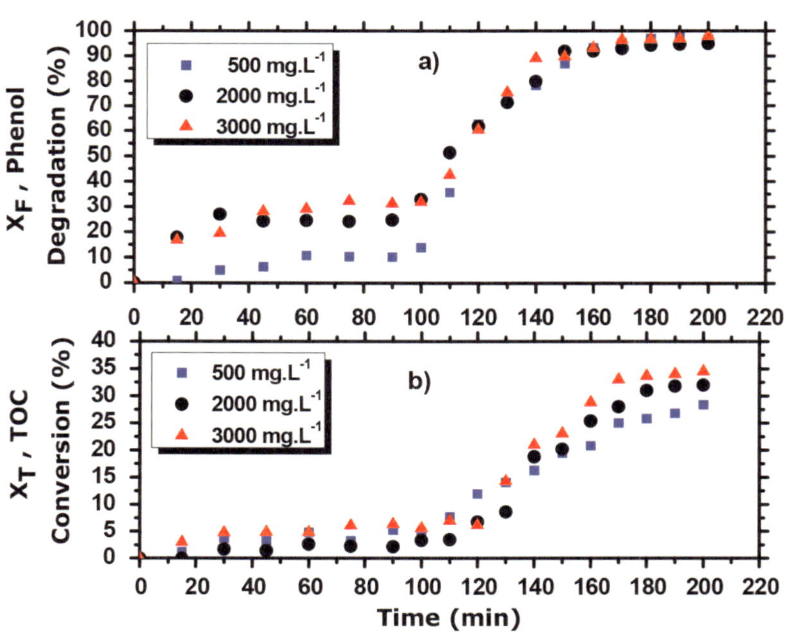

Figure 6. (a) Evolution of phenol degradation as a function of the operating time. (b) Evolution of TOC conversion as a function of the operating time. E = 40%, Q_{GN} = 4 m^3 h^{-1}, Q_L = 170 L h^{-1}, $R_{P/H}$ = 50% and Q_{RG} = 50%.

Figure 7a and **b** show, respectively, the results obtained in the quantification of the concentrations of hydroquinone and catechol formation. The evolution of the concentration profiles of hydroquinone and catechol in function of the time confirm clearly the induction time of reaction, around 110 min, identified initially by the curves of time evolution of the phenol degradation (**Figure 6a**) and TOC conversion (**Figure 6b**). It can be that maximum values of hydroquinone and catechol concentration formed in approximately 140 min operating time, reaching the maximum speed of phenol degradation and TOC conversion and that the catechol concentrations are always higher than the hydroquinone concentration.

After 140 min, both hydroquinone and catechol concentrations decrease, thus allowing the formation of other organic compounds that are not acids, because the pH becomes practically constant (pH = 3) after 140 min of operation (**Figure 5b**). It can be that the products resulting from the oxidation of hydroquinone and catechol are possibly aldehydes (Glyoxal, for example, in the case of hydroquinone and catechol) and alkenes (1,4-dioxo-2-butene, for example, in the case of hydroquinone).

The phenol oxidation produces catechol and hydroquinone [20]. Analyses indicated a higher catechol production then hydroquinone. This may be explained by the mesomeric effect. This signifies an electron re-distribution to the ortho position, which increases its reactivity at this position of the molecule due to the proximity of opposing charges [4, 27].

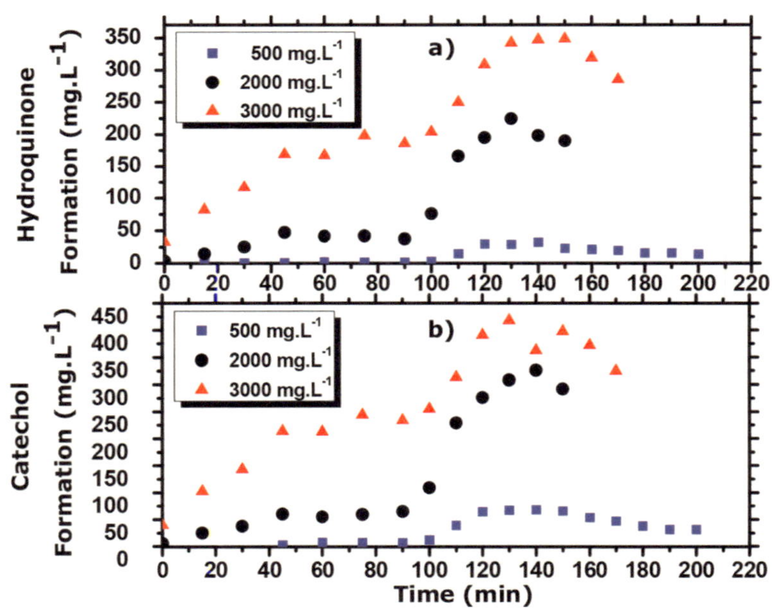

Figure 7. (a) Evolution of hydroquinone formation as a function of the operating time. (b) Evolution of catechol formation as a function of the operating time. E = 40%, Q_{GN} = 4 m^3 h^{-1}, Q_L = 170 L h^{-1}, $R_{P/H}$ = 50% and Q_{RG} = 50%.

http://dx.doi.org/10.5772/66366

4. Conclusions and recommendations

The Advanced Oxidation Processes (AOPs) are found to be an environmental friendly process for the degradation and mineralization of refractory compounds. The limitations of conventional processes in wastewater treatment necessitate study on the AOPs. Thus, applications of the Direct Contact Thermal Treatment (DiCTT) process is a promising technique increasingly used to remove phenolic compounds in water. The method advantages are operational and capital costs lower than other process and ability to allow total degradation and higher mineralization of target compounds when compared to conventional AOPs. On the other hand, further efforts are conducted to overcome the empirical aspect by studying the operating parameters, as well as the optimization of the process.

The complete degradation of phenol (almost 100%) was obtained independently of the flows of liquid effluent, 100 and 170 L h^{-1}, and, of the initial phenol concentrations, 500, 2000 and 3000 mg L^{-1} over a 180-min period. A TOC conversion of almost 35% was observed corresponding to an operational time of approximately 210 min at a Q_L of 170 L h^{-1}, which allows the speed of phenol mineralization is faster, but without interfering in the final value of almost 28% TOC conversion after 210 min of operation process. The flows of liquid effluent of 170 L h^{-1} was considered to be the best operating condition for the DiCTT process. An induction time of approximately 110 min was identified from the concentration profiles of hydroquinone and catechol. The concentrations of these intermediates tented to decrease independently of the flows of liquid effluent and of the initial phenol concentrations, indicating the formation of the other organic compounds, which were not acids (constant pH) according to data reported in the literature.

Acknowledgements

The authors wish to thank the Financiadora de Estudos e Projetos—FINEP/Ministry of Science and Technology—MCT—Brazil and PETROBRÁS for providing financial support during the development of this work and the Conselho Nacional de Desenvolvimento Científico e Tecnológico—CNPq for awarding the research grants.

Author details

Yana Batista Brandão[1]*, Julierme Gomes Correia de Oliveira[2] and Mohand Benachour[3]*

*Address all correspondence to: yanabatista@yahoo.com.br and mbena.br@gmail.com

1 Department of Engineering, University Federal Rural of Pernambuco, Cabo de Santo Agostinho, Pernambuco, Brazil

2 Department of Engineering, College of Boa Viagem, Recife, Pernambuco, Brazil

3 Department of Chemical Engineering, University Federal of Pernambuco, Recife, Pernambuco, Brazil

References

[1] Rodrigues, GD, Silva, LHM, Silva, MCH. Green alternatives for the preparation of sample and determination of phenolic pollutants in water, Química Nova, 2010: **33**, 1370–1378.

[2] Semrany, S, Favier, L, Djelal, H, Taha, S, Amrane, A. Bioaugmentation: possible solution in the treatment of Bio-Refractory Organic Compounds (Bio-ROCs). Biochem. Eng. J., 2012: **69**, 75–86.

[3] Mishra, VS, Mahajani, VV., Joshi, JB. Wet air oxidation. Ind. Eng. Chem. Res., 1995: **34**, 2-48.

[4] Brandão, Y, Teodosio, J, Dias, F, Eustáquio, W, Benachour, M. Treatment of phenolic effluents by a thermochemical oxidation process (DiCTT) and modelling by artificial neural networks. Fuel, 2013: **110**, 185-195.

[5] Luna, AJ, Rojas, LOA., Melo, DMA., Benachour, M, de Souza, JF. Total catalytic wet oxidation of phenol and its chlorinated derivates with MnO_2/CeO_2 catalyst in a slurry reactor. Brazil. J. Chem. Eng., 2009: **26**, 493–502.

[6] Experimental and kinetic diffusional resistance on wet oxidation of phenolic compounds using metal catalysts [Master of Science, Dissertation]. Natal, RN: UFRN; 2003.

[7] Hackbart LM. Liquid-liquid equilibrium of Systems Containing Phenol-Water-solvent:Obtaining and Thermodynamic Modelling [Master of Science, Dissertation]. Curitiba, PR: UFPR; 2007.

[8] Britto, JM, Rangel, MC. Advanced oxidation processes of phenolic compounds in industrial effluents. Química Nova, 2008: **31**, 114–122.

[9] National Council for the Environment (CONAMA). Resolution, 2011: **43**.

[10] Tisa, F, Raman, AAA, Daud, WMAW. Applicability of fluidized bed reactor in recalcitrant compound degradation through advanced oxidation processes: a review. J. Environ. Manage., 2014: **46**, 260–275.

[11] Coelho, A, Castro, AV, Dezotti, M. Treatment of petroleum refinery sour-water by advanced oxidation processes. J. Hazard Mater., 2006: **137**, 178–184.

[12] Ma, F, Guo, J-B, Zhao, L-J, Chang, C-C, Cui, D. Application of bioaugmentation to improve the activated sludge system into the contact oxidation system treating petrochemical wastewater. Bioresour. Technol., 2009: **100**, 597–602.

[13] Guo, J-B, Al-Dahhan, M. Catalytic wet air oxidation of phenol in concurrent downflow and upflow packed-bed reactors over pillared clay catalyst. Chem. Eng. Sci., 2005: **60**, 735–746.

[14] Oller, S, Malato, JA, Sánchez-Pérez. Combination of advanced oxidation processes and biological treatments for wastewater decontamination—a review. Sci. Total Environ., 2011: **409**, 4141–4166.

[15] Babuponnusami, A, Muthukumar, K. A review on Fenton and improvements to the Fenton process for wastewater treatment. J. Environ. Chem. Eng., 2014: **2**, 557–572.

[16] Hoigne, J. Inter-calibration of OH radical sources and water quality parameters. Water Sci. Technol., 1997: **35**, 1–8.

[17] Bach, A, Shemer, H, Semiat, R. Kinetics of phenol mineralization by Fenton-like oxidation. Desalination, 2010: **264**, 188–192.

[18] Garrido-Ramírez, EG, Theng, BKG, Mora, ML. Clays and oxide minerals as catalysts and nanocatalysts in Fenton-like reactions—a review. Appl. Clay Sci., 2010: **47**, 182–192.

[19] Lucas, MS, Peres, JA. Degradation of Reactive Black 5 by Fenton/UV-C and ferrioxalate/ H_2O_2/solar light processes. Dyes Pigments, 2007: **74**, 622–629.

[20] Esplugas, S, Gimenez, J, Contreras, S, Pascual, E, Rodrigues, M. Comparison of diferente advanced oxidation processes for phenol degradation. Water Res., 2001: **36**, 1034–1042.

[21] Ayoub, K, Nelieu, S, van Hullebusch, ED, Maia-Grondard, A, Cassir, M, Bermond, A. TNT oxidation by Fenton reaction: reagent ratio effect on kinetics and early stage degradation pathways. Chem. Eng. J., 2011: **173**, 309–317.

[22] Nidheesh, PV, Gandhimathi, R, Ramesh, ST. Degradation of dyes from aqueous solution by Fenton processes: a review. Environ. Sci. Pollut. Res. 2013: **20**, 2099–2132.

[23] Oppenländer T. Photochemical purification of water and air—Advanced Oxidation Processes (AOPs): principles, reaction mechanisms, reactor concepts. Wiley-VCH: Weinheim, 2003. 368p.

[24] Devlin, HR, Harris, IJ. Mechanism of the oxidation of aqueous phenol with dissolved oxygen. Ind. Eng. Chem. Res., 1984: **23**, 387–92.

[25] Guy, C, Benali, M and Ostiguy, E. Free radical oxidation process and installation for treating liquid effluents contaminated by organic substances. Canadian Patent, 2002, no. 2187982.

[26] Benali, M and Guy, C. Thermochemical oxidation of phenolic-laden liquid effluent models. J. Environ. Eng. Sci., 2007: **6**, 543-552.

[27] Brandão, Y, Teodosio, J, Benachour, M, Oliveira, J, Marinho, I, Figueirêdo, F, Anselmo-Filho, P. Study of the effect of excess air and dissipation of the burner on the capabilities of the DiCTT process in liquid phenolic wastewater treatment. Revista Iberoamericana Sistemática—Cibernética e Informática, 2010: **7**, 1–9.

[28] Fonseca, JCL, Silva, MRA, Bautitz, IR. Nogueira, RFP, Marchi, MRR. Evaluation of the reliability of analytical determinations of total organic carbon (TOC). Ecletica Quim, 2006: **31**, 47–52.

[29] Pernambuco Gas Company (COPERGÁS) [Internet]. 2009. Available from: http://www.copergas.com.br/site/ctudoconteudo.asp?idsecao=4 [Accessed: 2009-04-03].

[30] Oliveira, JGC. Theoretical study via CFD-computational and experimental study of the combustion of natural gas for liquid organic effluent treatment by applying the new technology DICTT [Master of Science, Dissertation]. Recife, PE: UFPE; 2007.

[31] Baukal, Jr. Charles, E, Schwartz, RE. The John Zink combustion handbook. Boca Raton, FL: CRC Press; 2001, 354 p.

Chapter 14

Removal of Phenolic Compounds from Water by Adsorption and Photocatalysis

Eduardo Enrique Pérez Ramírez,
Miguel de la Luz Asunción,
Veronica Saucedo Rivalcoba,
Ana Laura Martínez Hernández and
Carlos Velasco Santos

Additional information is available at the end of the chapter

http://dx.doi.org/10.5772/66895

Abstract

Phenolic compounds are important industrial wastes, and are classified as hazardous substances contaminating groundwater resources. Therefore, the removal or diminish of these organics compounds in order to reach the permitted levels before discharging becomes a challenging. Several processes have been developed to remove phenolic compounds from waters, including electrochemical oxidation, redox reactions, membrane separation and photocatalytic degradation. Recently, tendency of phenolic compounds removal involves adsorption and photocatalytic process, using synthetic or natural particles, such as carbon materials and clays. Actually, materials in nanometric scale play an important role in the processes previously mention due to their unique chemical and physical properties. In this book chapter, the first part shows the chemical properties of phenolic compounds that play an important role in the removal process. In the second part, different materials in macro, micro and nanosize used as adsorbents or photocatalysts are reviewed. In addition, other removal processes of phenolic compounds as electrochemistry and redox reactions are included. The removal conditions in these process, such as pH, adsorbate and adsorbent concentration are analyzed and discussed. Furthermore, special emphasis is included in micro and nanocarbon materials, used as adsorbents or photocatalyst to remove phenol from water in recently researches.

Keywords: phenolic compounds, water pollution, removal methods, adsorption, photocatalysis

1. Introduction

Water pollution is one of the most important problems in the world, which represents a risk to the human and environment. The increasing industrial and human activities have caused an increase on the discharge of wastewater into the water resources. Phenolic compounds from different industrial activities such as refineries, pesticides, insecticides, pharmaceutical, etc., are found among the main pollutants of water. These compounds are toxic and their degradation is difficult; thus, it is important the development of materials and effective methods that allow the removal of these pollutants from water.

Different methods have been used to assist with this problem. The adsorption and photocatalysis are two promising technologies related to the removal of phenol from the water. In the adsorption process, the molecules of the contaminant are retained on the surface of the adsorbent material and then these can be separated from the water. In the photocatalysis process, a semiconductor material is used to produce chemical species with high reactivity and it is possible degrading the molecules of the contaminant.

Some materials commonly used on the removal of phenolic compounds by adsorption are activated carbon, clays, zeolites, membranes and recently has emerged a new class of adsorbents, the nanomaterials. These materials are very promising in this area. On the other hand, the size reduction in the semiconductors particles increases the degradation of phenolic compounds due to the increments on their surface area. In addition, carbon nanomaterials as graphene and graphene-based materials have demonstrated an important performance on the degradation of phenolic compounds.

Thus, this chapter presents a review of researches where the adsorbents before mentioned have been used on the removal of different phenolic compounds. Phenolic compounds degraded by photocatalysis are also reviewed and presented. The influence of some parameters such as temperature, pH solution, dosage of photocatalyst/adsorbent, solution concentration on the process of adsorption and photocatalysis are also analyzed.

2. Phenolic compounds

In the last times, phenolic compounds have attracted a great interest, because they have several applications indispensable in our daily life. The phenolic compounds are present in adhesives, foams, emulsifiers and detergents, insecticides, dyes, explosives, flavors and rubber chemicals, self-assembly to nanomaterials, resins and so on, in other different applications. All these products have great economic importance in many industries such as food, medicine, petrochemical, agriculture, chemical synthesis and polymer chemistry, among others [1]. Thus, there is a great environmental interest in the removal of phenolic compounds. Phenolic compounds are among the most important contaminants present in the environment. On the other hand, phenolic compounds are not only generated by human activity, but they are also formed naturally, the phenols are present in soils and sediments and therefore these compounds produce the contamination of groundwater. Some organizations such as the European Union and the US Environmental Protection Agency (EPA) have included as priority contaminants, some of the phenolic compounds due to their high toxicity and persistence in the environment [2]. The structures of eleven phenols considered priority pollutants by the EPA are shown in **Figure 1**.

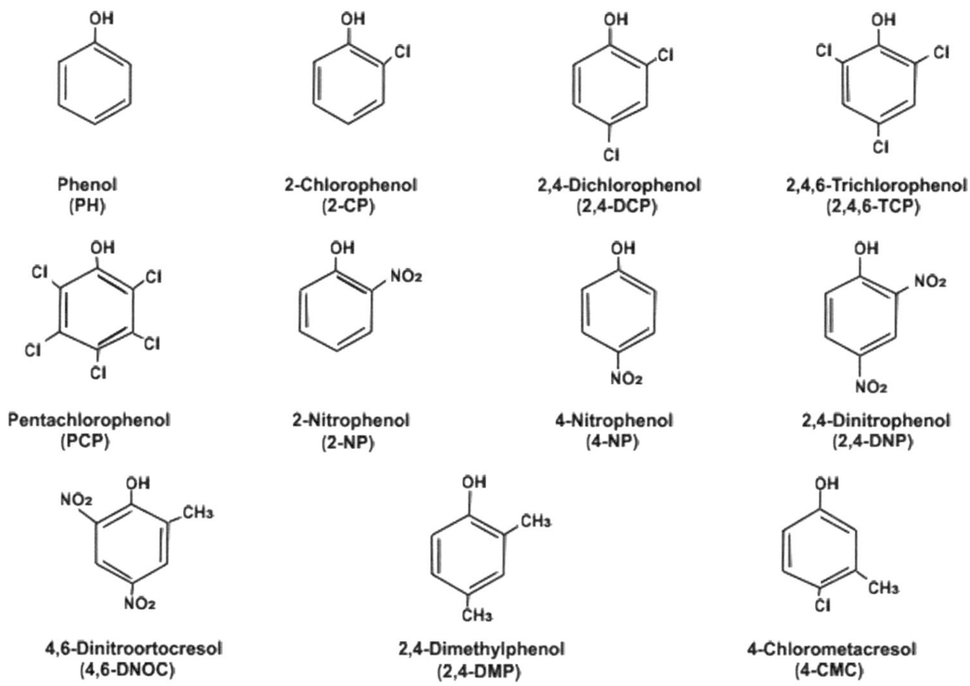

Figure 1. Phenolic compounds structures considered priority contaminants by US EPA. Reprinted with permission from Mahugo-Santana et al. [2]. Used under Creative Commons Attribution 3.0 Unported (CC BY 3.0) (https://creativecommons.org/licenses/by/3.0/).

2.1. Generalities

Industries generate large annual volumes of wastewater containing hazardous compounds to the environment, including natural water resources. Many of these compounds are in concentrations that are too low for recovery but high enough to act as contaminants; usually, they are nonbiodegradable substances. Therefore, conventional biological processes do not have the ability to remove all contaminants that are present in the industrial wastewater. Some organic compounds are hardy to biological degradation such as phenols, tannic acids, lignine, cellulose, chlorinated compounds, pesticides, aromatic hydrocarbons and so on; all of them are examples of nonbiodegradable organic compounds that are extensively employed in the chemical industry and, therefore, are common contaminants found in industrial wastewater discharges. In recent years, phenolic compounds have warranted more attention in the field of industrial wastewaters, because of their toxicity and the frequency of industrial processes producing waters contaminated by phenols [3].

Pollution of environment is one of the main problems facing humans today. Recently, the problem of environmental pollution has increased exponentially and reached worrying level in terms of its impact on the life of human beings. Among the contaminants that have harmful effects in animals and humans are considered the toxic organic compounds. As mentioned earlier, dissolved phenolic compounds that are present in industrial wastewater cause pollution of groundwater and owing to its harmful effect these compounds generate a serious problem in this type of water resources. Exposure to this type of chemical reagents, once they enter into human body can cause damage to the nervous and respiratory systems, kidney and blood system. Phenolic compounds have been classified as the top 45th in the list of priority hazardous substances by the Agency for Toxic substances and Disease Registry, USA, which require immediate treatment before disposal in the environment [4]. Consequently, removing organic compounds or reducing their concentrations to the permitted levels by environmental standards represents a big challenge.

2.2. Chemistry of phenolic compounds

The approach of this section is to present a summary of physicochemical properties of phenolic compounds.

2.2.1. Physical properties of phenols

Phenols are in some respects as alcohols, due to the presence of hydroxyl groups in their structures. They have the ability to form strong hydrogen bonds. Moreover, these compounds present higher boiling points than hydrocarbons of the same molecular weight. Phenols are also slightly soluble in water because of their ability to form strong hydrogen bonds with water molecules [5]. Phenols are stronger acids than alcohols. They react with bases like sodium hydroxide to form phenoxide ions. However, they are weaker acids than carboxylic acids and do not react with sodium hydrogen carbonate [6].

2.2.2. Reactions of phenols

Phenolic compounds behave as nucleophiles in most of their reactions and also the reagents that interact on them behave as electrophiles. In phenolic compounds, the site of nucleophilic reactivity may occur at the hydroxyl group or the aromatic ring. The reactions are carried out on the aromatic ring results in electrophilic aromatic substitution [7].

Halogenation: Bromination and chlorination of phenols occur easily even in the absence of a catalyst. Substitution occurs primarily at the para position to the hydroxyl group. When the para position is blocked, ortho-substitution is carried out.

Nitration: Phenol reacts with dilute nitric acid in either water or acetic acid. It is not necessary to use mixtures of nitric and sulfuric acids, due to the high reactivity of phenolic compounds. The o-nitrophenol is a phenolic compound ortho-substituted and therefore, this compound has considerably lower boiling point than the meta and para isomers. This is due to the hydrogen bond that is produced between the hydroxyl group and the substituent partially compensates for the energy required to go from the liquid to the vapor phase.

http://dx.doi.org/10.5772/66895

Nitrosation: The nitrosonium ion [N ≡ O^+] is obtained during the acidification of sodium nitrite, which is a weak electrophile and reacts with the strongly activated ring of a phenol. The resulting product is a nitrosophenol.

Sulfonation: The sulfonation of the ring can be carried out by reacting phenol with concentrated sulfuric acid.

Friedel-Crafts alkylation: Alcohols in combination with acids serve as sources of carbocations. Attack of a carbocation on the electron ring of a phenol results in the alkylation.

Oxidation of phenols: quinones: Phenols do not undergo oxidation in the same way that alcohols do because they do not have a hydrogen atom on the hydroxyl-bearing carbon. Instead, oxidation of a phenol yields a cyclohexa-2,5-diene-1,4-dione, or quinone [8].

3. Removal methods of phenolic compounds from water

Phenolic compounds are priority contaminants with high toxicity even at low concentrations. These compounds are present in industrial effluents, where increase biochemical and chemical oxygen demands resulting in detrimental effects on the environment. Some of them are highly toxic as well as carcinogenic and can remain in the environment for a long time due to their stability and bioaccumulation. Owing to the high toxicity of phenolic compounds, treatment of the organic wastewater has an important effect on the lives of human beings [9].

Many phenolic compounds can be removed efficiently by conventional treatments such as extraction, distillation, chemical oxidation, electrochemical oxidation and adsorption among others. On the other hand, some advanced treatments use less chemical reagents compared to the conventional processes, but they have the disadvantage of having high energy costs. Within the advanced treatments are as follows: Fenton, ozonation, wet air oxidation and photochemical method. Biological treatments have certain advantages compared to physico-chemical treatments; among these advantages may be mentioned: environmentally friendly and energy saving. However, it has the disadvantage that cannot treat high concentration of contaminants. One of the best ways to treat phenolic compounds under mild conditions is the enzymatic treatment, which uses different enzymes such as peroxidases, laccases and tyrosinases [10]. Thus, there is a need to treat waste contaminated with phenolic compounds at low and high concentrations before discharge. Some methods used today are described below:

Adsorption: Adsorption method for removal of phenols from water is effective from low concentrations to high concentrations, depending on the economics and recycling the required secondary material, adsorbent. Activated carbon (AC) is the most used in industry as adsorbent. It is expensive but has been shown to be effective for removal of trace organic compounds. Therefore, new options are being developed including impregnation with nanoparticles, different sources of carbon, different activation methods, carbon nanotubes (CNTs), graphene-based materials, as well as substitution with low cost biosorbents, such as chitin/chitosan which are promising alternatives to remove phenolic compounds [11–13].

Membrane processes: Membrane processes are applied in water and wastewater treatment to remove organic contaminants. At present, this technology has been investigated for the phenolic compounds removal. Low energy consumption, low operating cost and easy scale up by membrane modules are the main advantages of these technologies. Today, separation membranes have many uses with a growing potential for industrial applications in biotechnology, nanotechnology and purification processes.

Reverse osmosis and nanofiltration: Reverse osmosis (RO) is a membrane-based demineralization technique that is used to separate dissolved solids, especially ions, mainly from aqueous solutions. On the other hand, nanofiltration (NF) is widely used for removing organic pollutants, inorganic salts, color and hardness from aqueous solutions. NF is useful to use prior to an RO unit in order to decrease the pressures associated with organic matter [14].

Chemical oxidation: Chemical oxidants provide destructive methods of phenolic compounds. The processes have low consumption of reagents and energy costs, operating under mild conditions (temperature and pH). Ozone, chlorine, chlorine dioxide, chloramines, ferrate [Fe (VI)] and permanganate [Mn (VII)] are the most common chemicals applied in oxidative treatment of contaminated water.

Electrochemical oxidation: This technique can effectively oxidize many organic contaminants at high chloride concentration, usually larger than 3 g/L. Electrochemical oxidation is an alternative destructive of phenols which does not require addition of reagents. This technique is divided into direct and indirect oxidation. Direct or anodic treatment occurs through adsorption of the contaminants on the anode surface. Various anode materials are used with Pt, PbO_2, SnO_2, IrO_2 and BDD (boron-doped diamond) being the most investigated ones. Parameters such as current density, pH, anode material and electrolytes used have significant impact on process efficiency.

Advanced oxidation processes: Advanced oxidation processes (AOP) are techniques that present the common feature that they form hydroxyl radical (OH•) in situ and this free radical is capable of mineralizing most organics, including phenolics compounds. AOP are used mainly for the treatment of contaminated waters that contain recalcitrant organics (e.g., pesticides, surfactants, coloring matters, pharmaceuticals) [10].

Fenton and fenton-like treatment: An AOP with the capability to oxidize aromatic compounds is the Fenton reagent, which consists of hydrogen peroxide (H_2O_2) and ferrous ion at low pH. The iron (II) reacts with H_2O_2 to produce iron (III) and hydroxyl radicals. Then, Iron (III) is regenerated to Fe (II) by H_2O_2 in acid medium. Some of the variants of the Fenton process are as follows: Fenton-like, photo-Fenton and electro-Fenton [15].

Biological treatment: Biological treatment is the most commonly applied treatment for aqueous phenols. The treatment is an inexpensive method, simple design and maintenance, for transforming phenolic solutions into simple end products.

Wide research is carried out daily on phenolic compounds removal from water, from conventional methods to new technologies. Optimization and modification of conventional processes provide attractive alternatives on contaminants removal. Some other methods used in the

http://dx.doi.org/10.5772/66895

removal of phenol are as follows: wet air oxidation (WAO), catalytic wet air oxidation (CWAO), solvent extraction, extractive membrane bioreactors (EMBR), photocatalytic membrane reactors (PMR), UV/H_2O_2 treatment with microwave, etc.

3.1. Adsorption

Adsorption process is preferred over all methods because it is nondestructive and with this method is possible recover the organics pollutants through regeneration, relatively simple. Due to the high adsorption capacity of adsorbents, adsorption seems to be the best process, especially for the removal of moderate and low concentration phenolic compounds from an effluent [9].

The most common method for the removal of dissolved organic material is the adsorption with activated carbon, a product that is produced from a variety of carbonaceous materials, including wood, pulp mill char, peat, lignite, etc. Adsorption is the physical and/or chemical process in which a substance is accumulated at an interface between phases. The substance which is being removed from the liquid phase to the interface is called as adsorbate and the solid phase in the process is known as adsorbent. Physical adsorption (physisorption) is relatively nonspecific and is due to the operation of weak forces between molecules. Chemical adsorption (chemisorption) is also based on electrostatic forces, but much stronger forces act a major role on this process. In chemisorption, the attraction between adsorbent and adsorbate is due to a covalent bond or electrostatic forces among atoms [16].

3.1.1. Zeolites and clays

Zeolites and clays are two adsorbent materials commonly used on the adsorption process. Different investigations have shown interesting results on the phenolic compound removal.

Khalid [17] has carried out a research on phenol removal using four kinds of zeolites as adsorbents and adsorption properties were compared to those of an activated carbon. In this investigation, phenol diluted in water was used as contaminant and adsorption was carried out in batch and continuous flow. Siliceous BEA zeolite was successfully used; the adsorption capacity was slightly higher at low phenol concentration (1.6 g/L) than the one of activated carbon. Siliceous BEA zeolite showed to be efficient as adsorbent able to be easily regenerated.

Investigation of removal of 3-nitrophenol isomers (ortho, meta and para) was studied by Huong [18]. They used nano zeolite (NZ) as adsorbents. The adsorption of nitrophenols onto NZ reached equilibrium within 150 min at pH 6.0. The maximum adsorption capacities of NZ for meta-, ortho- and para-nitrophenols were 125.7, 143.8 and 156.7 mg/g, respectively. The removal percentages of nitrophenols were maintained at more than 70% of the initial values. The regeneration process showed that desorption efficiency of nitrophenols remained above 70% even after five adsorption-desorption cycles.

Adsorption capacity of a modified zeolite was evaluated by Xie [19] for the removal of ionizable phenolic compounds (phenol, p-chlorophenol and bisphenol A) and nonionizable organic compounds (aniline, nitrobenzene and naphthalene). The isotherm data of ionizable compounds fitted well to the Langmuir model but those of non-ionizable chemicals followed a linear equation. Adsorption capacity of ionizable compounds depended greatly on pH, increasing at alkaline pH conditions. On the other hand, adsorption of non-ionizable compounds was practically the same at all pH levels studied.

Djebbar et al. [20] employed as adsorbent, a natural clay, for the removal of phenol from aqueous solutions. This clay was easily activated. Some parameters such as pH solution, temperature, contact time and initial phenol concentration were studied. The adsorption experiments were carried out employing 100 mg of adsorbent and 100 mL of phenol solution at different initial concentrations of phenol at 23°C. The results indicated that up to 60 and 70% of phenol was removed by activated and natural clay after of 5 h of contact time. The activated process improved the adsorption of phenol onto natural clay. The adsorption capacity of phenol decreased when the temperature was increasing. The best results were obtained at pH 5. Adsorption equilibrium data were well fitted to both Freudlich and Langmuir isotherm indicating that the adsorption was favorable. The adsorption of phenol onto activated natural clay was exothermic.

Other investigation of phenol removal from water with clay of low cost was investigated by Nayak and Singh [21]. The influence of pH phenol solution, temperature and particles size was studied. Results indicated that the higher adsorption capacity of phenol was achieved when the particle size decreased from 140 to 50 μm, the pH decreased from 10 to 2 and the temperature increased from 30 to 50°C. The adsorption process was found to be spontaneous.

The removal of p-chlorophenol (PCP) and p-nitrophenol (PNP) from water with two types of organoclays prepared from different surfactants such as dodecyltrimethylammonium bromide (DDTMA) and didodecyldimethylammonium bromide (DDDMA) was investigated by Park [22]. In the experiments, 200 mg of adsorbents were dispersed into 30 mL of PNP and PCP solutions at initial concentration of 100 mg/L and pH 5–6. In the isotherm studies, the initial concentration of PNP and PCP was studied in the range of 5–250 g/L. The best adsorption results were found on the organoclays where DDDMA surfactant was used. The adsorption of PNP and PCP onto organoclays was more efficient than in unmodified clay which was attributed to hydrophobic behavior. The adsorption equilibrium data were well fitted to Freundlich isotherm, indicating the presence of multilayer sorption.

3.1.2. Membranes

Membranes are considered a process to separate two streams, a barrier to facilitate the selective mass transport between fluids; feed and permeate [23]. Before to select the optimal membrane to remove or recover a specific compound, it is important to know the macro and molecular separation level. Munirasu et al. [24], divided into two categories, inorganic; anion and cation and organic compounds; these later compounds are quite complex and due to its nature, they can be classified such as oil, grease, dissolved, disperse and emulsified organic forms, solids and/or particles, such as clays, waxes, bacteria, sand or any solids based on chemical productions. Membranes separation efficiency depends on diverse factors including physic-chemical composition; as type, weight, polarity and solute charge, operating parameters; as feed flow rate, transmembrane pressure, temperature, permeate flux; also it is important to contemplate the membrane characteristics, for example, membrane material, porous size and configuration of membranes (modules). On the other hand, these aspects play an important key role on specific phenomena related to the concentration polarization and membrane fouling, contributing directly to the solute retention and hydrophobic interactions between the solute and membrane surface [25].

The application of a driven force, measure as transmembrane pressure (TPM), divided the conventional membrane separation in microfiltration (MF), ultrafiltration (UF) and nanofiltration (NF) processes (**Figure 2**), the membranes separate the feed solution into permeate and retentate. The permeate stream contains the solvent passing through the membrane, this stream is rich in solutes with a nominal weight cut-off (NMWCO) below the porous size of membranes and the retentate stream are particles and dissolved compounds which are kept inside membrane. When a pressure force is applied, the membrane operation and the hydrodynamic resistance fluctuate, depending on the pore size; thus, the operating pressure increases while the pore size of the membrane decreases [24, 25]. Pressure driven force is a strategy to improve low weight molecules removal, for example, salt or organic compounds; however, despite the excellent rejection of salts, the process frequently present low rejection levels to organic molecules, which include aliphatic or aromatic chemical structures; with polar or nonpolar properties, as well as different kinds of alcoholic, amino, carboxyl, phenol, or hydroxyl functional groups, just to mention some examples [24, 26]. Bellona et al. [27] reported the factors affecting the permeation of solutes in high-pressure membrane and

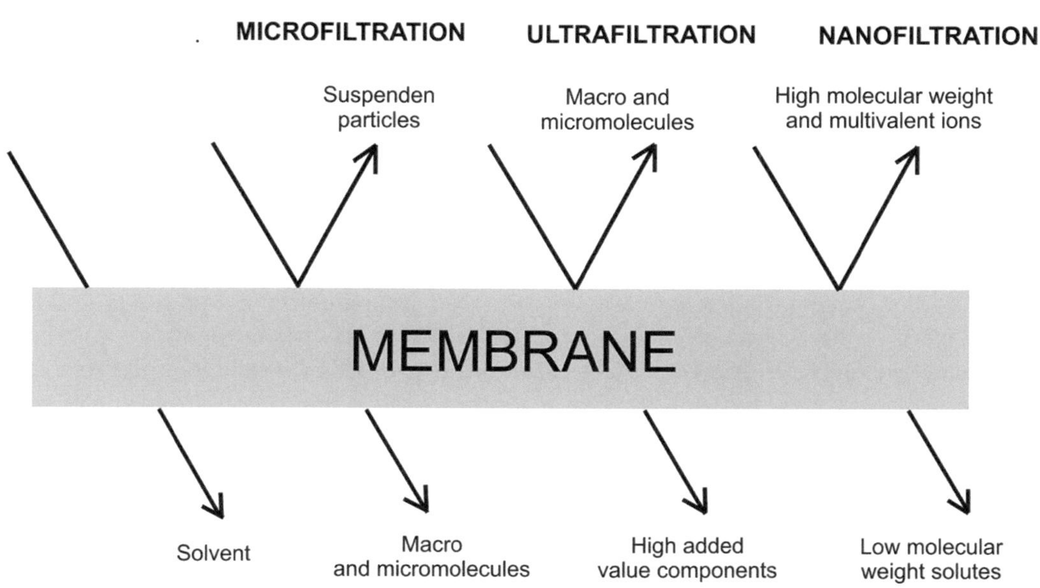

Figure 2. Scheme for microfiltration (MF), ultrafitltration (UF) and nanofiltration (NF) technologies. Adapted from Castro-Muñoz et al. [25].

some important solute parameters were identified to those that affect mainly solute rejection: molecular weight, molecular size (length and width), acid disassociation constant (pKa), hydrophobicity/hydrophilicity character and diffusion coefficient. Besides, membrane properties play an important role on the separation process: molecular weight cut-off, pore size, surface charge (zeta potential), hydrophobicity/hydrophilicity (contact angle) and surface morphology (roughness). Furthermore, water composition, such as pH, ionic strength, hardness and the presence of organic matter, was also associated to influence on the solute rejection.

Membrane process has been used to remove organic pollutants, among these technologies the liquid, anion exchange, nanofiltration, reverse osmosis and pervaporation membranes are precedents of this technology [28]. Hybrid processes is a recent tendency to get better removal of these compounds. These procedures are based on combination of diverse techniques, such as adsorption pretreatment process accomplish with reverse osmosis [29] and pervaporation with reverse osmosis [30], forward osmosis (FO) and reverse osmosis (RO) [31], NF/RO membranes [32], polymerization of phenolic compounds and UF/MF removal membranes [33].

Therefore, there is still a necessity to find advanced techniques to remove nonbiodegradable, high concentration organic substances from wastewater, not only those come from refineries but all complex wastewater from industry. In this sense, researchers are trying to design a combination of treatment methods for a complete and successful removal of such pollutants, due to variability of wastewater composition, the traditional methods become inadequate and could not be used individually in full scale [34]. Phenol and phenol stability usually offers difficulties to remove them, some of the main advantages of applying advanced techniques are the interfacial area; lower solvent losses, downstream phase separation and easy scale up. In the case of membrane process, the higher interfacial mass transfer, overcome the lower mass transfer rate in this kind of systems [25, 35–37].

In recent years, also composite membranes have been investigated to remove phenol compounds. In these materials, surface properties are controlled during membrane synthesis. Membrane aromatic recovery system (MARS) is a promising technology to recover phenol and aromatic amines. Composite membranes including poly(dimethylsiloxane) (PDMS) are commonly investigated to control porosity and the operational stability by synthetizing nonporous selective layer coated on a microporous support layer cast; reinforcement polymers could be poly(vinylidine fluoride), polyethersulfones, polyetherimides, polyacrylonitrile, polyester, polyphenylenesulphones. Xiao et al. [38] developed a pertraction membrane (pervaporation and extraction combine process), through plate composite polydimethylsiloxane/polyvinylidene fluoride polymers to recover phenol compound. Results show that mass transfer coefficient is five times higher compared to silicon rubber membranes nonreinforced (from 15×10^{-7} to 3.5×10^{-7} m/s), another property improved was the mass flux (2.38×10^{-2} kg/m^2 h), however, diminish the activation energy of permeation (9.7 kJ/mol), permeability (5.9×10^{-12} m^2/s) and diffusion coefficient (2.4×10^{-11} m^2/s). Lee et al. [39] also found a higher permeate flux using a wet phase inversion process polydimethylsioloxane/polysulfone (PDMS/PS) composite. Permeate flux is influenced by controlling the skin layer thickness of the asymmetric membrane during formation reaction, phenol concentration and recirculation rate. Additionally, the relatively hydrophilic nature of phenol and specificity of membranes are related to chemical moieties, both point of views allow making highly flexible polymer, with hydrophobicity and organophilicity properties and controlled free volume; this is the case of the poly(dimethylsiloxane) (PDMS) and poly(vinylmethylsiloxane) (PVMS) to synthesize extractive membrane bioreactor (EMBR). Main characteristic to the EMBR system is to permeate organic compounds, meaning it should have a high organic flux while being effectively impermeable to inorganic and water, such as silicon-based rubbers [40].

Due to complexity of phenol and phenolic compounds, researchers are focused on to develop new technologies to improve efficiency removal, mainly on the size/steric exclusion, electrostatic repulsion, fouling and energy consumption. Hybrid or combine process involving two or more steps, previously mention, could be the answer to the problem. Heo et al. [41] suggest that forward osmosis/reverse osmosis (FO/RO) provide the advantage to enhance the driven pressure force, diminishing the fouling property and making it less energy costly process. Due to, FO depends on the molar concentration of the solution instead of the nature of the solutes and RO offers higher selectivity characteristics, internal concentration polarization and low flux; the efficiency of both membranes (FO/RO) was attributed to porous, mesh fabric, hydrophobicity and steric hindrance. Surface fouling is one of the issues to overcome on the wastewater treatment; thus, forward osmosis alleviated the reverse osmosis membrane fouling as demonstrated by Choi et al. [31], after repeated cleaning membrane process where the permeate flux was recovered. Biopolymer-like substances were persistently accumulated on the membrane surface as seen in **Figure 3**.

Some authors have paid attention on the porosity as main property to develop integrated systems on phenol removal and/or recovery, taking into account the molecular weight (MW) of the species, such as the case of the combined nanofiltration and reverse osmosis [32, 42]. However, in case that a specific compound is desired to isolate, the complexity of the system could include more than two combine steps, for example, an innovative integrated process to recover an important food polyphenol, such as the Gallic acid. The proposal consisting of purification steps based on the MW of the specific molecule with UF-NF-RO and their further separation with an adsorption/desorption resins, where the final product had a phenol concentration of 378 g/L in Gallic acid equivalents and the initial quantity was 2.64 g/L [43].

The membranes contactor is an alternative technology, based on solvent extraction using hollow fiber membranes (membrane-based extraction method), to recover or remove low concentration of aromatic compounds. Two fluid phases flow in adjacent channels with the interface maintained in the intermediate membranes pores; in other words, the process involves extraction of the compound to a second phase stabilizing the aqueous and organic phases within the pores of the polymeric membrane. If compared to the conventional solvent extraction, membrane contactors offer a large interfacial area, a reduction in solvent by-products and lower solvent losses. Nevertheless, the operational pressure range limits the applications of the process. Research challenger is the interface stabilization of fluids in the membrane pores, which affect the operational conditions and the properties of both, membrane materials and fluids [44, 45]. Diverse factors to the design of this kind of membranes include modified surface properties. It can improve the gradual erosion caused by the shear forces generated by the aqueous phase

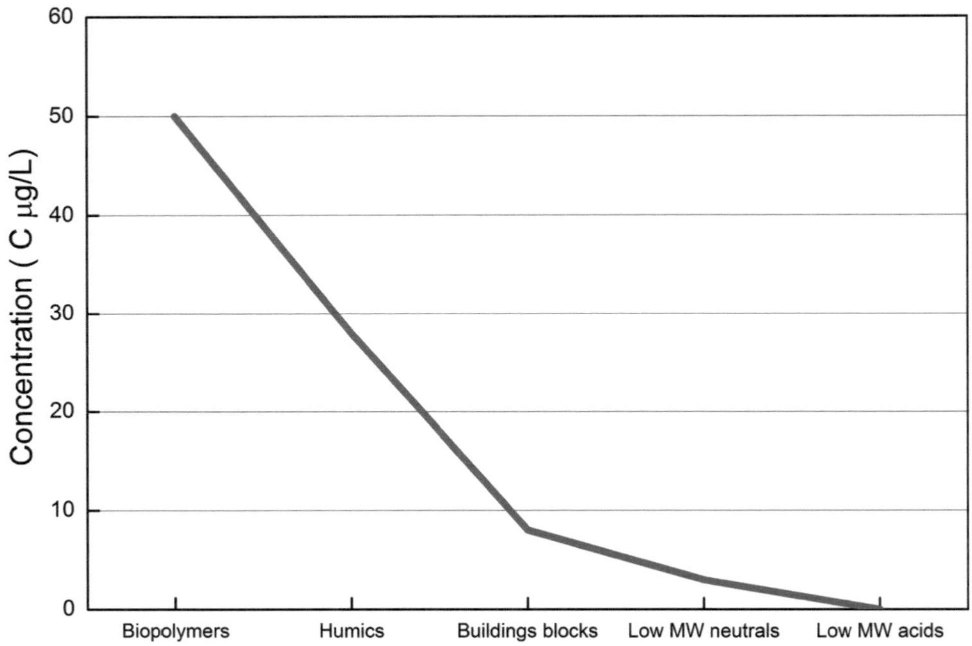

Figure 3. Concentration of DOMs (byopolymers, humic, buildings blocks, low-MW neutrals and low-MW acids) on the fouled membranes surface. Adapted from Choi et al. [31].

flowing over the liquid membrane. Alternatives, such as hollow fiber, supported liquid membrane (HFSLM), with solids-extractants immobilized on the pores of the hydrophobic surface, serve to dissolve a nonvolatile carrier solvent to maintain a high distribution coefficient for the solute. Trioctylphosphine oxide (TOPO) act as extractants in the liquid membranes; due to in solid crystalline powder can be impregnated on a polypropylene membrane, changing from a conventional liquid-liquid extraction into a solvent extraction liquid membrane of aromatic compounds; to study model molecules [46] or applied as osmotic membranes bioreactor (TPPOMBR) [47]. A variant of liquid-liquid phenol extraction and hollow fiber contactor membranes are the liquid membrane processes. Mainly, the process combines liquid-liquid extraction and stripping operations in a single unit operation; pertraction. In a pertraction system, the liquid membrane phase (organic solution phase) separates into two additional liquid phases (feed and stripping phase) which are immiscible (aqueous solution phase). There exist different types of liquid membranes, bulk liquid membranes (BLM), supported liquid membranes (SLM) and emulsion liquid membranes (ELM). The bulk liquid membrane is considering one of the simplest arrangements of liquid membrane systems. Mass transfer, solvent extractants, pH, temperature and rate migration are the foremost drawbacks to study [48–50].

3.1.3. Carbon-based materials (activated carbon and nanomaterials)

Due to structural and surface properties, carbon materials as activated carbon obtained of different sources have been used as effective adsorbents of pollutants from water. However, in addition to the structural and surface properties, the size of the adsorbents is an important factor that can improve the efficiency of the process, therefore, carbon nanomaterials as carbon nanotubes and graphene have been used as adsorbents of pollutants from water obtaining good results. Thus, investigations related to carbon activated, carbon nanotubes and graphene-based materials used as adsorbents of phenolic compounds from water are reviewed in this section.

Activate carbon (AC) is a carbon material very effective in the removal of pollutants from water. Its effectivity was probed on the removal of bisphenol A (BA). Liu et al. [51] investigated the effect of modification treatments of two AC (W20 and F20) onto removal of BA. The ACs treated with nitric acid were labeled as W20A and F20A and the AC modified with thermal treatment under a flow of N_2 was labeled as W20N and F20N. The highest adsorption capacities of BA were found on W20 (382.12 mg/g) and W20N (432.34 mg/g) samples. The thermal treatment favored more the BA adsorption on AC than the acid treatment. The surface charge density of the different ACs and their content of oxygenated groups are factors very important that affect the BA adsorption. Similarly than other experiments with phenolic compounds, the adsorption of BA onto ACs decreased when the temperature was increasing. The removal by adsorption of Bisphenol A also has been investigated with AC derivates of coconut and coal with good results [52]. The best results were found in the pH range from 3 to 9. The adsorption capacity of both ACs decreased in pH range (>10), it was attributed to the electrostatic repulsion between the negative charge surface of the ACs and the bisphenolate anion from the ionization of bisphenol A. Factors as surface area and pore volume were determinants for the good adsorption of BA onto ACs; however, surface polarity also played an important role in this adsorption process.

Fasfous et al. [53] studied the adsorption of tetrabromobisphenol (TBBPA) on multiwalled carbon nanotubes (MWCNTs). The results showed that MWCNTs have a high potential for removal of TBBPA from water. The removal of TBBPA after of 60 min was 90%. The adsorption capacity was increasing when the initial TBBPA concentration and contact time were increased. Oppositely, the adsorption capacity of TBBPA decreased when the temperature and pH (pH > 7) were increased. The experimental kinetic data were well adjusted to the pseudo-second-order model and the Freundlich and Langmuir models described well the experimental equilibrium data. The adsorption of TBBPA on MWCNTs was spontaneous with exothermic nature. In other investigation for TBBPA removal, Zhang et al. [54] employed graphene oxide (GO) as adsorbent. The experiments were conducted modifying the pH of TBBPA solution in the range from 2 to 12 and the temperature from 288 to 318 K. The results indicated that at 0.3 and 1 mg/L TBBPA concentrations, the maximum adsorption capacities were of 70–90% after of 120 min, indicating an influence of the initial concentration of TBBPA. The adsorption capacity of TBBPA on GO decreased when the temperature of the solutions was increased. The adsorption process of TBBPA on GO was exothermic. These effect of the temperature and type of process were similar when were used MWCNTs. The main interaction mechanisms between TBBPA and GO were π- π interaction and hydrogen bonding. Ji et al. [55] also investigated the removal of TBBPA but using Fe_3O_4 nanoparticles loaded on MWCNTs (MWCNTs-Fe_3O_4) composite and MWCNTs-Fe_3O_4 modified with 3-aminopropyltriethoxysilane (APTS) (MWCNTs/Fe_3O_4-NH_2). The solution concentration of the TBBPA was 10 mg/L and the adsorbent dosage was 0.5 g/L for all experiments. The solution pH was adjusted from 1.4 to 9 in order to optimize the pH for the maximum adsorption capacity of TBBPA on the composites. The results showed that the adsorption capacity of TBBPA on the two composites was increasing when the pH increasing from 1.5 to about 5.5, where the TBBPA is not dissociated. The donor-acceptor interactions between TBBPA and the magnetic nanocomposites through the graphene sheets of MWCNTs, the aromatic structure of TBBPA and π-π interactions between the benzene-ring structure on both of TBBPA and MWCNTs and Hydrogen bonding are the main possible adsorption mechanism of TBBPA on the two nanocomposites. The functionalization of the MWCNTs/Fe_3O_4 with amine groups improved the adsorption of TBBPA. The maximum adsorption capacity of TBBPA was found on MWCNTs/Fe_3O_4 nanocomposite (33.72 mg/g). The experimental kinetic data were well adjusted to the pseudo-second-order model.

The removal of phenol, 2-chlorophenol and 4-chlorophenol from aqueous solutions using as adsorbents activated carbon, multiwalled carbon nanotubes and carbon-encapsulated iron nanoparticles (CEINs) was investigated by Strachowski and Bystrzejewski [56]. The surface area of the different materials was found to be 1187, 156 and 36 m^2/g for AC, MWCNTs and CEINs, respectively. All adsorption kinetic experiments were carried out with a 150 mg/L initial concentration of the phenolic compounds and the relation of adsorbent mass and solution volume was 0.5 g/L. The results showed that the maximum adsorption capacity of the phenolic compounds was obtained for AC followed of activated carbon nanotubes (act-MWCNTs), MWCNTs and CEINs in this order. However, the maximum adsorption kinetic rate of the studied adsorbates was found in MWCNTs followed of act-MWCNTs = CEINs and AC, in this order. The highest adsorption capacity was found for 2-chlorophenol (549.5 mg/g). MWCNTs showed a rapid adsorption kinetic and the equilibrium concentration was achieved around of 5 min. The adsorption kinetic data were well fitted to the pseudo-second-order model.

Different studies have investigated the removal of phenol from water employing different carbon materials. de la Luz-Asunción et al. [13] realized the removal of phenol from aqueous solutions with carbon nanomaterials of 1D and 2D. The adsorbents used were MWCNTs and oxidized MWCNTs (O-MWCNTs), pristine single-walled carbon nanotubes (SWCNTs) and oxidized SWCNTs (O-SWCNTs), GO and reduced graphene oxide (rGO). The oxidation of the carbon nanotubes was carried out in microwave using H_2O_2 as oxidant agent. This method reduced the time of oxidation. GO and rGO were obtained by Hummer's method. The results showed that the pristine 1D nanomaterials (MWCNTs and SWCNTs) have a better adsorption capacity than rGO, however, GO presents a higher adsorption capacity than O-MWCNTs and O-SWCNTs. The best adsorption capacities of phenol were obtained by GO and O-SWCNTs. Oxygenated functional groups play an important role in the removal of phenol with carbon nanomaterials of 1D and 2D. The kinetic adsorption data and the adsorption equilibrium data were well fitted to the pseudo-second-order model and the Freundlich model, respectively. **Figure 4** shows the fit of Freundlich isotherms on the adsorption of phenol onto carbon nanomaterials. The main mechanism of adsorption of phenol onto the different carbon structure was due to π- π interactions between the aromatic structure of the graphitic layers and aromatic rings of the phenol structure. Other mechanism proposed in the adsorption of phenol onto carbon nanomaterials was hydrogen bonding. Li et al. [57] also studied the removal of phenol from aqueous solutions onto rGO. The effect of phenol solutions pH on the adsorption capacity of rGO was analyzed

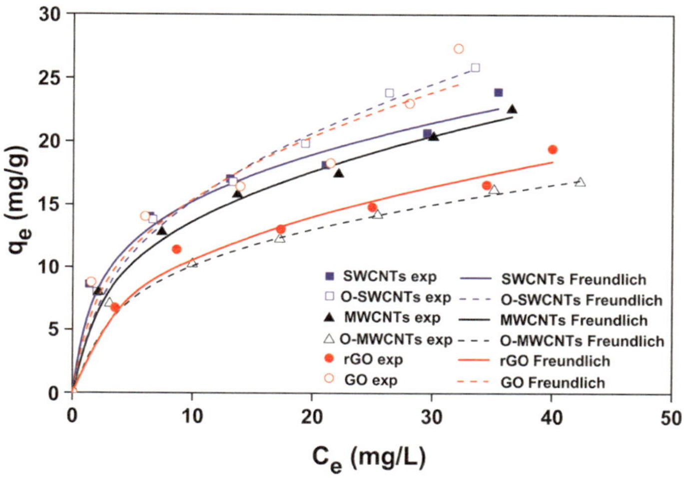

Figure 4. Freundlich isotherms of phenol adsorption onto carbon nanomaterials. Reprinted and adapted with permission from de la Luz-Asunción et al. [13]. Copyright © 2015, Hindawi Publishing Corporation.

when the pH increased from 2.3 to 11.5. The highest rGO adsorption capacities were found when the pH of phenol solutions was adjusted between 4 and 6.6. In this pH range, the complexation capability of the oxygenated groups present on the surface of rGO increased. Besides, the graphitic structure of rGO and the aromatic ring of phenol interact through of π–π interactions. The phenol removal increased when the rGO concentration was increasing gradually in the range from 0.5 to 1.7 g/L. Thermodynamic study revealed that the adsorption process of phenol onto rGO was endothermic and spontaneous.

Others phenolic compounds have been studied through of the adsorption process with carbon materials. Mehrizad and Gharbani [58] used rGO as adsorbent for the removal of 4-Chloro-2-nitrophenol (4C2NP) from aqueous solutions. The adsorption capacity of rGO onto 4C2NP removal decreased with increasing dosage of adsorbent from 0.2 to 0.8 g/L but in all cases the adsorption velocity was rapid (10 min) reaching the equilibrium at about 60 min. The best results of adsorption were found in the pH range from 3 to 7, 298 K and initial concentration of 10 mg/L. The pseudo-second-order model described well the adsorption kinetic data and the equilibrium adsorption data were well fitted to the Freundlich model. The adsorption of 4C2NP was found to be spontaneous and exothermic process in the temperature range from 298 to 328 K.

The adsorption of other phenol chloride compound onto carbon materials was studied by Pei et al. [59]. For the adsorption of 2,4,6-trichlorophenol (246TCP) from aqueous solutions were used rGO and GO as adsorbents. The adsorption of 246TCP onto rGO and GO was favored to pH range from 2.0 to 6.0. After of pH 6.0, the fraction of negatively charged 246TCP species increase in the solution causing electrostatic repulsion with negatively charged surfaces of rGO and GO. The adsorption capacity of rGO was higher than GO. The adsorption experimental data of 246TCP onto rGO and GO were well fitted to Freundlich equation.

Definitely, the application of carbon materials on the removal of phenolic compounds by adsorption process is a good alternative for the remediation of the contamination problem of water by this kind of compounds.

3.2. Photocatalysis

In advanced oxidation process, heterogeneous photocatalysis is a process with important potential for the degradation of recalcitrant organic contaminants in water. In this process, the use of a radiation source that generally is UV light and a semiconductor material as catalyst is necessary. The photon energy is converted into chemical energy which is capable of to degrade the organic pollutants. The photogenerated holes in the valence band diffuse to particle semiconductor surface and react with organic molecules present in aqueous solution forming hydroxyl radical (OH•). Meanwhile, electrons in the conduction band participate in reduction processes, reacting with molecular oxygen in the air to produce superoxide radical anions ($O_2 \cdot^-$). Titanium dioxide (TiO_2) is the most important semiconductor used in the heterogeneous photocatalysis due to different properties such as superhydrophilicity, chemical stability, long durability, nontoxicity, low cost and transparency to visible light; however, other semiconductor materials have been used also in this process with good performance in the contaminants degradation from water. The size of the semiconductors particles is an important factor that has influence on the efficiency of the photocatalytic process. Others important factors in the photocatalysis are the specific surface area, pore volume, pore structure, crystalline phase and the exposed surface. On the other hand, the charge separation is a problem that affects the efficiency of the photocatalytic process and whereby the improving of photocatalyts is a challenge. Thus, in this section, it is reviewed different works related to semiconductor particles and nanoparticles in the photocatalytic degradation of phenolic compounds. Also, we present some investigations associated with graphene nanomaterials focused to reduce the charges recombination in the photocatalytic degradation of phenolic compounds.

3.2.1. Semiconductor particles and nanoparticles

Although TiO_2 is the semiconductor more employed as catalyst in the heterogeneous photocatalysis, at present, there exist others semiconductor materials that have been developed and employed in the degradation of pollutants from water obtaining good results. The reduction in the particle size is an important factor that can improve the photocatalytic performance due to an increase in the surface area. With the aim to improve the results on the degradation of contaminants from water, also, the combination of semiconductor particles has been developed, obtaining composites and hybrids materials. A review of investigations about the degradation of phenolic compounds using semiconductor particles is presented.

Phenol is the phenolic compound more studied in the photocatalysis. Different semiconductor materials have been used in its degradation, but TiO_2 is the semiconductor more used in the photocatalytics process. Ye and Lu [60] synthesized anatase TiO_2 nanocrystals with exposed {0 0 1} facets; these materials were obtained in the presence of fluoride ions. The photocatalytic results indicated that the oxidation of phenol increased with the rise

http://dx.doi.org/10.5772/66895

in the percentage of {0 0 1} facets of TiO_2. The main intermediates produced during the photocatalytic degradation of phenol were catechol and hydroquinone. The improving in the photocatalytic performance of TiO_2 nanocrystales was attributed to the synergistic effects of the exposed {0 0 1} facets and surface fluorination. The TiO_2 doped have been used to improve the photocatalytic activity on the phenol degradation [61]. An important efficiency in the degradation of phenol was found. Best results of phenol degradation were found at high pH values. Composites of V_2O_5/N,S–TiO_2 were used as photocatalyst for phenol degradation under direct solar light [62]. The sample of V_2O_5/N,S–TiO_2 activated at 500°C showed the best photocatalytic performance, reaching a degradation of 88% of phenol solution (100 mg/L) in 4 h. The V_2O_5 component played a key role for the visible light activity of the composite system at longer wavelengths. The photocatalytic activity of the composite was mainly attributed to the acid sites present on the surface; however, other factors such as the surface area, anatase/rutile ratio and the absorption at longer wavelengths were important.

The phenol degradation also was investigated by Liu et al. [63]. In this investigation, $BiPO_4$ synthesized through hydrothermal process was used as photocatalyst and the process was assisted with H_2O_2. The initial concentration of phenol solution was 50 mg/L and the catalyst concentration was 0.5 g/L. The results indicated that the phenol could be mineralized after of 4 h with $BiPO_4$ but no by H_2O_2. The efficiency of $BiPO_4$ was attributed to the high potential photogenerated holes in the valence band and the high separation efficiency of electron hole pairs.

ZnO, other important semiconductor also have been used in the phenol degradation. Europium-doped flower like ZnO hierarchical [64], Ni-loaded ZnO nanorods [65], cerium-doped ZnO hierarchical micro/nanospheres [66] and ZnO nanosheets immobilized on montmorillonite [67] are some of the ZnO-based catalysts that have been employed successfully on the phenol degradation.

Although Al_2O_3 is known as insulator material, Tzompantzi et al. [68] synthesized an Al_2O_3 by the sol–gel method. The Al_2O_3 was dried and annealed at 400, 500, 600 and 700°C. The samples were tested in the degradation of phenol from water. The best results of degradation were obtained with the sample calcinated at 400°C. The photocatalytic activity of the Al_2O_3 can be due to the modification in the Al-O bonds distances and hydroxyl groups present in the Al_2O_3 structure, delaying the recombination process. The sample annealed to 400°C also was tested in the p-cresol and 4-chlorophenol degradation. The major degradation was obtained on 4-clorophenol, followed by phenol and p-cresol, in this order.

The degradation of 4-chlorophenol also have been studied by different investigation groups. A composite of anatase/titanate nanosheet was employed by Liu et al. [69] for phenol degradation from water. Titanate acted as the main adsorption site. The phenol degradation was carried out in a binary system. About of 99% of phenol was degraded within 120 min. The important photocatalytic efficiency was attributed to the synergetic effect on the photo-oxidation of 4-clorophenol and photoreduction in Cr (VI) due to the efficient separation of electron-hole pairs. In other investigation of phenol degradation, Naeem and Ouyang [70] investigated the degradation of 4-chlorophenol on TiO_2 supported on materials as activated carbon, silica

Proposal system	Type of photocatalytic material	Degraded component	References
One photocatalytic step	Bismute vanadate, BiV_4	Phenol	[71]
	Ag/ZnO	Nitrophenol	[72]
Photocatalytic-UV-C irradiation	$BiPO_4$ photocatalytic activity degradation under UV-C irradiation	Phenol	[73]
Heterogeneous photocatalytic-UV-laser irradiation	ZnO nanoparticles coupled under UV-laser irradiation	Phenol	[74]
Solar light photocatalytic degradation	Ag-core TiO_2 nanoparticles under solar light irradiation	Phenol	[75]

Table 1. Photocatalytic materials used on the removal of phenolic compounds.

(SiO_2) and zeolite (ZSM-5). All materials TiO_2-supported reached a better photocatalytic performance on the 4-chlorophenol degradation than TiO_2 alone. AC was found to be the best support followed by ZSM-5 and SiO_2. The maximum degradation of 4-chlorophenol using TiO_2-AC as photocatalyst was 89.7%. Others photocatalytic materials that have been studied on phenols degradation in water are shown in **Table 1**.

3.2.2. Graphene materials and graphene-based materials

Graphene materials have the capacity to transfer the charges rapidly which is very important to reduce the charge recombination. Besides this, the photocatalytic activity of graphene oxide, a functional form of graphene, has been proved [76]. Therefore, a review of the investigations where graphene materials have been used individually and in combination with others compounds for the removal of phenolic compounds is presented.

Bustos-Ramirez et al. [77] investigated the removal of phenol by photocatalysis using as photocatalyst GO synthesized under different conditions. The time of oxidation (2, 4 and 6 h) and the degassing units (55 and 65) were modified. The samples were labeled as GEO-2-55, GEO-2-65, GEO-4-55, GEO-4-65, GEO-6-55 and GEO-6-65. The experiment was carried out in a batch photoreactor containing 100 mL of initial concentration of 100 mg/L. The time of reaction was 2 h and as radiation source an UV lamp of 254 nm was used. The band gap obtained by all samples indicated that these materials could act as photocatalysts. The best results of phenol removal were found with the sample GO-2-55 (38.62%), followed by GO-6-55 (14.96%) and GO-4-55 (12.29%). **Figure 5** shows the absorption spectra of phenol before and after of the photocatalysis process using the GO-2-55 sample as photocatalyst. The degassing units and the oxygen functional groups played an important role in the preparation of GO and its photocatalytic activity, respectively. The obtained results indicated that GO under specific synthesis conditions is a very viable material for its use in photocatalytic processes for the contaminants degradation.

A better photocatalytic activity of GO was found on the degradation of 4-clorophenol (4-CP) [78]. The photocatalytic experiments were carried out using 30 mL of phenol solution with initial concentration of 30 mg/L, pH 7 and GO dosage of 0.8 g/L. As irradiation source was

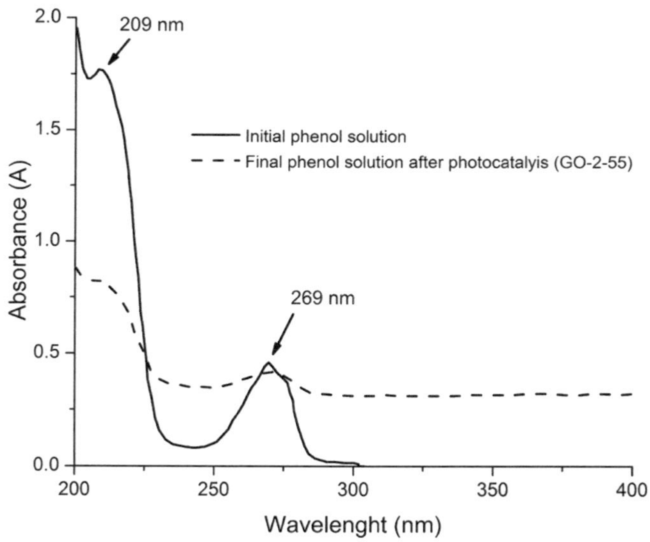

Figure 5. Phenol absorption spectra at initial and after of 2 h of photocatalytic reaction (reprinted with permission of Karina Bustos-Ramirez et al. [77]. Copyright © 2015, used under the Creative Commons Attribution License).

used an UV lamp (Pencil UV lamp, 254 nm and 5.5 W), which was introduced in the phenol solution. The photocatalytic results indicated that about of 80% of phenol was removed of the solution in a time of 100 min. About of 50% of removal was obtained in the first 20 min. The results of the chemical oxygen demand (COD) tests indicated that 97% of the organic matter was removed (**Figure 6**). Aromatic compounds and carboxylic acids are the main by-products generated in this photocatalytic process. The high efficiency of GO on the 4-CP degradation indicated that graphene materials have an important future in the photocatalysis area.

On the other hand, some investigations have studied the degradation of phenolic compounds employing composites of graphene and semiconductors particles. Some composites of graphene/TiO_2 have been synthesized by different routes for the degradation of phenol from water [79–81]. In all cases, the combination of graphene materials and TiO_2 particles improve the performance obtained with only TiO_2. This was attributed to an increase in the adsorption of phenol molecules, a better and more efficient charge separation and the improvement light absorption. The degradation of phenol was so well studied using other composites of graphene-based materials with good results [82–86].

Other phenolic compound that has been investigated was Bisphenol A. Wang et al. [87] synthesized a GO/ Ag_3PO_4 composite and proved its efficiency on the BA degradation. The degradation of BA was carried out using 75 mL of a solution of BA with initial concentration of 20 mg/L and 75 mg of catalyst. A 300 W Xe lamp with a 400 nm cutoff filter was used as irradiation source. The results indicated that the GO/Ag_3PO_4 (6 wt%) composite improved the degradation of BA with respect to the pure Ag_3PO_4. This enhancing of the photocatalytic activity of GO/Ag_3PO_4 was attributed to the presence of GO, which contributed to the separation electron-hole pairs in the composite. Similar results were found by Chen [88] on the

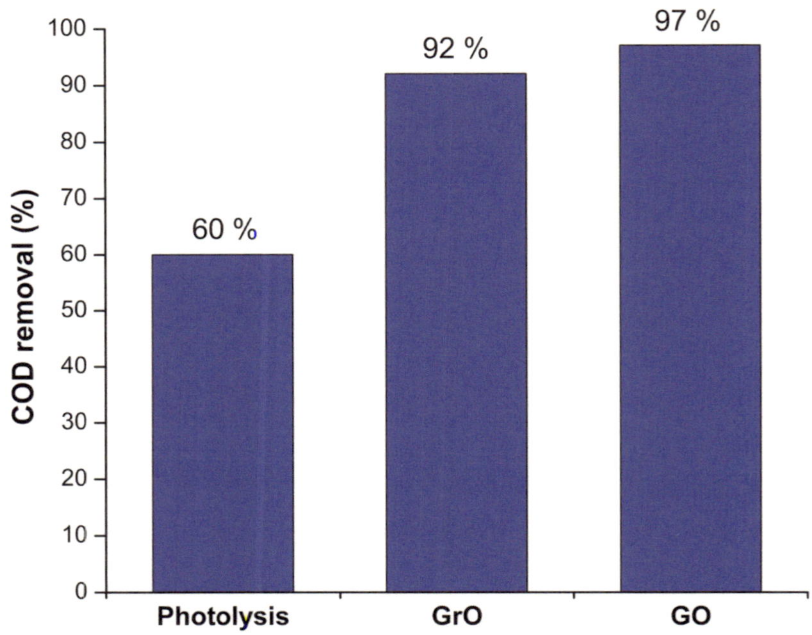

Figure 6. COD removal percentage obtained at the end of photolysis and photocatalysis tests using as catalysts (graphite oxide) GrO and GO on the 4-CP degradation. Adapted from Bustos-Ramírez et al. [78]; BioMed Central. 2015.

2,4-dichlorophenol degradation using as photocatalyst a GO/Ag_3PO_4 (5 wt%) composite. It is important to note that in both investigations was used visible light which is one of the most important challenges in the photocatalysis process.

In general, the revised investigations revealed that the combination of graphene materials with different semiconductors particles improve the degradation efficiency of the different phenolic compounds from water with respect to the individual particles. Besides, the graphene oxide showed an important photocatalytic activity capable of degrading the phenolic compounds.

3.3. Others

3.3.1. Electrochemistry

Currently, efforts have been made at developing more effective technologies to remove persistent organic pollutants. Advanced oxidation processes are based on the in situ production of reactive hydroxyl radicals (OH·) which non-selectively react with most organics, being able to degrade even resistant compounds. Although, OH· radicals are extremely reactive and cannot exist for a longtime, they can be used to decompose almost all organic to inorganic compounds [89, 90]. Combination process such as ozone, ultraviolet (UV) light, a semiconductor photocatalyst, hydrogen peroxide, ultrasound, Fenton reagent, photo-Fenton are widely studied to generate hydroxyl radicals (OH·). Recently, electrochemical advanced oxidation processes (EAOPs) are a promising derivative strategy that comes from AOPs. The easiest EAOPs method is the anodic oxidation (AO), where the organics can be directly oxidized at the anode surface by electron transfer and/or indirectly oxidized by OH· barely physisorbed at the anode surface whit the presence of agents at the bulk solution such as oxidizing reagents, O_3, chlorine species, persulfates or H_2O_2. On the other hand, when AO accomplish with cathodic electrogeneration of H_2O_2, the process is a cathodic oxidation. When H_2O_2 is electrochemically produced in the presence of Fe^{2+} at the bulk of reactions, as well as the OH·, it brings about the electro-Fenton process (EF). The phenomena originate a variety of techniques, for example, peroxi-coagulation (PC), Fered-Fenton, electrochemical peroxidation and sonoelectro-Fenton, or combine systems which include biological, chemical coagulation, electrocoagulation (EC) and membrane processes [90].

Phenol belongs to the recalcitrant pollutants commonly treated by conventional physicochemical and biological methods, so advanced oxidation (AO) represents an actual process for treatment of wastewater containing toxic persistent organic compounds. Pimentel et al. [91] applied a variant of advanced oxidation techniques to remove phenolic pollutants. They studied the oxidative degradation of aqueous phenol solutions in acidic medium by electro-Fenton technique using a carbon felt cathode and platinum anode in order to evaluate the mineralization efficiency, results evidenced that pH 3 enhance hydrogen peroxide electrochemical production, the most effective catalysts was ferrous iron ion at optimal concentration of 0.1 mM. Phenol oxidation by hydroxyl radical follows a pseudo-first-order kinetic with a rate constant of 0.037 min^{-1}. Additionally, phenol hydroxylation generate maleic, fumaric, succinic and glycolic acids in the beginning of the reaction; benzoquinone, catechol and hydroquinone as intermediate and oxalic and formic acids as final products. The total mineralization of phenol and its reactions intermediates put in context the effectiveness of the electro-Fenton process. If the process is combined, it could arises higher efficiencies as demonstrated Wang et al. [92], when combining electrocatalytic process and membrane bioreactor (MEBR), increases 11% the quality of the phenol removal, compared to the conventional and sum of the two individual processes; as result of the synergetic enhancement effect in one reactor. Also it was found that one of the degradation products is the benzoquinone (2,6-di.*tert*-butyl-*p*-benzoquinone).

Following with the electrochemical tendency process, Vasudevan [93] studied the peroxi-electrocoagulation method using mild steel as anode and graphite as cathode, obtaining 92% of removal from an initial phenol concentration of 2.5 mg/L and pH 2. The electro-coagulation (oxidation of sacrificial anode), amalgamates advantages from the separate procedures. Coagulants introduced without corresponding sulfate or chloride ions are more efficient to remove contaminants from waste, mainly when eliminate competitive anions and use a highly pure coagulant, it can be obtained lower metals residuals and less sludge as by-products if used metal salts. Moreover, when electrochemical reactors operate at high cell potential under acidic pH, the anodic process occurs in the potential region of water discharge and consequently hydroxyl radicals (OH·) are produced. This confirms that, ferrous ion generated in electrocoagulation function as coagulation materials and catalytically creates OH· radicals according to the conditions. Therefore, EAOPs can be even more effective than their chemical analogous, showing higher removal rates and greater reductions in organic toxic wastewater. EAOPs offer, the availability of higher amounts of H_2O_2 at the reaction beginning in the chemical processes; mainly in the presence of aromatic compounds; which is the case of phenol and phenolic molecules, demonstrated to induce faster initial removal rates for organic pollutants. Additionally, while increases the efficiency of the process the waste by-products diminish during the oxidation reaction to remove phenol compounds. Thus, EAOPs challenge needs to consider the implementation of high H_2O_2 quantities since the reaction initiation, as well as take into account aspects related to the investment costs, design of electrochemical cell; meaning less expensive hardware and electrodes materials and more versatile systems. In case of the light-assisted source, also should be considered, the UV lamps or photoreactors for natural sunlight capture. Even more, the operational costs include electrical energy for the electrochemical cell, plant operation, reagents and maintenance [90].

Nowadays, there is a growing interest to establish a great deal of attention to develop new strategies based on nanomaterials in conjunction with single and/or hybrid AOPs to remove or recover phenol species. One of the advantages of nanomaterial is the high surface area, where the volume/mass ratio will significantly improve the adsorption properties. Some nanomaterials studied are semiconductors, nanoclays, nanocatalyst, nanoclusters, nanorods, nanocomposites; for example, TiO_2, palladium, Fe_3O_2, Cerium oxide and magnetic chitosan, along or combined, $CoxFe_3$-xO_4, $CoFe_2O_4$ magnetic nanoparticles, $BiAg_xO_y$. From this point, can be synthesized nanoparticles, nanomembranes and nanopowders able to apply on the AOP technology [94].

Acknowledgements

The authors are grateful with the "Laboratorio Nacional de Materiales Grafénicos" located in the Centro de Investigación en Química Aplicada (CIQA) for supporting our research group, focused on the use of graphene materials in the process of removal of phenolic compounds.

Author details

Eduardo Enrique Pérez Ramírez[1], Miguel de la Luz Asunción[1], Veronica Saucedo Rivalcoba[2], Ana Laura Martínez Hernández[1] and Carlos Velasco Santos[1*]

*Address all correspondence to: cylaura@gmail.com

1 Division of Graduate Studies and Research, Technological Institute of Querétaro, Santiago de Querétaro, Querétaro, Mexico

2 Division of Postgraduate Studies and Research, Superior Technological Institute of Tierra Blanca, Tierra Blanca, Veracruz, Mexico

References

[1] Rappoport Z. The Chemistry of Phenols, Part 1: The Chemistry of Functional Groups. Hoboken, New Jersey: John Wiley & Sons. 2003.

[2] Mahugo-Santana C, Sosa Ferrera Z, Torres Padrón EM, Santana Rodríguez JJ. Methodologies for the extraction of phenolic compounds from environmental samples: new approaches. Molecules. 2009;**14(1)**: 298–320. doi:10.3390/molecules14010298.

[3] Santos A, Yustos P, Durban B, Garcia-Ochoa F. Catalytic wet oxidation of phenol: kinetics of the mineralization rate. Industrial & Engineering Chemistry Research. 2001;**40(13):**2773–2781. doi:10.1021/ie001129k.

[4] Atieh MA. Removal of phenol from water different types of carbon—a comparative analysis. APCBEE Procedia. 2014;**10:**136–141. doi:10.1016/j.apcbee.2014.10.031.

[5] Solomons TWG. Organic Chemistry. 11th ed. New York: John Wiley & Sons Incorporated. 2014.

[6] Coulter V. Advanced Organic Chemistry. 1st ed. Chandni Chowk, Delhi: Global Media. 2009.

[7] Carey FA. Organic Chemistry. 4th ed. Dubuque, Iowa: McGraw-Hill Companies. 2001.

[8] McMurry J. Organic Chemistry. 8th ed. Belmont, CA: Cengage Brooks/Cole. 2010.

[9] Baruah JB. Chemistry of phenolic compounds: State of the art. New York: Nova Science Publishers, Inc. 2011.

[10] Villegas LGC, Mashhadi N, Chen M, Mukherjee D, Taylor KE, Biswas NA. Short review of techniques for phenol removal from wastewater. Current Pollution Reports. 2016;1–11. doi:10.1007/s40726-016-0035-3.

[11] Park HS, Koduru JR, Choo KH, Lee B. Activated carbons impregnated with iron oxide nanoparticles for enhanced removal of bisphenol A and natural organic matter. Journal of Hazardous Materials. 2015;**286**: 15–324. doi:10.1016/j.jhazmat.2014.11.012.

[12] Tran VS, Ngo HH, Guo W, Zhang J, Liang S, Ton-That C, Zhang X. Typical low cost biosorbents for adsorptive removal of specific organic pollutants from water. Bioresource Technology. 2015;**182**:353–363. doi:10.1016/j.biortech.2015.02.003.

[13] de la Luz-Asunción M, Sánchez-Mendieta V, Martínez-Hernández AL, Castaño, VM, Velasco-Santos C. Adsorption of phenol from aqueous solutions by carbon nanomate-rials of one and two dimensions: kinetic and equilibrium studies. Journal of Nanomaterials. **2015**;2015:1–15. doi:10.1155/2015/405036.

[14] Mohammadi S, Kargari A, Sanaeepur H, Abbassian K, Najafi A, Mofarrah E. Phenol removal from industrial wastewaters: a short review. Desalination and Water Treatment. 2015;**53(8)**:2215–2234. doi:10.1080/19443994.2014.883327.

[15] Madani M, Aliabad, M, Nasernejad B, Kermanj Abdulrahman R, Yalili Kilic M, Kestioglu K. Treatment of olive mill wastewater using physico-chemical and Fenton processes. Desalination and Water Treatment. 2015;**53(8)**:2031–2040. doi:10.1080/19443994.2013.860882.

[16] Sawyer CN, McCarty PL, Parkin GF. Chemistry for environmental engineering and science. 5th ed. New York: McGraw-Hill. 2003.

[17] Khalid M, Joly G, Renaud A, Magnoux P. Removal of phenol from water by adsorption using zeolites. Industrial & Engineering Chemistry Research. 2004;**43(17)**:5275–5280. doi:10.1021/ie0400447.

[18] Huong PT, Lee BK, Kim J. Improved adsorption properties of a nano zeolite adsorbent toward toxic nitrophenols. Process Safety and Environmental Protection. 2016;**104**:314–322. doi:10.1016/j.psep.2016.08.018.

[19] Xie J, Meng W, Wu D, Zhang Z, Kong H. Removal of organic pollutants by surfactant modified zeolite: Comparison between ionizable phenolic compounds and non-ionizable organic compounds. Journal of Hazardous Materials. 2012;**231**:57–63. doi:10.1016/j.jhazmat.2012.06.035.

[20] Djebbar M, Djafri F, Bouchekara M, Djafri A. Adsorption of phenol on natural clay. Applied Water Science. 2012;**2(2)**:77–86. doi:10.1007/s13201-012-0031-8.

[21] Nayak PS, Singh BK. Removal of phenol from aqueous solutions by sorption on low cost clay. Desalination. 2007;**207(1)**:71–79. doi:10.1016/j.desal.2006.07.005.

[22] Park Y, Ayoko GA, Kurdi R, Horváth E, Kristóf J, Frost RL. Adsorption of phenolic compounds by organoclays: Implications for the removal of organic pollutants from aqueous media. Journal of Colloid and Interface Science. 2013;**406**:196–208. doi:10.1016/j.jcis.2013.05.027.

[23] Paidar M, Fateev V, Bouzek K. Membrane electrolysis-History, current status and perspective. Electrochimica Acta. 2011;**209**:737–756. doi:10.1016/j.electacta.2016.05.209.

[24] Munirasu S, Haija MA, Banat F. Use of membrane technology for oil field and refinery produced wáter treatment—a review. Process Safety Environmental Protection. 2016;**100**:182–202. doi:10.1016/j.psep.2016.01.010.

[25] Castro-Muñoz R, Yañez-Fernández J, Fila V, Phenolic compounds recovered from agro-food-by-products using membrane technologies: an overview. Food Chemistry. 2016;**213**:753–762. doi:10.1016/j.foodchem.2016.07.030.

[26] Bódalo A, Gómez JL, Gómez M, León G, Hidalgo AM, Ruíz MA. Phenol removal from water by hybrid processes: study of the membrane process step. In: Conference on Desalination and the Environment. 22–25 April 2007; Halkidiki, Greece: Desalination; 2008. pp. 323–329.

[27] Bellona C, Drewes JE, Xu P, Amy G. Factors affecting the rejection of organic solutes during NF/RO treatment—a literature review. Water Research. 2004;**38**:2795–2809. doi:10.1016/j.watres.2004.03.034.

[28] Bódalo A, Gómez E, Hidalgo AM, Gómez M, Murcia MD, López I. Nanofiltration membranes to reduce phenol concentration in wastewater. In: Proceedings Conference Engineering with Membranes 2008; 25–28 May 2008; Vale do Lobo, Algarve, Portugal. Desalination: 2009. pp. 680–686.

[29] Ipek U. Phenol removal capacity of RO with and without pre-treatment. Filtration & Separation. 2004;**41**:39–40. doi:10.1016/S0015-1882(04)00321-0.

[30] Sagehashi M, Nomura T, Shishido H, Sakoda A. Separation of phenols and furfural by pervaporation and reverse osmosis membranes from biomass-superheated steam pirolisis-derived aqueous solution. Bioresource Technology. 2007;**98**:2018–2026. doi:10.1016/j.biortech.2006.08.022.

[31] Choi BG, Kim DI, Hong S. Fouling evaluation and mechanisms in a FO-RO hybrid process for direct potable reuse. Journal of Membrane Science. 2016;**520**:89–98. doi:10.1016/j.memsci.2016.07.035.

[32] López-Muñoz MJ, Arsuaga JM, Sotto A. Separation of phenol and their advanced oxidation intermediate products in aqueous solution by NF/RO membranes. Separation and Purification Technology. 2010;**71**:246–251. doi:10.1016/j.seppur.2009.12.004.

[33] Ko Ch-H, Chen Sh-Sh. Enhanced removal of three phenols by laccase polymerization with MF/UF membranes. Bioresource Technology. 2008;**99**:2293–2298. doi:10.1016/j.biortech.2007.05.017.

[34] El-Nass MH, Alhaija MA, Al-Zuhair S. Evaluation of three-step process for the treatment of petroleum refinery wastewater. Journal of Environmental Chemical Engineering. 2014;**2**:56–62. doi:10.1016/j.jece.2013.11.024.

[35] Kulkarni SJ, Kaware JP. Review of research for removal of phenol from wastewater. International Journal of Scientific and Research Publications. 2013;**3(4):**1–5.

[36] Othman N, Noah NFM, Shu LY, Ooi Z-Y, Jusoh N, Idroas M, Goto M. Easy removing of phenol from wastewater using vegetable oil-based organic solvent in emulsion liquid membrane process. Chinese Journal of Chemical Engineering. 2016: 1–8. doi:10.1016/j.cjche.2016.06.002.

[37] Hemmati M, Nazari N, Hemmati A, Shirazian S. Phenol removal from wastewater by means of nanoporous membranes contactors. Journal of Industrial and Engineering Chemistry. 2015;**21**:1410–1416. doi:10.1016/j.jiec.2014.06.015.

[38] Xiao M, Zhou J, Zhang Y, Hu X, Li Sh. Pertraction performance of phenol through PDMS/PVDS composite membrane aromatic recovery system (MARS). Journal of Membrane Science. 2008;**428**:172–180. doi:10.1016/j.memsci.2012.10.030.

[39] Lee M-S, Huang Ch, Lee K-R, Pan JR, Chang W-K. Factors affecting phenol transfer through polydimethylsiloxane composite membrane. In: Fourth Conference of Aseanian Membrane Society (AMS 4), 16–18 August 2007, Taipei, Taiwan. Desalination; 2008. pp. 416–425.

[40] Loh ChH, Zhang Y, Goh Sh, Wang R, Fane AG. Composite hollow fiber membranes with different poly(dimethylsiloxane) intrusions into substrate for phenol removal via xtractive membrane bioreactor. Journal of Membrane Science. 2016;**500**:236–244. doi:10.1016/j.memsci.2015.12.001.

[41] Heo J, Boateng LK, Flora JRV, Lee H, Her N, Park Y-G, Yoon Y. Comparison of flux behavior and synthetic organic compound removal by forward osmosis and reverse

osmosis membranes. Journal of Membrane Science. 2013;**443**:69–82. doi:10.1016/j.memsci.2013.04.063.

[42] Yüksel S, Kabay N, Yüksel M. Removal of bisphenol A (BPA) from water by various nanofiltration (NF) and reverse osmosis (RO) membranes. Journal of Hazardous Materials. 2013;**263**:307–310. doi:10.1016/j.jhazmat.2013.05.020.

[43] Zagklis DP, Vavouraki AI, Kornaros ME, Paraskeva ChA. Purification of olive mill wastewater phenols through membrane filtration and resin adsorption/desorption. Journal of Hazardous Materials. 2015;**285**:69–76. doi:10.1016/j.jhazmat.2014.11.038.

[44] Shen Sh, Kentish SE, Stevens GW. Effects of operational conditions on the removal of phenol from wastewater by a hollow-fiber membranes contactor. Separation and Purification Technology. 2012;**95**:80–88. doi:10.1016/j.seppur.2012.04.023.

[45] Hasanoglu A. Removal of phenol from wastewaters using membrane contactors: comparative experimental analysis of emulsion pertraction. Desalination. 2013;**309**:171–180. doi:10.1016/j.desal.2012.10.004.

[46] Praveen P, Loh K-Ch. Trioctylphosphine oxide-impregnated hollow fiber membranes for removal of phenol from wastewater. Journal of Membrane Science. 2013;**437**:1–6. doi:10.1016/j.memsci.2013.02.057.

[47] Praveen P, Loh K-Ch. Osmotic membrane bioreactor for phenol biodegradation under continuous operation. Journal of hazardous materials. 2016;**305**:115–122. doi:10.1016/j.jhazmat.2015.11.034.

[48] Reis MTA, Freitas OMF, Agarwal Sh, Ferreira LM, Ismael MRC, Machado R, Carvalho JMR. Removal of phenols from aqueous solutions by emulsion liquid membranes. Journal of Hazardous Materials. 2011;**192(3):**986–994. doi:10.1016/j.jhazmat.2011.05.092.

[49] Ethash M, Fournier-Salaün M-Ch, Dimitrov K, Salaün P, Saboni A. Phenol removal from aqueous media by pertraction using vegetable oil as liquid membrane. Chemical Engineering Journal. 2014;**250**: 42–47. doi:10.1016/j.cej.2014.03.068.

[50] Lu Sh, Pei L. A study on phenol migration by coupling the liquid membrane in the ionic liquid. International Journal of Hydrogen Energy. 2016;**41**:15724–15732. doi:10.1016/j.ijhydene.2016.05.008.

[51] Liu G, Ma J, Li X, Qin Q. Adsorption of bisphenol A from aqueous solution onto activated carbons with different modification treatments. Journal of Hazardous Materials. 2009;**164**:1275–1280. doi:10.1016/j.jhazmat.2008.09.038.

[52] Tsai WT, Lai CW, Su TY. Adsorption of bisphenol-A from aqueous solution onto minerals and carbon adsorbents. Journal of Hazardous Material. 2006;**B134**:169–175. doi:10.1016/j.jhazmat.2005.10.055.

[53] Fasfous II, Radwan ES, Dawoud JN. Kinetics, equilibrium and thermodynamics of the sorption of tetrabromobisphenol A on multiwalled carbon nanotubes. Applied Surface Science. 2010;**256**:7246–7252. doi:10.1016/j.apsusc.2010.05.059.

[54] Zhang Y, Tang Y, Li S, Yu S. Sorption and removal of tetrabromobisphenol A from solution by graphene oxide. Chemical Engineering Journal. 2013;**222**:94–100. doi:10.1016/j.cej.2013.02.027.

[55] Ji L, Zhou L, Bai X, Shao Y, Zhao G, Qu Y, Wang C, Li Y. Facile synthesis of multiwall carbon nanotubes/iron oxides for removal of tetrabromobisphenol A and Pb(II). Journal of Materials Chemistry. 2012;**22**:15853–15862. doi:10.1039/c2jm32896h.

[56] Strachowski P, Bystrzejewski M. Comparative studies of sorption of phenolic compounds onto carbon-encapsulated iron nanoparticles, carbon nanotubes and activated carbon. Colloids and Surfaces A: Physicochemical Engineering Aspects. 2015;**467**:113–123. doi:10.1016/j.colsurfa.2014.11.044.

[57] Li Y, Du Q, Liu T, Sun J, Jiao Y, Xia Y. Equilibrium, kinetic and thermodynamic studies on the adsorption of phenol onto graphene. Materials Research Bulletin. 2012;**47(8)**:1898–1904. doi:10.1016/j.materresbull.2012.04.021.

[58] Mehrizad A, Gharbani P. Decontamination of 4-chloro-2-nitrophenol from aqueous solution by graphene adsorption: Equilibrium, kinetic and thermodynamic studies. Polish Journal of Environmental Studies. 2014;**23(6):**2111–2116. doi:10.15244/pjoes/26779.

[59] Pei Z, Li L, Sun L, Zhang S, Shan X, Yang S, Wen B. Adsorption characteristics of 1,2,4-trichlorobenzene, 2,4,6-trichlorophenol, 2-naphthol and naphthalene on graphene and graphene oxide. Carbon. 2013; **51**:156–163. doi:10.1016/j.carbon.2012.08.024.

[60] Ye H, Lu S. Photocatalytic selective oxidation of phenol in suspensions of titanium dioxide with exposed {0 0 1} facets. Applied Surface Science. 2013;**277**:94–99. doi:10.1016/j.apsusc.2013.04.008.

[61] Khraisheh M, Wu L, Al-Muhtaseb AH, Albadarin AB, Walker GM. Phenol degradation by powdered metal ion modified titanium dioxide photocatalysts. Chemical Engineering Journal. 2012;**213**:125–134. doi:10.1016/j.cej.2012.09.108.

[62] Satyabadi M, Das DP, Biswal N, Parida KM. Facile synthesis of visible light responsive V2O5/N,S–TiO_2 composite photocatalyst: enhanced hydrogen production and phenol degradation. Journals of Materials Chemistry. 2012;**22**:10695. doi:10.1039/c2jm30462g.

[63] Liu Y, Zhu Y, Xu J, Bai X, Zong R, Zhu Y. Degradation and mineralization mechanism of phenol by $BiPO_4$ photocatalysis assisted H_2O_2. Applied Catalysis B: Environmental. 2013;**142–143**:561–567. doi:10.1016/j.apcatb.2013.05.049.

[64] Sin JC, Lam SM. Hydrothermal synthesis of europium-doped flowerlike ZnO hierarchical structures with enhanced sunlight photocatalytic degradation of phenol. Materials Letters 2016;**182**:223–226. doi:10.1016/j.matlet.2016.06.126.

[65] Ahmed AB, Jibril B, Danwittayakul S, Dutta J. Microwave-enhanced degradation of phenol over Ni-loaded ZnO nanorods catalyst. Applied Catalysis B: Environmental. 2014;**156–157**:456–465. doi:10.1016/j.apcatb.2014.03.032.

[66] Sin JC, Lam SM, Lee KT, Mohamed AR. Preparation of cerium-doped ZnO hierarchical micro/nanospheres with enhanced photocatalytic performance for phenol degradation

under visible light. Journal of Molecular Catalysis A: Chemical. 2015;**409**:1–10. doi:10.1016/j.molcata.2015.08.004.

[67] Ye J, Li X, Hong J. Chen J, Fan Q. Photocatalytic degradation of phenol over ZnO nanosheets immobilized on montmorillonite. Materials Science in Semiconductor Processing. 2015;**39**:17–22. doi:10.1016/j.mssp.2015.04.039.

[68] Tzompantzi F, Piña Y, Mantilla A, Aguilar-Martinez O, Galindo-Hernandez F, Bokhimi Xim, Barrera A. Hydroxylated sol–gel Al_2O_3 as photocatalyst for the degradation of phenolic compounds in presence of UV light. Catalysis Today. 2014;**220**:49–55. doi:10.1016/j.cattod.2013.10.027.

[69] Liu W, Sun W, Alistair GL, Ting B, Li WF, Guan Y. Simultaneous removal of Cr(VI) and 4-chlorophenol through photocatalysis by a novel anatase/titanate nanosheet composite: synergetic promotion effect and autosynchronous doping. Journal of Hazardous Materials. 2016;**317**:385–393. doi:10.1016/j.jhazmat.2016.06.002.

[70] Naeem K, Ouyang F. Influence of supports on photocatalytic degradation of phenol and 4-chlorophenol in aqueous suspensions of titanium dioxide. Journal of Environmental Sciences. 2013;**25(2)**:399–404. doi:10.1016/S1001-0742(12)60055-2.

[71] Kunduz S, Soylu GSP. Highly active $BiVO_4$ nanoparticles: the enhanced photocatalytic properties under natural sunlight for removal of phenol from wastewater. Separation and Purification Technology. 2015;**141**:221–228. doi:10.1016/j.seppur.2014.11.036.

[72] Divband B, Khatamian M, Kazemi Eslamian GR, Darbandi M. Synthesis of Ag/ZnO nanostructures by different methods and investigation of their photocatalytic efficiency for 4-nitrophenol degradation. Applied Surface Science. 2013;**284**:80–86. doi:10.1016/j.apsusc.2013.07.015.

[73] Zhang Y, Selvaraj R, Sillanpää M, Kim Y, Tai ChW. The influence of operating parameters on heterogeneous photocatalytic mineralization of phenol over $BiPO_4$. Chemical Engineering Journal. 2014;**245**:117–123. doi:10.1016/j.cej.2014.02.028.

[74] Hayat K, Gondal MA, Khaled MM, Ahmed Sh, Shemsi AM. Nano ZnO synthesis by modified sol-gel method and its applications in heterogeneous photocatalytic removal of phenol from water. Applied Catalysis A: General. 2011;**393**:122–129. doi:10.1016/j.apcata.2010.11.032.

[75] Shet A, Vidya ShK. Solar light mediated photocatalytic degradation of phenol using Ag-core-TiO_2 shell (Ag@TiO_2) nanoparticles in batch and fluidized bed reactor. Solar Energy. 2016;**127**:67–78. doi:10.1016/j.solener.2015.12.049.

[76] Krishnamoorthy K, Mohan R, Kim S-J. Graphene oxide as a fotocatalytic material. Applied Physics Letters. 2011;**98**:244101 (1–3). doi:10.1063/1.3599453.

[77] Bustos-Ramirez K, Barrera-Diaz CE, De Icaza M, Martínez-Hernández AL, Velasco-Santos C. Photocatalytic activity in phenol removal of water from graphite and graphene oxides: effect of degassing and chemical oxidation in the synthesis process. Journal of Chemistry. 2015;**2015**:1–10. doi:10.1155/2015/254631.

[78] Bustos-Ramirez K, Barrera-Diaz CE, De Icaza M, Martínez-Hernández AL, Velasco-Santos C. 4-chlorophenol removal from water using graphite and graphene oxides as photocatalysts. Journal of Environmental Health Science & Engineering. 2015;**13**:1–10. doi:10.1186/s40201-015-0184-0.

[79] Adamu H, Dubey P anderson JA. Probing the role of thermally reduced graphene oxide in enhancing performance of TiO_2 in photocatalytic phenol removal from aqueous environments. Chemical Engineering Journal. 2016;**284**:380–388. doi:10.1016/j.cej.2015.08.147.

[80] Ren Y, Sun C, Li K, Wang L, Song M. Preparation of TiO_2–reduced graphene oxide–Pd nanocomposites for phenol photocatalytic degradation. Ceramics International. 2016;**42**:1339–134. doi:10.1016/j.ceramint.2015.09.072.

[81] Fu CC, Juang RS, Huq MM, Hsieh C-Te. Enhanced adsorption and photodegradation of phenol in aqueous suspensions of titania/graphene oxide composite catalysts. Journal of the Taiwan Institute of Chemical Engineers. 2016;1–8. doi:10.1016/j.jtice.2016.07.043.

[82] Gawande SB, Gawande KB, Thakare SR, Manwar NR, Mate VR. Photocatalytic degradation of phenol over novel rod shaped Graphene@$BiPO_4$ nanocomposite. Journal of Physics and Chemistry of Solids. 2015;**85**:132–137. doi:10.1016/j.jpcs.2015.04.012.

[83] Ma H, Shen J, Shi M, Lu X, Li Z, Long Y, Li N, Ye M.Significant enhanced performance for Rhodamine B, phenol and Cr(VI) removal by $Bi2WO_6$ nancomposites via reduced graphene oxide modification. Applied Catalysis B: Environmental. 2012;**121**–**122**:198–205. doi:10.1016/j.apcatb.2012.03.023.

[84] Liu H, Jin Z, Su Y, Wang Y. Visible light-driven $Bi_2Sn_2O_7$/reduced graphene oxide nanocomposite for efficient photocatalytic degradation of organic contaminants. Separation and Purification Technology. 2015; **142**:25–32. doi:10.1016/j.seppur.2014.12.027.

[85] Li Y, Tang Z, Zhang J, Zhang Z. Reduced graphene oxide three-dimensionally wrapped WO_3 hierarchical nanostructures as high-performance solar photocatalytic materials. Applied Catalysis A: General. 2016;**522**:90–100. doi:10.1016/j.apcata.2016.04.034.

[86] Ai B, Duan X, Sun H, Qiu X, Wang S. Metal-free graphene-carbon nitride hybrids for photodegradation of organic pollutants in water. Catalysis Today. 2015;**258**:668–675. doi:10.1016/j.cattod.2015.01.024.

[87] Wang C, Zhu J, Wu X, Xu H, Song Y, Yan J, Song Y, Ji H, Wang K, Li H. Photocatalytic degradation of bisphenol A and dye by graphene-oxide/ Ag_3PO_4 composite under visible light irradiation. Ceramics International. 2014;**40**:8061–8070. doi:10.1016/j.ceramint.2013.12.159.

[88] Chen X-J, Dai Y-Z, Wang X-Y, Guo J, Liu T-H, Li F-F. Synthesis and characterization of Ag_3PO_4 immobilized with graphene oxide (GO) for enhanced photocatalytic activity and stability over 2,4-dichlorophenol under visible light irradiation. Journal of Hazardous Materials. 2015;**292**:9–18. doi:10.1016/j.jhazmat.2015.01.032.

[89] Suzuki H, Araki A, Yamamoto H. Evaluation of advanced oxidation processes (AOP) using O_3, UV and TiO_2 for the degradation of phenol in water. Journal of Water Process Engineering. 2015;**7**:54–60. doi:10.1016/j.jwpe.2015.04.011.

[90] Moreira FC, Bonaventura RAR, Brillas E, Vilar VJP. Electrochemical advanced oxidation processes: a review on their application to synthetic and real wastewater. Applied Catalysis A: Environmental. 2017;**202**:217–261. doi:10.1016/j.apcatb.2016.08.037.

[91] Pimentel M, Oturan N, Dezontti M, Oturan MA. Phenol degradation by advanced electrochemical oxidation process electro-Fenton using a carbon felt cathode. Applied Catalysis B: Environmental. 2008;**83**:140–149. doi:10.1016/j.apcatb.2008.02.011.

[92] Wang T, Zhao H, Wang H, Liu B, Li Ch. Research on degradation production and reaction kinetics of membrane electro-bioreactor (MEBR) with catalytic electrodes for high concentration phenol wastewater treatment. Chemosphere. 2016;**155**:94–99. doi:10.1016/j.chemosphere.2016.03.140.

[93] Vasudevan S. An efficient removal of phenol from water by peroxi-electrocoagulation processes. Journal of Water Process Engineering. 2014;**2**:53–57. doi:10.1016/j.jwpe.2014.05.002.

[94] Bethi B, Sonawane SH, Bhanvase BA, Gumfekar SP. Nanomaterials-based advanced oxidation processes for wastewater treatment: a review. Chemical Engineering and Processing: Process Intensification. 2016;**109**:178–189. doi:10.1016/j.cep.2016.08.016.

Chapter 15

Phenolic Compounds Removal in Woodwaste Leachate by a Trickling Biofilter

Najat Kamal, Rosa Galvez, Gerardo Buelna and Rino Dubé

Additional information is available at the end of the chapter

http://dx.doi.org/10.5772/66755

Abstract

Woodwaste produces large volumes of leachate, which often contains high concentrations of phenolic compounds. These compounds necessitate appropriate management. Biological methods are efficient, innovative, and economic. In particular, biofiltration process has various advantages compared to CAS and MBR technologies. Two pilot filters, with and without biological activity, were designed for continuous mechanisms to follow. This study supposes that the three mechanisms of volatilization, sorption, and biodegradation are present, confirm these assumptions, and determine the contribution of each mechanism. Good efficiency was obtained in the biofilter and 97–98.2% of COD and BOD removal were observed, respectively. Excellent performances were achieved and reached 99.9% of initial concentrations removal for all the phenolic compounds.

Keywords: phenolic compounds, woodwaste leachate, sorption, volatilization, biofiltration, trickling biofilter

1. Introduction

The industry of paper is one of the principal economic sectors in Quebec. In 2011, 56.4% of total canadian production of paper was achieved by Québec and was evaluated at 14.9% of total world exportations [1]. In this year, approximately 373 million cubic meters were generated as water process and were discharged in the environment after only basic physicochemical treatment.

In 2006, the value of shipments of paper was 10.7 billion or 7.5% of the value of shipments of all Quebec's manufacturing industry. The value of exports in 2005 was estimated at 7.3 billion or 9.7% of all exports of Quebec and contributes to 46.1% of total newspaper production in Canada and 9% of global paper production.

Thus, considerable quantities of wood residue are generated, which produce large volumes of leachate by percolation of water and following industrial processes. Woodwaste leachate contains lignin and phenolic compounds at important concentrations and must be treated before its discharge in the environment.

Phenolic compounds are toxic to humans and ecosystem, persistent, and endocrine disruptor molecules. They are found in woodwaste leachate and pulp and paper industrial effluents. It is necessary to develop new and innovative technologies adapted for these toxic effluents' treatment and phenolic compounds' removal. Several processes were studied for phenol removal such as membrane processes, ozonation [2], advanced oxidation [3], and activated carbon [4]. Ozonation was proven to be effective but it requires energy, and produces gases and causes the formation of not yet identified by-products with an important risk to human health [5–8]. As phenolic compounds are persistent, it might be suitable to combine ozonation with biological treatment [9].

Biological methods are efficient, innovative, and economic. Biological treatment of organic pollutants was investigated in conventional activated sludge treatment plants (CAS) and in membrane bioreactors (MBR). Mailler et al. [10] compared emerging pollutants removal by both biofiltration and conventional activated sludge plants. This classical biological unit (CAS) has already been well documented and compared with membrane bioreactor [11–18]. No significant difference was observed between CAS and MBR when a critical solids retention time (SRT) necessary for nitrification (SRT > 10 days at 10°C) was exceeded and the suspended solids concentration in the effluent was low for CAS plants [19].

Biofiltration process has various advantages compared to CAS and MBR technologies [20, 21]. It is an economical, simple, and innovative solution for water and air pollution control [20, 22, 23].

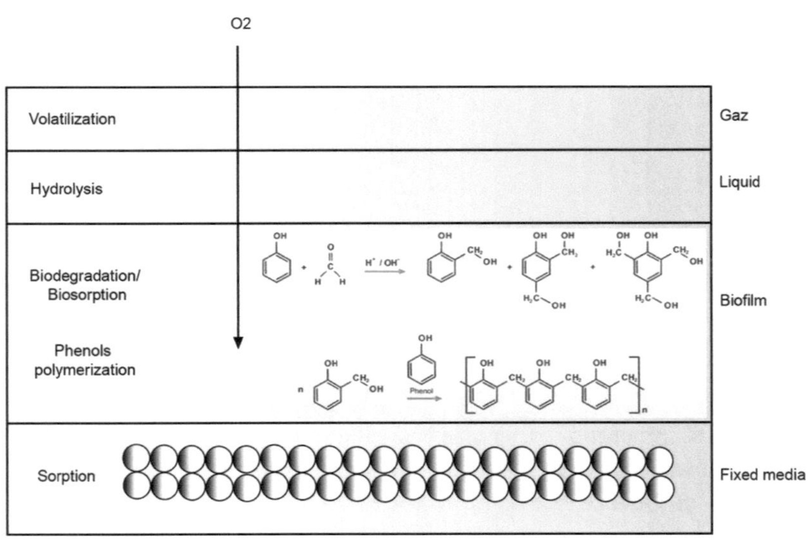

Figure 1. Possible mechanisms in operation during biofiltration process.

http://dx.doi.org/10.5772/66755

Phenolic compounds removal by biofiltration was investigated by [9, 24] but the involved mechanisms of this removal were not described before.

This work aims to study phenolic compounds removal in woodwaste leachate using a trickling biofilter and, second, to describe the mechanisms responsible for their transformation and removal. In otherwise, we aim to answer the following questions: Are phenolic compounds eliminated by volatilization, biodegradation, sorption, and biosorption or by all of these processes, and what is the contribution of each mechanism?

This study supposes that the mechanisms of volatilization, sorption, and biodegradation operate simultaneously during biofiltration process. We aim at confirming these assumptions and determine the contribution of each mechanism to the overall phenols removal efficiency (**Figure 1**).

2. Material and methods

2.1. Pilot scale unit and experiment set up

Two PVC columns of 0.15 m diameter and 2 m high were packed with a solid support (peat: perlite; 1:9) and used to follow phenols removal in woodwaste leachate. These units (**Figure 2**) are equipped with utilities, which are a storage leachate system composed of two mini-

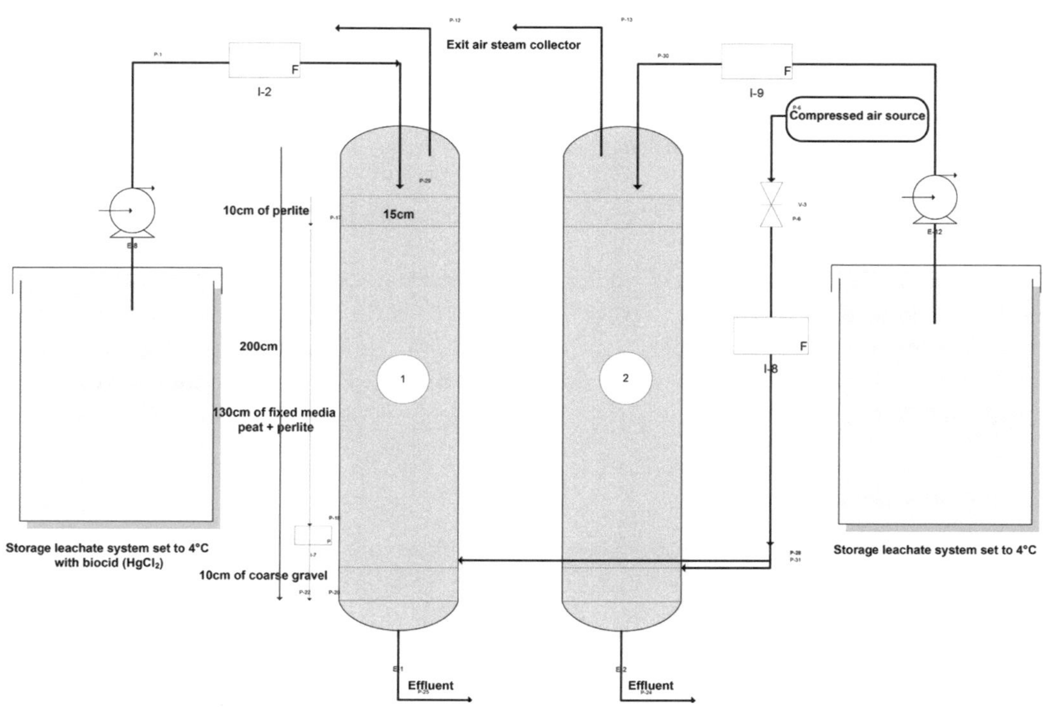

Figure 2. Pilot scale unit for phenolic compounds removal study.

refrigerators set to 4°C and aeration system connected to compressed air source with copper tubes in T-shape and connected to the bottom of each column using rigid tubes after passing in control air flow rate system. Columns are also equipped with exit air steam collector composed of four bottles containing each 100 ml of methanol, connected in series for each column via T-shape connection with the out vacuum evacuation. The output air steam was collected for 4 hours twice a week and if the air steam collection was not in operation, the valve connected to the out vacuum evacuation was opened. Woodwaste leachate was fed continuously to columns via Masterflex tubes at the top using a peristaltic pump, simulating trickling filtration.

2.2. Media composition and preparation

Columns were packed with solid support consisted of a mixture of peat (70% with 0.5 mm) and perlite (4–6 mm) at a 1:9 ratio. After characterization, the peat was dried at 103°C during 48 h to minimize bacterial activity. A layer of coarse gravel with 0.10 m of height was installed first at the bottom of the two columns and then these were packed with the media at a layer height of 1.30 m as described in **Figure 1**. A layer of perlite with 0.10 m of height was installed at the top of the two columns. For the conditioning, columns were fed slowly with tap water (1.5–2 h) in an ascending way to reach a maximum absorption capacity of water by peat [25], a good consistency of media and a resistance to compaction. The conditioning of columns removes also the retained air in the system.

2.3. Laboratory scale unit set up

Two mini-columns with 0.12 m of diameter and 0.20 m of height were set up as described in **Figure 3** to simulate the top of pilot columns and to follow efficiency and parameters variability with time at the top of the pilot scale units.

2.4. Leachate

Woodwaste leachate was collected from a sawmill site in the Quebec City area and characterized. This effluent is a complex matrix with highly variable chemical composition. Concentrations of phenolic compounds depend on the temperature, age, and industrial activities of the region, in particular, those of sawmills and paper industries. Physicochemical characteristics are given in **Table 1**.

2.5. Biomass acclimation

Activated sludge from the sewage treatment station Valcartier in Québec City area was used as a source of microorganisms for biomass acclimation. Volatile fatty acids (VFAs) and phenol were used as the carbon source for microorganisms. The acclimation was conducted progressively to avoid shock to the microorganisms due to the toxicity and to allow the biomass growth. At first, only VFAs were used as carbon source and were replaced gradually by phenol. **Table 2** shows the detailed composition for acclimation process. A synthetic mixture

http://dx.doi.org/10.5772/66755

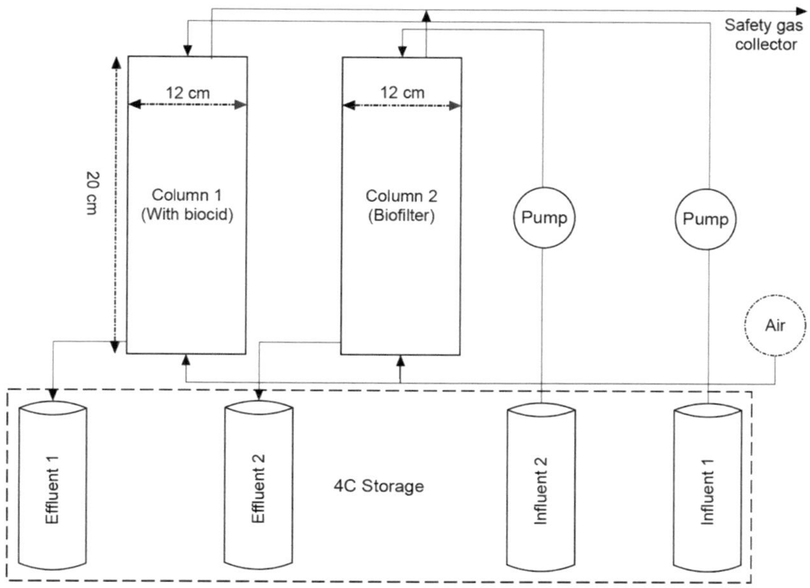

Figure 3. Laboratory scale unit simulating the top of pilot columns to follow phenolic compounds removal.

Parameter (mg L^{-1})	Winter 2006	Summer 2006	Winter 2007	Winter 2008	Autumn 2008	Summer 2009
pH	4.50	4.62	4.48	4.60	4.73	4.45
COD	1400	1100	1600	1700	1100	3792
BOD_5	600	400	750	760	450	1862
MES	415	850	410	480	600	1400
PO4	5.00	5.20	5.10	4.89	5.00	4.30
Total volatile fatty acids	80	50	120	150	40	425
Total phenolics (μg L^{-1})	1946	1625	2010	2080	1540	3810
Phenol	380.86	77.99	450.12	437.25	81.55	80.75
4-Nitrophenol	171.41	160.79	282.05	291.04	158.43	712.25
p, m-Cresol	111.55	95.28	107.38	117.28	88.97	1503.12
o-Cresol	133.86	77.92	141.57	139.50	99.75	801.15
2-Chlorophenol	177.23	175.57	191.35	211.47	178.98	162.17
2,4-Dinitrophenol	1102.03	35.75	78.15	81.75	28.67	576.85
2, 4-Dimethylphenol	3.87	1.35	4.07	5.73	1.55	239.72

Note: Total phenolic compounds were determined by spectrophotometric analysis. Individual phenolic compound values reported in μg L^{-1}.

Table 1. Physicochemical characterization and phenolic compounds quantification of woodwaste leachate.

		Acclimation steps				
		1	2	3	4	5
	V added (mL)	3	3	2	1	0
AGV	mg C	109.92	109.92	73.28	36.64	0.00
	(% C)	100.00	96.63	90.54	76.13	0.00
	ppm	0	2	4	6	8
Phenol	mg phenol	0	5	10	15	20
	V added (mL)	0	5	10	15	20
	mg C	0	3.83	7.66	11.49	15.32
	% (C)	0	3.37	9.46	23.87	100
Total	mg C	109.92	113.75	80.94	48.13	15.32

VFAs: 80 g L^{-1} (36.64 g-C/L); phenol: 1 g L^{-1} (0.766 g-C/L); nutrients: 65.86 g L^{-1} (0.052 g-N/L); volume of reactor: 2.5 L.

Table 2. Acclimation of biomass steps.

of the eight studied phenolic compounds was given as substrate when the biomass was acclimated and then the real woodwaste leachate was used for the biomass growth before use in the biofilter. The biomass growth and substrates utilization were followed by respirometric measurements during all the acclimation process.

2.6. Experiments

The two pilot-scale columns were followed for 6 months under continuous flow conditions. The liquid flow rate was maintained at 3 Ld^{-1}. Woodwaste leachate percolated filters with a hydraulic load of 0.169 m^3 m^{-2} d^{-1}. Air was applied upward at a flow rate of 5 m^3 m^{-2} h^{-1}. Columns were fed with real woodwaste leachate, which presented a pH between 4.5 and 4.7 spiked to 1 μg mL^{-1} with eight phenolic compounds and operated at room temperature (20–25°C). A flow sheet of the pilot scale unit is described in **Figure 2**.

2.6.1. Pilot 1 (with mercuric chloride)

The bacterial activity was inhibited in this column using mercuric chloride ($HgCl_2$) at 0.5 g L^{-1} in the fed leachate. Only physicochemical retention by sorption and volatilization are in operation and responsible for phenolic compounds removal in this column. Liquid influent and effluent were sampled each day for 6 months to establish the parameters variation profiles. pH and COD were determined daily for the first 3 months and each 2 days from the 4th month. Phenolic compounds were quantified each 2 days in the liquid influent and effluent and twice a week in the collected steam during all the operation period. Parameters profile variations are presented in **Figures 4** and **5**.

http://dx.doi.org/10.5772/66755

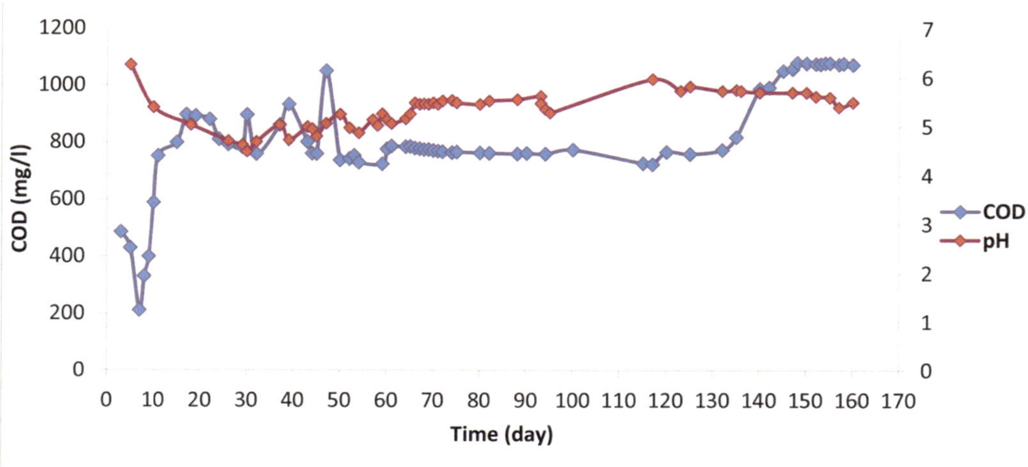

Figure 4. Physicochemical parameters variation in the effluent of column 1 (without bacterial activity).

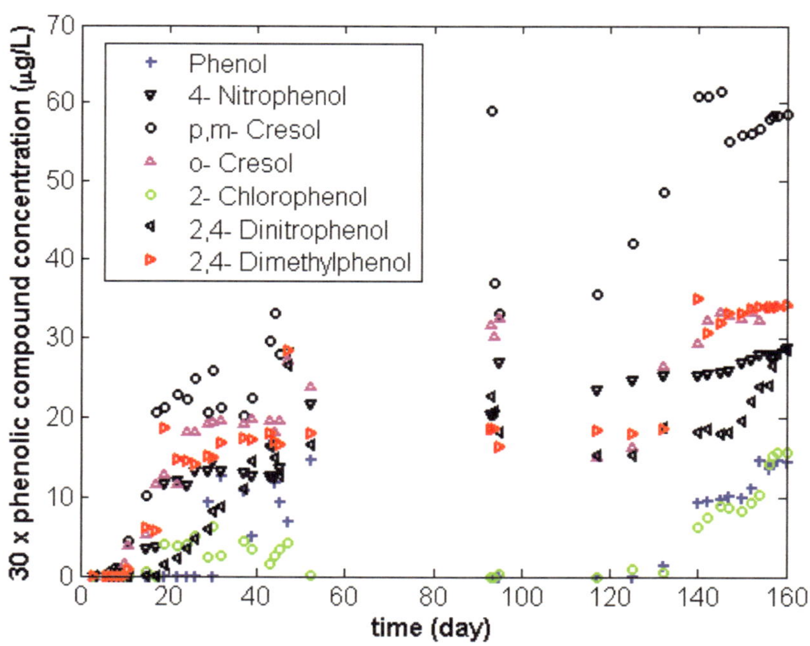

Figure 5. Phenolic compounds profiles variation in the column 1 (without biological activity).

2.6.1.1. Media without biological activity characterization

Both conventional scanning electron microscopy (SEM) and environmental scanning electron microscopy (ESEM) characterized media of pilot 1, with mercuric chloride. The environmental SEM offers all of the performance advantages of a conventional SEM, but allows high resolution and the analysis of wet and environmental samples in their current state without preliminary manipulations. Analysis was performed on JEOL 840-A (MEB) instrument for conventional

SEM and on JSM-6360 instrument at low voltage (30 V of acceleration voltage) and with different vacuum values for ESEM.

2.6.2. Pilot 1 (with mercuric chloride)

This pilot simulates the real behavior of a trickling biofilter. The mechanisms of biodegradation, sorption, and volatilization are to be present and in operation during biofiltration. For this column also, phenolic compounds were quantified each 2 days in the liquid influent and effluent and twice a week in the collected steam during all the operation period. The parameters that were monitored in the liquid influent and effluent were BOD, COD, pH, Nitrogen, and phenolic compounds and measured once a week for BOD, and each 2 days for all the other parameters during all the operation period. VFAs were determined sporadically. Parameters profile variations for the biofilter are presented in **Figures 6** and **7**.

2.6.2.1. Biomass characterization

Microbial aspects of fixed biofilm at different biofilter layers were shown by both conventional SEM and ESEM imaging performed on JEOL 840-A (MEB) and JSM-6360 instruments, respectively.

Biological characterization of bacterial diversity in the mixed biofilm was performed by DGGE of the 16 S ribosomal RNA genes molecular technique. PCR-DGGE is a rapid and sensitive method for the detection and investigation of active species in fixed media based on the DNA fragments extraction. Amplicons, with the same length, of dominant microbial organisms are separated in gradient gel based on their differential denaturation process and nucleotide compositions. Two samples were collected by coring the media, during and at the end of experiments. The first coring was done in the middle of the column. The second coring extraction was realized at eight different level heights of the biofilter. Gel preparation and analysis were carried out according to the protocol established by Ercolini [26] and run at Dr Duchaine's research laboratory at Centre de Recherche de l'Institut Universitaire de Cardiologie et de Pneumologie de l' Hôpital Laval (Québec, Canada).

2.6.3. Pilot 1 (with mercuric chloride)

The two laboratory-scale columns were set-up and followed for 4 months under continuous flow conditions. Real woodwaste leachate at pH of 4.5–4.7 and spiked with 1 $\mu g\ mL^{-1}$ of eight phenolic compounds was percolated under the same conditions applied to the pilot unit: hydraulic loading of 0.169 $m^3\ m^{-2}\ d^{-1}$ and flow rate aeration of 5 $m^3\ m^{-2}\ h^{-1}$. These laboratory units were followed to reproduce the top of pilot-scale columns and to describe the phenolic compounds removal at the top of these columns. A flow sheet of this mini-columns setup is described in **Figure 3**.

2.7. Parameters analysis

Analyses of phenolic compounds were performed on a Breeze HPLC system (Waters ©) consisting of a 717 plus auto sampler, a 1525 binary pump with internal degasser, and a code

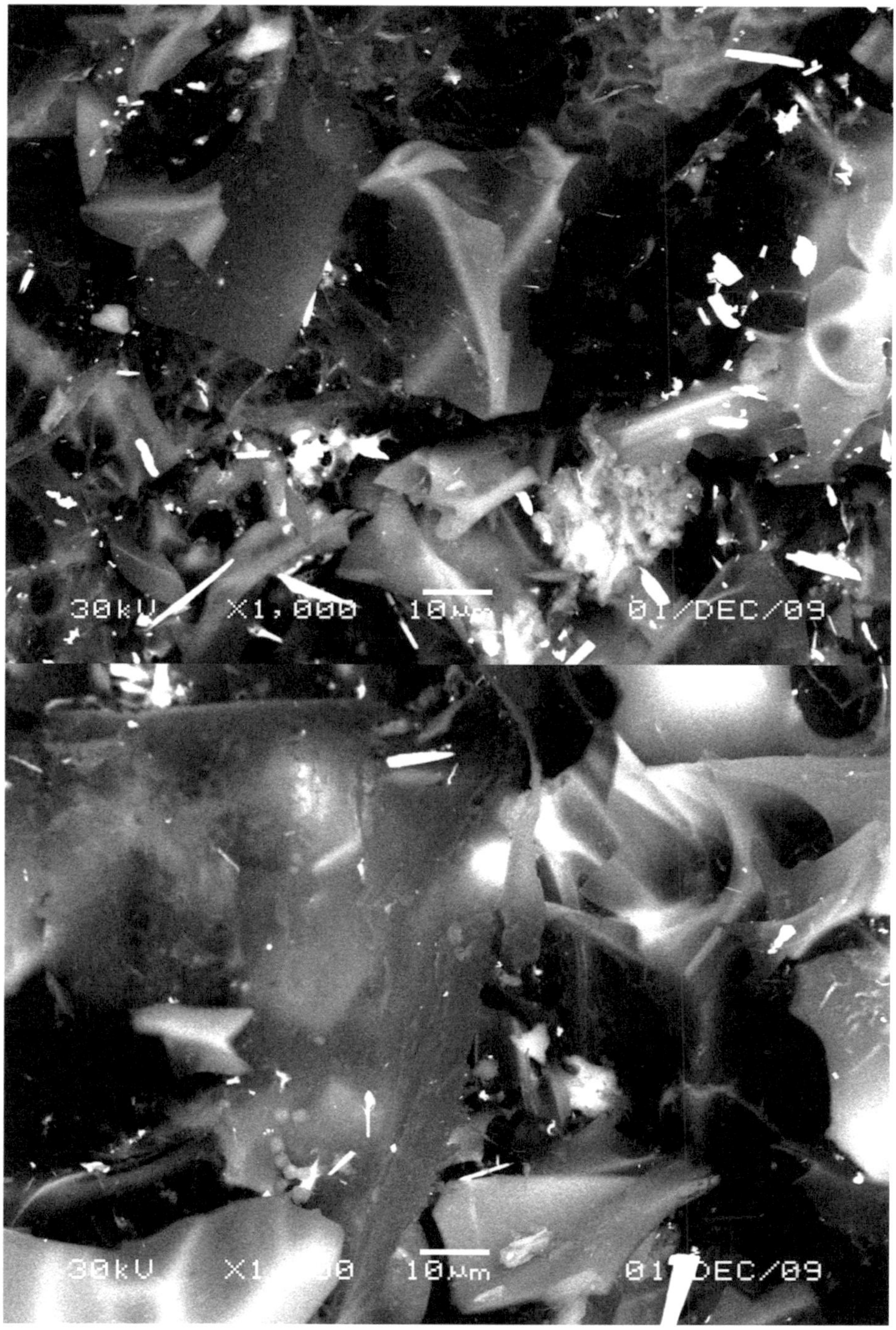

Figure 6. SEM and ESEM imaging of media of pilot 1 (with mercuric chloride).

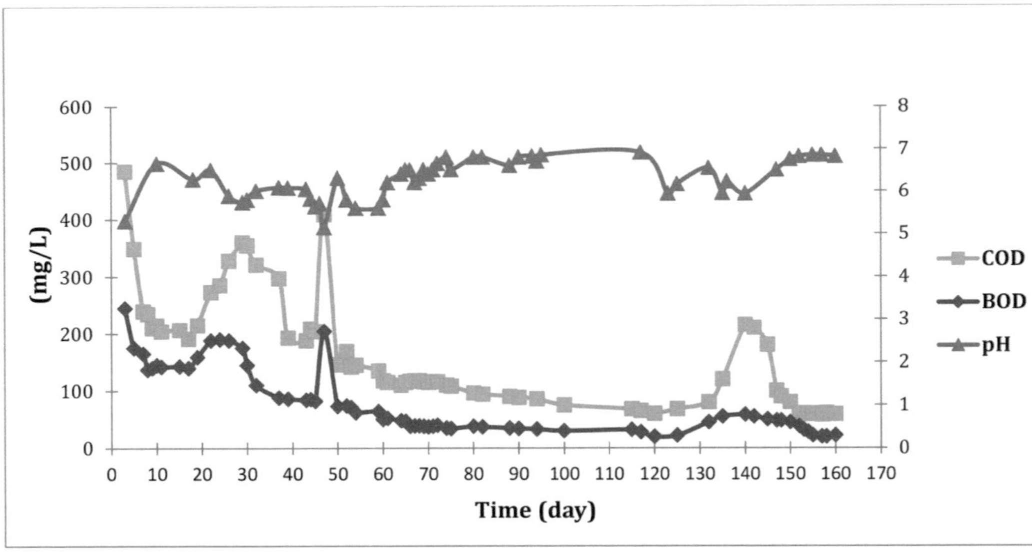

Figure 7. Physicochemical and biological parameters variation in the effluent of column 2 (biofilter).

model 5CH column oven. The separation was carried out on a Symmetry C_{18} (4.6 × 150 mm, 5 μm) column and detection by a 2487 dual wavelength detector. Samples were previously acidified to pH between 1.8 and 2, filtered on 0.45 μm membrane filter, concentrated by SPE on Oasis HLB (60 mg) cartridges and then analyzed by liquid chromatography according to the method described by Kamal et al. [27].

COD and nitrogen were determined according to standard methods using a DR3000 spectrophotometer. BOD was measured using a standard BOD Track system and a polyseed as bacterial consortium.

3. Results and discussion

3.1. Woodwaste leachate characteristics

Table 1 shows the real leachate physicochemical characterization and phenolic compounds quantification from winter of 2006 to summer of 2009. Phenols were found at concentrations varying between 0.38 μg mL^{-1} for phenol and 1.1 μg mL^{-1} for 2,4-dinitrophenol in real woodwaste leachate on winter of 2006.

3.2. Pilot 1 (with mercuric chloride) results

COD drastic decrease was observed in the effluent during the first 10 days of operation. Note that 84.41% of initial COD was removed during this operation period. From the second week,

http://dx.doi.org/10.5772/66755

a removal performance of 41.1% was obtained. A steady state with 43.82% efficiency was then reached and continued until the saturation of the media on the 135th day.

3.2.1. *Phenolic compounds profiles variation in the column 1 (without biological activity)*

For phenolic compounds also, a significant decrease was observed in the effluent during the first 10 days of operation. A removal performance of 33.33–40.1% was obtained at steady state for all the phenolic compounds except phenol and chlorophenol for which, 40.2–83.33% of initial concentrations was removed. On the other hand, 41.1% of phenol and chlorophenol initial concentrations was found in the collected gas. Thus, in this column, both physicochemical retention and volatilization are responsible for phenolic compounds removal. At the saturation of the media on the 135th day, a maximum of volatilization was obtained and reached 41.1% for more volatile compounds.

3.2.2. *Media without biological activity characterization results*

Both conventional scanning electron microscopy (SEM) and environmental scanning electron microscopy (ESEM) characterized media of pilot 1, with mercuric chloride. Images are presented in **Figure 6**. No bacterial activity was observed. It was noted that metal spots were present. The X-ray analysis confirmed a presence of mercury (Hg).

3.3. Pilot 2 (biofilter) results

A significant decrease was observed in the effluent during the first 10 days of operation for both COD and BOD. These parameters increased during the second and third week. This can be explained by a resistance and often observed inhibitory effect of organisms during the acclimation process. Good performances of the biofilter were obtained after 30 days. On the 47th day, the system received a high organic load and toxicity due to a periodic increase at the sawmill providing the effluent. However, the biofilter was able to manage this organic shock and fluctuations and to achieve excellent performances at steady state. Indeed, 97–98.2% of COD and BOD removal were observed, respectively. Another increase of the measured parameters was observed on the 135th operation day. This can be explained by the saturation of the media and biological phenomena become a predominant and responsible factor for the treatment. The system, in this situation also, was able to manage the change of physicochemical retention capacities of the media and to achieve the performances of the previous steady state in few days. **Figure 7** shows 59 and 22 mg L^{-1} values for COD and BOD, respectively, with superior performances in reducing organics (97% COD and (98.2% BOD reduction)).

3.3.1. *Phenolic compounds profiles variation in the biofilter*

Phenolic compounds profile variation, presented in **Figure 8**, followed absolutely COD and BOD variations in the biofilter. Indeed, a significant decrease was observed in the effluent during the first 10 days of operation and performance removal of 99.9% was obtained at steady state for all the phenolic compounds. A maximum of 5–7% of volatilization was observed for phenol and chlorophenol in the biofilter.

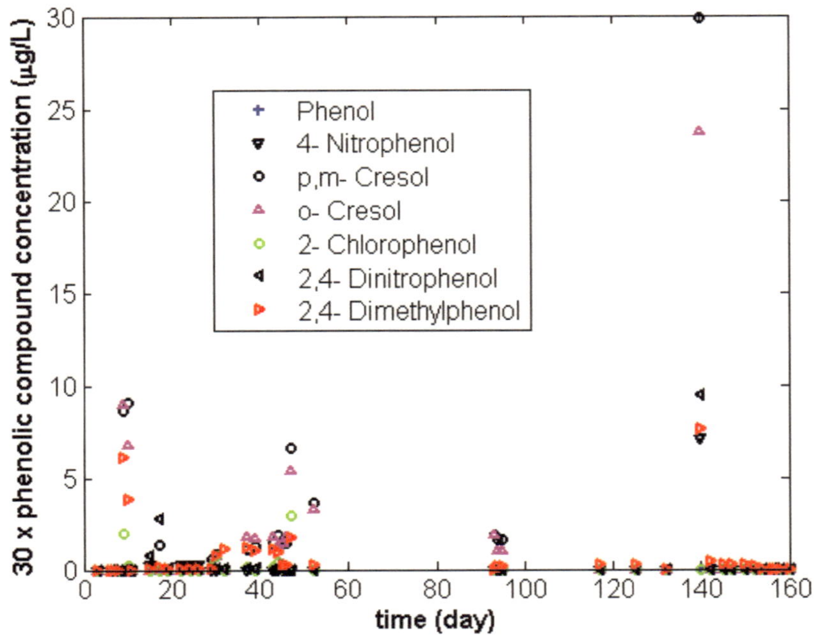

Figure 8. Phenolic compounds profiles variation in the column 2 (biofilter).

3.3.2. Biomass acclimation results

Table 2 shows the detailed steps for acclimation process. Activated sludge from the sewage treatment station Valcartier in Québec City area was used as a source of microorganisms for biomass acclimation. Volatile fatty acids (VFAs) and phenol were used as the carbon source for microorganisms.

3.3.3. Biomass characterization results

Both conventional SEM and ESEM imaging showed microbial aspects of fixed biofilm at different biofilter layers. SEM and ESEM images are given in **Figure 9**.

A sampling at different column heights was done, by coring of the biofilter in the last week of operation period. **Figure 10** shows the column sampling. Media samples were analyzed by DGGE technique with gel sequence and sequences amplified to determine the composition of the bacterial community. Samples 7 and 8 represent the 10 cm perlite layer at the top of the biofilter. Results of DGGE banding patterns of 16S rDNA fragments presented by **Figure 11** show the presence of generally similar bacterial species. Two intense bands present in perlite layer disappeared in the depth of biofilter to be replaced by other bands. This may be explained by the development of other microorganisms in the media mixture of peat and perlite. The bands representing the different heights of the mixed media are in general similar; this indicates that the same active species are present in all the peat/perlite mixture height of the biofilter and are involved in the phenolic compounds biodegradation process. The list of

http://dx.doi.org/10.5772/66755

Figure 9. SEM and ESEM imaging of microbial aspects of fixed biofilm at different biofilter layers.

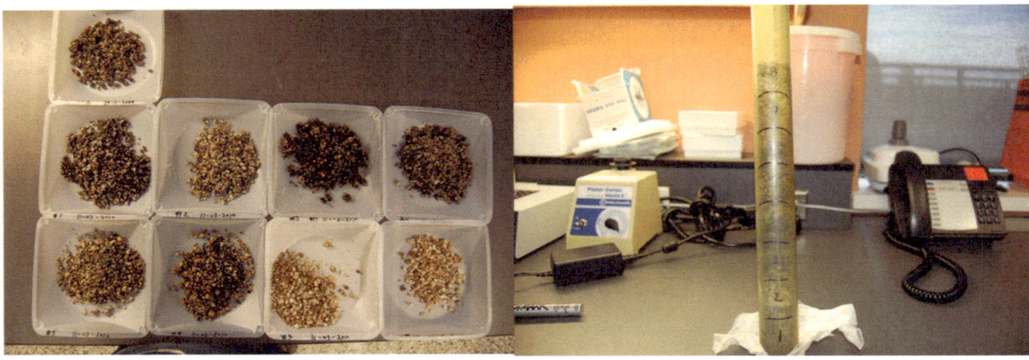

Figure 10. Media samples from different heights of the biofilter.

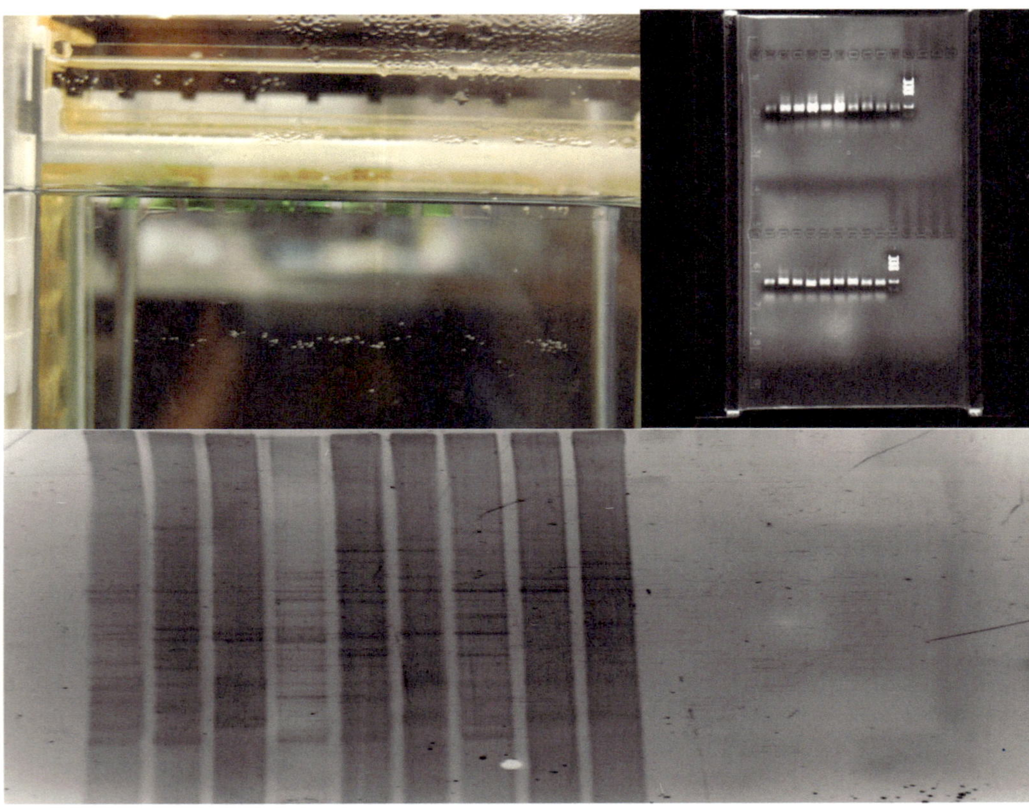

Figure 11. Gel sequence and DGGE banding results of active spices in different biofilter levels. Names of referred bacteria are listed in Table 3.

most abundant active species is given in **Table 3**. Both heterotrophic and autotrophic bacterial species are present in the biofilter with a predominance of heterotrophic population.

Number	Active species	GenBank	Group	Accession number
1	Alpha proteobacterium	AF256029.1	–	–
2	Gamma proteobacterium clone	EU665097.1	–	–
3	Gamma proteobacterium clone	EU665097.1	–	–
4	Bacillus sp.	GU935303.1		
5	Oxalobacteraceae bacterium	GU473126.1	Herbaspirillum sp	GQ355286.1
6	Oxalobacteraceae bacterium	GU473092.1	Herbaspirillum sp	GQ355286.1
7	Rhizobiales bacterium	FN794225.1		
8	Aucune corespondance	–	–	–
9	Aucune corespondance	–	–	–
10	Alpha proteobacterium	EF665816.1	Uncultured Rhizobiales bacterium	FJ475425.1
11	Alpha proteobacterium	EU051984.1	–	–
12	Séquence double	–	–	–
13	Alpha proteobacterium	GU553130.1	Methylosinus sp	GU556372.1
14	Alpha proteobacterium	–	–	–
15	Séquence double	–	–	–
16	Aucune corespondance	–	–	–
17	Aucune corespondance	–	–	–
18	Acidobacteriaceae bacterium	EU449736.1	–	–
19	Nitrobacter sp.	GU556337.1	–	–

Table 3. Most abundant active species and results of PCR re-amplified and sequenced 16S rDNA DGGE bands.

3.3.4. Toxicity evaluation

Toxicity evaluation of leachate before and after treatment was performed in the Centre d'Expertise en Analyses Environnementales de Québec using the Microtox test on two species susceptible to the toxicity, which are daphnia and algae. Results are presented in **Table 4**. Toxicity evaluation confirmed that effluent was not toxic for algae and daphnia. It was confirmed also that algae were stimulated in the presence of diluted (5.6% V/V) effluent.

3.4. Pilot unit profiles comparison

Figures 12 and **13** illustrate the profiles performance comparison. COD (feed) was the influent COD and COD (Out) was the effluent COD in column 1 without bacterial activity. The variation of BOD in the woodwaste influent and in the biofilter effluent showed high biofilter performances.

The pH distribution in column 1 (without bacterial activity) and in column 2 (biofilter) is presented in **Figure 14**.

Concentration % V/V	Influent			Effluent		
	Daphnia		Algae	Daphnia		Algae
	% Mortality	% Immobility	% Inhibition	% Mortality	% Immobility	% Inhibition
56	100	100	99	0	1	0
42	100	100	99	0	0	0
24	99	100	99	0	0	0
18	30	40	99	0	0	0
13.5	25	30	99	0	0	0
10	25	30	99	0	0	0
7.5	20	20	99	0	0	0
5.6	10	5	80	0	0	–1.7
Control	0	0	0	0	0	0

Table 4. Toxicity evaluation of leachate before and after treatment.

3.5. Mini-columns COD profile comparison

The mini-columns COD decreased in the first 10 days of operation from 1896 mg L^{-1} as a fed COD influent to 968 mg L^{-1} for the first column (without bacterial activity) and 737 mg L^{-1} for

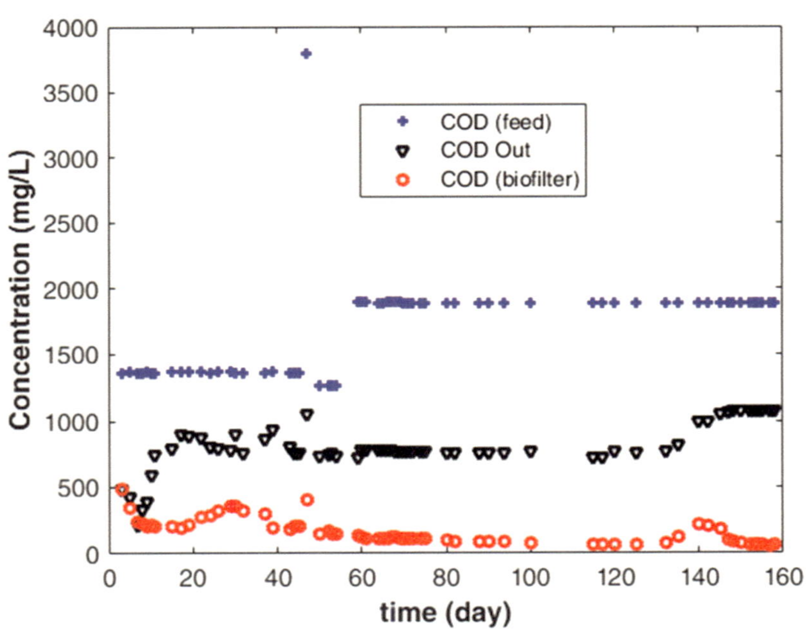

Figure 12. COD profiles comparison in woodwate leachate influent and in both columns with and without bacterial activity.

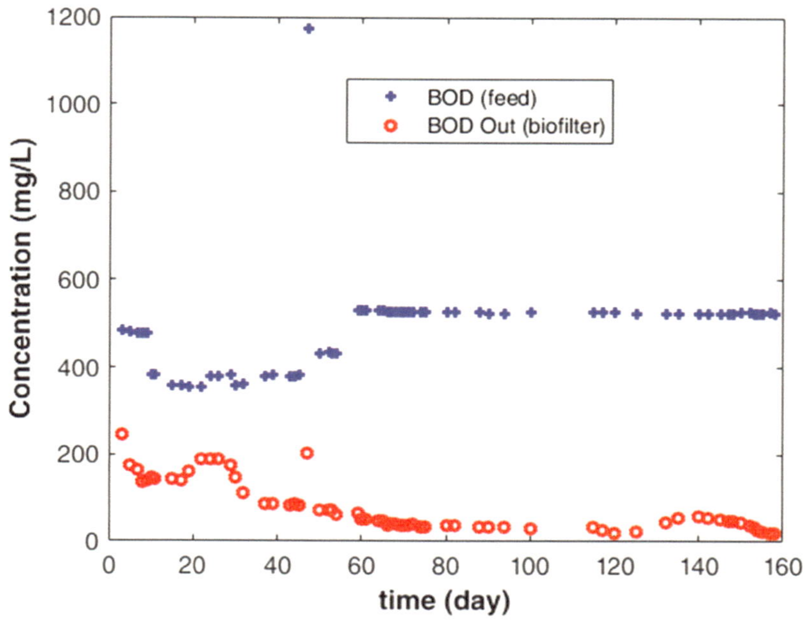

Figure 13. Treatment efficiency and BOD performance in the biofilter.

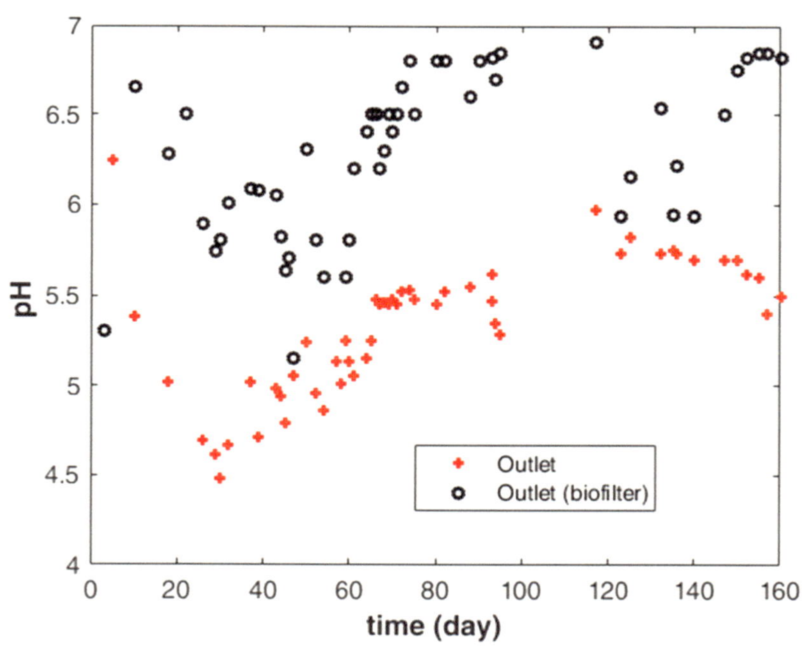

Figure 14. pH distribution in column 1 (without bacterial activity) and in column 2 (biofilter).

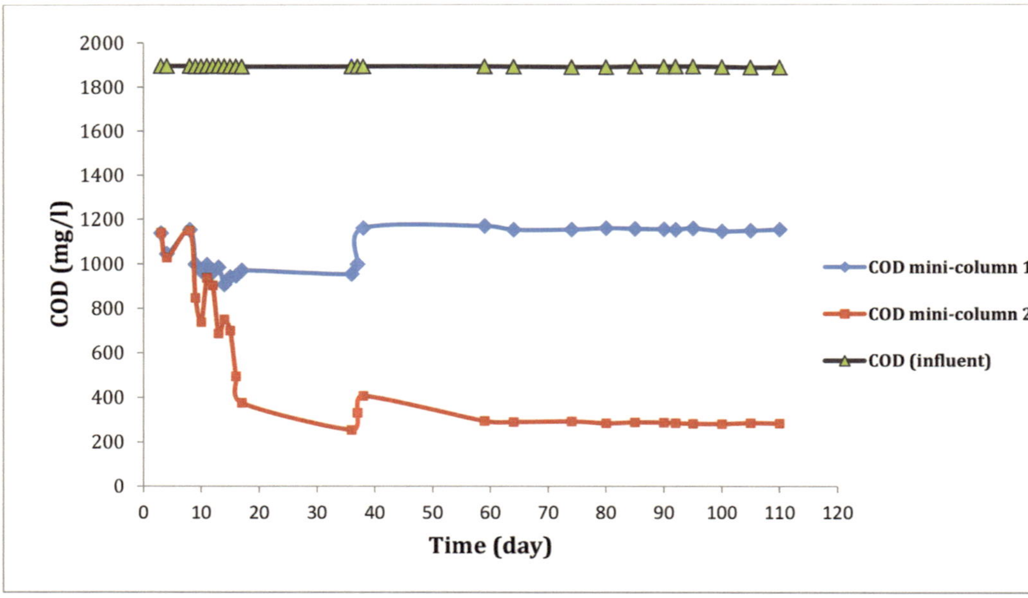

Figure 15. Mini-columns COD profiles comparison.

the second one (mini-biofilter). After 10 days of operation, the mini-biofilter COD fluctuated due to the biomass adaptation and then decrease in time at 254 mg L^{-1} (**Figure 15**). COD concentration of the two mini-columns increased at the 37th day of operation. This increase may due to the media saturation. COD of mini-column without bacterial activity was set at COD concentration influent while that of the mini-biofilter decreased to reach 285 mg L^{-1} values at the steady state with 84.9% of COD removal.

4. Conclusion

Two pilot filters, with and without biological activity, were designed for continuous mechanisms to follow. It has been shown that mechanisms of volatilization, sorption, and biodegradation operate simultaneously during the biofiltration process. For, column 1, in which bacterial activity was inhibited by mercuric chloride addition, a removal efficiency of 33.33–40.1% was obtained at steady state for all the phenolic compounds except phenol and chlorophenol for which 40.2–83.33% of initial concentrations was removed. On the other hand, 41.1% of phenol and chlorophenol initial concentrations were found in the collected gas. Thus, in this column, both physicochemical retention and volatilization are responsible for phenolic compounds removal. At the saturation of the media on the 135th day, a maximum of volatilization was obtained and reached 41.1% for more volatile compounds.

Good treatment efficiency was obtained in the biofilter (column 2). Note that 97–98.2% of COD and BOD removal were observed, respectively. Excellent performances were achieved and reached 99.9% of initial concentrations removal for all the phenolic compounds. Volatilization did not exceed 5–7% at the steady state for the more volatile compounds in this column simulating the real biofilter operation.

Woodwaste leachate was treated with an excellent efficiency by biofiltration process on the mixed media peat: perlite, although a slight problem with color due probably to peat. This is why the authors recommend improving aspect effluent quality by a final treatment if necessary. Toxicity evaluation confirmed that effluent was not toxic for algae and daphnia. It was confirmed also that algae were stimulated in the presence of diluted effluent.

Author details

Najat Kamal[1]*, Rosa Galvez[1], Gerardo Buelna[2] and Rino Dubé[2]

*Address all correspondence to: najat.kamal@gmail.com

1 Department of Civil Engineering, Laval University, Canada

2 Direction of Environment, Industrial Research Center of Québec, Canada

References

[1] Environment, Ministry of Environmental Assessment 2011. Report, Industrial water services; Water Policy Department. Québec: Ministry of Sustainable Development, Environment, Wildlife and Parks., 2013..

[2] Huber, M. M., Göbel, A., Joss, A., Hermann, N., Löffler, D., McArdell, C. S., et al. Oxidation of pharmaceuticals during ozonation of municipal wastewater effluents: a pilot study. *Environmental Science and Technology*. 2005; 39: 4290–4299.

[3] Huber, M. M., Canonica, S., Park, G. Y., & von Gunten, U. Oxidation of pharmaceuticals during ozonation and advanced oxidation processes. *Environmental Science and Technology*. 2003; 37: 1016–1024.

[4] Westerhoff, P., Yoon, Y., Snyder, S., & Wert, E. Fate of endocrine-disruptor, pharmaceutical, and personal care product chemicals during simulated drinking water treatment processes. *Environmental Science and Technology*. 2005; 39: 6649–6663.

[5] Dodd, M. C., Rentsch, D., Singer, H. P., Kohler, H. P., & von Gunten, U. Transformation of beta-lactam antibacterial agents during aqueous ozonation: reaction pathways and quantitative bioassay of biologically-active oxidation products. *Environmental Science and Technology*. 2010; 44 (15): 5940–5948.

[6] Stalter, D., Magdeburg, A., & Oehlmann, J. Comparative toxicity assessment of ozone and activated carbon treated sewage effluents using an in vivo test battery. *Water Research*. 2010; 44 (8): 2610–2620.

[7] Benner, J., & Ternes, T. A. Ozonation of metoprolol elucidation of oxidation pathways and major oxidation products. *Environmental Science and Technology*. 2009; 43 (14): 5472–5480.

[8] Radjenovic, J., Godehardt, M., Petrovic, M., Hein, A., Farre, M., Jekel, M., et al. Evidencing generation of persistent ozonation products of antibiotics roxithromycin and trimethoprim. *Environmental Science and Technology*. 2009; 43 (17): 6808–6815.

[9] Reungoat, J., Escher, B., Macova, M., & Keller, J. Biofiltration of wastewater treatment plant effluent: effective removal of pharmaceuticals and personal care products and reduction of toxicity.*Water Research*. 2011; 45: 2751–2762.

[10] Mailler, R., Gasperi, J., Rocher, V., Gilbert-Pawlik, S., Geara-Matta, D., Moilleron, R., et al. Biofiltration vs conventional activated sludge plants: what about priority and emerging pollutants removal?. *Environmental Science Pollution Research*. 2014; 5379–5390.

[11] Ruel, S. M., Esperanza, M., Choubert, J. M., Valor, I., Budzinski, H., & Coquery, M. On-site evaluation of the efficiency of conventional and advanced secondary processes for the removal of 60 organic micropollutants. *Water Science and Technology*. 2010; 62: 2970–2978.

[12] Sipma, J., Osuna, B., Collado, N., Monclus, H., Ferrero, G., Comas, J., et al. Comparison of removal of pharmaceuticals in MBR and activated sludge systems. *Desalination*. 2010; 250: 653–659.

[13] De Wever, H., Weiss, S., Reemtsma, T., Vereecken, J., Müller, J., Knepper, T., et al. Comparison of sulfonated and other micropollutants removal in membrane bioreactor and conventional wastewater teatment. *Water Research*. 2007; 41: 935–945.

[14] Gonzalez, S., Petrovic, M., & Barcelo, D. Removal of broad range of surfactants from municipal wastewater—comparison between membrane bioreactor and conventional activated sludge treatment. *Chemosphere*. 2007; 67: 335–343.

[15] Bernhard, M., Müler, J., & Knepper, T. Biodegradation of persistent polar pollutants in wastewater: comparison of an optimised lab-scale membrane bioreactor and activated sludge treatment. *Water Research*. 2006 ; 40: 3419–3428.

[16] Clara, M., Strenn, B., Gans, O., Martinez, E., Kreuzinger, N., & Kroiss, H. Removal of selected pharmaceuticals, fragrances and endocrine disrupting compounds in a membrane bioreactor and conventional wastewater treatment plants. *Water Research*. 2005; 39: 4797–4807.

[17] Joss, A., Keller, E., Alder, A., Göbel, A., McArdell, C., Ternes, T., et al. Removal of pharmaceuticals and fragrances in biological wastewater treatment. *Water Research*. 2005; 39: 3139–3152.

[18] Katsoyiannis, A., & Samara, C. Persistent organic pollutants (POPs) in the conventional activated sludge treatment process: fate and mass balance. *Environmental Research*. 2005; 97: 245–257.

[19] Schaar, H., Clara, M., Gans, O., & Kreuzinger, N. Micropollutant removal during biological wastewater treatment and a subsequent ozonation step. *Environmental Pollution*. 2010; 158: 1399–1404.

[20] Buelna, G., Dubé, R., & Turgeon, N. Pig manure treatment by organic bed biofilttration. *Desalination*. 2008; 231: 297–304.

[21] Elmrini, H., Bredin, N., Shareefdeen, Z., & Heitz, M. Biofiltration of xylene emmissions: bioreactor response to variation in the pollutant inlet concentration and gas flow rate. *Chemical Engineering Journal*. 2004; 100 : 149–158.

[22] Vijay kumar, K., Sridevi, V., Harsha, N., Chandana lakshmi, M., & Rani, K. Biofiltration and its application in treatment of air and water pollutants—A review. *International Journal of Application or Innovation in Engineering and Management. 2013; 2*: 226–231.

[23] Shareefdeen, Z., Herner, B., Webb, D., & Wilson, S. Biofiltration eliminates nuisance chemical odors from industrial air streams. *Journal of Industrial Microbiologyand Biotechnology.* 2003; *30*: 168–174.

[24] Zabneva, O., Smolin, S., Klymenko, N., Shvydenko, O., Synelnikova, A., & Nevynna, L. Biofiltration of the Chlorophenol Aqueous Solution through the Activated Carbon Bed. *Journal of Water Chemistry and Technology 2013; 35(1)*: 36–42.

[25] Garzon Marco, *Mécanisme d'enlèvement de l'azote du lisier de porc par biofiltration aérée sur tourbe* [Thesis] Département de génie civil, Université Laval, Québec, Canada; 2001.

[26] Ercolini, D. PCR-DGGE fingerprinting: novel strategies for detection of microbes in food. *Journalof Microbiology Methods*. 2004; *56*: 297–314.

[27] Kamal, N., Galvez, R., & Buelna, G. Application of a solid phase extraction-liquid chromatography to quantify phenolic compounds in woodwaste leachate. *Water Quality Research Journal of Canada. 2014; 49*: 210.

Chapter 16

Degradation of Phenolic Compounds Through UV and Visible-Light-Driven Photocatalysis: Technical and Economic Aspects

Pankaj Chowdhury, Sharmistha Nag and Ajay K. Ray

Additional information is available at the end of the chapter

http://dx.doi.org/10.5772/66134

Abstract

Phenolic compounds are found in surface and groundwater as well as wastewater from several industries. It is necessary to eliminate phenols and phenolic compounds from contaminated water before releasing into water bodies due to their toxicity to human beings. Photocatalytic degradation seems to be a promising technology for the degradation of several phenolic compounds. Complete mineralization of phenol and phenolic compound has been achieved with TiO_2-based photocatalysts under both UV and visible-light irradiation. This chapter will evaluate the conventional processes and advanced oxidation processes for the degradation of phenol and phenolic compounds. The process economics and efficiencies of different advanced oxidation processes will also be discussed. The main focus of the chapter is photocatalytic degradation processes under UV and visible light along with a detailed review of several factors affecting degradation of phenol and phenolic compounds. Photocatalytic degradation process is governed by reactions with hydroxyl radical or superoxide ion. The extent of degradation depends on light sources (UV, visible, and solar), the type of photocatalyst, and experimental conditions (pH, photocatalyst dosage, initial concentration of phenolic compounds, light intensity, electron donor concentration, etc.). Visible-light-active photocatalysts are applied by several researchers to exploit sunlight and to make the photocatalysis process sustainable. In the future, using sunlight in place of UV could make photocatalysis economically more efficient.

Keywords: AOPs, dye sensitization, photocatalyst, phenol, TiO_2, UV, visible

1. Introduction

Phenol and phenolic compounds (chlorophenols, nitrophenols, etc.) detected in water and wastewater are toxic in nature and treated as primary water pollutants as per different countries' regulations. Phenol is one of the first compounds included in the US EPA list of priority pollutants [1]. Chlorophenols and nitrophenols are even more toxic than phenol itself. The exposure, health effects, and regulatory limits of phenols are mentioned in **Table 1** [2]. High concentration of phenolic compounds is present in the effluents from different industries, such as textiles, plastics, paint, paper, petroleum refining, coal processing, wood products, pharmaceuticals, and steel manufacturing [3]. Phenols can be removed by conventional techniques such as (i) physicochemical processes and (ii) biological processes. A comparative study of different phenol degradation methods is presented in the following section. Because of several limitations of the conventional phenol degradation processes, researchers are now relying on advanced oxidation processes (AOPs) for the complete mineralization of phenols. AOPs provide much faster degradation rate with the participation of hydroxyl radical ($HO^•$), and phenols are mineralized to CO_2 and water instead of transferring the pollutants from one phase to another [4]. Heterogeneous photocatalysis process became very popular among the AOPs, and it requires mainly three components such as (i) semiconductor photocatalyst, (ii) light energy (UV or visible or solar), and (iii) electron donor or hole acceptor. When semiconductor photocatalyst is illuminated with light energy greater than the band gap of the photocatalyst, charge carriers (i.e., electron-hole pair) are produced which ultimately generate hydroxyl radicals ($HO^•$) in the system. Recently, photocatalytic degradation of phenol and phenolic compounds in wastewater has been extensively studied by several research groups [4–13]. Titanium dioxide (TiO_2) photocatalyst is frequently used in the degradation of phenols under ultraviolet light [14–16]. TiO_2 is nontoxic, photostable, insoluble under most conditions, and inexpensive and has exceptional chemical and biological inertness [17]. There are few other photocatalysts such as ZnO, CuO, and β-Ga_2O_3 which are also used for phenol degradation under UV light. TiO_2 shows highest efficiency among the photocatalysts [6]. However, the use of UV light is neither feasible nor economical for the degradation of a larger volume of industrial effluent containing phenols. Again, sunlight contains only a small fraction of UV light (4% of solar spectrum) in comparison to visible light (46% of solar spectrum) [18]. For this reason, visible-light-active photocatalyst development is necessary to utilize sunlight in photocatalytic degradation of phenols. There are two approaches to achieve the visible-light-active photocatalysts: (i) modifying existing photocatalysts (via techniques such as doping, composite semiconductor, and dye sensitization) and (ii) developing novel undoped single-phase mixed oxide photocatalysts [19]. Phenol degradation is achieved successfully under visible light with doped-TiO_2 photocatalysts, where different dopants such as iodine [20], nitrogen [21], sulfur [12], praseodymium [22], and iron [23] are used to expand their photoresponses into the visible spectrum. In the case of a composite semiconductor, a large band gap semiconductor is coupled with a small band gap semiconductor with a more negative conduction band level. Therefore, the conduction band electrons can be injected from the small band gap semiconductor to the large band gap semiconductor providing a better charge carrier separation [24]. There are few composite photocatalysts such as Co_3O_4/$BiVO_4$ [19], TiO_2/multiwalled carbon nanotubes [13], and coke-containing TiO_2 [25] that are reported for the degradation of phenols under visible light. In dye-sensitization process, electron injection occurs from the excited dye into the conduction band of the semiconductor photocatalyst, followed by interfacial electron transfer [7]. Dyes are naturally visible light active, and upon light illumination, they get excited. Vinu et al. [11] used eosin Y and fluorescein as sensitizers of TiO_2 to degrade 4-chlorophenol, 2,4-dichlorophenol, and 2,4,6-trichlorophenol under visible light. Chowdhury et al. [7] used eosin Y-sensitized Pt-loaded TiO_2 for phenol degradation. Qin et al. [8] applied N719 dye-sensitized TiO_2 for the degradation of 4-chlorophenol. Degradation of 4-nitrophenol is studied with Cu(II)-porphyrin and Cu(II)-phthalocyanine sensitized TiO_2 under visible light [26].

Phenol and phenolic compounds	Use/exposure	Health effects	Human health for the consumption of water and organism (maximum contaminant levels) [2]
Phenol	(i) Used in the production of aniline, phenolic resins, cresols, alkylphenols, dyes, pesticides, synthetic fiber, disinfectant, and antiseptic; (ii) found in industrial effluent from different industries such as pulp and paper, food processing, textile, pharmaceuticals, coal gasification, and petroleum refining	Headache, skin irritation, kidney damage, liver damage	4 mg L^{-1}
Chlorophenols	(i) Used in antiseptics and pesticide production, (ii) produced in chlorine-bleaching process during paper making, (iii) produced via chlorination of humic matter during the chlorination of drinking water, and (iv) also produced in textile, chemical, and pharmaceutical industry	Burning pain in the mouth, headache, lung damage, affects the digestive tract and immune system	2-Chlorophenol (0.03 mg L^{-1}), 3-methyl-4-chlorophenol (0.5 mg L^{-1}), 2,4-Dichlorophenol (0.01 mg L^{-1}), pentachlorophenol (0.03 μg L^{-1}), 2,4,6-Trichlorophenol (1.5 μg L^{-1})
Nitrophenols	(i) Formed via reaction of phenol and nitrite ions in water under light (UV or solar), (ii) also produced during production and degradation of pesticides, (iii) produced from metallurgic and electronic industry, and (iv) used in solvents, dyes, and plastic production	Weakness, muscles pain, anorexia, kidney damage	2,4-Dinitrophenol (0.01 mg L^{-1})

Table 1. Exposure and regulatory limits of phenol and phenolic compounds [1, 2].

In the first part of this chapter, conventional treatment methods for the degradation of phenol and phenolic compounds are presented, followed by the application of AOPs for such treatment. The process economics and efficiencies of different AOPs are also discussed for the degradation of phenolic compounds.

In the second part of the chapter, we focus on the photocatalytic degradation processes concerning different areas such as (i) basic principle of photocatalysis, (ii) experimental details of photocatalytic degradation of phenols, (iii) photocatalysis reaction mechanism for the degradation of phenols, and (iv) effect of different experimental parameters on degradation of phenol and phenolic compounds.

Finally, we demonstrate a dye-sensitized method to improve the photocatalytic activity and visible-light response of TiO_2-based photocatalyst to perform visible-light-driven phenol degradation.

2. Degradation methodologies of phenol and phenolic compounds

It is of utmost importance to treat wastewater containing phenols before disposal to the environment in order to save the aquatic life. Physical/chemical treatment methods such as activated carbon adsorption, ion exchange, liquid-liquid extraction, chlorine oxidation, chlorine dioxide oxidation, and hydrogen peroxide oxidation are mostly applied for the removal of phenols. However, these methods are expensive and produce hazardous by-products. On the other hand, biological treatment methods for phenol degradation are superior in the above aspects, but these are applicable only for low concentration of phenols [3]. AOPs normally produce hydroxyl radicals ($HO^{\bullet}$) as active species which have very low selectivity and can drive phenol degradation through complete mineralization [27]. A review of the conventional and advanced degradation methods of phenol and phenolic compounds is presented in **Table 2**.

No.	Technology	Target compound	Process details and significant factors	Experimental results	References
1.	Chemical oxidation with sulfatoferrate (VI)	Phenol	Solution pH 9, stirring for 1 h, $Fe^{(VI)}O_4^{2-}$ to phenol molar ratios (1:1, 10:1, and 15:1), phenol initial conc. 30 ppm	(i) Phenol oxidation follows a second-order reaction kinetics; (ii) at Fe(VI)/phenol of 10:1, phenol degrades 100%, TOC decreases 57%, and COD decreases 82%; (iii) oxidation reaction follows a radical pathway to undergo ring opening forming intermediates such as phenoxyphenol and benzoquinone	[32]
2.	Chemical oxidation with potassium permanganate	Phenol and bisphenol A (BPA)	Solution pH (4–8.5), initial conc. of BPA 0–16.8 μM, phenol initial conc. 10 μM, permanganate conc. 181 μM	(i) Oxidation follows a second-order reaction kinetics; (ii) oxidation of both phenol and BPA improve under mildly acidic conditions (pH 4–6); (iii) oxidation of phenol delays at pH range 7.5–8.5; (iv) manganese intermediates such as Mn(V) and Mn(VI) form during the reactions	[33]
3.	Chemical oxidation with ozone (O_3)	BPA	O_3 concentration, solution pH, bicarbonate concentration, initial conc. of BPA 35 μM	(i) BPA oxidation with aqueous O_3 follows a second -order rate equation at pH 7; (ii) O_3 conc. and solution pH show a significant effect on BPA removal	[34]
4.	Adsorption	BPA	Adsorbent: powdered activated	(i) Reach adsorption equilibrium in 150 min; (ii) adsorption follows a Freundlich isotherm; (iii) iron oxide impregnation improves BPA removal	[35]

No.	Technology	Target compound	Process details and significant factors	Experimental results	References
			carbons impregnated with iron oxide nanoparticles		
5.	Reverse osmosis	Phenol	Polyamide thin-film composite reverse osmosis membrane, pH (3–12), phenol conc. (50–1000 ppm), ionic strength (NaCl conc. 0.1–0.001 $molL^{-1}$), transmembrane pressure (5–15 bar)	(i) Phenol retention depends on transmembrane pressure, feed concentration, solution pH, and ionic strength; (ii) phenol mainly diffuses through the membrane; (iii) phenol retention exceeds 85% at alkaline pH; (iv) electrostatic repulsion plays a more important role than size exclusion process	[36]
6.	Nanofiltration	Phenol	Composite polyamide nanofiltration membrane, transmembrane pressure, pH, recovery rate, volumetric cross flow rate, phenol initial conc. 132 ppm	(i) 97% phenol and 96% COD removal take place by cross flow nanofiltration; (ii) nanofiltration membrane successfully removes coke-oven wastewater containing phenol, oil and grease, cyanide, and ammonia	[37]
7.	Solvent extraction	Phenol	Cumene is the extractant, pH range for extraction 1–7, phase ratio between 0.5 and 4, extraction temperature (from 25 to 55°C), phenol initial conc. 500–5000 ppm	(i) Stripping efficiency for phenol is more than 99%; (ii) cumene shows excellent extraction performance on phenol in acidic solution	[38]
8.	Biodegradation by activated sludge	BPA	Operating temperature 20°C, DO 4 ppm, pH 7.5, sludge age 30 or 45 days, hydraulic retention time 48 h, BPA initial conc. 40 ppm	(i) Metabolic intermediates are 4-hydroxyacetophenone, 4-hydroxybenzaldehyde, and 4-hydroxybenzoic acid; (ii) biodegradation kinetics is influenced by the sludge age, BPA concentration, and the acclimation process	[39]

No.	Technology	Target compound	Process details and significant factors	Experimental results	References
9.	Enzymatic process	Phenol (in refinery effluent)	Packed bed bioreactor, biocatalyst weight 135 g, effluent pH 7, temperature (20–32°C), flow rate(3–6 ml min $^{-1}$), H_2O_2 conc. (1–9 mM), phenol initial conc. 100 ppm	(i) 97% phenol degradation is attained; (ii) H_2O_2 concentration, temperature, and flow rate have a positive effect on phenol degradation; (iii) immobilized enzyme shows better stability at broad pH range and at high temperature	[40]
10.	Advanced oxidation with UV, O_3, and TiO_2	Phenol	A low-pressure mercury lamp (λ, 184.9 and 253.7 nm), circulation flow rate for phenol 1000 ml min^{-1}, phenol initial conc. 50–200 ppm	(i) Phenol degradation follows a pseudo-first-order kinetics; (ii) O_3-UV-TiO_2 process achieves complete degradation of phenol within 2 h; (iii) formic acid, acetic acid, propionic acid, and fumaric acid are the reaction intermediate	[41]
11.	Advanced oxidation by Fenton process	Phenol	Concentration of iron (II) sulfate 0.001 mol L^{-1}, flow rate of H_2O_2 (0.075, 0.15, and 0.3 mol per 30 min), pH 3, temperature 30°C, phenol initial conc. 0.012 mol L^{-1}	(i) About 94% organic degradation possible in 2 h; (ii) excess iron(II) is responsible for the lower efficiency of Fenton process; (iii) higher H_2O_2 flow rate provides best results	[42]
12.	Photocatalysis with UV-TiO_2	4-Nitrophenol (4-NP)	Initial concentration of 4-NP, light intensity, partial pressure of oxygen, photocatalyst concentration, pH, chloride ion, and temperature	(i) Degradation rate of 4-NP follows pseudo-first-order kinetics with respect to its concentration; (ii) Cl^- ion shows a negative effect on the degradation	[14]
13.	Photocatalysis -visible-light $BiVO_4$	Phenol	Chelating agents (ascorbic acid or citric acid), solvent volumetric ratio, electron scavenger	(i) Three-dimension ordered macroporous (3D-OM) bismuth vanadates successfully remove phenol (94% removal) from wastewater under visible light; (ii)Bi(+V)/chelating agent optimum molar ratio is 2:1	[43]

No.	Technology	Target compound	Process details and significant factors	Experimental results	References
			(H_2O_2), phenol initial conc. 0.1–0.4 mmol L^{-1}		
14.	Photocatalysis -visible-light-S-doped TiO_2	Phenol	Thiourea to TiO_2 mass ratio, photocatalyst calcination temperature	(i) Photocatalyst activity depends on the doping amount of S; (ii) maximum activity is observed when the photocatalyst is calcinated at 600°C with the mass ratio of thiourea/TiO_2 of 1:1	[12]
15.	Photocatalysis -visible-light-multiwalled carbon nanotubes (MWNT)-TiO_2 composite	Phenol	MWNT/TiO_2 ratio, MWNT surface area, photocatalyst preparation method, phenol initial conc. 50 ppm	(i) MWNT-TiO_2 composite photocatalysts are synthesized via modified sol-gel method; (ii) the increase of MWNT/TiO_2 ratio from 5 to 20% favors the enhancement of the synergetic effect on phenol disappearance	[13]
16.	Photocatalysis -visible-light-dye-sensitized TiO_2/Pt	Phenol	pH, photocatalyst loading, triethanolamine concentration, Pt loading on TiO_2, visible solar light intensity, phenol initial conc. 20–100 ppm	(i) Superoxide ion is the active species for phenol degradation; (ii) complete phenol degradation is achieved in 1 h with initial phenol concentration of 20 ppm (pH 7.0, light intensity 100 mWcm^{-2})	[7]
17.	Photocatalysis -visible-light-dye-sentitized TiO_2	Phenolic compounds	Concentration of sensitizing dye	(i) Degradation order of the different phenolic compounds: chlorophenol > trichlorophenol > dichlorophenol > phenol	[11]

Table 2. Different treatment methods for removal/degradation of phenol and phenolic compounds.

There are two major constraints that need to be considered for industrial applications: (i) technical feasibility and (ii) economic feasibility. The overall costs of the processes are calculated by summing up the capital cost, operating cost, and maintenance cost [28]. In the following section, we compare the costs associated with different advanced oxidation methodologies. The treatment costs of the AOPs are ranked on a 0–5 scale, 0 being the most expensive and 5 being the least. In between 0 and 5, the ranking is evaluated based on Eq. (1) [29]:

$$Rank_{cost,\, i} = \left(\frac{Cost_{\max} - Cost_i}{Cost_{max} - Cost_{\min}} \right) \times 5 \tag{1}$$

where $Rank_{cost,\,i}$ is the cost rank of AOP, *i*. $Cost_{max}$ is the most expensive AOP, and $Cost_{min}$ is the least expensive AOP.

The different process costs are compared by few authors. Saritha et al. [30] compare UV, UV/TiO_2, UV/H_2O_2, UV/Fenton, Fenton, and H_2O_2-based AOPs for the degradation of 4-chloro-2 nitrophenol (4C-2-NP). Based on the overall costs, we find that AOP carried out using H_2O_2 and Fenton are least expensive having ranks of ~5, while UV is the most expensive, assigned a rank of 0. **Figure 1** shows the cost ranking of the different AOPs for the degradation of 4C-2-NP. Esplugas et al. [31] compare UV, O_3/H_2O_2, O_3/UV, O_3/UV/H_2O_2, UV/H_2O_2, and O_3 processes for phenol degradation. Again, based on the overall costs, the different O_3-based AOPs are least costly, while UV is the most expensive, as evident from **Figure 2**. We can infer from the cost comparison that incorporating a photocatalyst such as TiO_2 with UV lowers the overall cost by one-third [30]. In the future, using sunlight in place of UV could make AOPs economically more efficient.

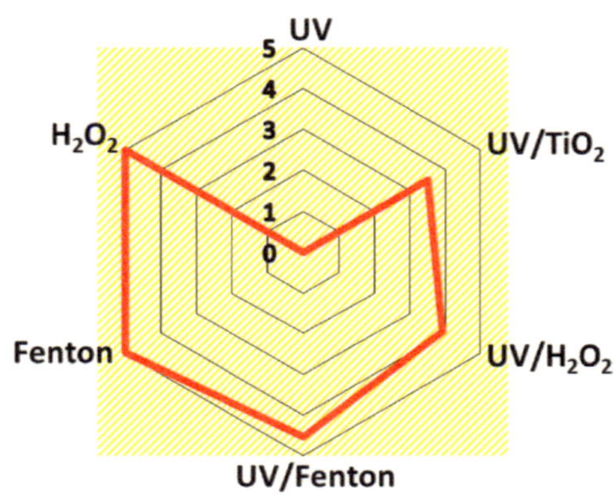

Figure 1. Cost comparison on AOPs for the degradation of 4C-2-NP.

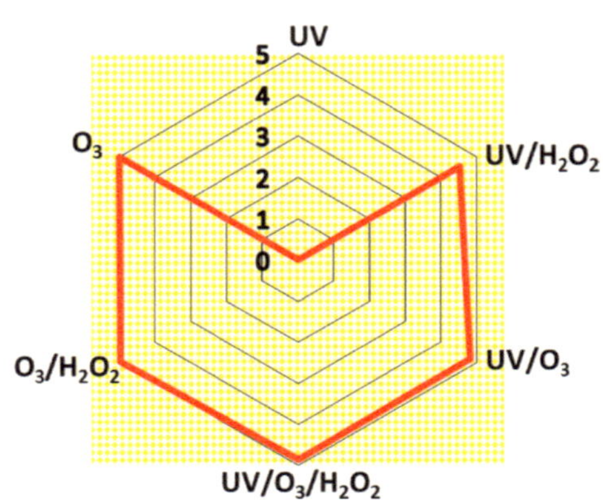

Figure 2. Cost comparison on AOPs for the degradation of phenol.

3. Photocatalytic degradation of phenol and phenolic compounds

3.1. Basic principle of photocatalysis

The precise definition of heterogeneous photocatalysis is a tricky one; particularly as in many cases, the complete mechanism of the reactions is uncertain [27]. In photocatalytic reactions, liquid or gas phase reactants and/or products come into contact with the light-absorbing semiconductor photocatalyst [44]. The semiconductor material can be activated by photons with sufficient energy equal to or greater than the band gap energy (E_g) between the conduction band and valence band of the material [45]. A photocatalytic reaction initiates with the formation of electron-hole pairs followed by oxidation and reduction reactions [46]. In the presence of hole scavenger, the reduction reactions become predominant, whereas in the presence of electron scavenger, the oxidation reactions are the key reactions. However, there are some unwanted reactions such as recombination of electron-hole pairs which reduces the photocatalysis efficiency [47]. **Figure 3** provides a detailed mechanism of photocatalytic reactions.

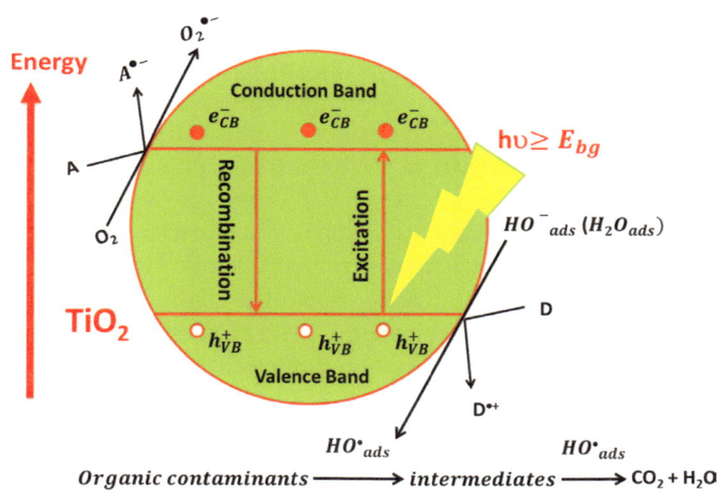

Figure 3. Schematic diagram of photocatalytic reactions for the degradation of organic contaminants.

3.2. Experimental details of photocatalytic degradation of phenols

3.2.1. Photoreactors

Photocatalytic degradation of phenols is performed with either slurry or immobilized photoreactors [15]. Slurry photoreactors provide a high photocatalytic surface area to reactor volume ratio but require filtration after the reaction. On the other hand, immobilized photoreactor can be used continuously without any photocatalyst separation step. However, immobilized reactors suffer from mass transfer limitations and high light scattering [10]. Slurry photocatalyst provides much higher phenol degradation efficiency than immobilized photocatalysts [15]. Our research group used few different types of photoreactors for the degradation of phenol and phenolic compounds. Chen and Ray [14] used two-phase monolithic-type photoreactor for the photodegradation of 4-nitrophenol under UV light. Sengupta et al. [48] used a Taylor vortex reactor (TVR) for the degradation of phenol under UV light. Chowdhury et al. [7] used slurry photoreactor with dye-sensitized photocatalyst, and Malekshoar et al. [5] used slurry photoreactor with graphene-based photocatalyst for phenol degradation under solar light. Studies by other researchers reported degradation of phenols with (i) TiO_2-coated-fiber-optic cable reactor [16], (ii) tubular photoreactor [49], (iii) continuous flow photoreactor [50], and (iv) solar photoreactor (CPC modules and flat reactor) [51].

3.2.2. Light sources

Photocatalysis process efficiency largely depends on photocatalyst surface area and incident photons. Ray [17] combined these two factors and came up with a parameter called illuminated photocatalyst surface area (κ, $m^2\ m^{-3}$) which mainly represents the illuminated photocatalyst inside the photoreactor to undergo the photocatalysis process. Therefore, distribution of light inside the photoreactor is a crucial factor. In majority cases of phenol degradation, photoreactors use an external light source (UV or solar) with a slurry reactor. Chen and Ray [14] used 125 W high-pressure Hg vapor lamp (Philips) in a swirl-flow reactor. However, such external-type photoreactors are limited by the low value of κ, and thus scale up is not possible. Sengupta et al. [48] used a TVR with immersion-type lamp and immobilized photocatalyst for phenol degradation and achieved a κ value of 80 $m^2\ m^{-3}$. In such case, photoreactor, scale up is possible with larger reactor volume [52]. Chowdhury et al. [7] used a solar simulator (1000 W Xe arc lamp with AM 1.5 G filter) in an external-type slurry photoreactor for dye-sensitized phenol degradation under solar-visible light. Gimenez et al. [51] studied the photocatalytic degradation of phenol and 2, 4-dichlorophenol under natural sunlight using compound parabolic collectors (CPCs) and the flat reactor (cylindrical tank). CPCs showed higher phenol degradation efficiency, but it is technologically more complicated than the flat reactor.

3.2.3. Photocatalysts

Degussa P25 TiO_2 (DP25) is the most common photocatalyst used for phenol degradation under UV light. Some other commercial TiO_2 photocatalysts such as Hombikat UV100, TTP, and PC500 are also used for the same. Among them, DP25 provides the highest photocatalytic activity due to slow electron-hole recombination during photocatalysis [53]. Several visible-light-active photocatalysts such as eosin Y-sensitized TiO_2/Pt [7], dye-sensitized TiO_2[11], MWNT-TiO_2 composite [13], S-doped TiO_2 [12], and BiO_4 [43] are also used for degradation of phenol and phenolic compounds.

3.2.4. Experimental procedure

Aqueous solutions of target compounds (phenol and/or phenolic compounds) are prepared at a desired initial concentration. Solution pH is adjusted with HCl or HNO_3 or NaOH solutions. In some cases, buffer solutions are used to maintain the solution pH. In the case of

http://dx.doi.org/10.5772/66134

slurry photocatalyst, the powered photocatalysts are dispersed in the solution with ultrasonication and then mixed with a magnetic stirrer. Sometimes the photocatalyst slurry is circulated through a peristaltic pump. Appropriate electron acceptor or hole scavenger is used in the reaction mixture. At first, the dark reaction is performed to study the adsorption behavior of phenols over the photocatalyst. Then photocatalytic reactions are performed under illuminated conditions, and aqueous samples are collected at regular time interval to check the residual concentration of phenols [7, 14].

3.2.5. Analyses of phenols

Chowdhury et al. [7] used high-pressure liquid chromatography (HPLC) to quantify the concentration of phenol and phenolic compounds in aqueous medium The instrument is equipped with a column oven and a diode array detector. AC18 column (5 μm × 150 mm × 4.6 mm) and a mobile phase of methanol and water (67/33% v/v) at a flow rate of 0.5 ml min^{-1} are used. The temperature of the column oven is kept at 25°C throughout the analysis. The wavelengths of analyses for phenol and reaction intermediates catechol, hydroquinone, and 1,4-benzoquinone are done at 270, 290, 275, and 255 nm, respectively.

3.3. Photocatalysis reaction mechanism for the degradation of phenols

Here we are discussing the photocatalysis reaction mechanism for TiO_2 photocatalyst only. The first step in the photocatalytic degradation is the formation of electron-hole pairs within the TiO_2 photocatalyst. Most of the electron-hole pairs are recombined producing heat energy. However, hydroxyl radicals ($HO^{\bullet}$) are formed in the presence of electron acceptor (dissolved O_2) while hole (h^+) oxidizes water or TiO_2 surface active —OH group. Dissolved O_2 reacts with the electron (e^-) and generates superoxide ion ($O_2^{-\bullet}$). Finally, the $HO^{\bullet}$ reacts with either phenol or phenolic compounds until complete mineralization. Photodegradation mechanism of 4-nitrophenol (4-NP) under UV light is presented as follows [14]:

$$\boldsymbol{TiO_2} \xrightarrow{\boldsymbol{h\vartheta,\ \lambda \leq 388\ nm}} \boldsymbol{TiO_2(e_{CB}^- + h_{VB}^+)} \tag{2}$$

$$h_{VB}^+ + H_2O_{ad} \rightarrow HO_{ad}^{\bullet} + H^+ \tag{3}$$

$$h_{VB}^+ + HO_{ad}^- \rightarrow HO_{ad}^{\bullet} \tag{4}$$

$$e_{CB}^- + O_2 \rightarrow O_2^{-\bullet} \tag{5}$$

$$C_6H_4OHNO_2 + 7O_2 \rightarrow 6CO_2 + 2H_2O + HNO_3 \tag{6}$$

Overall reaction stoichiometry shows complete mineralization of 4-NP with the involvement of $HO^{\bullet}$ (Eq. (6)). Devi and Rajashekhar [9] described the possible degradation mechanism for phenol under natural sunlight/UV light using nitrogen-doped TiO_2. Phenol mineralization went through the formation of dihydroxybenzene (catechol or resorcinol), pent 2-enedioic acid, and oxalic acid. In a parallel reaction path, benzoquinone and maleic acid were formed during the mineralization (**Figure 4**).

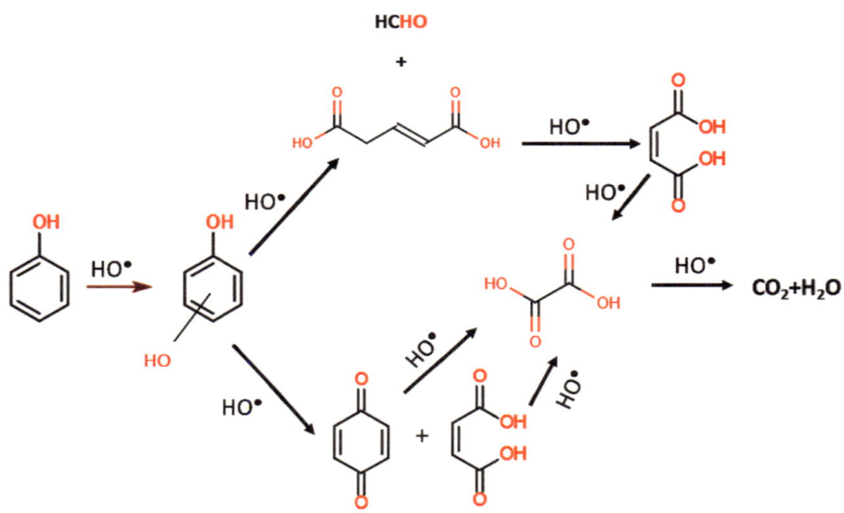

Figure 4. Phenol degradation mechanism (adapted from Ref. [9]).

3.4. Effect of different experimental parameters on degradation of phenol and phenolic compounds

Different parameters such as solution pH, light intensity, initial concentration of target compounds, photocatalyst concentration, and electron acceptors play a significant role on photocatalytic degradation of phenol and phenolic compounds. The following section will provide a review of recent studies on the degradation of phenol and phenol derivatives.

3.4.1. Effect of solution pH

Solution pH plays a vital role in the photocatalytic degradation of phenol and phenolic compounds since it influences two surface properties of the photocatalyst: (i) band edge position and (ii) surface charge. TiO_2 P25 shows a point zero charge at pH 6.8. Thus at pH < 6.8, TiO_2 surface attains positive charge and can easily adsorb anionic species at the photocatalyst surface [54]. Again, the protonation and deprotonation of phenols greatly depend on solution pH. Different phenolic compounds show different optimum pH during photodegradation. Venkatachalam et al. [55] studied the photocatalytic activity of Mg^{2+} and Ba^{2+}-doped TiO_2 nanoparticles for the degradation of 4-chlorophenol (4-CP). In the acidic pH (pH 5), 4-CP was well adsorbed on the photocatalyst surface and showed higher degradation rate than alkaline pH. Lathasree et al. [56] reported the photocatalytic degradation of phenol and chlorophenols with ZnO under UV light. Significant phenol degradation was achieved at neutral and mildly acidic pH. The zero point charge for ZnO was 8, and at alkaline pH, chlorophenols exist as negatively charged chlorophenolate anion. Thus the photodegradation rate was higher at acidic pH ($pH < 8$).

3.4.2. *Effect of light intensity*

Photodegradation rates of different organic compounds improve with increasing light intensity. At high light intensity when mass transfer limitation is low, the reaction rate is found to be proportional to the square root of light intensity. However, at the low-intensity level, the photodegradation rate is directly proportional to the light intensity [14, 54]. Al-Sayyed et al. [57] observed a similar rate shift from first order to half order in intensity while they studied photocatalytic degradation rate of 4-CP in the light intensity range of 2–50 mW cm^{-2}. Chen and Ray [14] correlated 4-CP degradation rate constant (k) with light intensity (I): $k \propto I^{0.84}$ indicating that the degradation was independent of mass transfer limitation.

3.4.3. *Effect of initial concentration of phenols*

As the effect of phenol concentration is of importance in the process of treatment of phenolic wastewater, it is necessary to investigate its dependence. Different concentration profiles can be seen during phenol degradation at different initial concentrations. The degradation rates at same concentration with different initial concentrations are different. However, all the concentration profiles could be correlated with an exponential function as follows [14]:

$$C = C_0 exp\left(-kt\right) \tag{7}$$

where C_0 and C, respectively, are the initial concentration and concentration of phenol at time t and k is the apparent rate constant. Chen and Ray [14] studied the photocatalytic degradation of 4-nitrophenol (4-NP) with varying initial concentration between 10 and 120 ppm. The degradation rate mainly followed a pseudo-first-order kinetics with respect to the initial concentration 4-NP.

3.4.4. *Effect of photocatalyst concentration*

Photocatalyst concentration is a crucial parameter that has been widely studied for photocatalytic processes. The optimum photocatalyst concentrations usually vary between 0.15 g l^{-1} and 8 g l^{-1} for different photocatalyst systems and photoreactors. A large difference in optimum photocatalyst concentration (0.15–2.5 g l^{-1}) was reported even for the same photocatalyst (DP25). Chen and Ray [15] expressed the photocatalytic degradation rate as follows:

$$r_i = K\left[1 - exp\left(-\epsilon\beta C_{photocat}H\right)\right] \quad (8)$$

$$\text{again, } K = \frac{k_1 A f(C_i) I_0^{\beta}}{\epsilon\beta} \quad (9)$$

where k_1 is a proportionality constant, A is the illuminated area of the photoreactor window, C_{photocat} is the photocatalyst concentration, ε is the light absorption coefficient of the photocatalyst, I_0 is the incident light intensity, and β is a constant. They reported β values of 0.84, 0.72, and 0.82 for the degradation of 4-NP, 4-CP, and phenol, respectively.

3.4.5. Effect of electron acceptor

Photocatalytic degradation reaction requires the use of electron acceptor to reduce the charge carrier recombination. Oxygen is the most common electron acceptor because of its availability, higher solubility, and nontoxic nature. The partial pressure of oxygen is adjusted by mixing the oxygen stream with nitrogen stream by maintaining the total flow rate of gas at a constant value. The photocatalytic reaction of phenols will terminate if sufficient oxygen is not available in the solution [15]. Chen and Ray [14] showed the improvement of 4-NP photodegradation rate with increasing oxygen partial pressure. The photodegradation rate constant reached approximately 70% of its maximum value at oxygen partial pressure of 0.2 atm. The effect of oxygen partial pressure on the photodegradation of 4-NP is described by a noncompetitive Langmuir kinetic equation as follows:

$$k_p \infty \frac{K_{O_2} p_{O_2}}{1 + K_{O_2} P_{O_2}} \quad (10)$$

where k_p is the kinetic constant for 4-NP degradation, K_{O2} is the adsorption constant of dissolved oxygen on photocatalyst, and p_{O2} is the partial pressure of dissolved oxygen.

4. Dye-sensitized photocatalytic degradation of phenol and phenolic compounds

4.1. Theory of dye sensitization

The process of expanding the spectral sensitivity of semiconductor materials with a dye to the visible spectra is known as dye sensitization. Dye is typically adsorbed onto the semiconductor surface by chemical adsorption process. Chemisorbed dye molecules act as spectral sensitizer that upon excitation with visible light inject an electron into the conduction band of the semiconductor [27]. To undergo successful electron injection, the dye molecule should include few basic properties regarding energy levels, ground-state redox potential, and surface anchoring group. Carboxylic and phosphoric acid groups form strong covalent bonds with semiconductor and provide fast electron transfer rate [58]. Recent studies mention the use of a group of sensitizers such as poly(aniline), poly(thiophene), porphyrins, coumarin, phthalocyanines, eosin Y, alizarin red S, and carboxylate derivatives of anthracene [11, 26, 27]. Among the photosensitizers, transition metal-based sensitizers have shown best results in dye-sensitization process. Transition metals such as Fe (II), Ru (II), and Os (II) form d^6 complex and undergo intense charge transfer absorption across the entire visible range [59]. However, metal-based sensitizers are not environment friendly, and thus researchers are now focusing on the use of natural dyes as an alternative for dye-sensitization process [59–62]. Several semiconductor photocatalysts have been studied for dye sensitization including TiO_2, $SrTiO_3$, ZnO, SnO_2, and Cu_2O [63, 64]. Among them, TiO_2 is the best photocatalyst in terms of (i) cost, (ii) availability, (iii) toxicity, (iv) stability against photocorrosion, and (v) electronic energy band structure [65, 66].

4.2. Dye-sensitized photocatalytic phenol degradation mechanism

Dye-sensitized photodegradation of phenol under visible light began through excitation of the dye molecule from its ground state to the excited state, which then assists the electron transfer to the conduction band of the semiconductor. The oxidized dye molecule (dye^+) can interact with either phenol or an electron donor to return back to its ground state [7]. Chowdhury et al. [67] used eosin Y (EY) as the sensitizer which provided TiO_2/Pt a significant visible-light activity via dye sensitization. Eosin Y contains both hydroxyl and carboxyl end groups, which actually assists the dissociative surface adsorption of eosin Y onto the surface hydroxyl (Ti–OH_2^+) sites of EY-sensitized TiO_2/Pt [11]. Triethanolamine (TEOA) was used as an electron donor, which was consumed through an irreversible oxidation by extending the lifetime of eosin Y during phenol degradation (Eqs. (11) and (12)) [7].

$$(CH_2-CH_2-OH)_3N: + EY^+ = (CH_2-CH_2-OH)_3N^+ + EY \tag{11}$$

$$\left(CH_2-CH_2-OH\right)_3 N^+ = \left(CH_2-CH_2-OH\right)_2 - N - CH_2 - CH_2 - OH + H^+ \tag{12}$$

(acid-base equilibrium of TEOA)

Eosin Y-sensitized phenol degradation mechanism under visible light is described below [7, 68]:

$$TiO_2 - \mathbf{EY} \overset{h\nu \leq 520\,nm}{\rightarrow} TiO_2 - EY^* \tag{13}$$

$$TiO_2 - EY^* \rightarrow TiO_2 - \left(EY^{\oplus} + e_{CB}^-\right) \tag{14}$$

$$TiO_2 - \left(EY^{\oplus} + e_{CB}^-\right) + PhOH \rightarrow TiO_2 - \left(EY + e_{CB}^-\right) + PhOH^{\oplus} \tag{15}$$

$$TiO_2 - \left(EY^{\oplus} + e^-_{CB}\right) + TEOA \rightarrow TiO_2 - \left(EY + e^-_{CB}\right) + TEOA^{\oplus} \quad (16)$$

$$TiO_2 - \left(EY + e^-_{CB}\right) + O_2 \rightarrow TiO_2 - \mathrm{EY} + O_2^{\bullet -} \quad (17)$$

$$O_2^{\bullet -} + H^+ \rightarrow HOO^{\bullet} \rightarrow\rightarrow HO^{\bullet} \quad (18)$$

$$HO^{\bullet} + C_6H_5OH \rightarrow intermediates + CO_2 \quad (19)$$

$$HO^{\bullet} + intermediates \rightarrow\rightarrow H_2O + CO_2 \quad (20)$$

4.3. Dye-sensitized photocatalytic phenol degradation kinetics

In dye-sensitized photodegradation under the visible light, the dye molecule is first activated by visible light ($\lambda > 420$ nm) and then injects electrons into the conduction band of the semiconductor. Chowdhury et al. [7] described the kinetics of phenol degradation using eosin Y-sensitized TiO_2/Pt with a modified Langmuir-Hinshelwood equation as follows:

$$-\left(\frac{dC_{Ph}}{dt}\right) = \frac{W\left(k_r K_A C_{ph}\right)}{V\left(1 + K_A C_{Ph0}\right)} \times I^{\beta} \quad (21)$$

where W is the mass of photocatalyst, C_{Ph} is the phenol concentration at time t, C_{Ph0} is the initial phenol concentration, V is the volume of the reaction mixture, K_A is the adsorption equilibrium constant, k_r is the kinetic rate constant, I is the light intensity, and β is a constant. The apparent kinetic constant as is defined as follows:

$$K_{app} = \frac{W}{V}\left(k_r K_A\right) \quad (22)$$

Combining Eqs. (21) and (22), they obtained Eq. (23) as follows:

$$-\left(\frac{dC_{Ph}}{dt}\right) = \frac{K_{app} C_{Ph}}{1 + K_A C_{Ph0}} \times I^{\beta} \quad (23)$$

Eq. (23) was used to predict the kinetic parameters of phenol photodegradation at different irradiation intensities (range, 25–100 mW cm^{-2}). Based on a parameter estimation using the experimental data, the values of K_{app}, K_A, and β were obtained for the degradation of phenol. The values of K_{app} was 8.02×10^{-6} min^{-1}, K_A was 0.13 L mg^{-1}, and β was 2.15 [7].

4.4. Application of dye-sensitized photocatalyst for the degradation of phenols

Dye-sensitized photodegradation of phenol and phenolic compounds showed promising results under visible-light irradiation [7, 11, 26, 69–71]. Vinu et al. [11] reported degradation of phenol, 4-chlorophenol, 2,4-dichlorophenol, and 2,4,6-trichlorophenol using eosin Y and fluorescein-sensitized combustion-synthesized nano TiO_2 under visible light. Eosin Y-sensitized photocatalyst showed better performance than fluorescein-sensitized photocatalyst. Phenol degradation rate was slowest among the phenolic compounds. Chowdhury et al [7] used eosin Y-sensitized TiO_2/Pt for the degradation of phenol under the visible solar light in the presence of triethanolamine as an electron donor. About 93% phenol degradation (initial concentration 40 ppm) was achieved within 90 min using eosin Y-sensitized TiO_2/Pt photocatalyst under optimum experimental conditions (pH 7.0, photocatalyst concentration of 0.8 g L^{-1}, triethanolamine concentration of 0.2 M, 0.5% Pt loading on TiO_2, visible solar light intensity of 100 mW cm^2). Mele et al. [26] studied photocatalytic degradation of 4-nitrophenol with polycrystalline TiO_2 impregnated with functionalized Cu(II)-porphyrin or Cu(II)-phthalocyanine. Cu(II)-based sensitizers provided better results for the degradation of 4-nitrophenol in comparison with metal-free sensitizers. Iliev [69] studied photooxidation of phenol with phthalocyanine-modified TiO_2 and Al_2O_3 under visible light. The degree of photodegradation of phenol in the presence of phthalocyanine-modified TiO_2 is much higher than phthalocyanine-modified Al_2O_3. Superoxide ion was considered as the active species during photodegradation. Grandos et al. [70] used Co(II) and Zn(II) tetracarboxyphthalocyanine (TcPcM)-sensitized TiO_2 for the degradation of phenol under visible light. The photodegradation efficiencies were reported to be 4.3 and 3.3% for TcPcCo/TiO_2 and TcPcZn/TiO_2, respectively. Ghosh et al. [71] demonstrated the photocatalytic degradation of 4-chlorophenol with coumarin-sensitized TiO_2 under visible LED light. The degradation rate followed a first-order kinetics and moved toward a limiting value at a photocatalyst concentration of 0.3 g L^{-1}.

5. Conclusions

Phenol and phenolic compounds are considered as priority pollutants by US EPA because of their high toxicity. They impose severe short-term and long-term health problems to human beings. In this review, we discussed different phenol degradation methods such as physical treatments, biological treatments, and advanced oxidation processes (AOPs). AOPs provide much faster degradation rate than conventional treatment methods and undergo complete mineralization instead of transferring the pollutants from one phase to another.

Heterogeneous photocatalysis is such an AOP that limits the use of oxidizing chemicals (e.g., ozone and hydrogen peroxide) and only utilizes light (UV or solar) and photocatalyst to generate hydroxyl radicals ($HO^•$). Several photoreactors are used with UV lamp, namely, swirl-flow reactor, Taylor vortex reactor, and two-phase monolithic-type reactor for the photocatalytic degradation of phenolic compounds. In UV-light-driven photocatalysis, hydroxyl radicals are the active species which react with either phenol or phenolic compounds until complete mineralization. Different parameters such as solution pH, light intensity, initial concentration of target compounds, photocatalyst concentration, and electron acceptors play a significant role on photocatalytic degradation of phenol and phenolic compounds. However, our economic assessment indicates that the use of UV light significantly increases the overall process cost.

Visible-light-active photocatalysts are developed to utilize the most abundant sunlight to make the photocatalysis economically feasible. Compound parabolic collectors (CPCs) are commonly used for solar photocatalytic degradation of phenol and phenolic compounds. Photocatalysts are modified via doping, dye sensitization, and coupling method to expand the photoresponse to the visible region. Among these, dye-sensitized photocatalysis is shown to be an efficient method for phenol degradation under the visible solar light. The process involves electron transfer to the conduction band of semiconductor initiated by dye sensitization under the visible solar light. Dye-sensitized photocatalysis processes are shown to be Efficient methods for the degradation of phenol and phenolic compounds under the visible solar light.

Author details

Pankaj Chowdhury[1], Sharmistha Nag[2*] and Ajay K. Ray[3]

*Address all correspondence to: snag@fdu.edu

1 Department of Chemistry, Simon Fraser University, Burnaby, Canada

2 Department of Economics, Finance and International Business, Fairleigh Dickinson University, Vancouver, Canada

3 Department of Chemical and Biochemical Engineering, Western University, London, Canada

References

[1] Michalowicz J, Duda W. Phenols–sources and toxicity. Pol. J. Environ. Stud. 2007;16: 347–362.

[2] US EPA. National Recommended Water Quality Criteria–Human Health Criteria Table. 2016. Available from: https://www.epa.gov/wqc/national-recommended-water-quality-criteria-human-health-criteria-table [Accessed: 2016-08-20].

[3] Villegas LGC, Mashhadi N, Chen M, Mukherjee D, Taylor KE, Biswas N. A short review of techniques for phenol removal from wastewater. Curr. Pollut. Rep. 2016;2: 157–167.

[4] Choquette-Labbe M, Shewa WA, Lalman JA, Shanmugam SR. Photocatalytic degradation of phenol and phenol derivatives using a nano-TiO_2 catalyst: Integrating quantitative and qualitative factors using response surface methodology. Water 2014;6: 1785–1806.

[5] Malekshoar G, Pal K, He Q, Yu A, Ray AK. Enhanced solar photocatalytic degradation of phenol with coupled graphene-based titanium dioxide and zinc oxide. Ind. Eng. Chem. Res. 2014;53: 18824–18832.

[6] Paschoalino FCS, Paschoalino MP, Jord E, de Figueiredo Jardim W. Evaluation of TiO_2, ZnO, CuO and Ga_2O_3 on the photocatalytic degradation of phenol using an annular-flow photocatalytic reactor. Open J. Phys. Chem. 2012;2: 135–140.

[7] Chowdhury P, Moreira J, Gomaa H, Ray AK. Visible-solar-light-driven photocatalytic degradation of phenol with dye-sensitized TiO_2: Parametric and kinetic study. Ind. Eng. Chem. Res. 2012;51: 4523–4532.

[8] Qin G, Sun Z, Wu Q, Lin L, Liang M, Xue S. Dye-sensitized TiO_2 film with bi-functionalized zones for photocatalytic degradation of 4-chlorophenol. J. Hazard. Mater. 2011;192: 599–604.

[9] Devi LG, Rajashekhar KE. A kinetic model based on non-linear regression analysis is proposed for the degradation of phenol under UV/solar light using nitrogen-doped TiO_2. J. Mol. Catal. A: Chem. 2011;334: 65–76.

[10] del Rio JM. Photocatalytic degradation of phenolic compounds in water: irradiation and kinetic modelling [thesis]. Chemical and Biochemical Engineering. London: The University of Western Ontario; 2011.

[11] Vinu R, Polisetti S, Madras G. Dye-sensitized visible light degradation of phenolic compounds. Chem. Eng. J. 2010;165: 784–797.

[12] Liu S, Chen X. A visible light response TiO_2 photocatalyst realized by cationic S-doping and its application for phenol degradation. J. Hazard. Mater. 2008;152: 48–55.

[13] Wang W, Serp P, Kalck P, Faria JL. Visible light photo-degradation of phenol on MWNT-TiO_2 composite catalysts prepared by a modified sol-gel method. J. Mol. Catal. A: Chem. 2005;235: 194–199.

[14] Chen D, Ray AK. Photo-degradation kinetics of 4-nitrophenol in TiO_2 suspension. Water Res. 1998;32: 3223–3234.

[15] Chen D, Ray AK. Photocatalytic kinetics of phenol and its derivatives over UV-irradiated TiO_2. Appl. Catal., B 1999;23: 143–157.

[16] Peill NJ, Hoffmann MR. Development and optimization of a TiO_2-coated fiber-optic cable reactor: Photocatalytic degradation of 4-chlorophenol. Environ. Sci. Technol. 1995;29: 2974–2981.

[17] Ray AK. Design, modelling and experimentation of a new large-scale photocatalytic reactor for water treatment. Chem. Eng. Sci. 1999;54: 3113–3125.

[18] Pardeshi SK, Patil AB. A simple route for photocatalytic degradation of phenol in aqueous zinc oxide suspension using solar energy. Sol. Energy 2008;82: 700–705.

[19] Long M, Cai W, Cai J, Zhou B, Chai X, Wu Y. Efficient photocatalytic degradation of phenol over $Co_3O_4/BiVO_4$ composite under visible light irradiation. J. Phys. Chem. B 2006;110: 20211–20216.

[20] Hong X, Wang Z, Cai W, Lu F, Zhang J, Yang Y, Ma N, Liu, Y. Visible-light-activated nanoparticle photocatalyst of iodine-doped titanium dioxide. Chem. Mater. 2005;17: 1548–1552.

[21] Wang Z, Cai W, Hong X, Zhao X, Xu F, Cai C. Photocatalytic degradation of phenol in aqueous nitrogen-doped TiO_2 suspensions with various light sources. Appl. Catal., B 2005;57: 223–231.

[22] Chiou CH, Juang RS. Photocatalytic degradation of phenol in aqueous solutions by Pr-doped TiO_2 nanoparticles. J. Hazard. Mater. 2007;149: 1–7.

[23] Nahar MS, Hasegawa K, Kagaya S. Photocatalytic degradation of phenol by visible light-responsive iron-doped TiO_2 and spontaneous sedimentation of the TiO_2 particles. Chemosphere 2006;65: 1976–1982.

[24] Ni M, Leung MKH, Leung DYC, Sumathy K. A review and recent developments in photocatalytic water-splitting using TiO_2 for hydrogen production. Renew. Sustain. Energy Rev. 2007;11: 401–425.

[25] Lettmann C, Hildenbrand K, Kisch H, Macyk W, Maier WF. Visible light photodegradation of 4-chlorophenol with a coke-containing titanium dioxide photocatalyst. Appl. Catal., B 2001;32: 215–227.

[26] Mele G, Del Sole R, Vasapollo G, Garcia-Lopez E, Palmisano L, Schiavello M. Photocatalytic degradation of 4-nitrophenol in aqueous suspension by using polycrystalline TiO_2 impregnated with functionalized Cu(II)-porphyrin or Cu(II)-phthalocyanine. J. Catal. 2003;217: 334–342.

[27] Chowdhury P. Solar and visible light driven photocatalysis for sacrificial hydrogen generation and water detoxification with chemically modified TiO_2 [thesis]. Chemical and Biochemical Engineering. London: The University of Western Ontario; 2012.

[28] Mahamuni NN, Adewuyi YG. Advanced oxidation processes (AOPs) involving ultrasound for wastewater treatment: A review with emphasis on cost estimation. Ultrason. Sonochem. 2010;17: 990–1003.

[29] Acar C, Dincer I, Zamfirescu C. A review on selected heterogeneous photocatalysts for hydrogen production. Int. J. Energy Res. 2014;38: 1903–1920.

[30] Saritha P, Aparna C, Himabindu V, Anjaneyulu Y. Comparison of various advanced oxidation processes for the degradation of 4-chloro-2 nitrophenol. J. Hazard. Mater. 2007;149: 609–614.

[31] Esplugas S, Gimenez J, Contreras S, Pascual E, Rodriguez M. Comparison of different advanced oxidation processes for phenol degradation. Water Res. 2002;36: 1034–1042.

[32] Peings V, Frayret J, Pigot T. Mechanism for the oxidation of phenol by sulfatoferrate (VI): Comparison with various oxidants. J. Environ. Manage. 2015;157: 287–296.

[33] Yang J, Zhao J, Ma J, Liu ZQ. Simultaneous oxidation of phenol and bisphenol A by permanganate: Synergetic or competitive effect. Sep. Purif. Technol. 2013;116: 271–276.

[34] Garoma T, Matsumoto S. Ozonation of aqueous solution containing bisphenol A: Effect of operational parameters. J. Hazard. Mater. 2009;167: 1185–1191.

[35] Park HS, Koduru JR, Choo KH, Lee B. Activated carbons impregnated with iron oxide nanoparticles for enhanced removal of bisphenol A and natural organic matter. J. Hazard. Mater. 2015;286: 315–324.

[36] Mnif A, Tabassi D, Ben Sik Ali M, Hamrouni B. Phenol removal from water by AG reverse osmosis membrane. Environ. Prog. Sustain. Energy 2015;34: 982–989

[37] Kumar R, Pal P. Removal of phenol from coke-oven wastewater by cross-flow nanofiltration membranes. Water Environ. Res. 2013;85: 447–455.

[38] Liu J, Xie J, Ren Z, Zhang W. Solvent extraction of phenol with cumene from wastewater. Desalin. Water Treat. 2013;51: 3826–3831.

[39] Orozco AMF, Contreras EM, Zaritzky NE. Biodegradation of bisphenol A and its metabolic intermediates by activated sludge: Stoichiometry and kinetics analysis. Int. Biodeterior. Biodegradation 2016;106: 1–9.

[40] Rezvani F, Azargoshasb H, Jamialahmadi O, Hashemi-Najafabadi S, Mousavi SM, Shojaosadati SA. Experimental study and CFD simulation of phenol removal by immobilization of soybean seed coat in a packed-bed bioreactor. Biochem. Eng. J. 2015;101: 32–43.

[41] Suzuki H, Araki S, Yamamoto H. Evaluation of advanced oxidation processes (AOP) using O_3, UV, and TiO_2 for the degradation of phenol in water. J. Water Process Eng. 2015;7: 54–60.

[42] Pontes RFF, Moraes JEF, Machulek A, Pinto JM. A mechanistic kinetic model for phenol degradation by the Fenton process. J. Hazard. Mater. 2010;176: 402–413.

[43] Liu Y, Dai H, Deng J, Zhang L, Au CT. Three-dimensional ordered macroporous bismuth vanadates: PMMA-templating fabrication and excellent visible light-driven photocatalytic performance for phenol degradation. Nanoscale 2012;4: 2317–2325.

[44] Schiavello M. Heterogeneous photocatalysis, Wiley Series in Photoscience and Photoengineering. John Wiley & Sons 1997. pp. 87–107.

[45] Chowdhury P, Elkamel A, Ray AK. Photocatalytic processes for the removal of toxic metal ions. In: Sharma SK editor. Heavy Metals In Water: Presence, Removal and Safety. Cambridge: Royal Society of Chemistry; 2014. pp. 25–43.

[46] Chen DW, Sivakumar M, Ray AK. Semiconductor photocatalysis in environmental remediation. Dev. Chem. Eng. Min. Process 2000;8: 505–550.

[47] Chen D, Ray AK. Removal of toxic metal ions from wastewater by semiconductor photocatalysis. Chem. Eng. Sci. 2001;56: 1561–1570.

[48] Sengupta TK, Kabir MF, Ray AK. A Taylor vortex photocatalytic reactor for water purification. Ind. Eng. Chem. Res. 2001;40: 5268–5281.

[49] Ling CM, Mohamed AR, Bhatia S. Performance of photocatalytic reactors using immobilized TiO_2 film for the degradation of phenol and methylene blue dye present in water stream. Chemosphere 2004;57: 547–554.

[50] Meshram S, Limaye R, Ghodke S, Nigam S, Sonawane S, Chikate R. Continuous flow photocatalytic reactor using ZnO-bentonite nanocomposite for degradation of phenol. Chem. Eng. J. 2011;172: 1008–1015.

[51] Gimenez J, Curco D, Queral MA. Photocatalytic treatment of phenol and 2, 4-dichlorophenol in a solar plant in the way to scaling-up. Catal. Today 1999;54: 229–243.

[52] Ray AK. Photocatalytic reactor configurations for water purification: Experimentation and modeling. In: de Lasa H, Rosales BS editors. Advances in Chemical Engineering Photocatalytic Technologies. Elsevier Science; 2009. pp. 145–184.

[53] Ahmed S, Rasul MG, Martens WN, Brown R, Hashib MA. Heterogeneous photocatalytic degradation of phenols in wastewater: A review on current status and developments. Desalination 2010;261: 3–18.

[54] Chowdhury P, Elkamel A, Ray AK. Photocatalytic processes for the removal of dye. In: Sharma SK editor. Green Chemistry for Dyes Removal from Wastewater: Research Trends and Applications Beverly: Scrivener Publishing; 2015. pp. 119–137.

[55] Venkatachalam N, Palanichamy M, Murugesan V. Sol-gel preparation and characterization of alkaline earth metal doped nano TiO_2: Efficient photocatalytic degradation of 4-chlorophenol. J. Mol. Catal. A: Chem. 2007;273: 177–185.

[56] Lathasree S, Rao AN, SivaSankar B, Sadasivam V, Rengaraj K. Heterogeneous photocatalytic mineralisation of phenols in aqueous solutions. J. Mol. Catal. A: Chem. 2004;223: 101–105.

[57] Al-Sayyed G, D'Oliveira JC, Pichat P. Semiconductor-sensitized photo-degradation of 4-chlorophenol in water. J. Photochem. Photobiol., A 1991;58: 99–114.

[58] Chowdhury P, Gomaa H, Ray AK. Dye-sensitized photocatalyst: A breakthrough in green energy and environmental detoxification. In: Shamim N, Sharma VK, editors. Sustainable Nanotechnology and the Environment: Advances and Achievements Washington, DC: American Chemical Society; 2013. pp. 231–266.

[59] Polo AS, Itokazu MK, Iha NYM. Metal complex sensitizers in dye-sensitized solar cells. Coord. Chem. Rev. 2004;248: 1343–1361.

[60] Wongcharee K, Meeyoo V, Chavadej S. Dye-sensitized solar cell using natural dyes extracted from rosella and blue pea flowers. Sol. Energy Mater. Sol. Cells 2007;91: 566–571.

[61] Polo AS, Iha NYM. Blue sensitizers for solar cells: Natural dyes from Calafate and Jaboticaba. Sol. Energy Mater. Sol. Cells 2006;90: 1936–1944.

[62] Garcia CG, Polo AS, Iha NYM. Fruit extracts and ruthenium polypyridinic dyes for sensitization of TiO_2 in photoelectrochemical solar cells. J. Photochem. Photobiol., A 2003;160: 87–91.

[63] Chowdhury P, Gomaa H, Ray AK. Factorial design analysis for dye-sensitized hydrogen generation from water. Int. J. Hydrogen Energy 2011;36: 13442–13451.

[64] Nian JN, Hu CC, Teng H. Electrodeposited p-type Cu_2O for H_2 evolution from photoelectrolysis of water under visible light illumination. Int. J. Hydrogen Energy 2008;33: 2897–2903.

[65] Gratzel M. Dye-sensitized solar cells. J. Photochem. Photobiol., C 2003;4: 145–153.

[66] Nada AA, Hamed HA, Barakat MH, Mohamed NR, Veziroglu TN. Enhancement of photocatalytic hydrogen production rate using photosensitized TiO_2/RuO_2-MV^{2+}. Int. J. Hydrogen Energy 2008;33: 3264–3269.

[67] Chowdhury P, Gomaa H, Ray AK. Sacrificial hydrogen generation from aqueous triethanolamine with Eosin Y-sensitized Pt/TiO_2 photocatalyst in UV, visible and solar light irradiation. Chemosphere 2015; 121: 54–61.

[68] Chatterjee D, Dasgupta S, Rao NN. Visible light assisted photo-degradation of halocarbons on the dye modified TiO_2 surface using visible light. Sol. Energy Mater. Sol. Cells 2006;90: 1013–1020.

[69] Iliev V. Phthalocyanine-modified titania-catalyst for photooxidation of phenols by irradiation with visible light. J. Photochem. Photobiol., A 2002;151: 195–199.

[70] Granados G, Martinez F, Paez-Mozo EA. Photocatalytic degradation of phenol on TiO_2 and TiO_2/Pt sensitized with metallophthalocyanines. Catal. Today 2005;107–108: 589–594.

[71] Ghosh JP, Langford CH, Achari G. Characterization of an LED-based photo-reactor to degrade 4-chlorophenol in an aqueous medium using coumarin (C-343) sensitized TiO_2. J. Phys. Chem. A 2008;112: 10310–10314.

Chapter 17

Phenolic Compounds in Water: Sources, Reactivity, Toxicity and Treatment Methods

William W. Anku, Messai A. Mamo and
Penny P. Govender

Additional information is available at the end of the chapter

http://dx.doi.org/10.5772/66927

Abstract

Phenolic compounds exist in water bodies due to the discharge of polluted wastewater from industrial, agricultural and domestic activities into water bodies. They also occur as a result of natural phenomena. These compounds are known to be toxic and inflict both severe and long-lasting effects on both humans and animals. They act as carcinogens and cause damage to the red blood cells and the liver, even at low concentrations. Interaction of these compounds with microorganisms, inorganic and other organic compounds in water can produce substituted compounds or other moieties, which may be as toxic as the original phenolic compounds. This chapter dwells on the sources and reactivity of phenolic compounds in water, their toxic effects on humans, and methods of their removal from water. Specific emphasis is placed on the techniques of their removal from water with attention on both conventional and advanced methods. Among these methods are ozonation, adsorption, extraction, photocatalytic degradation, biological, electro-Fenton, adsorption and ion exchange and membrane-based separation.

Keywords: phenolic compounds, toxicity, wastewater treatment, photodecomposition, membrane-based separation

1. Introduction

There has recently been heightened concern among policymakers and scientists with regard to the effects of human and wildlife exposure to chemical compounds in the environment, particularly the aquatic environment. Phenolic compounds are among the chemicals of major concern in this regard as they tend to persist in the environment over a long period of time, accumulate and exert toxic effects on humans and animals [1]. Some phenolic compounds are abundant in nature and are associated with the colours of flowers and fruits [2]. Others are synthesised and are used in varied aspects of mankind's everyday life.

The entrance of phenolic compounds into the aquatic environment results from natural, industrial, domestic and agricultural activities. Their presence may be due to the degradation or decomposition of natural organic matter present in the water, through the disposal of industrial and domestic wastes into water bodies and through runoffs from agricultural lands [3]. These chemicals, upon entry into the water, have the tendency of undergoing transformations into other moieties that can even be more harmful than the original compounds. This transformation is normally due to their interaction with physical, chemical and biological or microbial factors in the water [4].

Phenolic compounds have been enlisted by the United States Environmental Protection Agency (USEPA) and the European Union (EU) as pollutants of priority concern. This enlistment is due to the fact that these chemicals are noted to be toxic and have severe short- and long-term effects on humans and animals [5]. The occurrence of phenolic compounds in the aquatic environment is therefore not only objectionable and undesirable but also poses a danger as far as human health and wildlife are concerned. As a result, a number of wastewater treatment techniques have been developed and used for the removal of phenolic compounds from industrial, domestic and municipal wastewaters prior to their disposal into water bodies so as to minimise the devastating effects of these chemicals on human and aquatic lives. Some of these techniques include extraction, polymerisation, electro-Fenton process, photocatalytic degradation and so on.

This chapter presents a general overview of selected topics in relation to phenolic compounds. It dwells on the sources and reactivity of phenolic compounds in water, their toxic effects on humans and methods of their removal from water. Specific emphasis is placed on the techniques of their removal from water with attention on both conventional and advanced methods. Among these methods are adsorption, extraction, polymerisation, electro-coagulation, photocatalytic degradation, biological methods, electro-Fenton method, advanced oxidation processes, adsorption and ion exchange and membrane-based separation techniques.

2. Classification of phenolic compounds

Phenolic compounds are a class of organic compounds with a hydroxyl group(s) directly bonded to one or more aromatic rings. The first member of chemicals belonging to this category of organic compounds is called phenol, also known as carbolic acid, benzophenol or hydroxybenzene with the chemical formula of C_6H_5OH (**Figure 1**). All other members of the group are derivatives of phenol [6].

Phenolic compounds are classified into different groups based on different factors. Among these factors of categorisation are the carbon chain, basic phenolic skeleton or the number of phenol units present in the molecule, distribution in nature and location in plants.

http://dx.doi.org/10.5772/66927

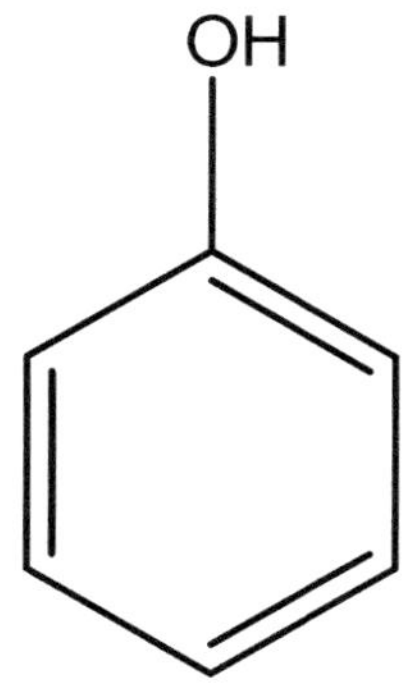

Figure 1. Structure of phenol.

2.1. Classification based on carbon chain

The classes of phenolic compounds identified under this category are listed in **Table 1**. The integers attached to the carbon atoms bonded to the aromatic ring (C6) represent the number of carbon atoms directly or indirectly bonded to the aromatic ring.

Carbon chain	Number of phenolic units	Class	Examples
C6	1	Simple phenols	Catechol
C6	1	Benzoquinones	Hydroquinone
C6-C1	1	Phenolic acids	Gallic
C6-C2	1	Acetophenones	3-Acetyl-6-methoxy benzaldehyde
C6-C2	1	Phenylacetic acids	p-Hydroxypheylacetic acid
C6-C3	1	Hydroxycinnamic acids	Caffeic acid
C6-C3	1	Phenylpropenes	Eugenol
C6-C3	1	Coumarins/isocoumarins	Umbelliferone
C6-C3	1	Chromones	Cromolyn
C6-C4	1	Naphthoquinones	Plumgagin, juglone
C6-C1-C6	2	Xanthones	Mangiferin
C6-C2-C6	2	Stilbenes	Resveratrol
C6-C3-C6	2	Flavonoids	Amentoflavone
$(C6\text{-}C3)_2$	2	Lignans and neolignans	Pinoresinol, eusiderin
$(C6\text{-}C3)_n$	n > 12	Lignins	Tannic acid

Table 1. Classification of phenolic compounds based on carbon chains.

2.2. Classification based on the number of phenol units present in the molecule

Phenolic compounds can be grouped as simple, bi and polyphenols depending on the number of phenol groups present in a particular molecule. Simple phenols are phenols with only one substituted phenolic ring. In other words, they have substituted phenols. Examples of simple phenols include phenolic acid, hydroquinone, resorcinol, thymol, etc. [7]. Biphenols contain two phenolic units while polyphenols consist of multiple units of the phenolic structure. Based on the number of phenol rings present, and the type of elements binding rings together, polyphenols can be subdivided into various groups such as flavonoids, phenolic acids, tannins, stilbenes and lignans (**Figure 2**) [8].

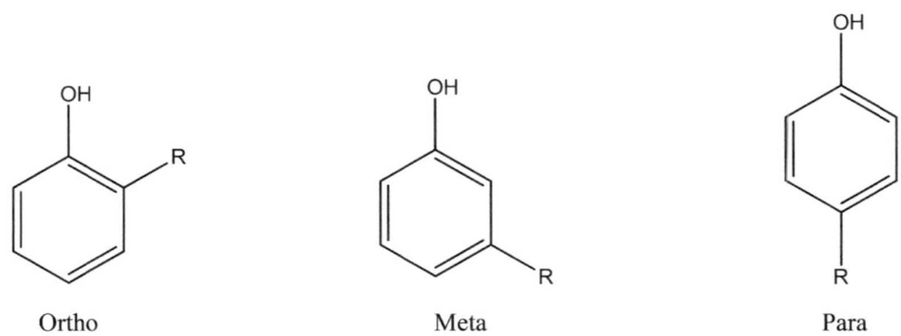

Figure 2. Structure of simple phenols.

2.3. Classification based on distribution in nature

Based on their extent of distribution in nature, phenolic compounds have been classified as being shortly distributed, widely distributed and as polymers. Widely distributed phenols are the types that are present or available in all plants, or are of high significance in specific plants. Examples include flavonoids and/or flavonoid derivatives, coumarins, phenolic acid including benzoic acid and cinnamic acid. Those that are shortly or less widely distributed are limited and include simple phenols, pyrocatechol, hydroquinone and resorcinol. Examples of the polymer class of phenolic compounds are tannin and lignin [9].

2.4. Classification based on location in plants

Phenolic compounds that are free in soluble forms in cells are categorised as being in solution while those bound as complexes in the cell wall, as insoluble. Soluble phenolic compounds include low and medium molecular weight phenolics such as simple phenol, flavonoids and tannins. Insoluble phenolics include high molecular weight tannins and phenolic acids. Other insoluble phenolics include those bound to polysaccharides in the cell walls (usually with low molecular weight), and proteins [10].

3. Sources of phenolic compounds in water

The existence of phenolic compounds in water can be attributed to natural and anthropogenic activities. Natural sources of phenolic compounds in water pollution include decomposition of dead plants and animals (organic matter) in the water. They are also synthesised by microorganisms and plants in the aquatic environment. Industrial, domestic, agricultural and municipal activities constitute the anthropogenic sources of water pollution with phenolic compounds. This section provides details on various sources of introduction of phenolic compounds in water bodies.

3.1. Natural sources

3.1.1. Decomposition of organic matter

The occurrence of phenolic compounds in water is sometimes due to the decomposition of dead plants and animals in the water bodies; or as a result of runoff from the land where the decomposing materials are washed into water bodies. Phenolic compounds are components of many plant species, aquatic or terrestrial. Some of these compounds are formed from amino acids, present in hemicelluloses of some plants, under ultraviolet light irradiation. For instance, willow bark is known to contain a certain amount of salicylic acid [11]. Green and red marine algae also contain macromolecules of phenolic compounds. Hydroxybenzene, for example, is produced due to decomposition of organic matter [12]. The body of humans and animals, without any external exposure, produces phenol which finally gets excreted. Thus, the metabolic waste products of humans and animals also contain phenol [13]. Phenol is produced in the gut of mammals as a result of the transformation of tyrosine in their digestive tract [14]. The compounds are also components of many food stuff including fruits and vegetables. Phenol is found to be present naturally in coal tar and creosote. It is also produced during natural fires, and through benzene degradation in the atmosphere under the influence of ultraviolet light radiation [14]. Direct decomposition of these materials in water or indirect introduction from runoffs and rainfall result in the pollution of the water bodies with these compounds.

3.1.2. Synthesis by microorganisms

The potential of microorganisms to degrade naturally occurring substrates into phenolic compounds, particularly hydroxybenzoate, is well established [15]. Debaryomyces hansenii is noted for its ability to convert ferulic acid into varied phenolic compounds in the presence of glucose and nitrogen. Debaryomyces hansenii metabolism of ferulic acid was identified by Max et al. [16] to produce 4-vinyl guaiacol, vanillic acid, acetovanillone, ferulic acid, vanillin and 4-ethylguaiacol, while Shashwati et al. [17] noted the ability of Streptomyces sannanensis to convert ferulic acid to vanillic acid. Fermentation of plant extracts by microorganisms is also known to result in the formation of different types of phenolic compounds. A study involving the use of Lentinus edodes for the fermentation of cranberry pomace identified the formation of ellagic acid [18], while fermentation of ethanolic acid (obtained from oat) by a fungus, Aspergillus oryzae, produced caffeic and ferulic acids [19].

3.1.3. Synthesis by plants

There is a vast distribution of phenolic compounds in the plant kingdom. Synthesis of phenolic compounds in plants occurs in the chlorophyll under the influence of certain external stimuli or factors including ultraviolet radiation from sunlight, chemical stressors (cations, pesticides), and microbial infections. Phenylalanine is a precursor for phenolic compound synthesis in plants. Typically, phenylalanine is deaminated to cinnamate through phenylalanine ammonia-lysase catalysis. A hydroxylation process catalysed by cinnamate-4-hydroxylase then results in the conversion of cinnamate to coumaric acid which becomes the starting material for the synthesis of stilbenes, flavonoids and furanocoumarines including other classes of phenolics catalysed by different enzymes [20]. These compounds are stored in the leaves, roots and stems of plants. Roots and leave exudates of plants contain these phenolic compounds, which are finally introduced into the soil by the exudates. Decomposition of dead leaves, roots and plants also introduces the phenolic compounds into the soil. Runoff from land then washes these compounds into the nearby water bodies.

3.2. Anthropogenic sources

3.2.1. *Industrial waste*

Phenolic compounds have varied applications in the day-to-day lives of human beings. Phenol is widely used in different industries such as the chemical industry where it is used in the production of other derivatives such as alkylphenols, cresols, aniline and resins [21]. Its application in the oil and gas and coal industries is also noteworthy [1]. Phenolic resins, which are also produced from phenol, are used heavily in appliance, wood and construction industries for various purposes. Dyes, textiles and explosive industries all depend on phenol as raw material. Other phenolic compounds such as bisphenol A serves as the raw material from which non-polymer additives, polycarbonate plastics and epoxy resins are manufactured. Nylon 6 (polycaprolactam) and some fibres (synthetic) are produced using caprolactam as raw material [22]. Phenolic compounds are also constituents of some pesticides and other insecticides. Other industrial activities such as wood distillation, use of chlorine for water disinfection, cooking processes and paper production all result in the formation of chlorophenols [23]. Direct or indirect discharge of effluents and/or influents from these industrial activities into water bodies culminate in their pollution with phenolic compounds. Some of these compounds are also released into the atmosphere through vehicular activities and are finally washed into water bodies as rain water.

3.2.2. *Agricultural waste*

Application of pesticides, insecticides and herbicides constitutes the main source of water pollution with phenolic compounds through the agricultural source. Availability or detection of phenol and some chlorophenols such as 2-chlorophenol, 2,4-dichlorophenol and some catechols in the aquatic environment have been attributed to biodegradation of some of these pesticides. Among these pesticides are 2,4-dichlorophenoxyacetic acid, 4-chloro-2-methylphenoxyacetic acid, 2,4,5-richloro-phenoxy acetic acid [24, 25]. Another pesticide vastly used in the agricultural sector is pentachlorophenol which finally degrades to other chlorophenols with lower chlorine substituents [26]. These herbicides, fungicides, and pesticides with their degradation by-products are washed into the water bodies through agricultural runoff.

3.2.3. *Domestic waste*

Phenol is a component of many household chemicals. It is present in disinfectants, antiseptics and slimicides. Medical or pharmaceutical products including body lotions, ointments, mouthwashes and some oral sprays meant for anaesthetic purposes or for sore throat treatment all contain phenol. Phenol is also present in other household products such as soaps, toys, paints, lacquers, perfumes and varnish removers [13]. Household wastewater, which invariably contains these products, is usually drained through the sinks or gutter, and finally, enter nearby water bodies contaminating them.

3.2.4. Municipal waste

Effluents and influents emanating from municipal waste treatment plants, and leachates from municipal solid waste landfill sites, are another source of phenolic compounds into water bodies. p-Cresols have been identified in leachates from a municipal waste landfill site and are believed to originate from incineration residues. Similarly, 2,4,6-trichlorophenol, 4-tertrabutylphenol and bisphenol A were also found in leachates and are believed to emanate from fly ash while combustibles were deemed to be the source of 4-tert-octylphenol in landfill leachates. Some chlorophenols, 4-nonylphenol and phenol have all been identified in municipal waste landfill sites [27]. Thus, the release of untreated leachates from landfill sites, the release of residues of incineration such as solid fly ash and the release of incombustible materials into nearby water bodies result in the pollution of the aquatic environments with phenolic compounds.

4. Reactivity of phenolic compounds in aquatic environment

As a result of their high reactivity, phenolic compounds present in water have high tendencies of interacting or reacting with other components of the aquatic environment. Among these components are inorganic compounds and microorganisms.

4.1. Interaction with microorganisms

In the aquatic environment, certain bacteria are known to degrade non-ionic surfactants into other phenolic compounds such as alkylphenolic compounds, most of which tend to be more toxic or harmful compared to the original compounds. For example, bacterial degradation of nonylphenol polyethoxylate surfactant results in the formation of nonylphenol. There is also evidence of microbial degradation of nonylphenoxyacetic acid into nitrophenol. Bacteria conversion of nonylphenol polyehoxylaes into other intermediate forms is also identified to occur under certain specific environmental and chemical conditions. Nonylphenol polyehoxylaes are converted to nitrophenols under anaerobic conditions [28], and additional conversion of nitrophenol by iso-substitution occurs under aerobic conditions [29]. Microbial interaction with 4-chlorophenoxyacetic acid results in the production of 4-chlorophenol. Pentachlorophenol degradation by bacteria produces tetrachlorocatechol which can undergo further degradation to form chlorinated catechols. Chlorocatechol was also identified to be the microbial degradation product of chlorobenzenes [30].

4.2. Interaction with inorganic compounds

Ultraviolet radiation from the sun initiates a reaction between phenol and nitrite ions in the aquatic environment resulting in the formation of 2-nitrophenol and 4-nitrophenol [31]. In a similar manner, the interaction between hydroxyl radical and phenol in water produces 2-nitrophenol [32]. Conversion of phenol to nitrophenol also occurs with the availability of nitric ions. Photolysis of phenol in the presence of charge transfer complexes results in the formation of hydroquinone, while the formation of chlorophenol occurs through chlorination of aromatic compounds in water [33]. Some phenolic compounds also coordinate with metal cations of water enhancing their ionisation with the subsequent increase in their solubility in water [34].

5. Toxic effects of phenolic compounds on humans

Most phenolic compounds can easily penetrate the skin through absorption and can readily be absorbed from the gastrointestinal tract of humans. Once in the system, they undergo metabolism and transform to various reactive intermediate forms particularly quinone moieties, which can easily form covalent bonds with proteins, resulting in their ability to exert toxic effects on humans [35].

Chlorophenols, aminophenols, chlorocatechols, nitrophenols, methylphenols and other phenolic compounds have all been characterised as exerting toxic influence on humans [36]. Bisphenol A and some alkylphenols have been identified to exert endocrine disrupting effects on humans by altering the development of mammary glands in exposed animals [37]. Similar work also discloses the tendency of bisphenol A to delay the onset of puberty in girls [38]. Consumption of liquids, including drinking water, containing a high concentration of phenol results in problems with the gastrointestinal tract and muscle tremor with difficulty in walking. Application of products containing a high concentration of phenol to the skin causes blisters and burns on the skin; heart, kidneys, and liver damage may occur with exposure to high levels of phenol [39]. Because of their tendency to readily oxidise to quinone radicals, which tend to be more reactive, catechols have the tendency to cause DNA damage or arylation, destroy some proteins in the body and disrupt transportation of electrons in energy transducing membranes [35]. Caffeic and dihydrocaffeic acids, in the presence of copper, also cause damage to DNA [40]. Chlorophenol poisoning causes mouth burning, throat burning and necrotic lesions in the mouth, stomach and oesophagus. It also induces abnormal temperature and pulse fluctuation, weak muscles and convulsions [41]. Other effects of chlorophenol poisoning include damage to the liver, kidneys, lungs, skin and the digestive tract [42]. Hydroquinone also damages chromosomes. Para-cresol and 2,4-dimethyl phenol have been classified as a chemical with the potential of inducing carcinogenic effects [43].

6. Techniques for the removal of phenolic compounds from water

Recovery of phenolic compounds from the aquatic environment is a mandatory requirement in order to safeguard the life of humans and aquatic organisms through possible contamination of these toxic chemicals. Deployment of appropriate technologies for effective removal of these compounds will not only eliminate problems of possible harm associated with pollutants, as well as waste disposal problems, but also allow the attainment of value-added phenolic compounds as by-products. This section, therefore, discusses various techniques being employed to effectively eliminate phenolic compounds from wastewater prior to their final discharge into water bodies.

6.1. Photocatalytic degradation of phenolic compounds

Photocatalytic degradation is the use of metal oxide catalysts to degrade pollutants where the catalyst is usually activated by absorption of a photon of appropriate energy and is capable of speeding up the reaction without being used up [44]. Photocatalytic properties of metal oxide catalysts are due to the fact that excitation of electrons from the valence to the conduction band of the catalyst occurs upon its irradiation with a light of appropriate wavelength. Promotion of the electrons (e^-) creates positive charges or holes (h^+) on the valence band, and accumulation of electrons on the conduction band of the catalyst. Generation of these charge carriers (e^- and h^+) initiates the photocatalytic degradation process. The valence band holes attack and the oxidised surface absorbs water molecules to form hydroxyl radicals ($OH^\bullet$). Conduction band electrons reduce oxygen molecules and produce oxygen radicals or superoxide radicals ($O_2^\bullet$). These highly reactive radicals then attack and convert the pollutants to harmless products such as carbon dioxide and water [45, 46].

Photocatalytic degradation is regarded as an efficient technique for the elimination of pollutants from polluted water as a result of its ability to completely degrade the pollutant instead of their transformation into other products. The degree of effectiveness of the degradation process is known to rely heavily on the catalyst dose, exposure time, solution pH and light intensity [47].

There have been a number of reports where photocatalytic degradation techniques have been utilised effectively to degrade phenol and its derivative from the water. Natural clinoptilolite zeolite and FeO-based nanoparticles were used by Mirian and Nezamzadeh-Ejhieh [48] in photocatalytic degradation of phenol in polluted water under simulated solar light irradiation. The results confirmed that using zeolite as a support for FeO enhanced its photocatalytic degradation efficiency. The improved photocatalytic activity of the FeO-zeolite composite was attributed to the fact that the zeolite prevented agglomeration of the FeO nanoparticles and minimised the charge carrier recombination rate. In their study, Shahrezaei et al. [49] explored the photocatalytic degradation ability of TiO_2 in the degradation of phenolic compounds present in wastewater from a refinery. Highest degradation efficiency of the phenol and its derivatives was identified at an optimum temperature of 318 K, pH 3 and 100 mg/l catalyst concentration. A 90% degradation efficiency of phenol was achieved within 2 hours at these optimum conditions. Photocatalytic degradation of phenolic compounds from wastewater has also been demonstrated by many researchers using various catalysts including TiO_2/reduced graphene [50], ZnO [51], Fe_2O_3 decorated on carbon nanotubes [52] and CuO [53].

6.2. Ozonation

Ozone (O_3) is formed naturally when ultraviolet (UV) rays from the sun enter the earth's atmosphere. It is also formed whenever lightning strikes during a thunderstorm. Under these conditions, oxygen molecules (O_2) split to form highly reactive oxygen radicals ($O^•$), which in turn react with O_2 to form ozone. Ozone has a very high oxidising potential (-2.74 V) which is much higher than that of hypochlorite ion (-1.49 V) and chlorine (-1.36 V) [54] which are all employed as oxidants for pollutant removal from water. This high oxidation potential forms the basis of the use of ozone as an oxidant for removal of organic pollutants from water.

Ozonation process begins with the formation of ozone through corona discharge simulation of lightning, or the use of UV-type ozone generator for simulation of ultraviolet radiation from the sun, by passing clean and dry air through high voltage ozone generators. The wastewater is then allowed to flow along a venture throat, which generates a vacuum and pulls the ozone into the wastewater, or the ozone is simply bubbled up through the wastewater. The ozone then oxidises and decomposes the pollutants leading to their elimination from the water. UV ozonation is mostly used for small-volume wastewater treatment while the corona discharge method is employed in large-scale wastewater treatment processes. Some advantages of ozonation include [55]:

(1) The process is completely natural with no inclusion of chemicals and produces no chemical waste.

(2) High microorganism elimination efficiency with the microorganisms having no potential to developing resistance against ozone.

(3) The process is self-sustaining as the main source of ozone is oxygen from the air.

(4) Ability to remove organic, inorganic, microorganism and improve taste and odour of the water.

Based on the above advantages, several research works have been performed on the use of ozonation technique for phenolic compounds removal from wastewater. Treatment of olive mill wastewaters containing garlic, p-hydroxybenzoic and p-coumaric acids based on ozonation was studied by Chedeville et al. [56]. They identified that the highest ozonation process was attained when the gas-liquid contractor was adopted to HaN_3 regime. The gas/liquid contractor used permitted a comprehensive removal of the phenolic compounds within a short time. A maximum of 80% of the pollutants was eliminated with up to 95% ozone mass transfer. Ozonation was also used to treat ethylene glycol containing wastewaters with emphasis on the impacts of pollutant dose, process time, and pH on the decontamination efficacy. After 180 min, ethylene glycol removal efficiencies were 93.31, 89.96 and 85.01% at 10, 20 and 50 mg/l pollutant concentrations, respectively. Removal efficiency was observed to be highest in alkaline medium [57].

6.3. Extraction method

Among the techniques used for removal of phenolics from water is extraction using polar organic solvents. The extraction method of phenolics removal from wastewater can be categorised as liquid-liquid extraction and solid phase extraction.

6.3.1. The liquid-liquid extraction method

This technique separates compounds on the basis of their solubilities in two immiscible liquids. The compounds are normally separated from one liquid phase to another. The immiscible liquids usually consist of water and an organic solvent.

The liquid-liquid extraction method was used by Rao et al. [58] to remove phenol from sebacic wastewater. The experiment was performed using batch, and serial approaches through the application of different solvents such as Aliquat-336, 1-hexanol, 1-octanol, 1-heptanol and castor oil. The extinction experiment which involved shaking and allowing for phase separation resulted in 75–96% extraction of phenol with the different solvents, where the solvent and the wastewater were mixed in a ratio of 5:250. The extraction efficiency was observed to vary according to the ratio of the solvent used. The best solvent for phenol extraction was Aliquat 336. Liu et al. [59] also used cumene to extract phenol from water. Cumene showed excellent extraction performance on phenol in acidic solution. The distribution coefficient was observed to be directly proportional to the temperature but decreased with increasing pH value.

6.3.2. Solid phase extraction method

Despite the efficiency of liquid-liquid extraction method of removal of chemicals from wastewater, the technique comparatively consumes a lot of time and is expensive with possible associated injuries from the large quantity of organic solvents (some toxic) used in the process. Solid phase extraction technique, which requires minimal time and organic solvents, highly selective and environmentally friendly, is therefore regarded as an appropriate alternative for liquid-liquid extraction method [5].

Solid phase extraction system consists of a syringe containing a merged silica fibre, which is coated with an immobilised phase. The aqueous solution containing the analyte is exposed to the fibre with the subsequent accumulation of the analyte on the stationary phase. The fibre is then removed from the aqueous solution followed by desorption of the extracted analyte in a column injector or gas chromatography. Polydimethylsiloxane is normally used as the stationary phase for removal of halogenated and polycyclic aromatic hydrocarbons and polychlorinated biphenyls [60].

Möder et al. [61] used a polyacrylate-coated fibre as a solid phase extractant for phenolic compound elimination from wastewater. Effects of humic acid and surfactant concentrations on the extraction efficiency were analysed. They attributed the successful extraction of naphthols, alkylated phenols and Tetra-ols from the wastewater to the fact that the polyacrylate coating demonstrated high specificity for polar hydroxylated aromatic compounds. Non-polar molecules hardly interacted with the extractant within the optimum 45 min extraction time. Another experiment on solid-phase extraction of phenols from wastewater was carried out by Tavallali and Shiri [62]. Their study involved the use of iron oxide nanoparticles modified with activated carbon as the solid adsorbent. They demonstrated that development of solid phase extraction method based on magnetised activated carbon prior to their spectrophotometric determination is an appropriate technique. Their result showed 98% removal of phenol from water, indicating the effectiveness of the iron oxide nanoparticle modified activated carbon solid adsorbent.

6.4. Biological method

Biological method of phenolic compound removal from wastewater is subdivided into microbial and enzymatic methods. The microbial method involves the deployment of bacteria, yeast and fungi in breaking down the phenolics into harmless products such as carbon dioxide and water. This method of phenolics removal is feasible as a result of the fact that some microorganisms are known to depend on aromatic compounds, including phenolics, as their source of carbon or nutrient [63, 64]. It has the advantage of a comparatively low operational cost. Microbial removal of phenolic compounds occurs through either aerobic or anaerobic processes and begins with hydroxylation (introduction of hydroxyl groups) of the aromatic ring [65]. Hydroxylation through aerobic degradation involves two steps with catechol being the end product [66]:

(1) Reduction of one of the molecular oxygen to water under the influence of a hydrogen donor (reduced pyrimidine nucleotide), and devouring of the other oxygen atom.

(2) The second step of the hydroxylation process occurs in the presence of dioxygenase enzyme with the subsequent formation of catechols.

Cleavage of the catechol aromatic rings then passes through various stages with specific enzymes, based on the type of microorganisms, resulting in the conversion of the phenolic compounds to compounds such as carbon dioxide and water [67]. Anaerobic degradation occurs whenever oxidising agents such as sulphates, nitrates and CO_2, or light are present. This process is believed to be initiated by carboxylation of phenol to 4-hydroxybenzoate [68].

In general, the aerobic process is known to be better suited for the degradation of phenolics with minimal substituents consisting of halogens. On the other hand, the anaerobic process is mostly appropriate for reduction of chlorinated phenolic compounds [69]. The anaerobic system produces methane in addition to carbon dioxide and water. A major advantage of the anaerobic system of degradation is the absence of aeration cost, recovery of methane and minimum excess biomass generation [70].

Kukadiya et al. [71] studied the effectiveness of using a moving bed biofilm reactor for phenolic compounds removal from wastewater. The laboratory scale model moving bed biofilm reactor was observed to be effective against the removal of phenol with about 98% efficiency. In their experiment, Sinha et al. [72] studied the p-chlorophenol and phenol microbial degradation as a single and mixed substrates by using *Rhodococcus* sp. RSP8 bacteria strain. The experiment was performed with a liquid mineral salt medium in a shake flask experiment at a neutral pH and a temperature of 37 °C. The two compounds (p-chlorophenol and phenol) served as the main source of carbon and energy for the cells and were consumed completely as individual solutions by the cells. The two pollutants, however, repressed each other's degradation by the cells in the mixed substrate experiment.

The enzymatic method of degradation, however, employs enzymes (biological catalysts). Enzymes can be used effectively to selectively eliminate pollutants in water since they catalyse specific reactions under modest temperature, pH and ionic strengths [73]. In addition, the enzymatic reaction is known to occur at much faster rates compared to other types of reactions [74]. As an advantage over the microbial system of pollutant degradation, the enzymatic system of pollutant removal can occur under conditions, which are unfavourable or toxic to bacteria. This system can operate under different pollutant concentration (high or low), eliminates the time requirement for biomass acclimatisation, involves no shock loading effect and with no generated biomass [74]. This method receives a high level of consideration due to its high pollutant removal efficiency, operation in wide temperature and pressure ranges and formation of harmless end products [75, 76]. The enzyme with a high promise for dephenolisation of phenolics in water is tyrosinase (KF1.14.18.1). This enzyme oxidises the phenolics to quinones, which are further broken down into the non-toxic intermediate product. The intermediate products are then removed through the addition of binding agents [77].

There has been a series of reported research works where enzymes have been used for the removal of phenolic compounds from wastewater [78, 79]. Among these reports is the work done by Shesterenko and co-workers [79]. They used tyrosinase isolated from *Agaricus bisporus* and immobilised it on polymer carriers, and inorganic coagulants to remove phenols from water. Peroxidase extracted from horseradish, hydrogen peroxide and polyethylene glycol (PEG) was also used to catalyse phenol removal from simulated wastewater [80]. Optimum degradation of 1 mM phenol (80%) was attained at 0.3 U/ml horseradish peroxidase and 3.0 mM hydrogen peroxide concentrations at pH 7 and 273 mg/l of PEG.

6.5. Adsorption

Adsorption is considered as one of the appropriate techniques for removal of phenolics from water because the technique is easy to design and operate. The technique produces no toxic wastes. The spent sorbent can serve as a source fuel to produce power [81]. Adsorption process involves the accumulation of the pollutant on the adsorbent's surface (usually solid material). An appropriate adsorbent must be porous with large surface area, possess high hydrophobicity and have the ability to selectively accumulate the pollutant from water onto its surface. Efficiency of the adsorption process is governed by [82]:

(1) Adsorbent's properties, i.e. its functional group composition, the size of its surface area together with the distribution of the pore size and the extent of its ash content.

(2) The solution chemistry including its pH, temperature, degree of polarity, availability of other solutes competing for the adsorbent surface area and the concentration of the adsorbate.

(3) Nature of the adsorbate. This includes its degree of solubility in water, hydrophobicity, size and molecular weight.

Reference [82] found phenol adsorption process to be solely dependent on the initial pollutant concentration and speciation, which in turn depends on pH of the solution. Adsorption of pollutants from water is believed to be based on the following steps [83]:

(1) Movement of the pollutant molecules towards the adsorbent across the external boundary layer.

(2) Movement of the pollutant molecules to the adsorbent's active surface sites.

(3) Adsorption of the pollutants on the active surface sites.

(4) Migration of the adsorbed pollutants through diffusion onto the pores' surfaces.

Various researchers have studied phenol adsorption from polluted water with different types of adsorbents. Phenol adsorption efficiency of different adsorbents including bagasse ash, activated carbon and charcoal from wastewater was studied by [84]. The adsorption efficiency was assessed based on the influence of pH, concentration of EDTA, anions and adsorbent dose. Their result showed 98, 90 and 90% phenol removal efficiencies by activated carbon, wood charcoal and bagasse ash systems, respectively. Removal efficiency was observed to increase with a decrease in the pH of the system. Effects of EDTA and nitrate ion content of the solution were identified as the factors that influenced the adsorption process. Chloride ion, on the other hand, exerted a significant adverse effect on the efficiency of bagasse ash system. Film diffusion was noted to control the adsorption efficiencies of all the adsorbents used. Similarly, the use of sugarcane bagasse-based activated carbons for effective phenol adsorption from aqueous medium was assessed by Akl et al. [85]. The result of the study proposed sugarcane bagasse-based activated carbon (SCBAC) as a viable adsorbent for phenol elimination from water. The pollutant eradication process depended solely on its concentration, solution pH and temperature.

6.6. Membrane-based separation method

A membrane is a specific type of a barrier that enables the separation of species in a gas or liquid through various mechanisms such as diffusion, sieving or sorption. The selective separation occurs as a result of the semipermeable nature of the membranes. This is the ability of the membrane to allow the passage of certain substances through it while it prevents the passage of others based on their sizes and/or molecular weights. Thus, in membrane-based separation method of water purification, the water usually passes through the membrane while the suspended pollutant, usually with comparatively larger sizes and molecular weights, is unable to pass through the membrane. They get retained in the medium or on the membrane and later removed.

Membrane separation is a general term used to encompass different types of separation processes that are characteristically the same or similar since they all use membranes. The difference lies in the pore size of the membranes and the driving force involved in the separation process. The driving forces for separation may include high pressure application, the creation of concentration gradient and the use of electric potential [86]. These processes are categorised as microfiltration, ultrafiltration, nanofiltration and reverse osmosis [87]:

- Microfiltration: The membrane's pore size of this technique ranges from 0.1 to 1.0 μm. It is normally used to filter suspended particles or colloidal solutions with large particles and bacteria.

- Ultrafiltration: The pore diameter of this type of membrane ranges from 0.01 to 0.1 μm and can be used for filtration macromolecules such as polymers and proteins from solution.

- Nanofiltration: The pore size range of this type of membrane is 1–10 nm. It is used for brackish water desalination and removal of micropollutants or metal ions.
- Reverse osmosis: This refers to a membrane with pore diameters in the range of 0.0001–0.0001 μm and may be used for filtration of solutions of polymer chains, ultrapure water production and desalination of sea and brackish water.

An ideal membrane system must have good fluxes and be highly selective. It must have excellent thermal, chemical and mechanical stability with low tendency of foul formation. Some advantages of membrane system of water purification include the following [87]:

(1) It has comparatively low energy requirements.

(2) It is simple, easy to use with low maintenance requirements.

(3) It is highly selective to the material to be separated.

(4) It is environmentally friendly as the system works without the addition of chemicals.

The technology is also not without disadvantages. Some of these disadvantages include [88]:

(1) Many membranes composed of polymeric materials can decompose, or swell or become weak under harsh conditions, thereby weakening the selectivity, and shortening the lifespan of the membrane.

(2) Some of the membranes based on polymeric material have temperature limitation as most of these polymers are unable to maintain their properties at temperatures above 100°C.

(3) Membrane fouling also occurs and interferes with the permeability of the membrane.

Phenol has been separated from water with membrane-based separation technique by using non-modified, and ionically, and covalently cross-linked ethylene methacrylic acid copolymer-based membranes [89]. They found out that the total flux increased with increasing phenol content in the feed while the enrichment factor decreased. They, however, observed lesser fluxes and higher enrichment factors when non-modified membrane containing a higher amount of methacrylic acid monomer was used. Ionic cross-linked membrane proved to be the most efficient membrane against the feed containing a high concentration of phenol.

Use of ionic liquids in the form of bulk liquid membranes for the elimination of phenol from water has also been studied by Ng et al. [90]. High hydrophobic ionic liquids including 1-butyl-3-methylimidazolium hexafluorophosphate, 1-butyl-3-methylimidazolium bis (trifluoromethylsulfonyl) imide and 1-butyl-3-methylimidazolium tris(pentafluoroethyl) trifluorophosphate were used for the experiment. The stability, membrane loss and phenol elimination efficiency of these liquids were compared. Their results identified 1-butyl-3-methylimidazolium bis (trifluoromethylsulfonyl) imide as the best performing liquid in terms of phenol elimination and stripping efficiencies. This liquid exhibited phenol extraction efficiency of 96.21% and stripping efficiency of 98.10%. These values were attained at optimum conditions of 225 and 135 rpm aqueous and membrane stirring speed, respectively.

This solvent was identified to possess higher hydrogen bonding, basicity and low viscosity compared to the other two solvents used.

6.7. Electro-Fenton method

Fenton reaction is a reaction between iron (II) and hydrogen peroxide resulting in the formation of hydroxyl radicals ($OH^{\bullet}$), a non-selective and strong oxidising agent [91]. As a result, this process has been used to oxidise organic pollutants in aqueous solution to carbon dioxide and water. However, Fenton process has the disadvantage of the high cost of procuring the reactants (hydrogen peroxide and iron (II)) and sludge generation. An improved technique, electro-Fenton process, which is capable of overcoming the above-mentioned hindrances and permits improved control of hydroxyl radical generation [92], has therefore been devised. Electro-Fenton process involves cathodic reduction of iron (III) in solution to iron (II) with a potential of E^{o} = 0.77 V/SHE. This process, often referred to as electrochemical catalysis, produces the iron (II) at a faster rate and thus promotes the production of hydroxyl radical for enhanced pollutant oxidation process [92].

Evaluation of the electro-Fenton process as an appropriate substitute technique for elimination of phenol from a phenol simulated wastewater was conducted by Abdelaziz et al. [93]. They used a sacrificial iron anode as the source of iron (II), added hydrogen peroxide to the system externally and used nitrogen gas sparging to stir the batch reactor. They also examined the influence of some factors on the effectiveness of the electro-Fenton process. Their results revealed 97% overall COD reduction of 50 mg/l pollutant concentration at optimum conditions of pH 3, the superficial gas velocity of 1.18 cm/s, current density (1.7 mA/cm^2) and hydrogen peroxide concentration of 1500 mg/l. They observed that COD removal percentage increased with increasing current density, hydrogen peroxide concentration and sodium chloride concentrations but started decreasing, in all cases, beyond their respective optimum values. In addition, consumption of energy and iron decreased as the initial pollutant (phenol) and sodium hydroxide concentrations were increased, but increased with an increase in the current density.

In their study, where they investigated the effectiveness of the electron-Fenton process for removal of COD from paper mill wastewater, Un et al. [94] concluded that electro-Fenton process is effective for removal of COD from tissue paper waste water. They achieved a COD removal efficiency of 80% within 60 min at 20 mA/cm^2 current density with 0.1 M hydrogen peroxide concentration at pH 2. They noticed that though increasing the current density caused a corresponding increase in the COD removal, it also resulted in increasing energy consumption. Similarly, Rahmani et al. [95] have effectively applied the electro-Fenton technique to degrade phenol from aqueous solution. In this study, they applied a disposable iron anode as the source of ferrous iron and added the hydrogen peroxide manually. Highest degradation efficiency of 100% was obtained within 30 minutes at pH 3, 100 mg/l hydrogen peroxide concentration and 5 mA/cm^2 optimum conditions.

6.8. Adsorption and ion exchange

In an ion exchange process, an interchange of ions between two phases, usually a solid and a liquid phase, occurs. The ion change resin forms the solid phase while the sample under consideration is normally in the liquid phase. A typical ion exchange resin comprises a cross-linked polymer network with its surface covered with a uniform distribution of ions. Ions with charges similar to that of the resin are exchanged once the solution containing the ions comes into contact with the ion exchanger. The degree to which the ions are exchanged is governed by the concentration of the ions in solution (ions to be exchanged with the resin) and their degree of affinity for the ion exchange resin. The ion exchange process is reversible and is deemed as one of the unique technologies for wastewater treatment because of its high degree of recyclability and environmentally friendly nature [96]. The use of polymer-based anionic resins can allow successful elimination of phenolic compounds from polluted water through adsorption in conjunction with ion exchange. The ions of the resin provide sites for ion exchange while the porous nature permits adsorption through hydrophobic interaction [97].

The efficiency of phenol exclusion from polluted water involving the use of Amberlyst A26 and Amberlite IRA-67 as strong and weak base ion exchangers, respectively, has been conducted. As an observation, phenol removal efficiency of both the strong and weak base ion exchange resins decreased significantly with increased initial phenol concentration from 99.6% (1 mg/l phenol concentration) to 74.2% of 200 mg/l (strong base), and from 65.7% 1 mg/l phenol concentration to 22.1% of 200 mg/l phenol concentration (weak acid). The strong base ion exchange (Amberlyst A26) was considered to be a better ion exchange medium for phenol removal compared to Amberlite IRA-67 [98].

The use of Amberlite IRA-420, a strong base ion exchange resin, for the removal of phenol from water was also carried out by Carmona et al. [99]. Both theoretical and empirical approaches were used to identify the parameters that governed the rate at which the Amberlite IRA-420 removed phenol from the solution. According to their results, the phenol elimination process with Amberlite IRA-420 occurred in the acidic medium through adsorption, and in the alkaline medium through both adsorption and ion exchange. The process increased with increasing pH values from 9 to 14. Ahmed et al. [100] also compared the phenolic compound removal effectiveness of four different ion exchange resins, namely, Ameberlite XAD-4, Ameberlite XAD-7, Ameberlite IRA-94 and Ionac AFP-329, and observed that the effectiveness of the degradation process depended on the size of the resins' surface area and temperature. Ionac AFP-329 was the most efficient sample. However, Ameberlite XAD-4 demonstrated the highest desorption ability with almost complete pollutant recovery at 60 °C.

7. Conclusion

The rapid increase in industrial and domestic activities as a result of the desire to meet the demands of the ever-increasing human population creates the possibility of phenolic compounds introduction into water bodies. Extensive research has been performed on these compounds resulting in the elucidation of their structure or classification, their sources of entry into the aquatic environment and their reactivity or interaction with other components of the aquatic environment. Research has also unveiled the significant toxic effects that these compounds exert on humans and wildlife upon exposure. Significant efforts have been made for the total elimination of phenolic compounds from water before use. This resulted in the development of water treatment technologies including the conventional methods such as activated carbon adsorption, solvent extraction and advanced technologies such as electro-Fenton method, membrane-based separation method, photocatalysis and so on, which have all been successfully used for removal of phenolic compounds from water.

Author details

William W. Anku, Messai A. Mamo* and Penny P. Govender*

*Address all correspondence to: messaim@uj.ac.za and pennyg@uj.ac.za

Department of Applied Chemistry, University of Johannesburg, Doornfontein, Johannesburg, South Africa

References

[1] Bruce R. M., Santodonato J. and Neal M. W. Summary review of the health effects associated with phenol. *Toxicology and Industrial Health 3*, pp. 535–568, 1987.

[2] Campanella L., Beone T., Sammartino M. and Tomassetti M. Determination of phenol in wastes and water using an enzyme sensor. *Analyst 118*, pp. 979–986, 1993.

[3] Wallace J. *"Phenol" In: Kirk-Othmer encyclopedia of chemical technology*. 4th Ed.: John Wiley and Sons, New York, pp. 592–602, 1996.

[4] Kulkarni S. J. and Kaware D. J. P. Review on research for removal of phenol from wastewater. *International Journal of Scientific and Research Publications 3*, pp. 1–4, 2013.

[5] Mahugo-Santana C., Sosa-Ferrera Z., Torres-Padrón M. E. and Santana-Rodríguez J. J. Analytical methodologies for the determination of nitroimidazole residues in biological and environmental liquid samples: a review. *Analytica Chimica Acta 665*, pp. 113–122, 2010.

[6] Dixon R. A. and Paiva N. L. Stress-induced phenylpropanoid metabolism. *The Plant Cell 7*, p. 1085, 1995.

[7] Ignat I., Volf I. and Popa V. I. A critical review of methods for characterisation of polyphenolic compounds in fruits and vegetables. *Food Chemistry 126*, pp. 1821–1835, 2011.

[8] D Archivio M., Filesi C., Di Benedetto R., Gargiulo R., Giovannini C. and Masella R. Polyphenols, dietary sources and bioavailability. *Annali-Istituto Superiore di Sanita 43*, p. 348, 2007.

[9] Bravo L. Polyphenols: chemistry, dietary sources, metabolism, and nutritional significance. *Nutrition Reviews 56*, pp. 317–333, 1998.

[10] Sánchez-Moreno, C. (2002). Polyphenolic compounds: Structure and classification. Presence in food and consumption. Bioavailability and metabolism. *Food* 329, PP. 19-27, 2002.

[11] Davidson R. S. The photodegradation of some naturally occurring polymers. *Journal of Photochemistry and Photobiology B: Biology 33*, pp. 3–25, 1996.

[12] USEPA. *Ambient Water Quality Criteria DOC: Phenol. US EPA-440/5-80-066 (PB 81-117772).* pp. 1100–1156, 1980.

[13] Sheets Fact. Breast Cancer & The Environment Research Centers. 2007.

[14] Tsuruta Y., Watanabe S. and Inoue H. Fluorometric determination of phenol and p-cresol in urine by precolumn high-performance liquid chromatography using 4-(N-phthalimidinyl) benzenesulfonyl chloride. *Analytical Biochemistry 243*, pp. 86–91, 1996.

[15] Toms A. and Wood J. M. Degradation of trans-ferulic acid by Pseudomonas acidovorans. *Biochemistry 9*, pp. 337–343, 1970.

[16] Max B., Tugores F., Cortés-Diéguez S. and Domínguez J. M. Bioprocess design for the microbial production of natural phenolic compounds by Debaryomyces hansenii. *Applied Biochemistry and Biotechnology 168*, pp. 2268–2284, 2012.

[17] Ghosh S., Ashish Sachan S. K. S. and Mitra A. Microbial transformation of ferulic acid to vanillic acid by Streptomyces sannanensis MTCC 6637. *Journal of Industrial Microbiology and Biotechnology 34*, pp. 131–138, 2007.

[18] Vattem D. A. and Shetty K. Ellagic acid production and phenolic antioxidant activity in cranberry pomace (Vaccinium macrocarpon) mediated by Lentinus edodes using a solid-state system. *Process Biochemistry 39*, pp. 367–379, 2003.

[19] Cai S., Wang O., Wu W., Zhu S., Zhou F., Ji B., Gao F., Zhang D., Liu J. and Cheng Q. Comparative study of the effects of solid-state fermentation with three filamentous fungi on the total phenolics content (TPC), flavonoids, and antioxidant activities of sub-fractions from oats (*Avena sativa* L.). *Journal of Agricultural and Food Chemistry 60*, pp. 507–513, 2011.

[20] Daniel O., Meier M. S., Schlatter J. and Frischknecht P. Selected phenolic compounds in cultivated plants: ecologic functions, health implications, and modulation by pesticides. *Environmental Health Perspectives 107*, p. 109, 1999.

[21] Careghini A., Mastorgio A. F., Saponaro S. and Sezenna E. Bisphenol A, nonylphenols, benzophenones, and benzotriazoles in soils, groundwater, surface water, sediments, and food: a review. *Environmental Science and Pollution Research 22*, pp. 5711–5741, 2015.

[22] Budavari S. *The Merck Index, 13th ed. Whitehouse station.* NJ: Merck Co., pp. 1299–1367, 2001.

[23] Paasivirta J. H. K., Humppi T., Karjalainen A., Knuutinen J., Maëntykoski K., Paukku R., Piilola T., Surma-Aho K., Tarhanen J., Welling L. and Vihonen H. Polychlorinated phenols, guaiacols and catechols in environment. *Chemosphere 14*, pp. 469–491, 1985.

[24] McBain A., Senior E., Paterson A., Plessis C. and Watson-Craik I. Bioremediation of soil contaminated with 4-chloro-2-methylphenoxyacetic acid (MCPA): essential laboratory studies. *South African Journal of Science 92*, pp. 426–430, 1996.

[25] Różański L. Transformation of pesticides in the living organisms and the environment. *Agra-Enviro Lab, Poznań*, pp. 290–292, 1998.

[26] Laine M. M. and Jorgensen K. S. Straw compost and bioremediated soil as inocula for the bioremediation of chlorophenol-contaminated soil. *Applied and Environmental Microbiology 62*, pp. 1507–1513, 1996.

[27] Kurata Y., Ono Y. and Ono Y. Occurrence of phenols in leachates from municipal solid waste landfill sites in Japan. *Journal of Material Cycles and Waste Management 10*, pp. 144–152, 2008.

[28] Montgomery-Brown J. and Reinhard M. Occurrence and behavior of alkylphenol polyethoxylates in the environment. *Environmental Engineering Science 20*, pp. 471–486, 2003.

[29] Kohler H.-P. E., Gabriel F. L. and Giger W. ipso-Substitution–A novel pathway for microbial metabolism of endocrine-disrupting 4-nonylphenols, 4-alkoxyphenols, and bisphenol A. *Chimia International Journal for Chemistry 62*, pp. 358–363, 2008.

[30] Saure T-Ig Nazi G. G. J., Begui N C., Barelle M., Markowicz Y., Pelmo Nt J. and Tous-Saint T. Characterization of a chromosomally encoded catechol 1.2-dioksygenase [E.C. 1.13.11.1.] from Alicagenes eutrophus CH34. *Archives of Microbiology 166*,1996.

[31] Patnaik P. and Khoury J. N. Reaction of phenol with nitrite ion: Pathways of formation of nitrophenols in environmental waters. *Water Research 38*, pp. 206–210, 2004.

[32] Moussavi M. Effect of polar substituents on autoxidation of phenols. *Water Research 13*, pp. 1125–1128, 1979.

[33] Kinney L. C., Laboratory-XIII. R. A. T. W. R. C. A. W. T. R., Ivanuski V. R., Administration U. S. F. W. P. and Center R. A. T. W. R. *Photolysis Mechanisms for Pollution Abatement*. Federal Water Pollution Control Administartion, 1969.

[34] EPA U. *Water-Related Environmental Fate of 129 Priority Pollutants*. Office of Water Planning and Standards, Office of Water and Waste Management, US Environmental Protection Agency, Washington, DC, 1979.

[35] Schweigert N., Zehnder A. J. and Eggen R. I. Chemical properties of catechols and their molecular modes of toxic action in cells, from microorganisms to mammals. *Environmental Microbiology 3*, pp. 81–91, 2001.

[36] Schweigert N., Hunziker R. W., Escher B. I. and Eggen R. I. Acute toxicity of (chloro-) catechols and (chloro-) catechol-copper combinations in Escherichia coli corresponds to their membrane toxicity in vitro. *Environmental Toxicology and Chemistry 20*, pp. 239–247, 2001.

[37] Muñoz-de-Toro M., Markey C. M., Wadia P. R., Luque E. H., Rubin B. S., Sonnenschein C. and Soto A. M. Perinatal exposure to bisphenol-A alters peripubertal mammary gland development in mice. *Endocrinology 146*, pp. 4138–4147, 2005.

[38] Vom Saal F. S. and Hughes C. An extensive new literature concerning low-dose effects of bisphenol A shows the need for a new risk assessment. *Environmental Health Perspectives, 133*, pp. 926–933, 2005.

[39] Health U. D. O. and Services H. Agency for Toxic Substances and Disease Registry. *Toxicological Profile for Asbestos. Update.* Agency for Toxic Substances and Disease Registry, Atlanta, US, 1999.

[40] Li Y. and Trush M. A. Reactive oxygen-dependent DNA damage resulting from the oxidation of phenolic compounds by a copper-redox cycle mechanism. *Cancer Research 54*, pp. 1895s–1898s, 1994.

[41] Gosselin R. E., Smith R. P. and Hodge H. C. *Clinical Toxicology Of Commercial Products.* Williams & Wilkins, Baltimore, 1984.

[42] Schweigert N., Belkin S., Leong-Morgenthaler P., Zehnder A. J. and Eggen R. I. Combinations of chlorocatechols and heavy metals cause DNA degradation in vitro but must not result in increased mutation rates in vivo. *Environmental and Molecular Mutagenesis 33*, pp. 202–210, 1999.

[43] Zhang L., Wang Y., Shang N. and Smith M. T. Benzene metabolites induce the loss and long arm deletion of chromosomes 5 and 7 in human lymphocytes. *Leukemia Research 22*, pp. 105–113, 1998.

[44] Fox M. *Photocatalysis and Environment: Trends and Applications.* New York: Academic Publishers, pp. 445–467, 1988.

[45] Dursun G., Cicek H. and Dursun A. Y. Adsorption of phenol from aqueous solution by using carbonised beet pulp. *Journal of Hazardous Materials 125*, pp. 175–182, 2005.

[46] Ugurlu M., Gurses A., Yalcin M. and Dogar C. Removal of phenolic and lignin compounds from bleached kraft mill effluent by fly ash and sepiolite. *Adsorption 11*, pp. 87–97, 2005.

[47] Laoufi N., Tassalit D. and Bentahar F. The degradation of phenol in water solution by TiO_2 photocatalysis in a helical reactor. *Global NEST Journal 10*, pp. 404–418, 2008.

[48] Mirian Z.-A. and Nezamzadeh-Ejhieh A. Removal of phenol content of an industrial wastewater via a heterogeneous photodegradation process using supported FeO onto nanoparticles of Iranian clinoptilolite. *Desalination and Water Treatment 57*, pp. 16483–16494, 2016.

[49] Shahrezaei F., Akhbari A. and Rostami A. Photodegradation and removal of phenol and phenolic derivatives from petroleum refinery wastewater using nanoparticles of TiO_2. *International Journal of Energy and Environment 3*, pp. 267–274, 2012.

[50] Al-Kandari H., Abdullah A., Mohamed A. and Al-Kandari S. Enhanced photocatalytic degradation of a phenolic compounds' mixture using a highly efficient TiO_2/reduced graphene oxide nanocomposite. *Journal of Materials Science 114*, pp. 1–15, 2016.

[51] Abdollahi Y., Abdullah A. H., Zainal Z. and Yusof N. A. Photocatalytic degradation of p-Cresol by zinc oxide under UV irradiation. *International Journal of Molecular Sciences 13*, pp. 302–315, 2011.

[52] Asmaly H. A., Abussaud B., Saleh T. A., Gupta V. K. and Atieh M. A. Ferric oxide nanoparticles decorated carbon nanotubes and carbon nanofibers: From synthesis to enhanced removal of phenol. *Journal of Saudi Chemical Society 19*, pp. 511–520, 2015.

[53] Feng Y.-B., Hong L., Liu A.-L., Chen W.-D., Li G.-W., Chen W. and Xia X.-H. High-efficiency catalytic degradation of phenol based on the peroxidase-like activity of cupric oxide nanoparticles. *International Journal of Environmental Science and Technology 12*, pp. 653–660, 2015.

[54] Trapido M., Veressinina Y. and Munter R. The application of ozonation and advanced oxidation processes for degradation and detoxification of phenolic compounds, in Proc. 14th Ozone World Congress, Dearborn, MI, USA, 1999, pp. 235–249.

[55] Beltran F. J., García-Araya J. F. and Acedo B. Advanced oxidation of atrazine in water—II. Ozonation combined with ultraviolet radiation. *Water Research 28*, pp. 2165–2174, 1994.

[56] Chedeville O., Debacq M. and Porte C. Removal of phenolic compounds present in olive mill wastewaters by ozonation. *Desalination 249*, pp. 865–869, 2009.

[57] Hassani A., Borghei S., Samadyar H., Mirbagheri S. and Javid A. Treatment of waste water containing ethylene glycol using ozonation: Kinetic and performance study. *Bulletin of Environment, Pharmacology and Life Sciences 2*, pp. 78–82, 2013.

[58] Rao N., Singh J. R., Misra R. and Nandy T. Liquid-liquid extraction of phenol from simulated sebacic acid wastewater. *Journal of Scientific and Industrial Research 68*, pp. 823–828, 2009.

[59] Liu J., Xie J., Ren Z. and Zhang W. Solvent extraction of phenol with cumene from wastewater. *Desalination and Water Treatment 51*, pp. 3826–3831, 2013.

[60] Potter D. W. and Pawliszyn J. Rapid determination of polyaromatic hydrocarbons and polychlorinated biphenyls in water using solid-phase microextraction and GC/MS. *Environmental Science & Technology 28*, pp. 298–305, 1994.

[61] Möder M., Schrader S., Franck U. and Popp P. Determination of phenolic compounds in waste water by solid-phase micro extraction. *Fresenius' Journal of Analytical Chemistry 357*, pp. 326–332, 1997.

[62] Tavallali H. and Shiri M. Solid phase extraction of phenol from wastewater by magnetic iron oxide nanoparticles. *International Journal of ChemTech Research 4*, pp. 311–318, 2012.

[63] Paller G., Hommel R. K. and Kleber H. P. Phenol degradation by Acinetobacter calcoaceticus NCIB 8250. *Journal of Basic Microbiology 35*, pp. 325–335, 1995.

[64] Kanekar P., Sarnaik S. and Kelkar A. Bioremediation of phenol by alkaliphilic bacteria isolated from alkaline lake of Lonar, India. *Journal of Applied Microbiology 85*, pp. 128–133, 1998.

[65] Xiaojian Z., Zhansheng W. and Xiasheng G. Simple combination of biodegradation and carbon adsorption—The mechanism of the biological activated carbon process. *Water Research 25*, pp. 165–172, 1991.

[66] Müller R. and Babel W. Phenol and its derivatives as heterotrophic substrates for microbial growth—An energetic comparison. *Applied Microbiology and Biotechnology 42*, pp. 446–451, 1994.

[67] Bajpai P. and Bajpai P. K. Reduction of organochlorine compounds in bleach plant effluents, in Biotechnology in the Pulp and Paper Industry, K.E.L. Eriksson, et al. (Editors), Springer, Berlin Heidelberg, 1997, pp. 213–259.

[68] Sahm H., Brunner M. and Schoberth S. Anaerobic degradation of halogenated aromatic compounds. *Microbial Ecology 12*, pp. 147–153, 1986.

[69] Reinke W. and Knackmess H. J. Anaerobic degradation of halogenated aromatic compounds. *Microbial Ecology 12*, pp. 147–153, 1986.

[70] Chan Y. J., Chong M. F., Law C. L. and Hassell D. A review on anaerobic–aerobic treatment of industrial and municipal wastewater. *Chemical Engineering Journal 155*, pp. 1–18, 2009.

[71] Kukadiya A. V. Study of removal of phenol by biological treatment methods—With reference to moving bed biofilm reactor & activated sludge process. *International Journal of Engineering Research & Technology 5*, pp. 400–403, 2016.

[72] Sinha P. K., Sinha A. and Das M. Microbial removal of phenol and p-chlorophenol from industrial waste water using Rhodococcus sp. RSP8 and its growth kinetic modeling. *Journal of Water Resource and Protection 3*, pp. 634, 2011.

[73] Demarche P., Junghanns C., Nair R. R. and Agathos S. N. Harnessing the power of enzymes for environmental stewardship. *Biotechnology Advances 30*, pp. 933–953, 2012.

[74] Bailey J. and Ollis D. *Biochemical Engineering Fundamentals*, 2nd ed. McGraw-Hill, Maidenheach, 1986.

[75] Bevilaqua J., Commarota, M. and Freire, D. Phenol removal through combined biological and enzymatic treatments. *Brazilian Journal of Chemical Engineering 19*, pp. 151–158, 2002.

[76] Ikehata K. and Nicell J. A. Color and toxicity removal following tyrosinase-catalyzed oxidation of phenols. *Biotechnology Progress 16*, pp. 533–540, 2000.

[77] Halaouli S., Asther M., Sigoillot J. C., Hamdi M. and Lomascolo A. Fungal tyrosinases: New prospects in molecular characteristics, bioengineering and biotechnological applications. *Journal of Applied Microbiology 100*, pp. 219–232, 2006.

[78] Wilberg K. D. Q., Nunes D. G. and Rubio J. Removal of phenol by enzymatic oxidation and flotation. *Brazilian Journal of Chemical Engineering 17*, pp. 907–914, 2000.

[79] Shesterenko Y. A., Sevast'yanov O. and Romanovskaya I. Removal of phenols from aqueous solutions using tyrosinase immobilized on polymer carriers and inorganic coagulants. *Journal of Water Chemistry and Technology 34*, pp. 107–111, 2012.

[80] Khaled F. and Mossallam F. M. S. A. N. A. S. Peroxidase catalyzed the removal of phenol from synthetic waste water. *Thirteenth International Water Technology Conference, IWTC*, pp. 363–376, 2009.

[81] Aksu Z. and Yener J. A comparative adsorption/biosorption study of mono-chlorinated phenols onto various sorbents. *Waste Management 21*, pp. 695–702, 2001.

[82] Bhatnagar A. and Minocha A. Conventional and non-conventional adsorbents for removal of pollutants from water-a review. *Indian Journal of Chemical Technology 13*, pp. 203–217, 2006.

[83] Mohamed E. F., Andriantsiferana C., Wilhelm A.-M. and Delmas H. Competitive adsorption of phenolic compounds from aqueous solution using sludge-based activated carbon. *Environmental Technology 32*, pp. 1325–1336, 2011.

[84] Mukherjee S., Kumar S., Misra A. K. and Fan M. Removal of phenols from water environment by activated carbon, bagasse ash and wood charcoal. *Chemical Engineering Journal 129*, pp. 133–142, 2007.

[85] Akl M. A., Dawy M. B. and Serage A. A. Efficient removal of phenol from water samples using sugarcane bagasse based activated carbon. *Journal of Analytical & Bioanalytical Techniques 2014*, 2014. Doi: 10.4172/2155-9872.1000189.

[86] Peter-Varbanets M., Zurbrügg C., Swartz C. and Pronk W. Decentralized systems for potable water and the potential of membrane technology. *Water Research 43*, pp. 245–265, 2009.

[87] Bahareh Asadollahi Esfahani B. A. E., Koupaei M. S. and Ghasemi S. Z. Industrial waste water treatment by membrane systems. *Indian Journal of Fundamental and Applied Life Sciences 4*, pp. 1168–1177, 2014.

[88] Guo H., Wyart Y., Perot J., Nauleau F. and Moulin P. Low-pressure membrane integrity tests for drinking water treatment: A review. *Water Research 44*, pp. 41–57, 2010.

[89] Mixa A. and Staudt C. Membrane-based separation of phenol/water mixtures using ionically and covalently cross-linked ethylene-methacrylic acid copolymers. *International Journal of Chemical Engineering 2008*, 2009. Doi:10.1155/2008/319392.

[90] Ng Y. S., Jayakumar N. and Hashim M. A. Behavior of hydrophobic ionic liquids as liquid membranes on phenol removal: Experimental study and optimization. *Desalination 278*, pp. 250–258, 2011.

[91] Argun M. E., Dursun S., Karatas M. and Gürü M. Activation of pine cone using Fenton oxidation for Cd (II) and Pb (II) removal. *Bioresource Technology 99*, pp. 8691–8698, 2008.

[92] Brillas E., Sirés I. and Oturan M. A. Electro-Fenton process and related electrochemical technologies based on Fenton's reaction chemistry. *Chemical Reviews 109*, pp. 6570–6631, 2009.

[93] Abdelaziz A. A. M., Nosier S. A. and Hussien M. Treatment of industrial wastewater containing phenol using the electro-Fenton technique in gas sparged cell. *American Journal of Environmental Engineering and Science 2*, pp. 47–59, 2015.

[94] Un U. T., Topal S., Oduncu E. and Ogutveren U. B. Treatment of tissue paper wastewater: Application of electro-Fenton method. *International Journal of Environmental Science and Development 6*, p. 415, 2015.

[95] Rahmani A. R., Shabanloo A., Mehralipour J., Fazlzadeh M. and Poureshgh Y. Degradation of phenol in aqueous solutions using electro-Fenton process. *Research Journal of Environmental Sciences 9*, p. 332, 2015.

[96] Alexandratos S. D. Ion-exchange resins: a retrospective from industrial and engineering chemistry research. *Industrial & Engineering Chemistry Research 48*, pp. 388–398, 2008.

[97] Ioannidis S. and Anderko A. Equilibrium modeling of combined ion-exchange and molecular adsorption phenomena. *Industrial & Engineering Chemistry Research 40*, pp. 714–720, 2001.

[98] Maria D. V.-O., Javier M. O.-P. and Antonio M.-F. Equilibrium studies on phenol removal from industrial wastewater through polymeric resins. *Chemical Engineering Transactions*, pp. 253–258, 2016.

[99] Carmona M., De Lucas A., Valverde J. L., Velasco B. and Rodríguez J. F. Combined adsorption and ion exchange equilibrium of phenol on Amberlite IRA-420. *Chemical Engineering Journal 117*, pp. 155–160, 2006.

[100] Ahmed Z., Lyne S. and Shahrabani R. Removal and recovery of phenol from phenolic wastewater via ion exchange and polymeric resins. *Environmental Engineering Science 17*, pp. 245–255, 2000.